T0402726

Springer Series in Biophysics

Volume 24

The "Springer Series in Biophysics" spans all areas of modern biophysics, such as molecular, membrane, cellular or single molecule biophysics. More than that, it is one of the few series of its kind to present biophysical research material from a biological perspective.

All postgraduates, researchers and scientists working in biophysical research will benefit from the comprehensive and timely volumes of this well-structured series.

Mauricio Comas-Garcia · Sergio Rosales-Mendoza
Editors

Physical Virology

From the State-of-the-Art Research to the Future of Applied Virology

Editors
Mauricio Comas-Garcia
Autonomous University of San Luis Potosi
San Luis Potosí, Mexico

Sergio Rosales-Mendoza
Autonomous University of San Luis Potosi
San Luis Potosí, Mexico

ISSN 0932-2353 ISSN 1868-2561 (electronic)
Springer Series in Biophysics
ISBN 978-3-031-36814-1 ISBN 978-3-031-36815-8 (eBook)
https://doi.org/10.1007/978-3-031-36815-8

This Springer imprint is published by the registered company Springer Nature Switzerland AG
The registered company address is: Gewerbestrasse 11, 6330 Cham, Switzerland

Preface

The field of virology originates from the discoveries made by microbiologists, veterinarians, and physicians. For at least the first half of the XX century, virology focused on understanding a pathogen and the disease it causes. Furthermore, studying viruses has led us to a deep understanding of cellular and molecular biology at such degree that it would be impossible to think of modern cellular and molecular biology without virology. However, virologists, chemists, biochemists, and physicists realized that virions could have high-order symmetry and thus could be treated as physical objects. The success of W. M. Stanley in crystalizing tobacco mosaic virus opened the door to the study of virion symmetry and assembly, and it could be considered the beginning of the field of physical virology. Then, this field started to grow when J. B. Bancroft showed that the plant virus cowpea chlorotic mottle virus could be reconstituted from its purified components (i.e., capsid protein and RNA).

A long time has passed since the key experiments of Stanley and Bancroft, and the field of physical virology has evolved into a multidisciplinary science where physicians, experimental and theoretical physicists, chemists, virologists, structural biologists, and engineers have come together to study viruses as biological entities, physical objects, and as tools for biomedicine and nanotechnology. This book has tried to combine the experience of senior and junior independent researchers that are leaders in different areas of physical virology.

RNA was considered for a long time just as a messenger molecule; however, RNA viruses use this molecule as a genome, and thus, its role during assembly is far more complicated than a passive cargo. Chapters 1 and 2 focus on theoretical approaches to understanding viral RNAs and assembly. In Chap. 1, Vaupotic and co-workers explain how the viral RNA is a branched polymer and how this feature has tremendous consequences on the length of the packaged viral genome and on virions assembly. Chapter 2 by Cruz-Leon and co-workers focuses on atom and coarse-grained simulations to understand the viral RNA and the assembly process. Duran-Meza and co-workers show in Chap. 3 how the length of the genomic RNA controls the assembly process, and Stockley and Twarock explain the implications of the role of the cis-acting RNA sequences during assembly and infection.

The assembly and disassembly of virions is a necessary process of the viral infectious cycle. Chapters 5 through 9 focus on how artificial viruses and virus-like particles (VLPs) can be assembled. Chapter 5, by Moreno-Gutierrez and co-workers, explains how artificial viruses can be assembled by following well-defined rules. Hegawama et al., in Chap. 6, cover the self-assembly of VLPs to create hierarchical 3D materials. Chapter 7 by Cadena-Lopez and co-workers focuses on understanding the assembly of highly pathogenic viruses by producing VLPs in mammalian cultures. Thiede et al. analyze in Chap. 8 how norovirus capsid can alter its shape when interacting with metal ions and other molecules. Finally, in Chap. 9, Azad and co-workers discuss how different viruses can disassemble.

Viruses can also be seen as physical objects; thus, their physical properties can be linked to their biological activity. In Chaps. 10 and 11, Rodriguez-Espinosa and co-workers and M. G. Mateu, respectively, present analyses by atomic force microscopy to understand the mechanical properties of virions and correlate them with their biological functions. Luque et al., in Chap. 12, explain how cryo-electron microscopy and tomography can be used to understand the architecture of viruses.

Finally, the last two chapters focus on using viruses as therapeutic and diagnostic tools. C. Catalano, in Chap. 13, describes the potential application of the bacteriophage lambda as a tool for the targeted delivery of therapeutics and for vaccine development. In Chap. 14, S. Chaturvedi explains the novel concept of therapeutic interfering particles that were initially developed against SARS-CoV-2.

We greatly appreciate the tremendous amount of work done by all authors to assemble this book, and we apologize for any omissions. Unfortunately, it is impossible to edit a book that showcases all the areas and key works on physical virology. Nonetheless, this edition has brought together scientists with different backgrounds that push the limits of their respective areas and aims to explain the present and future of this field.

San Luis Potosí, Mexico

Prof. Mauricio Comas-Garcia, Ph.D.
Prof. Sergio Rosales-Mendoza, Ph.D.

Contents

Chapter 1
Viral RNA as a Branched Polymer

Domen Vaupotič, Angelo Rosa, Rudolf Podgornik, Luca Tubiana, and Anže Božič

Abstract Myriad viruses use positive-strand RNA molecules as their genomes. Far from being only a repository of genetic material, viral RNA performs numerous other functions mediated by its physical structure and chemical properties. In this chapter, we focus on its structure and discuss how long RNA molecules can be treated as branched polymers through planar graphs. We describe the major results that can be obtained by this approach, in particular the observation that viral RNA genomes have a characteristic compactness that sets them aside from similar random RNAs. We also discuss how different parameters used in the current RNA folding software influence the resulting structures and how they can be related to experimentally observable quantities. Finally, we show how the connection to branched polymers can be extended to take advantage of known results from polymer physics and can be further moulded to include additional interactions, such as excluded volume or electrostatics.

Keywords +ssRNA viruses · RNA secondary structure · Branched polymers · Scaling exponents · Graph theory

D. Vaupotič · A. Božič (✉)
Department of Theoretical Physics, Jožef Stefan Institute, Ljubljana, Slovenia
e-mail: anze.bozic@ijs.si

A. Rosa
Scuola Internazionale Superiore di Studi Avanzati (SISSA), Trieste, Italy

R. Podgornik
School of Physical Sciences and Kavli Institute of Theoretical Science, University of Chinese Academy of Sciences, Beijing, China

L. Tubiana
Physics Department, University of Trento, Trento, Italy

INFN-TIFPA, Trento Institute for Fundamental Physics and Applications, Trento, Italy

M. Comas-Garcia and S. Rosales-Mendoza (eds.), *Physical Virology*, Springer Series in Biophysics 24, https://doi.org/10.1007/978-3-031-36815-8_1

Introduction

RNA is an incredibly versatile biological macromolecule: not only does it act as a messenger between the DNA genome and the protein product, but it also assumes various roles in the form of transfer RNA, ribosomal RNA, microRNA, guide RNA, and long non-coding RNA, to name just a few [1, 2]. Its function is carried out both on the level of its primary sequence of nucleotides and by the local and global structures that are formed when the constituent nucleotides form base pairs with each other [3, 4]. Many RNA structures are thus involved in translational control, RNA localization, gene regulation, RNA stability, and more [5]. RNA structure folding is hierarchical, with the formation of base pairs—described by *secondary* structure—dominating the contribution to the folding energy and leading into its embedding in three-dimensional space, described by *tertiary* structure [6, 7]. In spite of recent improvements in the prediction of the tertiary structure of RNA molecules, it remains restricted to relatively short, individual sequences [8–10]. It is therefore of great advantage that RNA structure and its function can often be understood well by modelling it on the level of secondary structure, which can be further complemented by experimental methods such as SHAPE and its derivations [11–13].

In a large number of bacterial, plant, animal, and human *viruses*, positive-strand RNA (+ssRNA) takes on the role of their genomes [14]. Far from simply coding for the protein products, both local structural elements as well as long-range structural interactions in the genomes of +ssRNA viruses are involved in many fundamental viral processes such as virus disassembly, translation, genome replication, and packaging, and are thus in general important for viral fitness [15–18]. In particular, the genomes self-assemble together with capsid proteins to form a functional virion in an interplay of RNA sequence, length, and structure, further influenced by environmental variables such as pH and salt concentration [19–22]. For instance, in certain viruses, local structural elements called packaging signals—typically one or several hairpin loops with a more or less defined structure and nucleotide pattern—are responsible for specific interactions with the capsid proteins, initiating assembly through several possible pathways [22–24].

At the same time, non-specific electrostatic interactions between highly negatively charged RNA and positively charged domains of capsid proteins dominate the self-assembly of many +ssRNA viruses [25, 26]. Here, a number of experiments have demonstrated that viral capsids can assemble not only with their native RNA genomes but also with non-cognate RNA genomes of other viruses, other RNA molecules, and even linear polyelectrolytes [27, 28]. Success of the self-assembly and the resulting capsid(-like) structure, however, both depend on the *length and structure of the cargo* as well as on environmental variables [29–31]. Varying the salt concentration of the solution, for instance, changes the strength of RNA-protein interaction [32], and varying the strength of the interaction between RNA and an adsorbing substrate can change the latter's preference for adsorbing either single- or double-stranded RNA [33].

The *branching structure of viral RNA*, in particular, plays an important role in RNA-capsid interaction and virus assembly. Experiments have demonstrated that RNA structure and topology influence both packaging efficiency and the resulting capsid size and shape [29, 30, 34], while theoretical studies have shown that the degree of branching can greatly increase the amount of RNA that can be packaged into a capsid [35, 36]. Moreover, branching patterns of different RNAs have been shown to influence their size [37, 38], with genomes of +ssRNA viruses with icosahedral capsids being significantly more compact compared to those with helical capsids [39, 40]—with the former capsid type providing more severe spatial restrictions than the latter. This characteristic compactness appears to be a global structural property, and while even $\sim$5% of *synonymous* mutations were shown to destroy it [40], the question remains of where in the genome sequence these topological and structural properties are encoded [41]. Understanding the topological properties of the genomes of +ssRNA viruses is thus essential to understand their ability to self-assemble and consequently to design strategies to modify or interfere with their function [42].

In this chapter, we describe how the secondary structure of viral RNA can be mapped to a branched polymer, which properties can be extracted, what are some of the major results that can be obtained using this approach, and some pitfalls to be considered. To this purpose, we first introduce the main properties of branched polymers and demonstrate how RNA can be treated as one by being mapped onto a graph. We then describe some of the topological and structural properties that can be gleaned from this approach. Next, we illustrate this approach on random RNA sequences of different length and nucleotide composition, which provides a baseline for comparison of different biological RNAs. Focusing on the genomes of +ssRNA viruses, we explore the differences among them by comparing them to random RNAs as well as shuffled versions of themselves. We also show how model parameters used in the prediction of RNA secondary structure—specifically, multiloop energy and maximum base pair span—influence these predictions. Lastly, we briefly overview the field-theoretical description of RNA as a branched polymer, which makes use of the derived topological parameters and allows for a self-consistent inclusion of additional short- and long-range interactions in the analysis of interactions between the RNA genome and the capsid proteins.

RNA as a Branched Polymer

Secondary Structure of RNA as a Graph

Secondary structure prediction
Description of RNA structure on the intermediate level of its secondary structure forms a conceptually important step and explains the dominant part of the free energy of structure formation [43]. Modelling RNA on this level allows for analysis of large numbers of very long RNA sequences—which would be prohibitively expensive

to model on the level of their tertiary structure—while retaining the majority of the pertinent information about its local and global structure resulting from base-pairing. Numerous software packages exist for the prediction of RNA secondary structure, the most popular ones being ViennaRNA [44] and RNAstructure [45], based on energy models of base-pairing, and CONTRAfold [46] and EternaFold [47], which learn model parameters using stochastic context-free grammar. All of these algorithms necessarily come with limitations [47–49], but due to the complexity of structure prediction for long RNA molecules, they remain the tool of choice for studies of +ssRNA viral genomes, which can range anywhere from ~1000–30,000 nt in length (section "Branching Properties of Viral RNAs"). While some of the uncertainty in the prediction of RNA secondary structure can be alleviated by taking into account experimental data [12], such data is not widely available for most viral genomes.

Since the energy landscape of RNA structures is very shallow, predicting only the minimum free energy structure is typically insufficient, as the RNA can sample different conformations and several functional structures can co-exist in vivo [50]. The benefit of using energy-based folding algorithms for the prediction of secondary structure is that they enable generation and sampling of *thermal ensembles* of representative structures at a given temperature [51]. In the examples presented in this chapter, we use ViennaRNA v2.4 [44] to predict thermal ensembles of 500 structures at $T = 37°$ C for each RNA sequence and denote any quantity $\mathcal{O}$ averaged over this thermal ensemble of structures by $\langle \mathcal{O} \rangle$. As we show later on, this sample size produces sufficient statistics for each quantity we consider.

RNA as a graph

The idea that the complexity of base pairs and sequence-structure patterns in a folded RNA sequence can be reduced by mapping its secondary structure onto a graph is not new (Ref. [52] provides a detailed overview of the topic). In the absence of pseudoknots—a typical simplification which drastically reduces the computational complexity of structure prediction—the secondary structure can be described as a planar tree (Fig. 1.1). The simplest way to construct such a tree is by mapping double-stranded regions (base pairs) to *edges* with weights corresponding to the stem lengths,

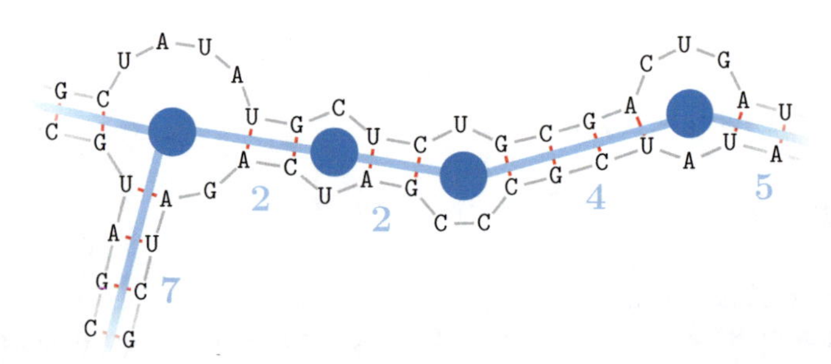

Fig. 1.1 Mapping (a part of) RNA secondary structure onto a planar tree. Double-stranded (base-paired) regions are mapped to graph edges, weighted with the stem length, while single-stranded regions are mapped to graph nodes

while single-stranded regions (unpaired nucleotides) are mapped to *nodes* connecting the edges. In the rest of the chapter, when we will refer to RNA trees, we will have in mind this procedure. This mapping is independent of the base-pairing model used to predict the structure—apart from the assumption of the absence of pseudoknots—and we discuss some differences that arise from using different model parameters in section "Influence of Model Parameters".

Once RNA secondary structure is mapped onto a tree composed of $N+1$ nodes $v_i \in \mathbb{V}$, $i = 0, \ldots, N$, connected by N (undirected) edges $e_i = (v_j, v_k) \in \mathbb{E}$ with weights (stem lengths) b_i, it is possible to derive various parameters describing its topology and structure [53–55], including:

- The distribution of ladder distances $p(\ell)$, where the ladder distance $\ell(v_i, v_j)$ is defined as the shortest path between a pair of nodes v_i and v_j. The most important derived measures are the maximum ladder distance (MLD),

$$\mathrm{MLD} = \max_{v_i, v_j \in \mathbb{V}} \ell(v_i, v_j), \tag{1.1}$$

corresponding to the diameter of the graph, and the average ladder distance (ALD),

$$\mathrm{ALD} = \frac{1}{(N+1)N} \sum_{v_i \neq v_j \in \mathbb{V}} \ell(v_i, v_j), \tag{1.2}$$

and its related quantity, the Wiener index $W = \sum_{v_i \neq v_j \in \mathbb{V}} \ell(v_i, v_j)$.
- The distribution of branch weights $p(N_{\mathrm{br}})$, obtained by cutting the expanded tree at each edge and taking the smaller of the two total weights of the resulting trees.
- The distribution of node degrees $p(d_i)$, indicating the presence of multiloops (nodes of degree $d_i \geqslant 3$), with the total number of nodes with degree k given by D_k. Some derived quantities are, e.g., Zagreb indices $M_1 = \sum_{v_i \in \mathbb{V}} d_i^2$ and $M_2 = \sum_{(v_i, v_j) \in \mathbb{E}} d_i d_j$.
- The Laplacian spectrum, the eigenvalues λ_i of the Laplacian matrix $\mathsf{L} = \mathsf{D} - \mathsf{A}$, where D is the matrix of node degrees and A is the node adjacency matrix. The second smallest eigenvalue $\lambda_2 \leqslant 1$ describes the connectivity of the graph, with larger values indicating better connectivity or a more star-like structure.

These quantities have the ability to distinguish, to various extents, between polymers with different types of tree topology, as illustrated in Fig. 1.2. Figure 1.3 further illustrates some of these quantities on an example of a (uniformly) random RNA sequence, $N_{\mathrm{nt}} = 2700$ nt in length. From the distribution of node degrees (panel (b)), one can for instance determine the Zagreb indices of the RNA tree, and from the distribution of ladder distances (panel (e)), one can determine both the MLD and the ALD. Panels (f) and (g) further show thermal ensemble distributions of MLD and the total number of base pairs B, demonstrating that their averages are well-defined quantities. As we will see in the following, combining different topological properties of RNA graphs with statistical mechanics of branched polymers can be used to gain insight into their physical properties.

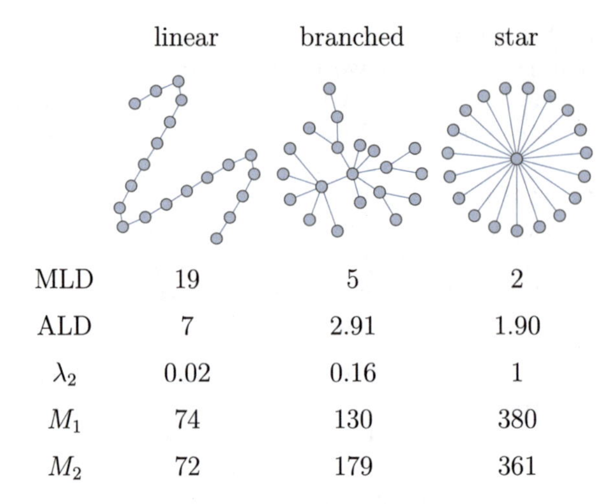

	linear	branched	star
MLD	19	5	2
ALD	7	2.91	1.90
λ_2	0.02	0.16	1
M_1	74	130	380
M_2	72	179	361

Fig. 1.2 Illustration of some topological quantities described in the main text: maximum (MLD) and average ladder distance (ALD), second Laplacian eigenvalue λ_2, and Zagreb indices M_1 and M_2. The values of these quantities are shown for three different types of polymers—linear, branched, and star polymer—all of them having the same total number of monomers N and unit edge weight

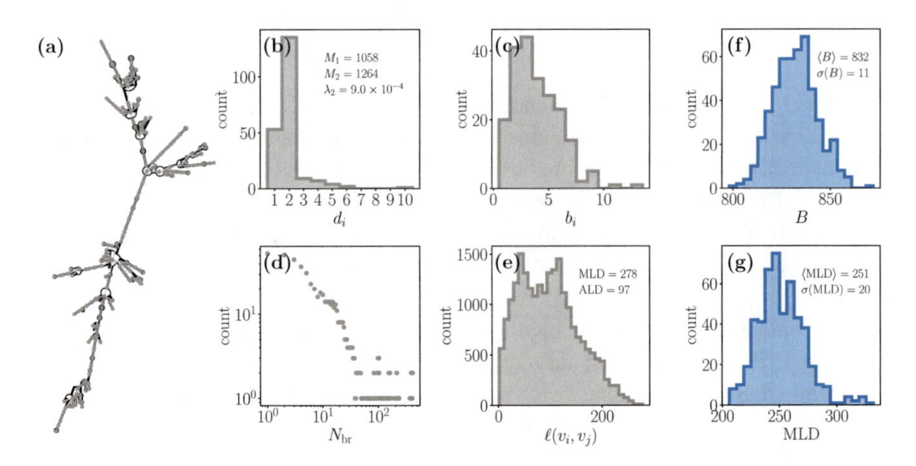

Fig. 1.3 Topological properties of RNA as a graph. **a** Representative secondary structure of a random RNA ($N_{\rm nt} = 2700$ nt), overlaid with its graph representation. **b–e** Distributions of node degrees d_i, edge weights b_i, branch weights $N_{\rm br}$, and path lengths $\ell(v_i, v_j)$ for the RNA structure shown in panel **a**. Also shown are the values of M_1 and M_2, λ_2, MLD, and ALD. **f–g** Distributions of the total number of base pairs B and of MLD for 500 structures of the same RNA sequence drawn from a thermal ensemble. Also shown are the values of the thermal averages and their standard deviation

Properties of Branched Polymers

Statistical mechanics of polymers is a very powerful theoretical tool with wide applications across biophysics, including, for instance, the scaling laws in the large-scale eukaryotic chromosome organization [56] and liquid-liquid phase separation in cells [57]. An important conceptual peculiarity of systems of *branched* polymers (trees)—such as RNA—concerns the necessity to distinguish between *annealed* (or, randomly *branching*) [58] and *quenched* (or, randomly *branched*) polymers [59]. Quenched trees are those whose topology of branches (tree connectivity) is fixed during the process of chemical synthesis and does not change afterwards. In contrast, the topology of annealed trees is not fixed at synthesis but instead can vary, typically in response to interactions (e.g., in the case of viral RNA, with capsid proteins) and/or changes in external conditions, and may fluctuate due to thermal motion. The class of annealed trees is particularly important as it is likely the most relevant for RNA molecules [60].

Scaling exponents
A physical description of polymer conformations has to be formulated by adopting the probabilistic language of statistical mechanics. In particular, since the total number of accessible conformations of a polymer chain increases exponentially with the number N of bonds (edges) [61], polymers are best described in terms of averages of corresponding observables.

The most distinct feature of a polymer conformation is its linear size, which can be expressed in terms of the radius of gyration, defined as

$$R_{\mathrm{g}}^2 \equiv \frac{1}{N}\sum_{i=1}^{N}(\mathbf{r}_i - \mathbf{r}_{\mathrm{cm}})^2, \tag{1.3}$$

where $\mathbf{r}_i$ is the spatial coordinate of the i-th monomer and $\mathbf{r}_{\mathrm{cm}} \equiv N^{-1}\sum_{i=1}^{N}\mathbf{r}_i$ is the centre-of-mass of the chain. The characteristic mean polymer size (i.e., its *mean* gyration radius) is given by the square root of the statistical average of Eq. (1.3) over the ensemble of all accessible conformations,

$$\langle R_{\mathrm{g}}(N)\rangle \equiv \sqrt{\langle R_{\mathrm{g}}^2\rangle} \approx bN^{\nu}. \tag{1.4}$$

The quantity b is the mean bond length, while the *scaling exponent* ν is a fundamental parameter which—as we will shortly see—depends on several factors, particularly on monomer-monomer interactions [62]. While many fundamental works in polymer physics have dealt with determining the exponent ν for various polymer ensembles, exact values are known only for a very few cases [60]. In most—and often the most relevant—cases, approximate (albeit accurate) results can be obtained by computationally extensive numerical methods or sophisticated mathematical tools [60]. The values of ν for polymer ensembles most relevant in the context of RNA are shown in Table 1.1; for other contexts, see the review by Everaers et al. [60].

Table 1.1 Best known values for scaling exponents of common polymer models in three dimensions

Polymer model	ν	ρ	ε	ν_{Flory}	ρ_{Flory}	References
Ideal linear	1/2	1	1	1/2	1	[63]
Self-avoiding linear	0.5877	1	1	3/5	1	[64]
Ideal branching	1/4	1/2	1/2	1/4	1/2	[63]
Self-avoiding branching	1/2	0.654	0.651	7/13	9/13	[65, 66]

Here, the "ideal" and "self-avoiding" refer to either a complete neglect or inclusion of excluded-volume effects, respectively (cf. Flory theory in section "Properties of Branched Polymers"). The values shown as fractions are exact, while others are approximate (obtained either from numerical simulations or by analytical methods)

While the exponent ν is sufficient to understand the physical properties of *linear* polymers (see Fig. 1.2), to completely understand an ensemble of *branching* polymers, such as viral RNAs, it is also necessary to characterize the topology of branching (or, equivalently, the *tree connectivity*) [60]. This is a particularly central problem for RNA secondary structure, since its mean gyration radius (Eq. 1.4) and hence the exponent ν are not easily accessible.

The problem of characterizing the connectivity of various ensembles of branching polymers has been theoretically addressed numerous times [66–68]. In the context of RNA, it is useful to introduce as a proper measure of chain connectivity the ensemble average of either the MLD or the ALD (Eqs. 1.1 and 1.2) as a function of the number of monomers N,

$$\langle \text{MLD}(N)\rangle \sim \langle \text{ALD}(N)\rangle \sim bN^{\rho}, \tag{1.5}$$

both of which account for the average length of linear paths on the tree. While originally defined for characterizing the connectivity and introduced independently from ν, the exponent ρ in Eq. (1.5) is related to it, and consequently also provides a fundamental insight into RNA folding *in physical space*.

Last but not least, we can also consider the average branch weight [66]:

$$\langle N_{\text{br}}(N)\rangle \sim N^{\varepsilon}, \tag{1.6}$$

which is defined as the average weight of the smallest of the two sub-trees obtained by systematically removing—one at time—the edges connecting two neighbouring nodes of the original tree [67]. Note that while the two scaling exponents ρ and ϵ describe very different quantities, they are not independent from each other. In fact, the relation

$$\varepsilon = \rho \tag{1.7}$$

is expected to hold for randomly branching polymers in general [66]. Equation (1.7) is particularly appealing because it can be used to support *a posteriori* the initial hypothesis that RNA behaves as a randomly branching polymer: In fact, it is "sufficient" to measure $\langle \text{ALD} \rangle$ and $\langle N_{\text{br}} \rangle$ as a function of N and compare the estimates for the corresponding scaling exponents. Table 1.1 again summarizes the known values of ρ and ε for selected polymer ensembles.

Even the simplest theory of branching polymers thus has to deal with the three distinct observables introduced in Eqs. (1.4)–(1.6). Since we are primarily interested in the secondary structure of random and viral RNAs, we will focus on topological observables such as $\langle \text{MLD} \rangle$ and $\langle N_{\text{br}} \rangle$. Nonetheless, we will show that this has important consequences for how RNA molecules fold in space, i.e., on the average molecular size as given by $\langle R_{\text{g}} \rangle$.

Flory theory

Exact values for the scaling exponents ν and ρ are known only in few special polymer cases. In this respect, Flory theories of polymers [60, 62, 69] provide a simple framework for first—and yet remarkably accurate—approximations of both ν and ρ. Flory theory is formulated in terms of a balance between *(i)* an entropic (elastic) term, given by a sum of two contributions coming from the classical entropy of swelling (F_{sw}) and the entropy of reconfiguration of the tree architecture (F_{tree}) due to swelling and interaction, and *(ii)* an interaction term (F_{inter}) arising from monomer-monomer collisions. Taken together, the Flory free energy (in units of $\beta^{-1} = k_B T$) reads [60, 62, 69]:

$$\begin{aligned} F &= F_{\text{sw}}(N, \langle R_{\text{g}} \rangle, \langle \text{ALD} \rangle) + F_{\text{tree}}(N, \langle \text{ALD} \rangle) + F_{\text{inter}}(N, \langle R_{\text{g}} \rangle) \\ &= \frac{\langle R_{\text{g}} \rangle^2}{\langle \text{ALD} \rangle b} + \frac{\langle \text{ALD} \rangle^2}{N b^2} + \upsilon_2 \frac{N^2}{\langle R_{\text{g}} \rangle^3}, \end{aligned} \tag{1.8}$$

where $\upsilon_2 \sim b^3$ is on the order of the *second virial coefficient* [60], accounting for the excluded-volume interaction between any two monomers. Although physically appealing, this representation of the free energy is itself an approximation, since the terms in the free energy are not independent from one another. Nonetheless, Flory theories turn out to be quite accurate [60, 62, 69].

A key feature of the free energy in Eq. (1.8) is that the interaction term does not depend on $\langle \text{ALD} \rangle$, which likely remains valid even in other ensembles with different forms of interaction energy F_{inter} [60]. Consequently, we can balance the first two terms without worrying about the third, and connect $\langle R_{\text{g}} \rangle$, $\langle \text{ALD} \rangle$ and N:

$$\langle \text{ALD} \rangle \sim b^{1/3} N^{1/3} \langle R_{\text{g}} \rangle^{2/3}, \tag{1.9}$$

or, conversely (see Eqs. 1.4 and 1.5),

$$\rho = \frac{1 + 2\nu}{3} \Longleftrightarrow \nu = \frac{3\rho - 1}{2}. \tag{1.10}$$

By reinserting Eq. 1.10 into Eq. (1.8) and balancing the remaining terms, we finally get the estimates for ν and ρ shown in Table 1.1, which, when compared to the exact ones, are remarkably accurate. In general, the relation between ν and ρ in Eq. (1.10) has been compared in various ensembles of randomly branching polymers and has been found to be very accurate in all cases [60]. Its practical implications are quite remarkable, as it allows us to connect branching (ρ) to $3D$ conformations (ν) by determining either of the two exponents in terms of the other. In the context of RNA, this relation is particularly useful, since we can determine ρ from the topological properties of its structure and extract ν afterwards.

Branching Properties of Viral RNAs

Random RNAs

Unlike viral RNA genomes which have a well-defined sequence length, random RNAs can be used to generate sequences of (in principle) arbitrary length and nucleotide composition. This enables one to explore how their topological properties change with both length and composition and in this way obtain different scaling relationships (section "Properties of Branched Polymers"). It is important to note here that RNA sequence length N_{nt} is, on average, directly proportional to the tree size N of its structure, and the two quantities can be used interchangeably.

Figure 1.4 shows some examples of scaling laws for uniformly random RNA sequences, $f(\text{A}) = f(\text{C}) = f(\text{U}) = f(\text{G}) = 0.25$, where $f(n)$ is the frequency of a nucleotide in the sequence. While some properties, such as $\langle\text{MLD}\rangle$ and $\langle N_{\text{br}}\rangle$, follow a scaling law with a well-defined exponent, others, such as the ratio of the number of degree 1 nodes (leaves of the tree—corresponding to hairpin configurations of RNA) and degree 3 nodes, D_1/D_3, tend towards a constant value. The scaling exponents are of course *asymptotic* properties valid for large RNA structures, as seen in the insets in panels (a) and (b) of Fig. 1.4, which show how the fitted values of exponents change as shorter sequences are progressively removed from the fit.

Nucleotide composition can vary significantly between different viral species (and biological RNAs in general) [70, 71], affecting their properties. Still, changing the composition of random RNA sequences mainly influences the prefactor of the $\langle\text{MLD}\rangle$ scaling law (Fig. 1.5a) and only minimally its exponent (Fig. 1.5b), even when their composition deviates significantly from a uniformly random one, as evaluated by the Euclidean distance $\delta^2 = \sum_{n\in\{\text{A,C,G,U}\}}[f(n) - 0.25]^2$. The decrease in the prefactor appears to be related to a decrease in the base pair percentage $2B/N_{\text{nt}}$ (Fig. 1.5c), which is perhaps unsurprising, as this leads to a smaller size of the RNA graph N at the same sequence length N_{nt}.

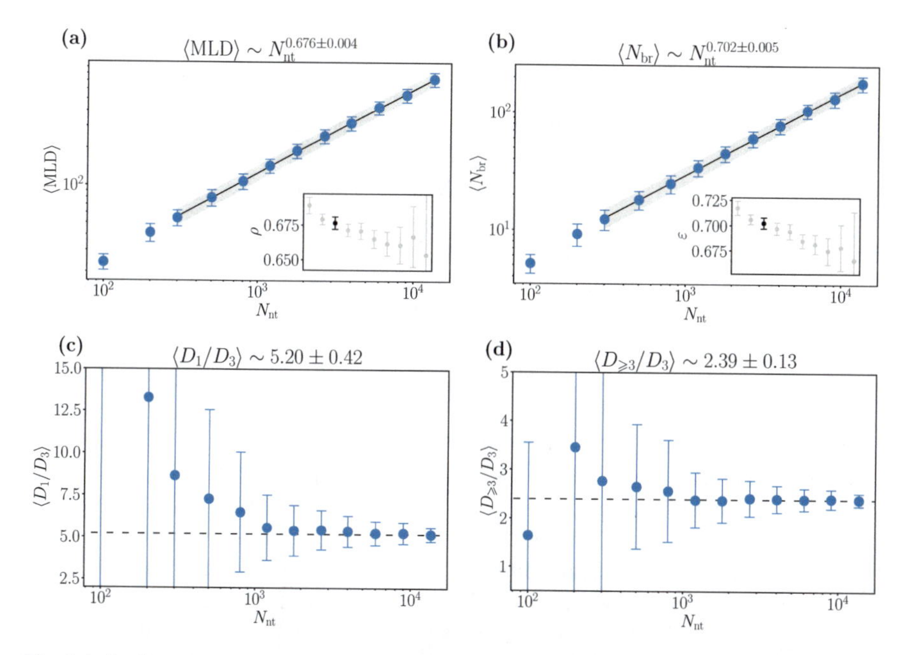

Fig. 1.4 Scaling of topological properties of (uniformly) random RNA sequences with their length: **a** maximum ladder distance ⟨MLD⟩, **b** branch weight ⟨N_{br}⟩, **c** ratio of the number of degree 1 and degree 3 nodes ⟨D_1/D_3⟩, and **d** ratio of the number of degree 3 and all multiloop (⩾ 3) nodes, ⟨$D_{\geqslant 3}/D_3$⟩. Each point in the plots represents an average over 200 random sequences and the error bars show the standard deviation. Insets in panels **a**, **b** show how the scaling exponents change with the starting point of the fit, with the black point corresponding to a fit over the shaded region in the panel

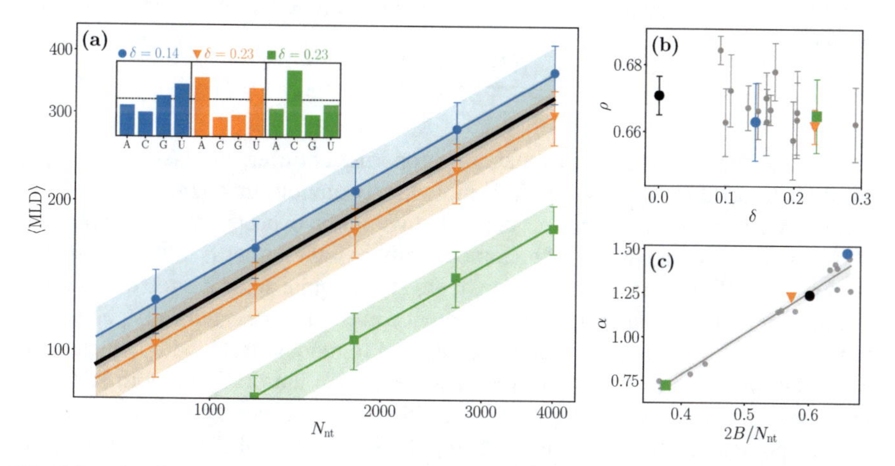

Fig. 1.5 **a** Scaling of ⟨MLD⟩ with sequence length, ⟨MLD⟩ $= \alpha N_{nt}^{\rho}$, for random RNA sequences with different nucleotide compositions. Each point in the plots represents an average over 200 random sequences with 500 thermal ensemble folds for each. **b** Scaling exponent ρ for random RNAs as a function of the Euclidean distance from the uniform composition δ. **c** Scaling prefactor α for RNAs with different compositions as a function of the base pair percentage, $2B/N_{nt}$

Viral RNAs

Random RNA sequences show what we can expect of biological and viral RNAs in general [72, 73]. An important example of this is the compactness of the viral RNA folds as captured by its proxy measure, the MLD (cf. section "Properties of Branched Polymers"). As already mentioned, the MLD of the genomes of +ssRNA viruses with icosahedral capsids, which need to pack the genome into a comparatively small volume, was found to be significantly smaller compared to random RNA sequences of viral-like composition [39, 40]. On the other hand, the MLD of genomes of viruses with helical capsids, which can in principle extend indefinitely, was indistinguishable from that of random RNAs. Figure 1.6a demonstrates these differences on an extended set of $\sim$1500 genomes of +ssRNA viruses from different families, obtained from the Virus Metadata Resource of ICTV [74], highlighted by capsid type.

Such an analysis opens up the possibility of comparing other topological properties of compact and non-compact viruses, as illustrated in panels (b)–(d) of Fig. 1.6 for genomes of beet virus Q (BVQ) and blueberry shock virus (BlShV), which are of comparable length but significantly less and more compact than random RNA, respectively (Fig. 1.6a). The second Laplacian eigenvalue, for instance, identifies the compact genome of BlShV as more star-like (cf. Fig. 1.2), but even more interesting is that the genome of BlShV forms *more* base pairs compared to the genome of BVQ and is at the same time located *lower* than the scaling law for uniformly random RNAs. This is in direct contrast to observations in random RNAs with different nucleotide composition, where the scaling prefactor was reduced for those RNAs which form fewer base pairs (Fig. 1.5), and indicates that the compactness of viral RNAs goes beyond simple differences in nucleotide composition.

Difference in a quantity $\mathcal{O}$ between viral and random RNAs can also be evaluated through the Z-score,

$$Z = \frac{\langle \mathcal{O} \rangle_{\mathrm{viral}} - \langle \mathcal{O} \rangle_{\mathrm{random}}}{\sigma(\mathcal{O})_{\mathrm{random}}}. \tag{1.11}$$

As Fig. 1.6e shows, this allows to study the properties of different sets of genomes, in this case grouped by viral family. It is immediately obvious that genomes in certain families are overall more compact than what would be expected of similar random RNAs, while the compactness of genomes in other viral families is indistinguishable from random RNAs. At the same time, there is also quite some degree of variation within viral families. Importantly, the difference in compactness typically persists no matter whether the genome $\langle$MLD$\rangle$ is compared to random RNA with a nucleotide composition similar to the one of the genome or to a uniformly random RNA, as indicated by the arrows in Fig. 1.6b. (Notable exception are Tymoviridae, which have a significantly different composition [39, 40].) This implies that the difference in nucleotide composition of various +ssRNA genomes does not suffice to explain the resulting differences in their compactness as measured by the $\langle$MLD$\rangle$.

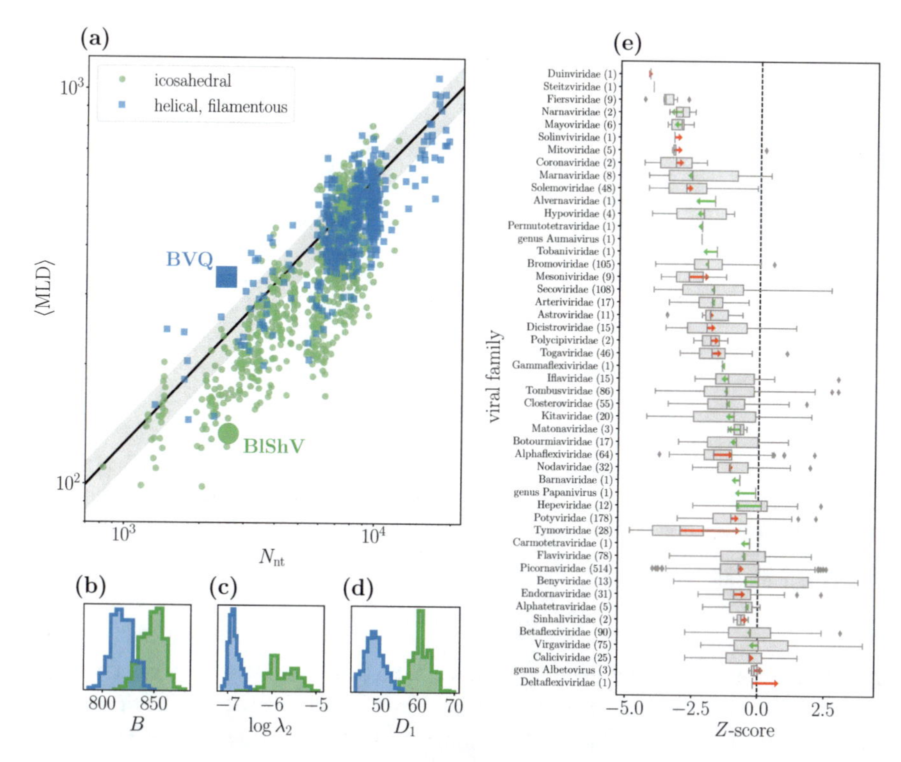

Fig. 1.6 **a** ⟨MLD⟩ of ∼1500 genomes of +ssRNA viruses of different lengths and capsid types. Black line shows the scaling for random RNA sequences with uniform composition, $\langle \text{MLD} \rangle \sim N_{\text{nt}}^{0.676}$, and the shaded area shows the region where $|Z| \leqslant 1$. **b–d** Number of base-pairs B, logarithm of the second Laplacian eigenvalue $\log \lambda_2$, and number of degree 1 nodes D_1 of genomes of two viruses indicated in panel **a**, BQV (large square) and BlShV (large circle). **e** Distribution of ⟨MLD⟩ Z-scores in different viral families, calculated with respect to random RNA sequences with uniform composition. Arrows indicate the shift in the median Z-score when the ⟨MLD⟩ is calculated with respect to random RNA with closest viral-like composition. The number of genomes included in each viral family is noted in parentheses next to its name

k-let shuffle of viral genomes

Both mononucleotide and dinucleotide frequencies of viral RNA genomes exhibit biases among different viral species [70, 75], even if they share the same host [76]. These biases are reflected in other properties—for instance, dinucleotide frequencies at codon position 2–3 were shown to explain the majority of codon usage bias [77]. Studies have made it clear that nucleotide composition alone does not suffice to explain the observed differences in the ⟨MLD⟩ of viral genomes [39, 40] (see also Fig. 1.6), which is further supported by computational observations that *synonymous* mutations preserving both mononucleotide and dinucleotide frequencies easily erase their characteristic compactness [40, 41].

One can thus take a step further and compare instead the topological properties of viral RNAs with their shuffled versions which conserve higher-order nucleotide

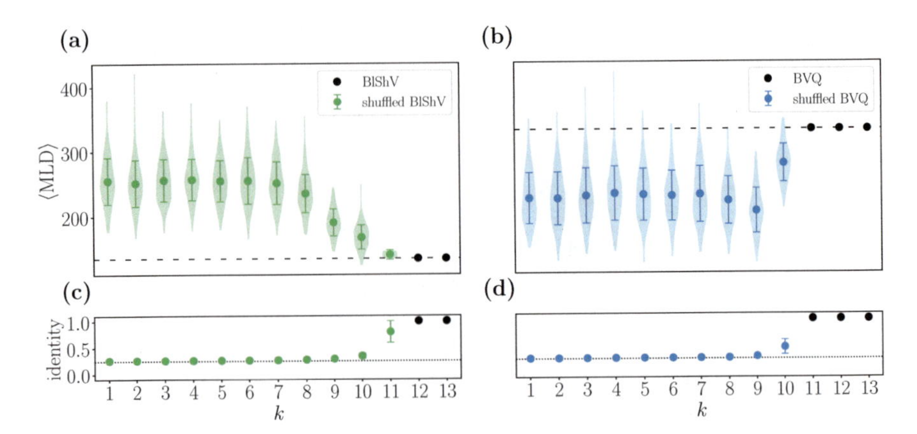

Fig. 1.7 ⟨MLD⟩ of shuffled sequences of **a** BlShV and **b** BVQ genomes with preserved k-let frequencies and **c** and **d** their corresponding sequence identities. The genomes of these two viruses are more and less compact than comparable random RNA, respectively (Fig. 1.6a). The distributions represent sequence ensembles, with 200 shuffled sequences used where possible (i.e., except for $k \geqslant 11$ and the wild-type genome)

frequencies of the original genomes. This can be achieved by using a k-let preserving shuffling algorithm, such as implemented by uShuffle [78], to shuffle the original genome sequences while *exactly* preserving k-let nucleotide frequencies. This means that for $k = 1$ we preserve mononucleotide frequencies (nucleotide composition), for $k = 2$ we preserve dinucleotide frequencies, and so on. Of course, as k increases, the number of possible shuffled sequences decreases. In the examples of BVQ and BlShV, there are still $\sim 10^3$–10^4 possible shuffled sequences available for $k = 10$, while the count drastically drops for $k = 11$ where only $\lesssim 10$ different shuffled sequences exist (Fig. 1.7). For the non-compact genome of BVQ, k-let shuffling does not produce drastic changes and the ⟨MLD⟩ of the shuffled sequences is comparable to that of random RNA for all possible values of k (Fig. 1.7b). On the other hand, for the compact genome of BlShV, there is a drastic change in the range of $k = 8$–10. For lower values of k, the shuffle completely destroys the compactness of the genome, as was previously seen for $k = 1$ and $k = 2$ [39–41]. For $k = 9$, however, the ⟨MLD⟩ remains close to the compactness of the original genome, and for $k = 10$ it is indistinguishable from it. There thus appears to be a particular length scale ($k \approx 10$) at which shuffling the compact viral genomes while preserving their k-let nucleotide frequency also preserves their compactness. This observation could hold important clues to the question of where at the sequence level the genome compactness is encoded, but of course needs to be explored more carefully.

Influence of Model Parameters

RNA secondary structure treated as a branched polymer provides a lot of information about the topological and structural nature of +ssRNA viral genomes and random RNAs alike. The inability to exactly predict the base pair patterns in a folded RNA sequence, however, can lead to differences in the predicted topology of the secondary structure, with consequences also for the subsequent tertiary structure prediction [79]. While prediction of thermal structure ensembles necessitates the use of energy-based models, there is nonetheless a wide variety of model parameters that can influence the resulting predicted structures. We will briefly comment on the effect that two of the most important ones—namely, energy of multiloop formation and the maximum allowed base pair span—have on the topological measures of secondary structure of viral RNA genomes.

Multiloop Energy Models

Energy-based folding algorithms predict both the minimum free energy fold of an RNA sequence as well as its pairing probability matrix, from which an ensemble of thermal folds can be obtained. These algorithms typically use a nearest neighbour energy model that breaks down the energy of an RNA structure into a sum of energies of its constituent loops. Commonly used sets of energy parameters are based on measurements provided by Turner [80], with two particular sets of parameters—Turner1999 and Turner2004—used as a basis by different versions of the most popular energy-based folding software such as ViennaRNA and RNAstructure. Several efforts have also been made to improve on these parameter sets by using various computational optimization techniques [81, 82].

Among the numerous energy parameters involved in structure prediction, *multiloop energies* are the least accurately known [83], even though occurrences of multiloops of degree 10 or higher are not uncommon in various RNAs [84]. Since allowing for an arbitrary size of a multiloop increases the computational complexity of structure prediction, earliest energy-based structure prediction models simply neglected multiloop contributions to the energy [85]. Most of the current structure prediction software assumes that the energy of a multiloop depends only on the amount of enclosed base pairs (number of branches) and the number of unpaired nucleotides in it, and uses a linear model of the form

$$E_{\mathrm{multiloop}} = E_0 + E_{\mathrm{br}} \times [\mathrm{branches}] + E_{\mathrm{un}} \times [\mathrm{unpaired\ nucleotides}], \tag{1.12}$$

where E_0 is the energy contribution for multiloop initiation, and E_{br} and E_{un} are the energy contributions for each enclosed base pair and unpaired nucleotide, respectively. While this form has been chosen mostly for its computational simplicity,

Table 1.2 Comparison of different energy parameters for the linear multiloop energy model (Eq. 1.12) used in prediction of RNA secondary structure

Energy model	E_0	E_{un}	E_{br}	References
Turner1999	10.1	−0.3	−0.3	[80]
Turner2004	9.25	NA	0.63	[80]
ViennaRNA (< v2.0)	3.4	0.0	0.4	[44]
ViennaRNA (v2.0+)	9.3	0.0	−0.9	[44]
RNAstructure	9.3	0.0	−0.6	[45]
Andronescu2007	4.4	0.04	0.03	[81]
Langdon2018	9.3	0.0	−0.8	[82]

improved models of multiloop energy that have been proposed seemingly do not lead to improved multiloop predictions compared to the linear one [86, 87].

Table 1.2 gives an overview of some of the most commonly used energy parameters for the linear multiloop model (Eq. 1.12). A notable difference between the models lies not only in the magnitude but in *the sign* of the parameter E_{br} which controls the number of branches stemming from the multiloop. Earlier versions of ViennaRNA (until v2.0), for instance, used a positive value of this parameter, penalizing high-degree nodes, while the latest versions of the software use a negative value, promoting high-degree nodes.

These differences will of course reflect in the predicted structures of long RNAs and their topological properties. Since the different energy models in Table 1.2 also differ in other aspects, it is easiest to compare multiloop energy parameters by modifying *only* the multiloop parameters in the current parameter set used by ViennaRNA (v2.4) with the ones from older versions (< v2.0), resulting in a modified set of parameters ViennaRNA-mod. Differences in their predictions are illustrated in Fig. 1.8. The opposite signs of the parameter E_{br} clearly lead to very different distributions of node degrees in the genome of BlShV, with the modified set of parameters predicting far fewer multiloops ($D_{\geqslant 3}$). Interestingly enough, however, the different multiloop energy parameters do not seem to lead to a different scaling behaviour of the $\langle$MLD$\rangle$ of uniformly random RNA (Fig. 1.8c), as the exponent ρ becomes indistinguishable between the two cases in the asymptotic limit of long sequences. On the other hand, the ratio of the number of degree 1 and degree 3 nodes is completely different (Fig. 1.8d). The choice of the multiloop energy parameters can thus lead to important differences in the predicted RNA topology [86, 88], which needs to be taken into account when comparing results obtained by existing studies on the branching properties of viral RNAs [37–40] that use different versions of folding software and thus potentially different energy models.

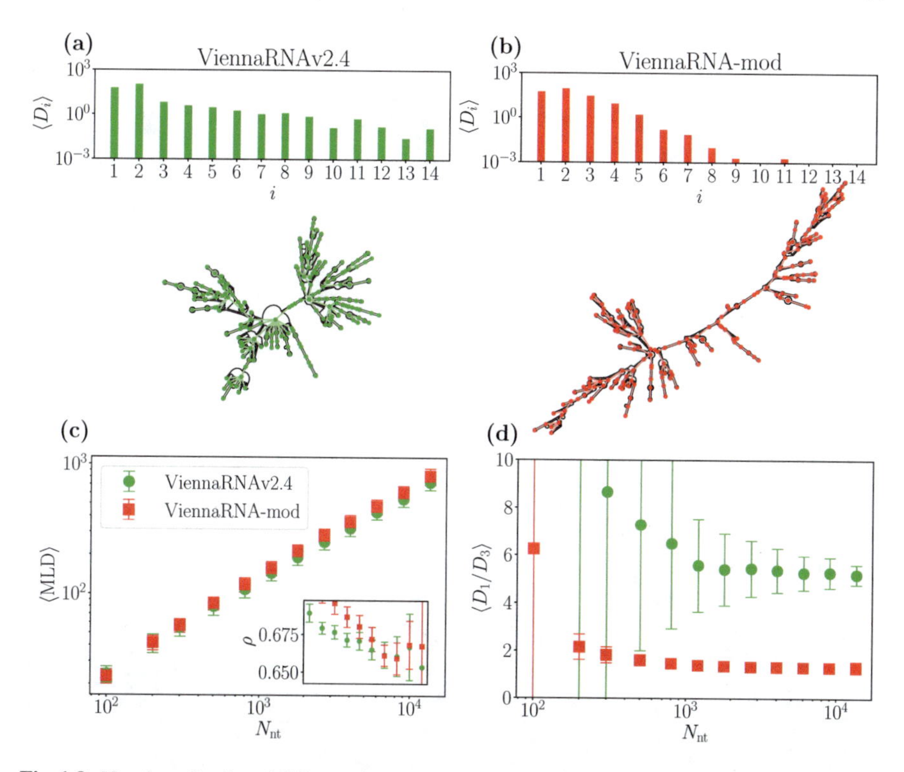

Fig. 1.8 Number of nodes of different degrees $\langle D_i \rangle$ in the BlShV genome obtained using the energy model in ViennaRNA (v2.4) with either **a** default multiloop energy parameters or **b** multiloop energy parameters from older versions of ViennaRNA (< v2.0). For the values of these parameters, see Table 1.2. Note the logarithmic scale in the histograms. Each panel also shows an example structure of the genome. Scaling of **c** ⟨MLD⟩ and **d** the ratio of the number of degree 1 and degree 3 nodes $\langle D_1/D_3 \rangle$ with the sequence length of uniformly random RNA as predicted by the two different sets of multiloop energy parameters

Maximum Base Pair Span

Parts of the folding process in very long RNA molecules (i.e., over several hundred nucleotides in length) are influenced by various factors such as co-transcriptional folding and the presence of other molecules in the cell [89, 90]. Consequently, the accuracy of RNA secondary structure prediction in general decreases with the span of a base pair—the length of the nucleotide sequence between two paired bases [91]. This effect can be incorporated in the folding prediction by restricting the maximum allowed base pair span [44, 91], which not only tends to yield more plausible local structure predictions but also drastically increases the computational efficiency. While restrictions on the maximum base pair span in the range of 200–600 nt are often made to improve the prediction of local structural elements [92, 93], this can neglect

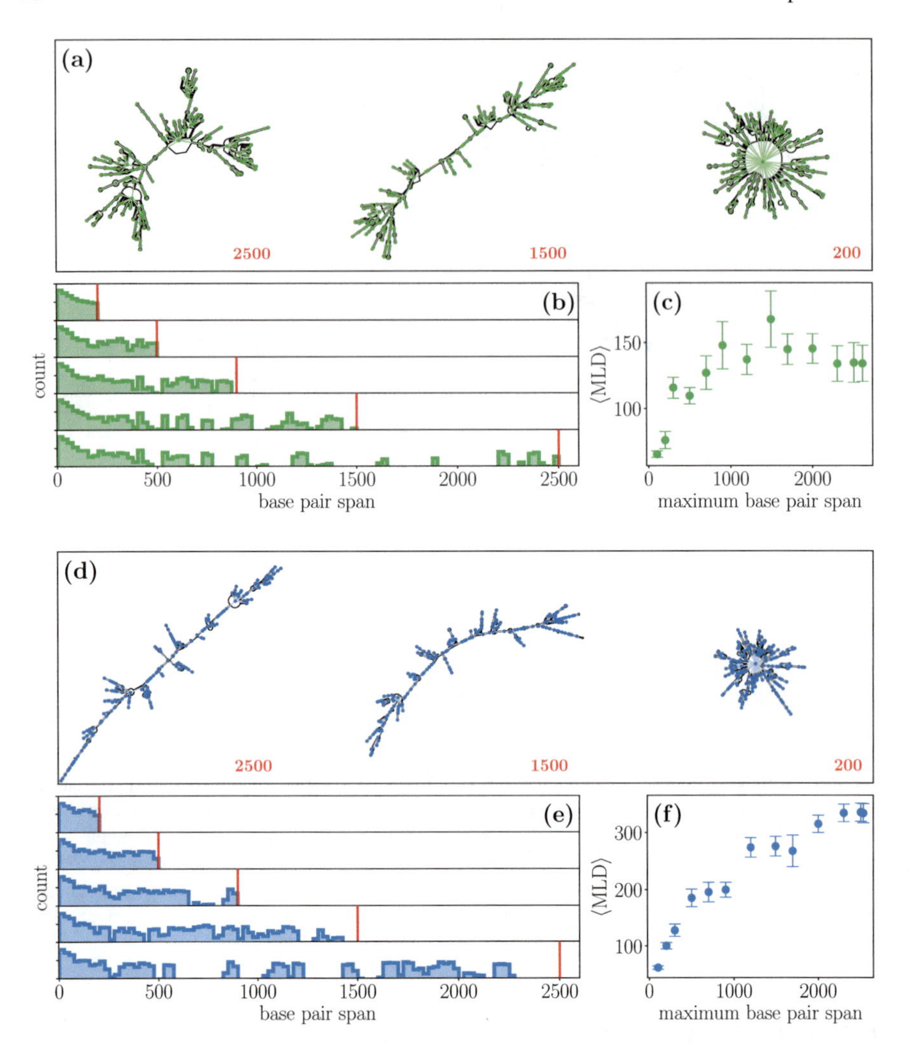

Fig. 1.9 **a** Representative structures of BlShV genome with different maximum base pair restrictions, indicated with each structure. **b** Distributions of base pair spans in the thermal ensemble of BlShV genome structures with different maximum base pair span restrictions (marked by vertical bars). **c** ⟨MLD⟩ of the BlShV genome as a function of the maximum base pair span restriction. **d–f** Same as panels **a–c** but for the BVQ genome

long-range base pairing and global structure which have been shown to be important in numerous +ssRNA genomes, including that of SARS-CoV-2 [17, 94, 95].

It is therefore natural to wonder to what extent restricting the maximum allowed base pair span influences the topological properties of the structure of viral RNAs. As panels (c) and (f) of Fig. 1.9 show, restricting the maximum base pair span to ~1000–2000 nt slightly changes the ⟨MLD⟩, regardless of whether the original genome belongs to a compact class or not, as long-range base pairs are progressively removed from the global structure (panels (b) and (e)). However, when the maximum base pair span is restricted to $\lesssim$1000 nt, $\langle$MLD$\rangle$ starts to decrease drastically. Inspection of the resulting structures (panels (a) and (d)) shows that significant maximum base pair span restrictions eventually result in RNA topology becoming more star-like because the individual hairpins are effectively being "strung" on a backbone of single-stranded RNA. This leads to a decrease in $\langle$MLD$\rangle$ which by definition does not take into account single-stranded regions of RNA (section "Secondary Structure of RNA as a Graph"), and could consequently affect the results showing that viral RNAs are more compact than random ones (section "Branching Properties of Viral RNAs"). From a topological and structural perspective, imposing a drastic restriction on the maximum base pair span in RNA would thus not only require a redefinition of the MLD but perhaps even a different mapping of its secondary structure onto a graph.

Field-Theoretical Description of Viral RNA as a Branched Polymer

RNA branching is also intimately connected with long-range interactions, such as electrostatic self-interaction and interactions between the RNA and capsid proteins. Since the strength of RNA self-interaction (base-pairing) is relatively weak and may easily be affected by either thermal fluctuations or electrostatic interactions [25], *annealed branched polymers* present a viable *coarse-grained* model system. Here, one starts with the grand canonical partition function [96]

$$\Xi(K, f_e, f_b; V) = \sum_{N, N_e, N_b} K^N f_e^{N_e} f_b^{N_b} \Omega(N, N_e, N_b; V) \tag{1.13}$$

where K are bonds (edges), and f_e and f_b are end- and branch point fugacities of the annealed polymer (with hairpins counted as the end points, cf. section "Secondary Structure of RNA as a Graph"). Branch points (nodes) of high degree ($d_i > 3$) can be considered as combinations of branch points of degree 3 and thus need not be treated separately in this description. The function $\Omega(N, N_e, N_b; V)$ is the number of ways to arrange N bonds, N_e end points, and N_b branch points on a lattice of volume V.

The grand canonical partition function can be obtained in the $n \to 0$ limit of the partition function of an $\mathcal{O}(n)$ model of a magnet [97] as a functional integral over a continuous field Ψ [58]

$$\Xi(K, f_e, f_b; V) \simeq \int \mathcal{D}[\Psi] e^{-\beta F_0[\psi]}, \tag{1.14}$$

with the square of Ψ proportional to the monomer density. The saddle-point (mean-field) free energy F_0—in absence of electrostatic interactions—is then

$$\beta F_0[\Psi] = \int_V \mathrm{d}^3r \left[\frac{a^2}{6}|\nabla\Psi|^2 + \frac{1}{2}\upsilon\Psi^4 - \frac{1}{\sqrt{a^3}}\left(f_e\Psi + \frac{a^3}{6} f_b\Psi^3 \right) \right], \tag{1.15}$$

where a is the statistical step length (Kuhn length; for RNA, $\sim$1 nm) and υ is the (repulsive) short-range excluded volume interaction term. Branch point and end point terms proportional to Ψ and Ψ^3 are negative (attractive), therefore increasing the local monomer density. The annealed numbers of end- (N_e) and branch points (N_b) of the RNA (corresponding to D_1 and $D_{\geqslant 3}$ in the graph representation, respectively) are related to the fugacities f_e and f_b by [58]

$$N_e = -f_e \frac{\partial \beta F_0}{\partial f_e} \quad \text{and} \quad N_b = -f_b \frac{\partial \beta F_0}{\partial f_b}. \tag{1.16}$$

If the total number of monomers is fixed, the number of end points for a single RNA molecule with no closed loops depends on the number of branch points as $N_e = N_b + 2$, meaning that f_e is not a free parameter. The polymer is linear if $f_b = 0$, and the number of branch points increases with f_b.

On this level of the description of RNA branching, *long-range interactions* can be straightforwardly implemented by adding additional terms to the free energy functional (Eq. 1.15), depending on local interaction fields and couplings with the Ψ field [35, 36, 98]. The free energy for, e.g., electrostatic interactions described by electrostatic potential field Φ would read

$$\beta F = \beta F_0[\Psi] + \beta F[\Phi] + \beta\tau \int_V \mathrm{d}^3r \; \Phi\Psi^2, \tag{1.17}$$

with τ being the charge per monomer and $\beta F[\Phi]$ the electrostatic interaction term [99], which can also describe the interactions between the genome and the capsid proteins. On the other hand, non-electrostatic interactions implied by *packaging signals* require a modified approach [100].

Modifying the topology of the genome by varying the fugacities f_e and f_b allows to use this methodology to simulate systems of viral RNAs with a fixed number of end- and branch points. This coarse-grained description of branching is complementary to the methodology based on explicit planar tree structure [42] and can, for instance, differentiate between the encapsulation behavior of RNA1 of brome mosaic virus (BMV) that has 65 branch points and of RNA1 of cowpea chlorotic mottle virus (CCMV) with 60.5 branch points—correctly confirming that BMV RNA1 is preferentially packaged over CCMV RNA1 by the CCMV capsid protein [35]. Furthermore, the straightforward implementation of electrostatic interactions on top of RNA topology is probably the most important forte of this methodology and allows one to assess the role of electrostatics in spontaneous co-assembly of the negatively charged genome and positively charged capsid proteins. Using this approach, it has been demonstrated that branching in fact allows viruses to maximize the amount of encapsulated genome and makes assembly more efficient [35], implies negative osmotic pressures across the capsid wall [36], and can explain the effect of number and location of charges in the capsid protein tails [101].

While the field-theoretical annealed-branching description is without doubt heavily coarse-grained, it not only provides an approximate implementation of topology, but also readily incorporates *short- and long-range interactions* on a level amenable to analytical calculations. Different approaches of standard polymer theory can then be transplanted into the statistical mechanics of RNA providing further insight into the coupling between topology and virion self-assembly.

Conclusions

Mapping RNA secondary structure onto a graph enables its description as a branched polymer and a subsequent study of its topological properties (e.g., MLD and node degree distribution; section "RNA as a Branched Polymer"). This, in turn, can be connected to the physical properties of the RNA, such as its size as given by its radius of gyration. This approach also provides insight into how RNA structure compares to other types of branched polymers in terms of, for instance, their scaling exponents. Branching properties of RNA also allow comparison of +ssRNA genomes of different viral families, both among themselves and with random RNA of similar length and composition (section "Branching Properties of Viral RNAs"). Such an analysis reveals the unusual compactness of genomes from certain viral families and may eventually provide an answer to the question of where in the sequence of viral RNA its physical compactness is encoded.

Describing RNA as a graph and in this way treating it as a branched polymer is, of course, an approximation. This description is made on the topological level of the RNA secondary structure, itself deriving from an energy-based base pair prediction, and thus depends on the model parameters used in it (section "Influence of Model Parameters"). Secondary structure prediction furthermore remains agnostic to steric interactions between different parts of the RNA, long-range interactions such as electrostatics, and other tertiary interactions. Some of these effects can, however, be included on a coarse-grained level by treating the branching RNA structure with a field-theoretical description, enabling, for instance, a coupling between topological parameters (such as node degree distribution) and electrostatic interactions (section "Field-theoretical Description of Viral RNA as a Branched Polymer").

Predictions obtained by treating viral RNA as a branched polymer can, to an extent, be verified experimentally, for instance by measuring its radius of gyration by gel electrophoresis [37] or by determining the distributions of node degrees and segment lengths from 2D projections of viral RNA molecules imaged by cryo-EM [37, 102]. Properly designed experiments on long RNA molecules of different lengths could thus, in principle, be compared with the predictions given by different multiloop energy models and in this way help determine the most appropriate model. Lastly, treating RNA as a graph and being able to understand how its sequence leads to its topological and structural properties can be beneficial not only in the ability to interfere with the function of viral genomes but also in the design of RNA molecules with specific topology for use in nanomedicine and synthetic biology [52, 103, 104].

Acknowledgements A.B. acknowledges support by Slovenian Research Agency (ARRS) under Contract No. P1-0055. L.T. acknowledges support by MIUR through the Rita Levi Montalcini grant and financial support from ICSC—Centro Nazionale di Ricerca in High Performance Computing, Big Data and Quantum Computing, funded by European Union—NextGenerationEU. R.P. acknowledges support from the Key Project No. 12034019 of the Natural Science Foundation of China. R.P. also thanks J.D. Farrell for his comments on an earlier version of the manuscript. The authors acknowledge networking support by the the COST Action No. CA17139 (EUTOPIA).

References

1. Eddy SR (2001) Non-coding RNA genes and the modern RNA world. Nat Rev Genet 2:919–929
2. Mattick JS, Makunin IV (2006) Non-coding RNA. Hum Mol Genet 15:R17–29
3. Gorodkin J, Ruzzo WL (2014) RNA sequence, structure, and function: computational and bioinformatic methods. Springer
4. Wang XW, Liu CX, Chen LL, Zhang QC (2021) RNA structure probing uncovers RNA structure-dependent biological functions. Nat Chem Biol 17:755–766
5. Mortimer SA, Kidwell MA, Doudna JA (2014) Insights into RNA structure and function from genome-wide studies. Nat Rev Genet 15:469–479
6. Brion P, Westhof E (1997) Hierarchy and dynamics of RNA folding. Annu Rev Biophys Biomol Struct 26:113–137
7. Mustoe AM, Brooks CL, Al-Hashimi HM (2014) Hierarchy of RNA functional dynamics. Annu Rev Biochem 83:441–466
8. Leontis N, Westhof E (eds) (2012) RNA 3D structure analysis and prediction. Springer, Berlin
9. Miao Z, Westhof E (2017) RNA structure: advances and assessment of 3D structure prediction. Annu Rev Biophys 46:483–503
10. Li J, Chen SJ (2021) RNA 3D structure prediction using coarse-grained models. Front Mol Biosci 8:720937
11. Low JT, Weeks KM (2010) SHAPE-directed RNA secondary structure prediction. Methods 52:150–158
12. Lorenz R, Hofacker IL, Stadler PF (2016) RNA folding with hard and soft constraints. Algorithms Mol Biol 11:1–13
13. Mitchell D III, Assmann SM, Bevilacqua PC (2019) Probing RNA structure in vivo. Curr Op Struct Biol 59:151–158
14. Holmes EC (2009) The evolution and emergence of RNA viruses. Oxford University Press, Oxford
15. Liu Y, Wimmer E, Paul AV (2009) Cis-acting RNA elements in human and animal plus-strand RNA viruses. Biochimi Biophys Acta 1789:495–517
16. Newburn LR, White KA (2015) Cis-acting RNA elements in positive-strand RNA plant virus genomes. Virology 479:434–443
17. Nicholson BL, White KA (2015) Exploring the architecture of viral RNA genomes. Curr Op Virol 12:66–74
18. Boerneke MA, Ehrhardt JE, Weeks KM (2019) Physical and functional analysis of viral RNA genomes by SHAPE. Annu Rev Virol 6:93–117
19. Schneemann A (2006) The structural and functional role of RNA in icosahedral virus assembly. Annu Rev Microbiol 60:51–67

20. Rao A (2006) Genome packaging by spherical plant RNA viruses. Annu Rev Phytopathol 44:61–87
21. Garmann RF, Comas-Garcia M, Knobler CM, Gelbart WM (2016) Physical principles in the self-assembly of a simple spherical virus. Acc Chem Res 49:48–55
22. Comas-Garcia M (2019) Packaging of genomic RNA in positive-sense single-stranded RNA viruses: a complex story. Viruses 11:253
23. Twarock R, Bingham RJ, Dykeman EC, Stockley PG (2018) A modelling paradigm for RNA virus assembly. Curr Op Virol 31:74–81
24. Stockley PG, Twarock R, Bakker SE, Barker AM, Borodavka A, Dykeman E, Ford RJ, Pearson AR, Phillips SE, Ranson NA et al (2013) Packaging signals in single-stranded RNA viruses: nature's alternative to a purely electrostatic assembly mechanism. J Biol Phys 39:277–287
25. Zandi R, Dragnea B, Travesset A, Podgornik R (2020) On virus growth and form. Phys Rep 847:1–102
26. Perlmutter JD, Hagan MF (2015) Mechanisms of virus assembly. Annu Rev Phys Chem 66:217
27. Hu Y, Zandi R, Anavitarte A, Knobler CM, Gelbart WM (2008) Packaging of a polymer by a viral capsid: the interplay between polymer length and capsid size. Biophys J 94:1428–1436
28. Comas-Garcia M, Cadena-Nava RD, Rao A, Knobler CM, Gelbart WM (2012) In vitro quantification of the relative packaging efficiencies of single-stranded RNA molecules by viral capsid protein. J Virol 86:12271–12282
29. Beren C, Dreesens LL, Liu KN, Knobler CM, Gelbart WM (2017) The effect of RNA secondary structure on the self-assembly of viral capsids. Biophys J 113:339–347
30. Marichal L, Gargowitsch L, Rubim RL, Sizun C, Kra K, Bressanelli S, Dong Y, Panahandeh S, Zandi R, Tresset G (2021) Relationships between RNA topology and nucleocapsid structure in a model icosahedral virus. Biophys J 120:3925–3936
31. Perlmutter JD, Qiao C, Hagan MF (2013) Viral genome structures are optimal for capsid assembly. eLife 2:e00632
32. Garmann RF, Goldfain AM, Tanimoto CR, Beren CE, Vasquez FF, Villarreal DA, Knobler CM, Gelbart WM, Manoharan VN (2022) Single-particle studies of the effects of RNA–protein interactions on the self-assembly of RNA virus particles. Proc Natl Acad Sci USA 119:e2206292119
33. Poblete S, Božič A, Kanduč M, Podgornik R, Vargas Guzmán HA (2021) RNA secondary structure regulates fragments adsorption onto flat substrates. ACS Omega 6:32823–32831
34. Singaram SW, Garmann RF, Knobler CM, Gelbart WM, Ben-Shaul A (2015) Role of RNA branchedness in the competition for viral capsid proteins. J Phys Chem B 119:13991–14002
35. Erdemci-Tandogan G, Wagner J, Van Der Schoot P, Podgornik R, Zandi R (2014) RNA topology remolds electrostatic stabilization of viruses. Phys Rev E 89:032707
36. Erdemci-Tandogan G, Wagner J, van der Schoot P, Podgornik R, Zandi R (2016) Effects of RNA branching on the electrostatic stabilization of viruses. Phys Rev E 94:022408
37. Gopal A, Egecioglu DE, Yoffe AM, Ben-Shaul A, Rao AL, Knobler CM, Gelbart WM (2014) Viral RNAs are unusually compact. PLoS One 9:e105875
38. Borodavka A, Singaram SW, Stockley PG, Gelbart WM, Ben-Shaul A, Tuma R (2016) Sizes of long RNA molecules are determined by the branching patterns of their secondary structures. Biophys J 111:2077–2085
39. Yoffe AM, Prinsen P, Gopal A, Knobler CM, Gelbart WM, Ben-Shaul A (2008) Predicting the sizes of large RNA molecules. Proc Natl Acad Sci USA 105:16153–16158
40. Tubiana L, Božič A, Micheletti C, Podgornik R (2015) Synonymous mutations reduce genome compactness in icosahedral ssRNA viruses. Biophys J 108:194–202
41. Božič A, Micheletti C, Podgornik R, Tubiana L (2018) Compactness of viral genomes: effect of disperse and localized random mutations. J Phys Condens Matter 30:084006
42. Farrell J, Dobnikar J, Podgornik R (2023) Role of genome topology in the stability of viral capsids. Phys Rev Res 5:L012040
43. Fallmann J, Will S, Engelhardt J, Grüning B, Backofen R, Stadler PF (2017) Recent advances in RNA folding. J Biotechnol 261:97–104

44. Lorenz R, Bernhart SH, Höner zu Siederdissen C, Tafer H, Flamm C, Stadler PF, Hofacker IL (2011) ViennaRNA Package 2.0. Algorithms Mol Biol 6:1–14
45. Reuter JS, Mathews DH (2010) RNAstructure: software for RNA secondary structure prediction and analysis. BMC Bioinform 11:1–9
46. Do CB, Woods DA, Batzoglou S (2006) CONTRAfold: RNA secondary structure prediction without physics-based models. Bioinformatics 22:e90–e98
47. Wayment-Steele HK, Kladwang W, Strom AI, Lee J, Treuille A, Becka A, Participants E, Das R (2022) RNA secondary structure packages evaluated and improved by high-throughput experiments. Nat Methods 19:1234–1242
48. Koodli RV, Rudolfs B, Wayment-Steele HK, Eterna Structure Designers, Das R (2021) Redesigning the Eterna100 for the Vienna 2 folding engine. bioRxiv. Available from: https://www.biorxiv.org/content/10.1101/2021.08.26.457839v1
49. Liu M, Poppleton E, Pedrielli G, Šulc P, Bertsekas DP (2022) ExpertRNA: a new framework for RNA structure prediction. INFORMS J Comput 34(5):2464–2484
50. Spasic A, Assmann SM, Bevilacqua PC, Mathews DH (2018) Modeling RNA secondary structure folding ensembles using SHAPE mapping data. Nucleic Acids Res 46:314–323
51. Mathews DH, Turner DH (2006) Prediction of RNA secondary structure by free energy minimization. Curr Op Struct Biol 16:270–278
52. Schlick T (2018) Adventures with RNA graphs. Methods 143:16–33
53. Gross JL, Yellen J, Anderson M (2018) Graph theory and its applications. Chapman and Hall/CRC, London
54. Todeschini R, Consonni V (2008) Handbook of molecular descriptors. John Wiley, Sons
55. Rouvray DH, King RB (2002) Topology in chemistry: discrete mathematics of molecules. Elsevier, Amsterdam
56. Sazer S, Schiessel H (2018) The biology and polymer physics underlying large-scale chromosome organization. Traffic 19:87–104
57. Perry SL (2019) Phase separation: bridging polymer physics and biology. Curr Op Colloid Interface Sci 39:86–97
58. Wagner J, Erdemci-Tandogan G, Zandi R (2015) Adsorption of annealed branched polymers on curved surfaces. J Phys Condens Matter 27:495101
59. Gutin AM, Grosberg AY, Shakhnovich EI (1993) Polymers with annealed and quenched branchings belong to different universality classes. Macromolecules 26(6):1293–1295
60. Everaers R, Grosberg AY, Rubinstein M, Rosa A (2017) Flory theory of randomly branched polymers. Soft Matter 13(6):1223–1234
61. Wang ZG (2017) 50th anniversary perspective: polymer conformation—a pedagogical review. Macromolecules 50(23):9073–9114
62. Bhattacharjee SM, Giacometti A, Maritan A (2013) Flory theory for polymers. J Phys Cond Matter 25(50):503101
63. Rubinstein M, Colby RH (2003) Polymer physics. Oxford University Press, New York
64. Li B, Madras N, Sokal AD (1995) Critical exponents, hyperscaling, and universal amplitude ratios for two-and three-dimensional self-avoiding walks. J Stat Phys 80(3):661–754
65. Parisi G, Sourlas N (1981) Critical behavior of branched polymers and the lee-yang edge singularity. Phys Rev Lett 46:871–874
66. Van Rensburg EJ, Madras N (1992) A nonlocal Monte Carlo algorithm for lattice trees. J Phys A Math Theor 25:303
67. Rosa A, Everaers R (2016) Computer simulations of randomly branching polymers: annealed versus quenched branching structures. J Phys A Math Theor 49:345001
68. Rosa A, Everaers R (2016) Computer simulations of melts of randomly branching polymers. J Chem Phys 145:164906
69. Flory PJ (1953) Principles of polymer chemistry. Cornell University Press, Ithaca (NY)
70. Simón D, Cristina J, Musto H (2021) Nucleotide composition and codon usage across viruses and their respective hosts. Front Microbiol 12:646300
71. Schultes E, Hraber PT, LaBean TH (1997) Global similarities in nucleotide base composition among disparate functional classes of single-stranded RNA imply adaptive evolutionary convergence. RNA 3:792–806

72. Higgs PG (1993) RNA secondary structure: a comparison of real and random sequences. J Phys I (3):43–59
73. Clote P, Ferré F, Kranakis E, Krizanc D (2005) Structural RNA has lower folding energy than random RNA of the same dinucleotide frequency. RNA 11:578–591
74. Lefkowitz EJ, Dempsey DM, Hendrickson RC, Orton RJ, Siddell SG, Smith DB (2018) Virus taxonomy: the database of the International Committee on Taxonomy of Viruses (ICTV). Nucleic Acids Res 46:D708
75. Gaunt ER, Digard P (2022) Compositional biases in RNA viruses: causes, consequences and applications. Wiley Interdiscip Rev RNA 13:e1679
76. Di Giallonardo F, Schlub TE, Shi M, Holmes EC (2017) Dinucleotide composition in animal RNA viruses is shaped more by virus family than by host species. J Virol 91:e02381–e023416
77. Belalov IS, Lukashev AN (2013) Causes and implications of codon usage bias in RNA viruses. PLOS One 8:e56642
78. Jiang M, Anderson J, Gillespie J, Mayne M (2008) uShuffle: a useful tool for shuffling biological sequences while preserving the k-let counts. BMC Bioinform 9:1–11
79. Zhao Y, Wang J, Zeng C, Xiao Y (2018) Evaluation of RNA secondary structure prediction for both base-pairing and topology. Biophys Rep 4:123–132
80. Turner DH, Mathews DH (2010) NNDB: the nearest neighbor parameter database for predicting stability of nucleic acid secondary structure. Nucleic Acids Res 38:D280–D282
81. Andronescu M, Condon A, Hoos HH, Mathews DH, Murphy KP (2010) Computational approaches for RNA energy parameter estimation. RNA 16:2304–2318
82. Langdon WB, Petke J, Lorenz R (2018) Evolving better RNAfold structure prediction. In: European conference on genetic programming, pp 220–236
83. Poznanović S, Wood C, Cloer M, Heitsch C (2021) Improving RNA branching predictions: advances and limitations. Genes 12:469
84. Wiedemann J, Kaczor J, Milostan M, Zok T, Blazewicz J, Szachniuk M, Antczak M (2022) RNAloops: a database of RNA multiloops. Bioinformatics 38(17):4200–4205
85. Zuker M, Stiegler P (1981) Optimal computer folding of large RNA sequences using thermodynamics and auxiliary information. Nucleic Acids Res 9:133–148
86. Ward M, Datta A, Wise M, Mathews DH (2017) Advanced multi-loop algorithms for RNA secondary structure prediction reveal that the simplest model is best. Nucleic Acids Res 45:8541–8550
87. Ward M, Sun H, Datta A, Wise M, Mathews DH (2019) Determining parameters for non-linear models of multi-loop free energy change. Bioinformatics 35:4298–4306
88. Poznanović S, Barrera-Cruz F, Kirkpatrick A, Ielusic M, Heitsch C (2020) The challenge of RNA branching prediction: a parametric analysis of multiloop initiation under thermodynamic optimization. J Struct Biol 210:107475
89. Amman F, Bernhart SH, Doose G, Hofacker IL, Qin J, Stadler PF, Will S (2013) The trouble with long-range base pairs in RNA folding. In: Brazilian symposium on bioinformatics. Springer, Berlin, pp 1–11
90. Pyle AM, Schlick T (2016) Challenges in RNA structural modeling and design. J Mol Biol 428:733
91. Lorenz R, Stadler PF (2020) RNA secondary structures with limited base pair span: exact backtracking and an application. Genes 12:14
92. Archer EJ, Simpson MA, Watts NJ, O'Kane R, Wang B, Erie DA, McPherson A, Weeks KM (2013) Long-range architecture in a viral RNA genome. Biochemistry 52:3182–3190
93. Lan TC, Allan MF, Malsick LE, Woo JZ, Zhu C, Zhang F, Khandwala S, Nyeo SS, Sun Y, Guo JU et al (2022) Secondary structural ensembles of the SARS-CoV-2 RNA genome in infected cells. Nat Comm 13:1–14
94. Simmonds P, Tuplin A, Evans DJ (2004) Detection of genome-scale ordered RNA structure (GORS) in genomes of positive-stranded RNA viruses: implications for virus evolution and host persistence. RNA 10:1337–1351
95. Cao C, Cai Z, Xiao X, Rao J, Chen J, Hu N, Yang M, Xing X, Wang Y, Li M et al (2021) The architecture of the SARS-CoV-2 RNA genome inside virion. Nat Comm 12:1–14

96. Lubensky TC, Isaacson J (1979) Statistics of lattice animals and dilute branched polymers. Phys Rev A 20:2130–2146
97. Lubensky TC, Isaacson J (1972) Statistics of lattice animals and dilute branched polymers. Phys Lett A 38:339–340
98. Li S, Erdemci-Tandogan G, Wagner J, van der Schoot P, Zandi R (2017) Impact of a nonuniform charge distribution on virus assembly. Phys Rev E 96:22401
99. Šiber A, Podgornik R (2008) Nonspecific interactions in spontaneous assembly of empty versus functional single-stranded RNA viruses. Phys Rev E 78:051915
100. Huang C, Podgornik R, Man X (2021) Selective adsorption of confined polymers: self-consistent field theory studies. Macromolecules 54:9602–9608
101. Dong Y, Li S, Zandi R (2020) Effect of the charge distribution of virus coat proteins on the length of packaged RNAs. Phys Rev E 102:062423
102. Garmann RF, Gopal A, Athavale SS, Knobler CM, Gelbart WM, Harvey SC (2015) Visualizing the global secondary structure of a viral RNA genome with cryo-electron microscopy. RNA 21:877–886
103. Jain S, Tao Y, Schlick T (2020) Inverse folding with RNA-As-Graphs produces a large pool of candidate sequences with target topologies. J Struct Biol 209:107438
104. Geary C, Grossi G, McRae EK, Rothemund PW, Andersen ES (2021) RNA origami design tools enable cotranscriptional folding of kilobase-sized nanoscaffolds. Nat Chem 13:549–558

Chapter 2
RNA Multiscale Simulations as an Interplay of Electrostatic, Mechanical Properties, and Structures Inside Viruses

Sergio Cruz-León, Salvatore Assenza, Simón Poblete, and Horacio V. Guzman

Abstract RNA is a functionally rich biomolecule with multiple hierarchical structures that emerge from phenomena occurring at different temporal and spatial scales. Multiscale simulations of RNA, including all-atom simulations, coarse-grained strategies, and continuum models, have enabled the quantification of electrostatic and mechanical interactions at the nanometer scale, providing physical insights and systematic interpretation of experiments. In this chapter, we present recent methodological developments, parametrization details, and applications for the study of nucleic acids at different scales. In particular, we explore RNA interactions with ions, substrates and virus capsids, compare RNA and DNA properties, and describe how these methods could contribute to the reconstruction of virus genome structures. Finally, we discuss future developments and challenges in this rapidly evolving field.

S. Cruz-León
Department of Theoretical Biophysics, Max Planck Institute of Biophysics, Max-von-Laue-Str. 3, 60438 Frankfurt am Main, Germany
e-mail: sergio.cruz@biophys.mpg.de

S. Assenza
Departamento de Física Teórica de la Materia Condensada, Universidad Autónoma de Madrid, 28049 Madrid, Spain
e-mail: salvatore.assenza@uam.es

Condensed Matter Physics Center (IFIMAC), Universidad Autónoma de Madrid, 28049 Madrid, Spain

Instituto Nicolás Cabrera, Universidad Autónoma de Madrid, 28049 Madrid, Spain

S. Poblete
Instituto de Ciencias Físicas y Matemáticas, Universidad Austral de Chile, 5090000 Valdivia, Chile
e-mail: spoblete@dlab.cl

Computational Biology Lab, Fundación Ciencia & Vida, 7780272 Santiago, Chile

H. V. Guzman (✉)
Department of Theoretical Physics, Jožef Stefan Institute, Jamova 39, 1000 Ljubljana, Slovenia
e-mail: horacio.guzman@uam.es

Departamento de Física Teórica de la Materia Condensada, Universidad Autónoma de Madrid, 28049 Madrid, Spain

M. Comas-Garcia and S. Rosales-Mendoza (eds.), *Physical Virology*, Springer Series in Biophysics 24, https://doi.org/10.1007/978-3-031-36815-8_2

Introduction

RNA research is thriving in physical virology. Several packing mechanisms of a few counted viruses and emerging functions of RNA structure inside virus shells have been already elucidated [1]. However, the assembly of RNA viruses is complex and involves the interaction between the genome and the capsid proteins. RNA viruses typically have genomes of several thousands of nucleobases, highly charged, structurally complex, and densely packed within the virion. Understanding the physics of viral RNA requires addressing diverse phenomena that occur on different temporal and spatial scales. In this chapter, we will show recent advances in the multiscale modeling of RNA. We will start by discussing the importance of electrostatic effects (Section "The Importance of RNA Electrostatic Interactions"). On the one hand, we will use continuum theories to describe the interaction of RNA with surfaces and the impact of nucleic acid secondary structure. On the other hand, we will describe electrostatics from the detailed insight of atomistic simulations. The simulations provide access to the thermodynamics and kinetics that govern interactions between RNA and its environment. In section "RNA Mechanical Properties", we focus on the elastic properties of nucleic acids. We review the experimental and computational methods to determine these properties and show advances in all-atom and coarse-grained models for an accurate description of the mechanical complexity of nucleic acids. In section "RNA Genome 3D Reconstruction", we offer algorithms that allow the assembly of the RNA tertiary structure using fragments. Finally, we discuss current challenges and the future role of realistic multiscale modeling in advancing our understanding of RNA biophysics.

The Importance of RNA Electrostatic Interactions

Electrostatic interactions are often prescribed as one of the generators, or the most important one in virus capsids [2], as well as full virion assembly [3]. The precise number of charges carried by the proteins and genome of virions depend strongly on the solution conditions and structural distribution of dissociable groups on the regions close to the solvent accessible surface. A major part of the volume of the virion is composed of DNA or RNA, which are macro-ions that carry massive amounts of charge, one phosphate group (negative charge) per nucleotide. Consequently, nucleic acids form an ionic atmosphere composed of excess of cations and depletion of anions around that also screen the electrostatic interactions. Without the ionic atmosphere, the confinement inside the viral capsid, folding, and function of nucleic acids would involve crossing enormous electrostatic barriers [4]. At physiological conditions, the ionic atmosphere for RNA viruses is commonly located close to the proteinaceous capsid, which also converges with the concept of neutral total charge inside the capsid required for spontaneous assembly [5].

The ionic atmosphere around nucleic acids is divided into diffusive and associated ions [6]. On the one hand, diffusive ions refer to a significant accumulation of mobile cations (and depletion of anions) around the polyelectrolyte determined largely by long range electrostatics. The diffusive layer contains the majority of the interacting ions, contributes significantly to the thermodynamic stability of DNA and RNA secondary structures [6, 7], and regulates the adsorption of RNA to substrates [8] (Section "Capsid Proteins and RNA-substrate Interactions with Coarse–Grained and Continuum Models"). On the other hand, the associated ions bind in specific sites of nucleic acids [7, 9]. Their effects strongly depend on the ion type and its understanding requires an atomistic level of description (Section "Ion-specific Effects in RNA Systems"). Associated ions induce conformational changes and affect the stability, melting temperature, folding kinetics, and biological activity of DNA and RNA (see reference [9] for an overview).

Furthermore, ions screen electrostatics, stabilize DNA and RNA and can be responsible for large conformational changes [10], such as the compactness of the genome inside a virus capsid. They can induce bending [11, 12], over- and underwinding [13–15], stabilize tertiary structures and assist the catalytic activity of RNAs [6, 7, 16, 17]. Therefore, to understand nucleic acids, sophisticated knowledge and modeling of the ions around them is crucial [4]. From the protein side, it is also crucial to understand the molecular and even atomistic identity [18], as well as the charged states [1].

Given the scales of virus genome, providing insight into the interpretation of electrostatics phenomenology requires an integrative multiscale approach. Atomistic simulations can provide a high level of detail but are limited to short time- and length scales. On the other hand, coarse-grained (CG) simulations or continuous models allow us to access larger systems, but their insight is more limited. For example, the SARS-CoV-2 reaches 30 kb and several millions of residues [19, 20]. The sampling of a system like SARS-CoV-2 in all-atom simulations is still computationally prohibitive. Here, coarse-grained methods have the opportunity of mediating between all-atom and continuum approaches [21, 22]. However, an integrative CG model for both proteins and RNA with tuned electrostatics is still a challenging topic [8, 23–25]. From a continuum Poisson-Boltzmann approach the ionic clouds surrounding RNA molecules have been also observed [9, 26]. Although continuous models may overlook associative ions at atomistic scale, they still provide *grosso modo* insight. Understanding viral electrostatics requires feedback loops between different modeling methods at different scales.

Capsid Proteins and RNA-Substrate Interactions with Coarse-Grained and Continuum Models

Recent research results show the electrostatic interactions of a nanoprobe with the proteins inside and at the surface of a Zika Virus capsid described as a function of the distance, molecular charge identity and variations of ionic strength [18]. RNA as flexible molecules go through structural changes to adsorb onto substrates. The problem of how the secondary structure regulates the adsorption mechanisms of RNA is unsolved from a generalized perspective. Here, we take a major step in its solution by creating a multiscale method, based on electrostatic Debye-Hückel theory, to calculate the adsorption free energies between archetypical RNA fragments and structureless flat substrates. The resulting model, taking into account with existing secondary structures obtained from chemical probing experiments, predicts the adsorption free energies and connects to RNA 3D structure [27, 28]. The latter modulates the interaction regimes, for "more or less" sounds colloquial. Better: "moderately" efficient adsorption. This works sets the stage for tackling RNA interactions to proteinaceous substrates, such as virus capsids and the effects those substrates may have on the elemental RNA secondary structure and vice versa.

Localized Proteinaceous Capsid Electrostatic Interactions in Variable Salinity Environments

A continuum description of electrostatic interactions is crucial for understanding function and phenomenology in biological systems. In particular, the interactions generated at the atomistic and molecular resolutions are of predominant interest for non-enveloped and enveloped viruses [1, 29–35]. For a few decades, the physical virology community has employed diverse experimental techniques, like force microscopy, to learn more about the mechanical and electrostatic properties of biomacromolecular systems [36–46]. However, the amount of viruses that are electrostatically characterized at the nanoscale are scarce [18, 38]. The most recent computational model [18] tackles the electrostatic force arising from the interaction between a nanoprobe and the virus capsid via Poisson-Boltzmann calculations.

There are several ways to calculate the electrostatic forces using Poisson-Boltzmann [47]. Here, the total force $\mathbf{F}$ is decomposed as,

$$\mathbf{F} = \int_{\Omega} \left[\rho_f \mathbf{E} - \frac{1}{2}\mathbf{E}^2 \nabla\epsilon - \frac{1}{2}\epsilon\kappa^2\phi^2\nabla\lambda \right] d\Omega = \mathbf{F}_{\text{qf}} + \mathbf{F}_{\text{db}} + \mathbf{F}_{\text{ib}} \tag{2.1}$$

where ρ_f is the charge distribution in the solute, $\mathbf{E}$ the electric field, and λ a unit step function that is 1 in Ω_w (the region containing the solvent-water) and 0 elsewhere, ϵ is the dielectric constant, κ is the inverse of the Debye length and ϕ is the electrostatic potential. The first term $\mathbf{F}_{\text{qf}}$ corresponds to the force on the solute's charges, then $\mathbf{F}_{\text{db}}$ corresponds to the pressure due to the dielectric jump on the interface, and the final term $\mathbf{F}_{\text{ib}}$ arises from the sudden jump in ionic concentration on the molecular surface. More details on the derivation of equations and definition of variables are contained elsewhere [18].

Figure 2.1 shows the interaction force (force magnitude normalized by the total magnitude $\bar{\mathbf{F}}_{\text{qf}}$), indicating that the atoms located closer to the tip are also concentrating most of the contribution to the total force (blue atoms in Fig. 2.1c and d). Note that the nanoprobe approaches from the right end in the x-axis. From the experimental AFM viewpoint, the dependence on the distance is a well-known observable of the sensed molecular interactions. In order to cover this aspect from the theoretical perspective, we analyzed the atoms corresponding to each amino acid based on 5 Å thick slices (normal to the x-direction), which resembles the tip-capsid distance in a similar form as AFM 'force tomography' or subsurface imaging [48]. Note that Fig. 2.1c and d differ only in the amount of ionic concentration in the solvent, namely, 20mM driving stronger electrostatic interactions and 150mM, which is rather screening most of the ions. The contribution of the atoms from specific amino acids is reflected in the force tomography profiles in Fig. 2.1a, b. In Fig. 2.1a, we identify 5 amino acids that are more sensitive to the nanoprobe, namely, LYS, ASP, GLU, ARG and ILE. Those amino acids have different net charges depending on the pH of the medium where they are located, whereby Fig. 2.1a shows that for LYS and ARG the interactions are attractive, while for both ASP and GLU are of repulsive nature. Nonetheless, the ILE residues are much more sensitive to the slice where they are located. At pH 7, ILE is an uncharged residue, however if it is sliced by the tomography-type analysis (the whole amino acid is distributed across several slices), the sensed polarity by the nanoprobe depends on the exact charges of the atoms of the specific slice (Fig. 2.1a). This explains the observed transitions from repulsive to attractive interactions and vice versa. Interestingly, the same quantification was performed for the higher salt concentration force, namely 150 mM (Fig. 2.1b). Here the amino acids with higher contribution to the force have changed drastically in order and also in species, as it is the case for LYS, THR and SER.

In summary of this subsection, we showed that a simple electrostatic model based on the Poisson-Boltzmann equation provides useful insights of fine-grain mechanisms in the interaction between an AFM tip and a virus capsid. In particular, we demonstrate that the electrostatic force originated from a proteinaceous capsid and detected by a nanoprobe depends on both the precise distribution and identity of charges inside the amino acids and the tip-capsid distance. The research highlights the strong locality of the total interaction force with the amino acids as it approaches the nanoprobe. These results also suggest that the protein shell may shield the RNA genome inside the capsid at higher ionic concentrations. However, further verification, of the interaction between the RNA and the capsid of the virus is currently investigated [1]. As a consequence, the electrostatic component of the force requires higher levels of resolution when modeling and interpreting heterogeneous surfaces, like the ones of virus capsids.

RNA Adsorption onto Flat Substrates

Molecular simulations have offered important insights into the adsorption of semi-flexible macromolecules to a molecular substrate [49–52], while the adsorption of macromolecules with either double-stranded or quenched internal structure onto a molecular substrate remains much less understood. The latter problem is particularly

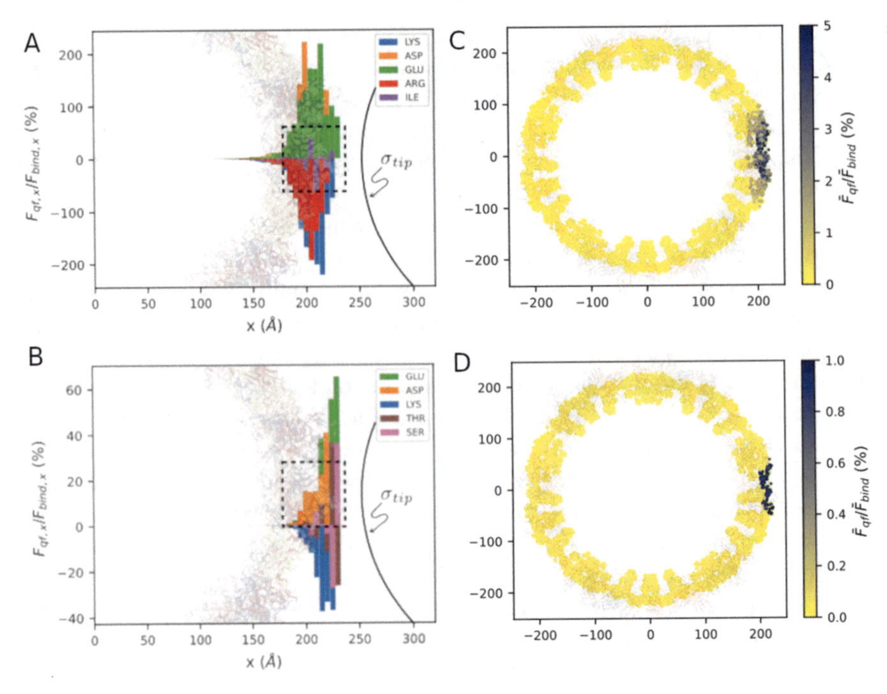

Fig. 2.1 **a** The top left panel shows the x-component of the interaction force (normalized) on atoms that belong to the 5 amino acids that contribute the most to the force (LYS, ASP, GLU, ARG, and ILE, in that order) on 5 Å-thick slices. As reference, we show the Zika capsid structure in the background, and the AFM tip, which is placed 2 Å away. **b** A similar plot as (**a**) with a remarking difference, now the ionic concentration is 150 mM and hence the indentity of the top 5 amino acids drastically changes in order and species (GLU, ASP, LYS, THR, and SER, in that order). **c** Relative magnitude of $\bar{\mathbf{F}}_{qf}$ on atoms located in a 5 Å slice centered at the $z = 0$ Å plane at ionic concentration is 20 mM and **d** 150 mM. *Source* Figure reproduced from [18] with licence CC-BY 3.0

relevant in the context of RNA-virus assembly phenomena [1, 3, 22, 53–57], where the soft and malleable RNA structure can respond to the adsorption process.

To investigate the general role of RNA secondary structure in adsorption processes, we model the substrate as a flat, featureless surface. Excluding topographical and molecular features of the substrate, such a scenario allows us to isolate the influence of the RNA secondary structure in its adsorption. The employed RNA model for unstructured fragments is similar to a single chain bead-spring polymer while the structured RNA uses a tractable CG scheme [58]. We model the attraction of the RNA phosphate groups to the adsorbing substrate by a Debye-Hückel-like interaction potential, which can be rationalized as stemming from the electrostatic interactions between the dissociated RNA phosphates and the substrate charges [59]. Combined with a generic short-range repulsive term [8], the surface potential acting on the RNA assumes the form,

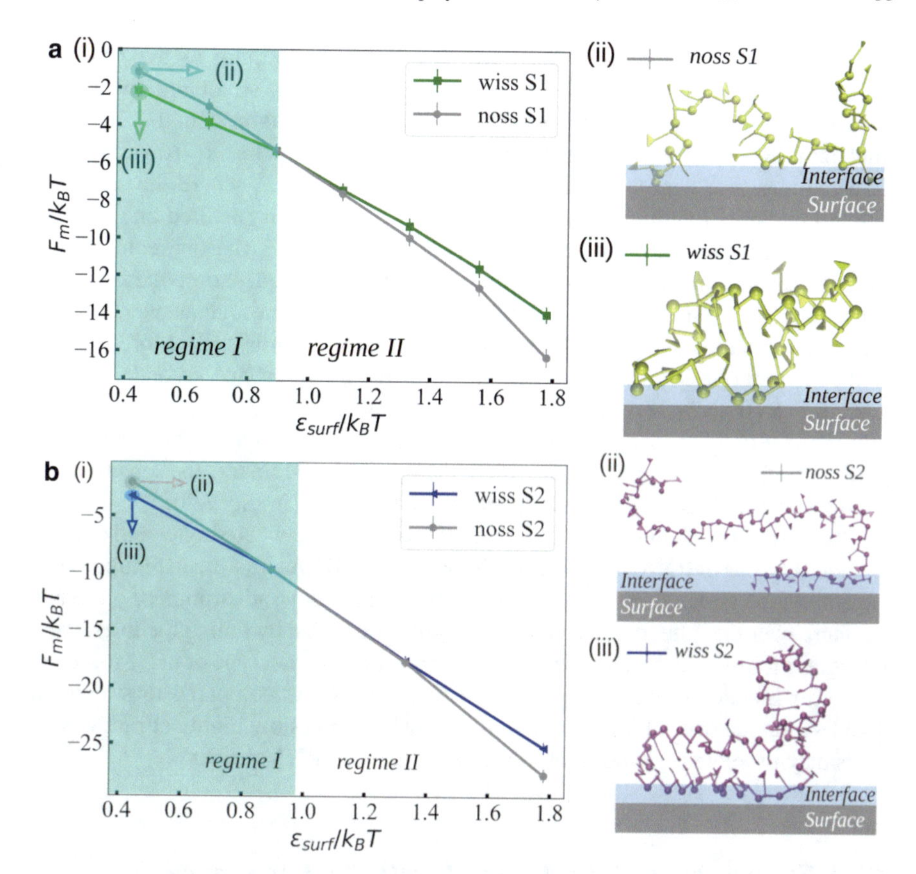

Fig. 2.2 **a** Adsorption free energy as a function of the substrate attraction strength, $\varepsilon_{\text{surf}}/k_{\text{B}}T$, for RNA fragments S1-wiss and S1-noss. For a weak attraction of $\varepsilon_{\text{surf}}/k_{\text{B}}T = 0.44$, we show simulation snapshots of the (ii) unstructured and (iii) structured fragments. The illustrated interface layer is a schematic definition of a distance slightly thicker than the diameter of phosphates. **b** Adsorption free energy F_m as a function of the substrate attraction strength $\varepsilon_{\text{surf}}$ for RNA fragments S2-wiss and S2-noss. Snapshots of (ii) unstructured and (iii) structured fragments at $\varepsilon_{\text{surf}} = 0.44\ k_{\text{B}}T$. *Source* Figure reproduced from [8] with licence CC-BY 4.0

$$U_{\text{surf}}(z) = \varepsilon_{\text{r}} \left(\frac{\lambda_{\text{r}}}{z}\right)^9 - \varepsilon_{\text{surf}} \exp\left(-\frac{z}{\lambda_{\text{D}}}\right), \tag{2.2}$$

where z is the distance to the surface, $\varepsilon_{\text{r}} = 1\ k_{\text{B}}T$, $\lambda_{\text{r}} = 0.1$ nm is the distance of activation of the repulsive Lennard-Jones term (well below the size of the RNA phosphates ≈ 0.3 nm, as defined in our model) and $\lambda_{\text{D}} = 1$ nm, considering RNA under typical physiological conditions as described in Refs. [60, 61]. The strength of the attractive potential, $\varepsilon_{\text{surf}}$, is varied in the range between $0.44\ k_{\text{B}}T$ and $1.78\ k_{\text{B}}T$, which are also observed experimentally for nucleic acid packaging [62].

Figure 2.2a shows the adsorption free energy F_m (minimum of the PMF) as a function of the surface attraction strength ε_{surf}, for both structured (wiss) and unstructured (noss) RNA systems. The tackled fragment is system 1 (S1), a 22 nucleotide RNA hairpin, with the following secondary structure *((.(((((......))))).))*. Two regimes are identifiable: the first for $\varepsilon_{surf} < 0.89\ k_B T$, where the fragment with secondary structure adsorbs more strongly than the unstructured one (regime I), while for $\varepsilon_{surf} > 0.89\ k_B T$, the unstructured fragment is the one exhibiting a stronger adsorption free energy (regime II). An intriguing question is how general is the existence of the two adsorption regimes. To provide a rough answer, we performed simulations using a different type of attractive potentials [8]. Moreover, we studied another archetypical RNA fragment of the STMV shape secondary structures [63]. System 2 (S2) contains 40 nucleotides and a slightly lower base-pair fraction *(((((...((.(((((......))))).)).....)))))*, namely 63% (S1), 60% (S2). This brings into play the shifting of the cross-over point to the right, which remarks the sensitivity of the molecule during the adsorption phenomena (as shown in Fig. 2.2b).

Our results, which indicate a selectivity in adsorption between single- and double-stranded regions of RNA, underline the importance of RNA structure in regulating its adsorption to various substrates. We expect that the selective adsorption of one RNA structure over the other could be experimentally controlled by tuning the interaction strength, for instance, by changing the salt concentration, salt type or pH. Moreover, the lower interaction free energies of unstructured RNAs compared to structured double-stranded ones at high attractions, suggest that possibly highly attractive surfaces may promote the unfolding of a double-stranded RNA structure.

RNA Electrostatics at the Atomic Level: The Importance of the Details

At the molecular level, the interaction between RNA and the surrounding ions and water completely determines the electrostatics. All-atom molecular dynamics simulation (MD) is a powerful computational technique that allows investigating physical systems with molecular resolution. MD is a "computational microscope" [64], and it is well suited to study RNA electrostatics because MD can account for the identity of ions, the critical role of water, the internal flexibility of nucleic acids, and the effect of the many-body problem with the required spatial and time resolution. MD has been successfully used in modeling many biological systems atom-by-atom. Recently and thanks to the effort of the community, MD models has proven also powerful in quantitatively reproducing the conformations dynamics and mechanical response of DNA and RNA [15, 65–71] (see section "RNA Mechanical Properties"). This section shows how MD provides molecular insight into the interactions of RNA with ions. This atomic view is essential to understand complex processes such as RNA packaging into viruses because ions can modify the physical and chemical properties of RNA and modulate its interaction with other molecules.

Ion-Specific Effects in RNA Systems

Multiple physical and chemical properties of RNA systems depend on the ion type present in the solution. In this section, we study the thermodynamics and kinetics of ion binding in specific sites of nucleic acids [7, 9] using an atomistic description. We used MD simulations using recently improved models [65, 71–74] combined with enhanced sampling techniques to investigate the interactions of eight mono- and divalent cations with the main binding sites of RNA [9]. The main binding sites on RNA correspond to the backbone (non-bridging oxygen atoms of the phosphate groups: atoms O1P and O2P) and nucleobase (atoms N7 and O6) binding sites. Using a dinucleotide (snapshots in Fig. 2.3), we investigated the binding mechanism, affinities, and exchange rates of Li^+, Na^+, K^+, Cs^+, Mg^{2+}, Ca^{2+}, Sr^{2+}, and Ba^{2+} to understand the molecular origin of ion-specific effects on RNA.

Figure 2.3 shows the free energy profiles $F(r)$ for the metal cations as a function of the distance to the backbone (Fig. 2.3a) and nucleobase binding sites (Fig. 2.3b, c).[1] In all cases, the free energy profiles exhibit two stable states separated by an energetic barrier that shows a common binding mechanism. The first minimum is an inner-sphere interaction where a partially dehydrated cation is in direct contact with the binding site (snapshots (i)). The second minimum is an outer-shell contact, where a fully hydrated ion interacts with the RNA site mediated by a water molecule (snapshots (o)). Therefore, ion-water interactions strongly influence ion binding to RNA.

The free energy profiles show opposing trends for binding affinities. First, at the backbone binding site, the first minimum in F(r) gets deeper with decreasing ion diameter (Fig. 2.3a). The depth of the first minimum indicates that ions with higher charge density (such as Li^+ or Mg^{2+}) interact stronger with the non-bridging oxygen atoms of the RNA backbone (O1P and O2P) and, consequently, have a higher binding affinity. This effect helps explaining changes in free energies [16, 71, 73] and nicely correlates with the ability of ions like Ca^{2+} and Mg^{2+} to stabilize RNA structures [7, 9, 75]. The trend is completely reversed for the nucleobase binding site N7. At the N7 site, ions with low charge density are preferred, which provides further evidence for the discussion on the binding of Mg^{2+} in nucleic acids [76].

Finally, the binding kinetics is also affected by ion type and RNA binding site. For example, at the backbone binding site, the energetic barrier separating the two stable states increases with decreasing ion diameter, therefore slowing exchange kinetics. We determined that ion-binding lifetimes span more than five orders of magnitude, from picoseconds for Cs^+ up to hundreds of nanoseconds for Ca^{2+} [9, 71].

Overall, the results from the dinucleotide reveal notorious ion-specific effects and a high selectivity of RNA binding sites. The site-specific affinities and the vast time-scale involved provide a microscopic explanation of the ion-specific effects observed in multiple macroscopic RNA properties. For example, the results from Fig. 2.3 help us to understand the central role of ion type in RNA folding stability and kinetics. RNA stability is largely determined by the binding affinity of the ions. On the other

[1] Details on the free energy profiles calculations can be found in Ref. [9].

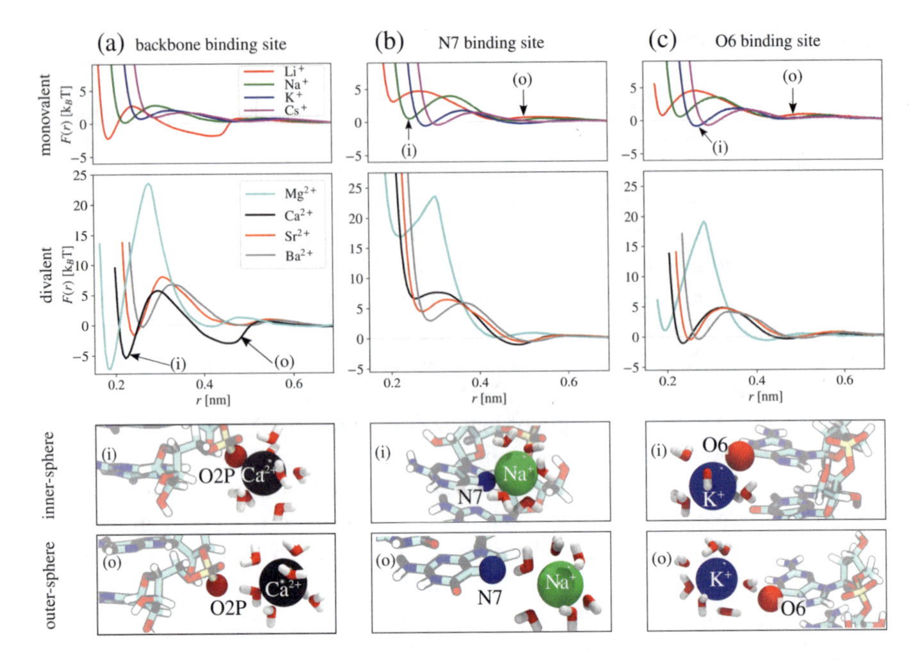

Fig. 2.3 Free energy profiles $F(r)$ for the binding of metal cations as a function of the distance to the O2P atom in the backbone binding site (**a**), and to the N7 (**b**) and the O6 binding site in the nucleobase (**c**). Simulation snapshots for inner-sphere (i) and outer-sphere (o) conformation at the different ion binding sites (bottom). The arrows indicate at which separation the snapshots were taken. For clarity, only water molecules in the first hydration shell of the cation are shown. *Source* Figure reproduced from [9] with licence CC-BY 4.0

hand, RNA folding times depend on binding kinetics, and therefore is affected by lifetimes on ion-binding. However, although conceptually useful, the dinucleotide results are limited when one tries to understand the complex structural landscape of nucleic acids. Consequently, in the following section, we use atomistic simulations to quantify the effects of ions in the more structurally complex nucleic acids.

Electrostatics Around DNA and RNA Duplexes

Ion distributions and nucleic acid structure are intertwined: ion distributions modify the nucleic acid structure [13–15]; conversely, nucleic acid structure modifies ion distributions [77]. In this section, we use MD simulations to show how the ion distributions, and consequently the electrostatics around dsDNA and dsRNA, change depending on ion type and the nucleic acid type. We focus on duplexes because they are the main form of DNA and the most common secondary fragment for RNA. For example, the crystal structure of the Satellite Tobacco Mosaic Virus reveals up to 60% of the genome folded into helices [78].

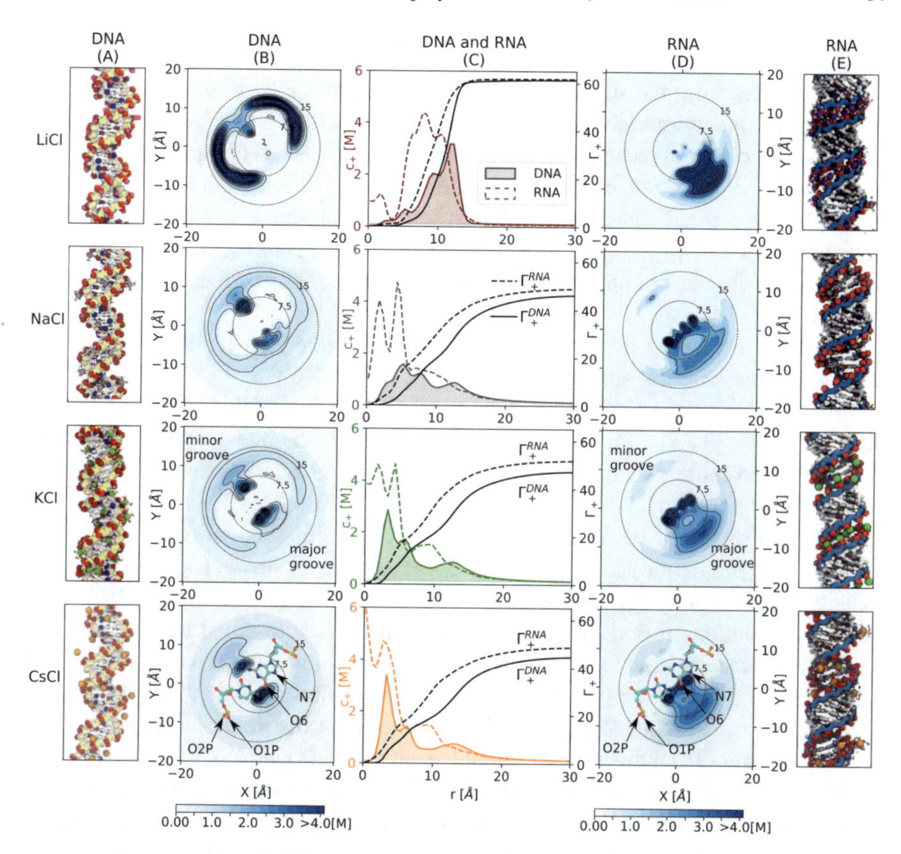

Fig. 2.4 Ion distribution and excess of monovalent cations around DNA and RNA. **a**, **e** Simulation snapshots. Backbone atoms are indicated in yellow and blue for DNA and RNA, respectively. The most frequent ion binding sites are highlighted: red (O1P, O2P, and O6 atoms) and blue (N7 atoms). **b**, **d** Top view of the untwisted helicoidal ion concentration obtained with the software canion [79, 80]. In this representation, the upper-left and lower-right corners correspond to the minor and major grooves as indicated by the superimposed molecular schemes of cytosine-guanine (bottom). In these schemes the most frequent ion binding sites are labeled. The dotted concentric circles indicate the distance to the center of the helix (radius in Å). **c** Ion concentration profiles c_+ as function of the distance r for DNA (solid line) and RNA (dashed line). Concentration profiles for DNA are filled for clarity. *Source* Figure reproduced from Ref. [77] with licence CC-BY 4.0

Figure 2.4 shows the ion distributions obtained from extensive unrestrained simulations of 33-base pairs (bp) DNA and 33-bp RNA duplexes in their B-helix and A-helix topologies, respectively (see details in Ref. [77]). To allow the identification and comparison of specific nucleic acid volumes such as the major groove, minor groove, and the non-bridging oxygen atoms in the phosphate groups along the backbone, we used the curvilinear helicoidal coordinate system introduced by Lavery et al. [79] (Fig. 2.4b, d).

The distributions of all the cations can be divided into two regimes: long and short distances from the surface of the DNA (Fig. 2.4c). Far away from the nucleic acid surface, the cylindrical cations concentration decay towards the bulk value, as predicted by classical mean-field theories [81, 82] (see also section "Introduction"). However, at short distances from the DNA helix ($r \lessapprox 15$ Å), the ion distributions are highly structured and unique for each ion type and nucleic acid type. The ion distributions close to the nucleic acid surface emerge from the direct interactions of the ions with the binding sites at the nucleic acids, as described in the previous section.

High and low charge density cations preferentially interact with the backbone and the nucleobase binding sites of nucleic acids, respectively. For example, all DNA profiles in Fig. 2.4 have a local maximum at the position of the backbone binding sites $r \approx 11 - 13$ Å, i.e., all the cations interact at the backbone of DNA. In addition, this peak is prominently higher for ions with high charge density, e.g., Li^+ compared to ions with intermediate or lower charge density, such as K^+ or Cs^+. In contrast, the innermost peaks ($r < 10$ Å), corresponding to the interaction with the nucleobases at the minor and major grooves (see Fig. 2.4b), increase with decreasing ion charge density ($Li^+ < Na^+ < K^+ < Cs^+$).

The simulations reveal that despite having helices with equal total charge and analogous sequences the ionic distributions depend both on ion type and the type of nucleic acid. Each cation type follows a distinct distribution when interacting with DNA or RNA. Simultaneously, for all ions, the ionic distribution around DNA is notably different from the distribution of the same ion around RNA. In DNA, which is normally a B-helix, ions are distributed between the major groove, the minor groove, and the backbone, whereas RNA, which typically is an A-helix, always favors ion localization in the major groove. As expected from the dinucleotide results (Section "Ion-specific Effects in RNA Systems"), ions with high charge density interact strongly with the nucleic acid backbone, whereas ions with low density interact with nucleobases. The ion-site affinities explain the preferential distributions around DNA because in the B-helix the binding sites are spatially segregated. However, in the case of RNA, the interactions between the individual binding site and ions are insufficient to describe the results. We need to consider the topology of RNA. The A-helix points the backbone binding sites toward the interior of the major groove, where the nucleobase binding sites are also located.

In summary, we used MD simulations and recent development in atomic models [65, 71–74] to quantify, in detail, the interaction of ions with nucleic acid binding sites [9, 71, 77]. The simulations showed high selectivity of ion binding to nucleic acids. Taken together, this detailed study allows us to understand a variety of experiments, to visualize how ions change the electrostatic environment around nucleic acids, and with it, the RNA properties and its interactions with other molecules.

RNA Mechanical Properties

The mechanical properties of nucleic acids play a key role for their functioning in vivo as they affect e.g. the affinity of binding proteins [83] or the packaging of genetic material in viral capsids. Experimental and simulation works have outlined a rich landscape for the elastic response of nucleic acids, unveiling a similar behavior of RNA and DNA in some properties, yet also showing marked differences in several other features. The current knowledge of nucleic-acids elasticity is mostly oriented towards duplexes. Several notable exceptions include experiments on single-stranded nucleic acids [84–93], which have highlighted a pronounced higher flexibility when compared to their double-stranded counterparts. Recent simulations [94, 95] and theory [90, 93, 96] have also provided fresh insights, although overall the state of the art is not as developed as for double-stranded nucleic acids, which will be the focus of the present section.

Local Elasticity: Base-Pair-Step Parameters

A powerful insight on molecular flexibility can be obtained by the study of fluctuations, since more rigid molecules are characterized by weaker thermally-induced conformational fluctuations. Based on this, atomistic simulations demonstrated that for various angular degrees of freedom double-stranded RNA (dsRNA) is stiffer than double-stranded DNA (dsDNA), including e.g. the sugar pucker and the glycosidic angle [97]. Nevertheless, later simulation work has shown that the relative flexibility between dsDNA and dsRNA depends on the particular mechanical stress begin considered, as well as the scale at which it is being applied [98, 99]. In order to characterize the flexibility of nucleic acids, one has first to introduce suitable theoretical frameworks, which we rapidly describe in this section.

In complex molecules such as nucleic acids, the conformational state is usually characterized by introducing multiple variables which are significantly coupled with each other [99, 100]. Considering a perturbative approach, the elastic energy U is harmonic around the ground state [101]:

$$U(\mathbf{q}) = \frac{1}{2}\Delta\mathbf{q}\mathbf{K}\,\Delta\mathbf{q}^T\,. \tag{2.3}$$

In the previous formula, $\mathbf{q} = \{q_1, \ldots, q_N\}$ is a vector containing the N conformational variables; $\Delta\mathbf{q} = \mathbf{q} - \mathbf{q}_0$, where $\mathbf{q}_0$ contains corresponds to the minimum-energy state; the apex ^{T} indicates transposition; and $\mathbf{K}$ is the stiffness matrix, where the diagonal elements are the elastic constants associated to the various modes and the off-diagonal elements are the coupling constants. Application of the equipartition theorem to this quadratic energy leads to

$$\mathbf{K} = k_B T\,\mathbf{V}^{-1}\,, \tag{2.4}$$

where $k_B T$ is the thermal energy and $\mathbf{V}$ is the covariance matrix, with elements $V_{ij} = \langle \Delta q_i \Delta q_j \rangle$ and $\langle \cdots \rangle$ denoting thermal averaging. The previous formula provides a handy method to determine the stiffness matrix $\mathbf{K}$ based on the knowledge of $\mathbf{q}$, acquired from different replicas of the system or measuring the same thermally-equilibrated sample at different times.

A popular choice to characterize local elasticity of duplexes is provided by the base-pair-step parameters [99, 101], which give the relative position (shift, slide, rise) and orientation (tilt, roll, twist) between the planes of consecutive base pairs. By setting $\mathbf{q} = \{\text{shift, slide, rise, tilt, roll, twist}\}$, the local elasticity of dsDNA and dsRNA has been assessed by database analysis of crystal structures [99, 101] as well as by atomistic simulations [99]. This analysis has shown that, on average, dsRNA is more rigid than dsDNA at the local level. Yet, there is a wide variability according to sequence, which can be exploited to easily device dsRNA molecules which are actually softer than dsDNA with the same sequence. Moreover, even for the same sequence, the relative flexibility of dsRNA and dsDNA sometimes depends on the particular conformational feature being addressed. For instance, for the base-pair step CG, dsRNA is stiffer than dsDNA for the slide, but it is softer for the rise [99].

Global Elasticity: Elastic Rod Model

For global features, a nucleic acid is usually ascribed to a homogeneous, continuous rod characterized by stretch, twist, and bend deformations. For molecules up to tens of base pairs, the bending fluctuations of dsRNA and dsDNA can be neglected. Moreover, also for long molecules (as the ones considered in single-molecule experiments) the bending effects can be uncoupled from the other elastic modes when large pulling forces are considered [102]. By focusing on stretch and twist conformational changes, one thus chooses $\mathbf{q} = \{L, \theta\}$ in Eq. (2.3), with L and θ being the overall contour length and torsion, respectively. In this case, Eq. (2.3) is usually written in the form

$$U(L, \theta) = \frac{1}{2}\frac{S}{L_0}\Delta L^2 + \frac{1}{2}\frac{C}{L_0}\Delta\theta^2 + \frac{g}{L_0}\Delta L \Delta\theta \,, \tag{2.5}$$

where L_0 is the equilibrium value for the contour length, ΔL and $\Delta\theta$ are the deviations for contour length and torsion, while the elastic constants S, C and g are the stretch modulus, the twist modulus and the twist-stretch coupling, respectively. Equation (2.5) characterizes the energy of the so-called elastic rod model. Based on Eqs. (2.4) and (2.5), one can obtain the elastic constants from the fluctuations of L and θ. An alternative route consists in applying a force f and/or a torque τ, and minimize the total energy $U - f\Delta L - \tau\Delta\theta$. Thanks to the formulas obtained, fit of the curves giving $\langle \Delta L \rangle$ and $\langle \Delta\theta \rangle$ as a function of f and τ enables computing the elastic constants. This approach implicitly assumes that the elastic constants are independent of the applied mechanical stress.

Table 2.1 Collection of elastic constants obtained in the literature for dsDNA and dsRNA

Experiments						
		S (pN)		C (pN·nm^2)		g (pN·nm)
dsDNA		900–1400 Refs. [11, 12, 104, 105]		390–460 Refs. [11, 100, 106–108]		−120 to −90 Refs. [11, 100, 102, 109]
dsRNA		350 to 700 Refs. [11, 12]		400 Refs. [11]		50 Refs. [11]
Atomistic simulations						
		Modification	S (pN)	C (pN·nm^2)	g (pN·nm)	Ref.
dsDNA		bsc0	1300	300	−220	[66]
		bsc1	1000 to 1500	200 to 400	−140	[65, 68]
		OL15	1700 to 1800	400 to 500	N.A.	[70, 110]
dsRNA		bsc0+χ_{OL3}	500 to 600	300	80 to 140	[66]
Coarse-grained Simulations (MADna [111])						
		S (pN)		C (pN·nm^2)		g (pN·nm)
dsDNA		1000		400		−120

The values of S, C and g obtained in single-molecule experiments are recapitulated in Table 2.1 top. While the twist response is similar for both duplexes, dsRNA is significantly softer than dsDNA for stretching. This is in contrast with the elasticity at the local level, where on average dsRNA was found to be stiffer, and provides a further example of the dependence of the relative flexibility between dsRNA and dsDNA on the deformation mode under consideration [99]. A remarkable qualitative difference between dsDNA and dsRNA emerges in the twist-stretch coupling, which has opposite sign for the two molecules [11]. This implies that, upon pulling, dsRNA relaxes its torsional state, while dsDNA overwinds. The sequence dependence of the elastic constants has hitherto not been experimentally characterized in detail, although it has been shown that dsDNA rich in A-tracts is generally stiffer than a random sequence [103]. Finally, it has been reported that for both dsRNA and dsDNA the stretching modulus increases with the ionic strength of the embedding solution, i.e. when the effect of electrostatics is weakened [12, 104].

Atomistic simulations have provided a fundamental contribution to the microscopic understanding of nucleic-acids elasticity. Indeed, the experimental results for S, C and g were reproduced with good accuracy by all-atom simulations performed on short sequences (Table 2.1 center). The simulations captured quantitatively the stretching modulus and, although with large variability, the twist modulus. Moreover, they correctly account for the opposite sign of g. A microscopic analysis of the simulations enabled to ascribe this qualitative difference to the different sugar pucker induced by the presence or absence of the hydroxyl group in the ribose [66]. Quantitatively, the magnitude of g is overestimated by the simulations [66, 68], although this can be partially ascribed to finite-size effects introduced by the short length of the simulated molecules ($\sim$20 base pairs) [111].

In order to simulate nucleic acids at larger scales, due to computational limits one needs to introduce coarse-grained descriptions. In this regard, until recently the mechanics of dsDNA was not satisfactorily captured by existing models, either due to the lack of sequence dependence or to the overestimation of the elastic constants [111–113]. Moreover, the sign of g was not captured, indicating inherent limits in the qualitative assessment of dsDNA mechanics. The accuracy of atomistic simulations has recently enabled the development of MADna, a sequence-dependent model focused on the conformational and mechanical properties of dsDNA, and entirely parameterized from atomistic simulations [111]. Despite its coarse-grained nature, MADna quantitatively reproduces the main sequence-dependent conformational and elastic properties obtained in atomistic simulations and experiments (see Table 2.1 bottom for average elastic constants). Moreover, it also accurately captures the sequence-dependent helical pitch as obtained from cyclization experiments [114], as well as the pronounced stiffness of A-tracts [103]. The development of coarse-grained models for the elasticity of dsRNA is still in its infancy, in comparison to dsDNA. At present, there is only a report of the stretch modulus for one such model, showing that S is significantly underestimated [115]. Given the importance of nucleic-acids elasticity, there is clearly a great need for further development of dsRNA models.

Global Elasticity: Persistence Length

For longer molecules, also thermally-induced bending has to be taken into account. In this regard, it is customary to resort to the tools provided by polymer theory. The self-correlation of the tangent vectors $\hat{t}$ along the chain is predicted to decay exponentially with the contour length Δs separating the points under inspection [116]: $\langle \hat{t}(s) \cdot \hat{t}(s + \Delta s) \rangle = \exp(-\Delta s / l_p)$, where the persistence length l_p gives the length scale of thermally-induced bending fluctuations. Experimental imaging via Atomic-Force Microscopy can be used to determine the correlation function, from which the value of l_p can be extracted [117]. Moreover, the wormlike chain model provides a handy formula for the response of the chain to a pulling force f [118]. Based on either of these approaches, experiments showed that $l_p = 45 - 55$ nm for dsDNA [11, 12, 104, 105, 110, 117], while $l_p = 57 - 63$ nm for dsRNA [11, 12, 119]. Therefore, dsRNA is stiffer than dsDNA when it comes to global bending. Moreover, at low pulling forces ($f < 1$ pN) the coupling between bending and twist results in an effective smaller twist modulus [120, 121], while for both dsDNA and dsRNA the persistence length decreases with the ionic strength of the solution [12, 104, 105]. Finally, cyclization experiments on short dsDNA fragments of repeating patterns investigated the sequence dependence of the persistence length [114].

Atomistic simulations for dsDNA based on AMBER force fields account with good accuracy for the experimental results. Particularly, it was found that $l_p = 43 - 51$ nm for parm99+bsc0 [122], $l_p = 40 - 60$ nm for parm99+bsc1 [65, 110, 123] and $l_p \simeq 60$ nm for parm99+OL15 [110]. In the case of dsRNA, it was estimated that $l_p = 70 - 80$ nm for bsc0 [99] and bsc0+χ_{OL3} [123]. Hence, all-atom simulations capture the relative bending flexibility of dsDNA and dsRNA, although slightly overestimating the quantitative estimations.

At the coarse-grained level, several models give a correct estimation of l_p for dsDNA [94, 111–113, 124–126], yet in various cases the parameters of the models were tuned for its optimization [112, 113, 124]. Notably, the TIS and 3SPN models capture the decrease of l_p with the ionic strength [94, 112]. As for the sequence dependence, the best performance has been shown for MADna and cgDNA [111, 126], which accurately account for the values of l_p obtained by cyclization experiments [114]. Finally, both MADna and oxDNA capture the effect of twist-bend coupling [111, 127]. In the case of dsRNA, there are only few works estimating l_p for coarse-grained models. Its value is usually underestimated [115, 128, 129], although a suitable parameterization of the MARTINI model can reproduce the correct result [128].

Force Dependence of Elastic Constants

Experimental evidence has pointed out that the elastic constants of nucleic acids depend on the pulling force f. For instance, stretched dsDNA overwinds until $f \simeq 35$ pN (which implies $g < 0$), while it underwinds for larger forces ($g > 0$) [100, 102]. On this matter, atomistic simulations have provided unique fundamental insight. However, a theoretical challenge first needed to be overcome, since neither of the approaches presented above is suitable to study stress-dependent elasticity: fitting full stress-dependent curves impedes by construction to extract the stress dependence of constants, while the formula for fluctuations (Eq. 2.4) is valid only for harmonic energies, which are perturbed by the linear terms in which the force or torque appear. In this regard, at a general level the energy can be written as [130]

$$U(\mathbf{q}) = \frac{1}{2}\Delta\mathbf{q}\mathbf{K}\,\Delta\mathbf{q}^T - \boldsymbol{\gamma}\cdot\mathbf{q}\,, \tag{2.6}$$

where $\boldsymbol{\gamma}$ is a vector containing the mechanical stresses associated to the various deformation modes. For instance, for $\mathbf{q} = \{L, \theta\}$ one has $\boldsymbol{\gamma} = \{f, \tau\}$, where f is the pulling force and τ the applied torque. Application of the generalized equipartition theorem gives [130]

$$\mathbf{K} = k_B T\,\mathbf{V}^{-1} + \mathbf{V}^{-1}\boldsymbol{\Gamma}\,, \tag{2.7}$$

where $\Gamma_{ij} = \langle\Delta q_i\rangle\gamma_j$. This approach was employed [130] to analyze the atomistic simulations from Refs. [66, 131, 132], where various dsDNA and dsRNA molecules with heterogeneous sequences were pulled with forces up to 20 pN. It was found that the stretching modulus of dsDNA increases with force, while that of dsRNA decreases. This unexpected result was rationalized by observing the opposite change in slide upon pulling the two molecules, which induces a strengthening and weakening of stacking interactions for dsDNA and dsRNA, respectively [130]. In contrast, the twist modulus was found to decrease for both kinds of duplexes. This behavior was ascribed to the progressive alignment of the normals to the base planes with the helical axis at increasing forces, which eventually identifies local torsion with the helical twist. A toy model based on this conjecture was capable of quantitatively reproducing the results for dsDNA without any fitting parameters, supporting this physical picture [130]. Finally, it was also found that, when considered with sign, g increases with force. Since for dsDNA $g < 0$ at low forces, this is in line with the experimental observations that for strong enough pulling the sign is reverted. All in all, these results add a further layer of complexity to the already rich landscape of nucleic-acids elasticity, where microscopic changes in the chemical structure result in strikingly different macroscopic mechanic behavior.

RNA Genome 3D Reconstruction

The assembly process in RNA viruses is a complex phenomenon which involves a cooperative interplay between the genome and the capsid proteins [1]. Consequently, the packaged structure of RNA in the virus' interior is of fundamental importance for its stability, and can shed light into the details of the assembly or disassembly processes [133]. As we saw in the sections above, mean-field theoretical approaches have pointed out the effect of persistence length, secondary structure, branching degree and electric charge into the assembly/disassembly of the packaged structure [134–137]. Other authors have stressed the importance of sequence-specific interactions between the genome and the capsid proteins, whose identification can lead to a sequential description of the assembly process [133]. A systematic and concrete picture of the geometrical conformation of the genome inside the virus can complement these results and be used for generating practical, all-atom intraviral RNA conformations. In fact, many experimentally determined structures contain some RNA fragments resolved, which have been used for proposing 3D models [138]. Such is the case of Satellite Tobacco Mosaic Virus (STMV) [78], Bacteriophage MS2 virus [139] and Cowpea Chlorotic Mottle Virus [140], among others. The case of STMV is of particular interest, since its crystal structure contains 30 double helices which comprise around 60% of its genome, although it lacks specificity about its sequence. It is therefore a good starting point for testing the 3D reconstruction in two aspects: a full model based in the secondary structure, and another one which considers the geometrical restraints imposed by the experimental crystallographic data of the STMV [27]. Given the size of the RNA involved, it is recommended to work, in an initial phase, with simpler descriptions. Coarse-grained models [69] offer a good trade-off between accuracy and efficiency, and can help to distinguish realistic models from spurious assemblies with clashes or unfeasible topological artifacts.

Fragment Assembly

Several well-established approaches generate three-dimensional RNA models from the assembly of experimentally determined fragments. Knowing the secondary structure, radius of gyration or tertiary contacts greatly helps to distinguish the most suitable structures among the large number of possibilities given a specific scoring function [141–146]. Inside the virus, the RNA might adopt non-trivial conformations mainly due to the confinement and the interaction with the capsid proteins, so the use of experimentally determined fragments could be too restrictive. We have then generated fragments of the secondary structure elements, hairpins, stems, and junctions, through on-the-fly short coarse-grained simulations using secondary structure restraints [27, 58]. Figure 2.5 shows an example on a fragment of STMV. Stems and hairpins are generated and later assembled with their closest junction, which is sampled from a simulation where the energy function contains only secondary structure and radius of gyration restraints, excluded volume and backbone connectivity. The conformation of the junctions can be sampled by performing several independent short simulations, in order to obtain a wide number of different geometries. After the assembly process, it is likely that the proximity of the assembled fragments will produce clashes between coarse-grained beads or, even worse, entanglements between the loops, stems, and junctions. These issues are circumvented by energy minimization and identifying and removing the links between the loops. The detection of these artifacts can be performed by known methods [147], and their removal can be done by applying repulsive energy terms on virtual sites situated at specific points of the loops, as illustrated in Fig. 2.5d [27, 148].

In a later step, the generated structures can be submitted to external forces which confine them to a spherical domain. Nevertheless, the structure might have additional restraints coming from the experimental data and must be treated in a specific way.

eRMSD Restraints

The imposition of geometric restraints requires a parameter to compare the structural differences between a given structure and a template. The eRMSD distance [149] is our choice since it is a specific metric for dealing with nucleic acids, taking into account the difference of the spatial arrangements of the nucleobases belonging to two structures. For this aim, it defines a local reference frame on each nucleobase, which is situated at the center of mass of their C2, C4 and C6 atoms. For each pair of bases i and j, the vector joining their centroids in the local reference frame of nucleobase i is denoted by $\mathbf{r}_{ij}$, which is rescaled in order to take into account the

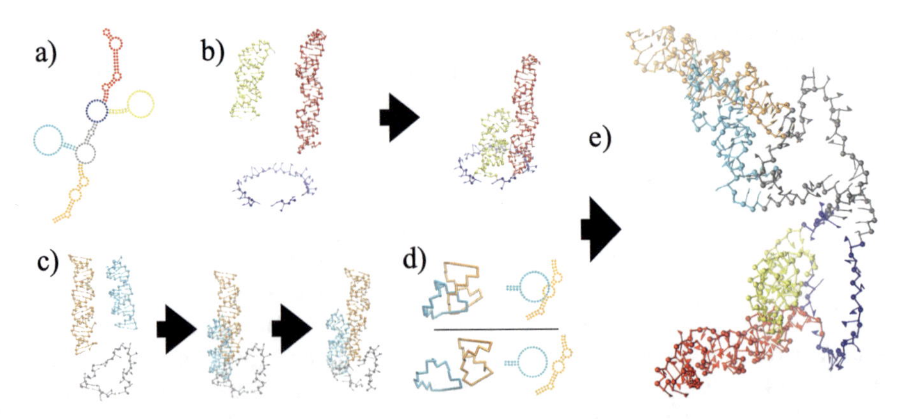

Fig. 2.5 Assembly process illustrated for nucleotides 330–575 of STMV, **a** in virio secondary structure. **b** Two loops containing hairpins are folded separately and attached to a junction, generated independently. **c** The procedure is repeated in another junction. In this case, the assembled fragments exhibit clashes and are entangled in a non-trivial manner. **d** Zoom of the entanglement, in backbone (left) and secondary structure (right) representation: each fragment forms a ring by joining the phosphate and sugar groups, closed by canonical base pairs. The rings are represented in the unrefined (top) and refined (bottom) conformations. **e** The whole structure is assembled, following energy minimization and removal of clashes and links between secondary structure elements [27]

nucleobase shape, by defining $\tilde{\mathbf{r}}_{ij} = (r_x/a, r_y/a, r_z/b)$, with $a = 5$ Å and $b = 3$ Å. The eRMSD between two structures α and β composed by N nucleotides is given by

$$\text{eRMSD}_{\alpha\beta} = \sqrt{\frac{1}{N}\sum\nolimits_{i\neq j}\left(\mathbf{G}(\mathbf{r}^{\alpha}_{ij}) - \mathbf{G}(\mathbf{r}^{\beta}_{ij})\right)},$$

where $\mathbf{G}(r)$ is a four-dimensional vector which satisfies $|\mathbf{G}(\tilde{\mathbf{r}}^{\alpha}) - \mathbf{G}(\tilde{\mathbf{r}}^{\beta})| \approx |\tilde{\mathbf{r}}^{\alpha} - \tilde{\mathbf{r}}^{\beta}|$ for $\tilde{\mathbf{r}}^{\alpha}, \tilde{\mathbf{r}}^{\beta} \ll D_c$ and $|\mathbf{G}(\tilde{\mathbf{r}}^{\alpha}) - \mathbf{G}(\tilde{\mathbf{r}}^{\beta})| = 0$ for $\tilde{\mathbf{r}}^{\alpha}, \tilde{\mathbf{r}}^{\beta} > D_c$, with D_c a dimensionless cutoff. The eRMSD restraint is defined as a harmonic potential energy term of the form

$$U(\{\mathbf{r}^{\alpha}\}, \{\mathbf{r}^{\beta}\}) = \frac{1}{2}K(\text{eRMSD}_{\alpha\beta})^2$$

where K is a constant with units of energy. The energy is imposed on a set of nucleotides with respect to its template, which can also be used for imposing secondary structure restraints. In our case, we have tested it on a set of double helices defined from the secondary structure proposed by Weeks and co-workers [63]. We took nucleotides 772–897, as illustrated in Fig. 2.6a, and chose five stems around a five-fold symmetry axis from PDB 4OQ9 (Fig. 2.6b) for defining two templates: one where the hairpins point outwards (case A, Fig. 2.6c) and another where they point inwards (case B, Fig. 2.6d). An initial condition can be easily generated using the procedure described in the previous section. The eRMSD restraints can be applied on any representation of RNA which is able to define the orientation of the nucleobases. In the present case, we have employed the SPQR coarse-grained model [58], which

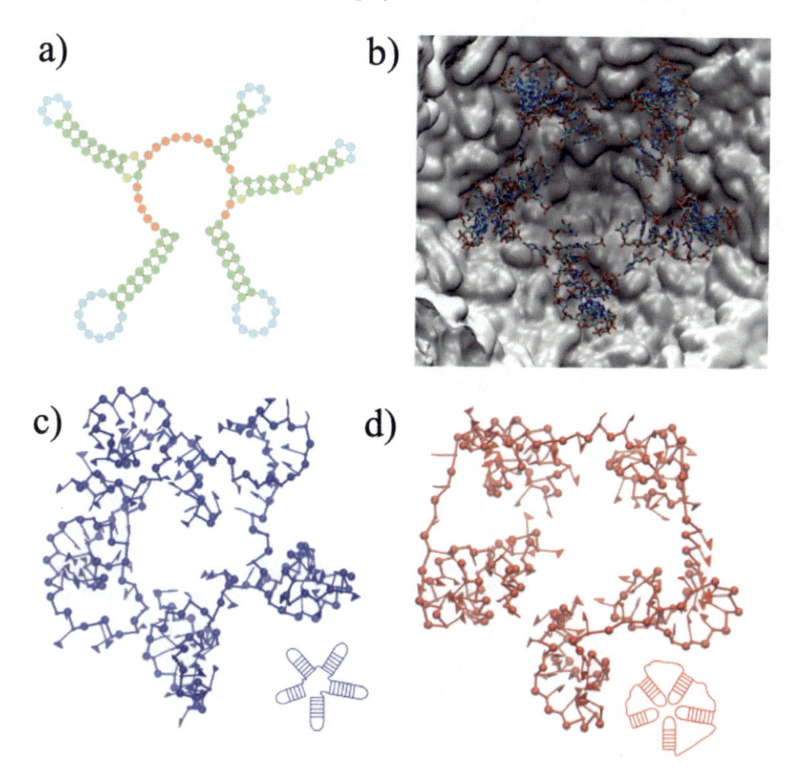

Fig. 2.6 Five consecutive hairpin loops are mapped according to different templates. **a** Secondary structure of the fragment. **b** Template taken from PDB structure. **c** Case A, after simulation **d** Case B, after simulation. *Source* Figure reproduced from [27] with licence CC-BY 4.0

has this feature incorporated into its code. The model represents each nucleotide by a triangular nucleobase and two beads for the sugar and phosphate group.

The value of the eRMSD and backbone energy are good descriptors of the suitability of the proposed configurations, as shown in Fig. 2.7. After running 3 short independent Monte Carlo simulations for each case, we see that these parameters can blindly distinguish that case A is more stable and favorable than case B, and therefore, selected for backmapping into atomistic detail. The procedure could be iterated over the rest of the structure in order to obtain a reduced set of possible conformations of the whole arrangement of the STMV genome inside the virus.

Discussion and Prospective Methods

The studies described in the previous sections outline the exquisite multiscale nature of RNA physics. The assembly of a viral capsid, as well as a virion requires to consider electrostatic interactions, determining feasible conformations of the genome

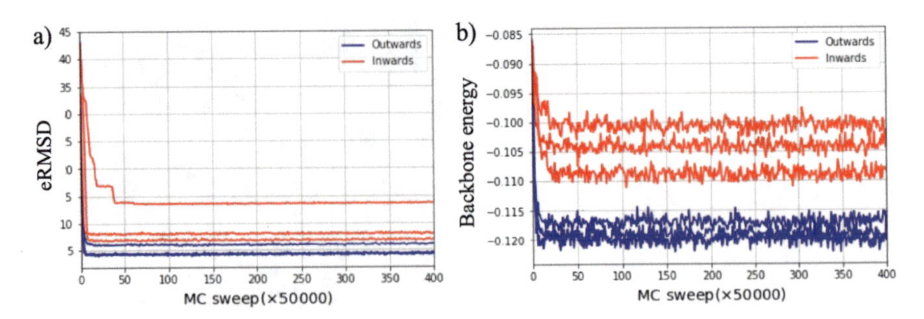

Fig. 2.7 **a** eRMSD between the 5 stems and their templates. **b** Backbone energy for both conformations as a function of MC sweep. *Source* Figure reproduced from [27] with licence CC-BY 4.0

and their effects on the nanomechanical stability of the capsid and genome itself. Here, we have shown the potential of atomistic, coarse-grained simulations and continuum models for modeling RNA and its interactions at different spatial and temporal scales. Still, there are several challenges ahead for RNA physics virology. Determining the folding of RNA and its dynamics within the virus remains a challenging open problem.Experimental techniques such as NMR, CryoEM, or recently CryoET have had dramatic progress recently and now allow us to visualize the nanoscopic details of the interior of the virions [150]. However, the 'dynamic structure' of RNA is commonly difficult to efficiently sample by pure experiments [151, 152]. Hence, high-resolution experiments are complemented with computational models. For example, computational approaches were used to refine 3D RNA reconstructions with additional restraints on tertiary contacts given by SAXS experiments [148, 153]. Furthermore, based on chemical probing and CryoEM, it was possible to reconstruct the frameshift stimulating element from SARS-CoV-2 all the way down to atomistic resolution [24]. Likewise, a general solution remains out of reach. However, after the astonishing success of Alphafold for protein folding [154] and similar developments for RNA [25, 155], machine learning-based methods hold a great promise for solving RNA folding.

Realistic models of whole virus dynamics require physics-informed models. Yet, a current problem with multiscale approaches is that they remain largely disconnected. No integrative model takes also into account detailed electrostatics and mechanical properties of both RNA and the proteins forming the non-enveloped viruses. The mechanics of RNA is vital for its packaging in a confined medium such as a viral capsid, yet scalable computational approaches focused on this feature are still at their infancy.

Similarly, electrostatics is central for the assembly, disassembly and stability of virus capsids [2], and virion assembly [3]. However, it remains unclear which is the required precision to describe the interactions between RNA and capsid-proteins, ions and short range interactions. Recent research has used free energy profiles extracted from atomic calculations as effective interactions in coarse-grained models [156] and also multiscale [8]. Similar approaches are already under discussion [1], which could include important interactions towards single amino-acids in proteins [18], water or surfaces into multiscale simulations [157]. In any case, there is little doubt that the exceptional challenge posed by RNA organization within viruses will be overcome only via a synergistic approach complementing computational approaches at multiple scales with the growing amount of experimental knowledge.

The integrative approach is complex itself, nonetheless, current developments to discern between molecular interactions can be used to test physical assumptions, guide the interpretation of experiments and possibly the direction of new experiments. Finally, a robust parametrization of the underlying RNA-protein molecular interactions (mechanical and electrostatics) would consequently permit the modification and/or control the capsid status from assembly to disassembly or vice versa, which are awaited features for the next generation of Virus-like Particles (VLPs) nanomedicine applications.

Acknowledgements H. V. G. acknowledges the core funding support from the Slovenian Research Agency, under grant No. P1-0055, and the financial support of the Community of Madrid and the European Union through the European Regional Development Fund (ERDF), financed as part of the Union response to Covid-19 pandemic. S. A. received the support of a fellowship from "la Caixa" Foundation (ID 100010434) and from the European Union's Horizon research and innovation programme under the Marie Sklodowska-Curie grant agreement No. 847648. The fellowship code is LCF/BQ/PI20/11760019. S. A. acknowledges support from the Ministerio de Ciencia e Innovación (MICINN) through the project PID2020-115864RB-I00 and the "María de Maeztu" Programme for Units of Excellence in R&D (grant No. CEX2018-000805-M). S. P. acknowledges the project FONDECYT Iniciación en Investigación 11181334 for financial support.

References

1. Zandi R, Dragnea B, Travesset A, Podgornik R (2020) On virus growth and form. Phys Rep 847:1–102
2. Xian Y, Karki CB, Silva SM, Li L, Xiao C (2019) The roles of electrostatic interactions in capsid assembly mechanisms of giant viruses. Int J Molecular Sci 20(8)
3. Garmann RF, Comas-Garcia M, Koay MST, Cornelissen JJLM, Knobler CM, Gelbart WM (2014) Role of electrostatics in the assembly pathway of a single-stranded RNA virus. J Virol 88:10472–10479
4. Lipfert J, Doniach S, Das R, Herschlag D (2014) Understanding nucleic acid-ion interactions. Annu Rev Biochem 83(1):813–841
5. Šiber A, Podgornik R (2008) Nonspecific interactions in spontaneous assembly of empty versus functional single-stranded rna viruses. Phys Rev E 78:051915
6. Draper DE (2004) A guide to ions and rna structure. RNA 10(3):335–343
7. Woodson SA (2005) Metal ions and rna folding: a highly charged topic with a dynamic future. Curr Opin Chem Biol 9(2):104–109
8. Poblete S, Božič A, Kanduč M, Podgornik R, Guzman HV (2021) Rna secondary structures regulate adsorption of fragments onto flat substrates. ACS Omega 6(48):32823–32831
9. Cruz-León S, Schwierz N (2020) Hofmeister series for metal-cation-rna interactions: the interplay of binding affinity and exchange kinetics. Langmuir 36(21):5979–5989
10. Sigel A, Sigel H, Sigel RKO (2012) Interplay between metal ions and nucleic acids, vol 10. Springer Science & Business Media

11. Lipfert J, Skinner GM, Keegstra JM, Hensgens T, Jager T, Dulin D, Köber M, Yu Z, Donkers SP, Chou F-C et al (2014) Double-stranded rna under force and torque: similarities to and striking differences from double-stranded dna. Proc Natl Acad Sci 111(43):15408–15413
12. Herrero-Galán E, Fuentes-Perez ME, Carrasco C, Valpuesta JM, Carrascosa JL, Moreno-Herrero F, Arias-Gonzalez JR (2012) Mechanical identities of rna and dna double helices unveiled at the single-molecule level. J Am Chem Soc 135(1):122–131
13. Paul A, William B (1978) Supercoiling in closed circular dna: dependence upon ion type and concentration. Biochemistry 17(4):594–601
14. You-Cheng X, Bremer H (1997) Winding of the DNA helix by divalent metal ions. Nucleic Acids Res 25(20):4067–4071
15. Cruz-León S, Vanderlinden W, Müller P, Forster T, Staudt G, Lin Y-Y, Lipfert J, Schwierz N (2022) Twisting DNA by salt. Nucleic Acids Res 50(10):5726–5738
16. Schnabl J, Sigel RKO (2010) Controlling ribozyme activity by metal ions. Curr Opin Chem Biol 14(2):269–275
17. Roychowdhury-Saha M, Burke DH (2006) Extraordinary rates of transition metal ion-mediated ribozyme catalysis. RNA 12(10):1846–1852
18. Cooper CD, Addison-Smith I, Guzman HV (2022) Quantitative electrostatic force tomography for virus capsids in interaction with an approaching nanoscale probe. Nanoscale 14(34):12232–12237
19. Wang B, Zhong C (2022) Tieleman DP (2022) Supramolecular organization of sars-cov and sars-cov-2 virions revealed by coarse-grained models of intact virus envelopes. J Chem Inf Model 62(1):176–186 PMID: 34911299
20. Pezeshkian W, Grünewald F, Narykov O, Lu S, Arkhipova V, Solodovnikov A, Wassenaar TA, Marrink SJ, Korkin D (2022) Molecular architecture and dynamics of sars-cov-2 envelope by integrative modeling. bioRxiv
21. Viso JF, Belelli P, Machado M, González H, Pantano S, Amundarain MJ, Zamarreño F, Branda MM, Guérin DMA, Costabel MD (2018) Multiscale modelization in a small virus: Mechanism of proton channeling and its role in triggering capsid disassembly. PLOS Comput Biol 14(4):e1006082
22. Martínez M, Cooper CD, Poma AB, Guzman HV (2019) Free energies of the disassembly of viral capsids from a multiscale molecular simulation approach. J Chem Inf Model 60(2):974–981
23. Ponce-Salvatierra A, Astha, KM, Nithin C, Ghosh P, Mukherjee S, Bujnicki JM (2019) Computational modeling of rna 3d structure based on experimental data. Biosci Rep 39(2):BSR20180430
24. Rangan R, Watkins AM, Chacon J, Kretsch R, Kladwang W, Zheludev IN, Townley J, Rynge M, Thain G, Das R (2021) De novo 3D models of SARS-CoV-2 RNA elements from consensus experimental secondary structures. Nucleic Acids Res 49(6):3092–3108
25. Kappel K, Zhang K, Zhaoming S, Watkins AM, Kladwang W, Li S, Pintilie G, Topkar VV, Rangan R, Zheludev IN, Yesselman JD, Chiu W, Das R (2020) Accelerated cryo-em-guided determination of three-dimensional rna-only structures. Nat Methods 17(7):699–707
26. Kirmizialtin S, Silalahi ARJ, Elber R, Fenley MO (2012) The ionic atmosphere around a-rna: poisson-boltzmann and molecular dynamics simulations. Biophys J 102(4):829–838
27. Poblete S, Guzman HV (2021) Structural 3d domain reconstruction of the rna genome from viruses with secondary structure models. Viruses 13(8)
28. Guo T, Modi OL, Hirano J, Guzman HV, Tsuboi T (2022) Single-chain models illustrate the 3d rna folding shape during translation. Biophys Rep 2(3)
29. Rotsch C, Radmacher M (1997) Mapping local electrostatic forces with the atomic force microscope. Langmuir 13(10):2825–2832
30. Butt H (1992) Measuring local surface charge densities in electrolyte solutions with a scanning force microscope. Biophys J 63(2):578–582
31. Yang Y, Mayer KM, Hafner JH (2007) Quantitative membrane electrostatics with the atomic force microscope. Biophys J 92(6):1966–1974

32. Adar RM, Andelman D, Diamant H (2017) Electrostatics of patchy surfaces. Adv Colloid Interface Sci 247:198–207. Dominique Langevin Festschrift: Four Decades Opening Gates in Colloid and Interface Science
33. Patel N, Wroblewski E, Leonov G, Phillips SEV, Tuma R, Twarock R, Stockley PG (2017) Rewriting nature's assembly manual for a ssrna virus. Proc Nat Acad Sci 114(46):12255–12260
34. Filippo V, Matteo G, Marta G, Laura C, Elisabetta B, Francesco F, Irene R (2020) Protein electrostatics: from computational and structural analysis to discovery of functional fingerprints and biotechnological design. Comput Struct Biotech J 18:1774–1789
35. Edwardson TGW, Levasseur MD, Tetter S, Steinauer A, Hori M, Hilvert D Protein cages: from fundamentals to advanced applications. Chem Rev 0(0):null. 0. PMID: 35394752
36. Garmann RF, Comas-Garcia M, Koay MST, Cornelissen JJLM, Knobler CM, Gelbart WM, Simon A (2014) Role of electrostatics in the assembly pathway of a single-stranded rna virus. J Virol 88(18):10472–10479
37. Ares P, Garcia-Doval C, Llauro A, Gomez-Herrero J, Van Raaij MJ, De Pablo PJ (2014) Interplay between the mechanics of bacteriophage fibers and the strength of virus-host links. Phys Rev E 89(5):052710
38. Hernando-Pérez M, Cartagena-Rivera AX, Božič AL, Carrillo PJP, Martín CS, Mateu MG, Raman A, Podgornik R, De Pablo PJ (2015) Nanoscale. Quantitative nanoscale electrostatics of viruses 7(41):17289–17298
39. Marchetti M, Wuite GJL, Roos WH (2016) Atomic force microscopy observation and characterization of single virions and virus-like particles by nano-indentation. Current Opin Virol 18:82–88. Antiviral strategies • Virus structure and expression
40. Llauró A, Luque D, Edwards E, Trus BL, Avera J, Reguera D, Douglas T, de Pablo Pedro J, Castón JR (2016) Cargo-shell and cargo-cargo couplings govern the mechanics of artificially loaded virus-derived cages. Nanoscale 8:9328–9336
41. Buzón P, Maity S, Roos WH (2020) Physical virology: from virus self-assembly to particle mechanics. WIREs Nanomed Nanobiotech 12(4):e1613
42. Ortega-Esteban Á, Mata CP, Rodríguez-Espinosa MJ, Luque D, Irigoyen N, Rodríguez JM, de Pablo PJ, Castón JR, López S (2020) Cryo-electron microscopy structure, assembly, and mechanics show morphogenesis and evolution of human picobirnavirus. J Virol 94(24):e01542-20
43. Martín-González N, Freire PI, Ortega-Esteban Á, Laguna-Castro M, San Martín C, Valbuena A, Delgado-Buscalioni R, de Pablo PJ (2021) Long-range cooperative disassembly and aging during adenovirus uncoating. Phys Rev X 11:021025
44. Krieg M, Fläschner G, Alsteens D, Gaub BM, Roos WH, Wuite GJL, Gaub HE, Gerber C, Dufrêne YF, Müller DJ (2019) Atomic force microscopy-based mechanobiology. Nat Rev Phys 1(1):41–57
45. Snijder J, Uetrecht C, Rose RJ, Sanchez-Eugenia R, Marti GA, Agirre J, Guérin DMA, Wuite GJL, Heck AJR, Roos WH (2013) Probing the biophysical interplay between a viral genome and its capsid. Nat Chem 5:502–509
46. de Pablo PJ (2018) Atomic force microscopy of virus shells. Seminars in Cell Devel Biol 73:199–208
47. Gilson MK, Davis ME, Luty BA, McCammon JA (1993) Computation of electrostatic forces on solvated molecules using the Poisson-Boltzmann equation. J Phys Chem 97(14):3591–3600
48. Ebeling D, Eslami B, De Jesus Santiago S (2013) Visualizing the subsurface of soft matter: simultaneous topographical imaging, depth modulation, and compositional mapping with triple frequency atomic force microscopy. ACS Nano 7(11):10387–10396 PMID: 24131492
49. Netz Roland R, David A (2003) Neutral and charged polymers at interfaces. Phys Rep 380(1):1–95
50. Egorov SA, Andrey M, Peter V, Kurt B (2016) Semiflexible polymers under good solvent conditions interacting with repulsive walls. J Chem Phys 144(17):174902

51. Nikoubashman A, Vega DA, Binder K, Milchev A (2017) Semiflexible polymers in spherical confinement: bipolar orientational order versus tennis ball states. Phys Rev Lett 118:217803
52. Milchev A, Binder K (2020) How does stiffness of polymer chains affect their adsorption transition? J Chem Phys 152(6):064901
53. Schneemann A (2006) The structural and functional role of RNA in icosahedral virus assembly. Annu Rev Microbiol 60:51–67
54. Stockley PG, Ranson NA, Twarock R (2013) A new paradigm for the roles of the genome in ssrna viruses. Fut Virol 8(6):531–543
55. Hernandez-Garcia A, Kraft DJ, Janssen AFJ, Bomans PHH, Sommerdijk NAJM, Thies-Weesie DME, Favretto ME, Brock R, de Wolf FA, Werten MWT, van der Schoot P, Stuart MC, de Vries R (2014) Design and self-assembly of simple coat proteins for artificial viruses. Nat Nanotechnol 9(9):698–702
56. Perlmutter JD, Hagan MF (2015) The role of packaging sites in efficient and specific virus assembly. J Mol Biol 427:2451–2467
57. Božič A, Micheletti C, Podgornik R, Tubiana L (2018) Compactness of viral genomes: effect of disperse and localized random mutations. J Phys Condens Matter 30(8):084006
58. Simón Poblete, Sandro Bottaro, Giovanni Bussi (2018) A nucleobase-centered coarse-grained representation for structure prediction of rna motifs. Nucleic acids research 46(4):1674–1683
59. Podgornik R, Harries D, DeRouchey J, Strey HH, Parsegian VA (2008) Interactions in macromolecular complexes used as nonviral vectors for gene delivery
60. Caliskan G, Hyeon C, Perez-Salas U, Briber RM, Woodson SA, Thirumalai D (2005) Persistence length changes dramatically as rna folds. Phys Rev Lett 95:268303
61. Singh N, Willson RC (1999) Boronate affinity adsorption of rna: Possible role of conformational changes. J Chromatogr A 840(2):205–213
62. Buzón P, Maity S, Christodoulis P, Wiertsema MJ, Dunkelbarger S, Kim C, Wuite GJL, Zlotnick A, Roos WH (2021) Virus self-assembly proceeds through contact-rich energy minima. Sci Adv 7(45):eabg0811, 2021
63. Larman Bridget C, Dethoff Elizabeth A, Weeks Kevin M (2017) Packaged and free satellite tobacco mosaic virus (stmv) rna genomes adopt distinct conformational states. Biochemistry 56(16):2175–2183
64. Dror RO, Dirks RM, Grossman JP, Xu H, Shaw DE (2012) Biomolecular simulation: a computational microscope for molecular biology. Annu Rev Biophys 41:429–452
65. Ivani I, Dans PD, Noy A, Pérez A, Faustino I, Hospital A, Walther J, Andrio P, Goñi R, Balaceanu A et al (2016) Parmbsc1: a refined force field for dna simulations. Nat Methods 13(1):55–58
66. Marin-Gonzalez A, Vilhena JG, Perez R, Moreno-Herrero F (2017) Understanding the mechanical response of double-stranded dna and rna under constant stretching forces using all-atom molecular dynamics. Proc Nat Acad Sci 114(27):7049–7054
67. Stelzl Lukas S, Nicole Erlenbach, Marcel Heinz, Prisner Thomas F, Gerhard Hummer (2017) Resolving the conformational dynamics of dna with Ångstrom resolution by pulsed electron-electron double resonance and molecular dynamics. J Am Chem Soc 139(34):11674–11677
68. Lei B, Xi Z, Ya-Zhou S, Wu Y-Y, Zhi-Jie T (2017) Understanding the relative flexibility of rna and dna duplexes: stretching and twist-stretch coupling. Biophys J 112(6):1094–1104
69. Sponer J, Bussi G, Krepl M, Baáš P, Bottaro S, Cunha RA, Gil-Ley A, Pinamonti G, Poblete S, Jurecka P et al (2018) Rna structural dynamics as captured by molecular simulations: a comprehensive overview. Chem Rev 118(8):4177–4338
70. Dohnalová H, Lankaš F (2022) Deciphering the mechanical properties of b-dna duplex. Wiley Int Rev Comput Molecular Sci 12(3):e1575
71. Cruz-León S, Grotz Kara K, Schwierz N (2021) Extended magnesium and calcium force field parameters for accurate ion-nucleic acid interactions in biomolecular simulations. J Chem Phys 154(17):171102
72. Mamatkulov S, Schwierz N (2018) Force fields for monovalent and divalent metal cations in tip3p water based on thermodynamic and kinetic properties. J Chem Phys 148(7):074504

73. Grotz Kara K, Cruz-León S, Schwierz N (2021) Optimized magnesium force field parameters for biomolecular simulations with accurate solvation, ion-binding, and water-exchange properties. J Chem Theory Comput 17(4):2530–2540
74. Zgarbová M, Otyepka M, Šponer J, Mládek A, Banáš P, Cheatham TE, Jurečka P (2011) Refinement of the cornell et al. nucleic acids force field based on reference quantum chemical calculations of glycosidic torsion profiles. J Chem Theory Comput 7(9):2886–2902
75. Koculi E, Hyeon C, Thirumalai D, Woodson SA (2007) Charge density of divalent metal cations determines rna stability. J Am Chem Soc 129(9):2676–2682
76. Leonarski F, D'Ascenzo L, Auffinger P (2016) Mg^{2+} ions: do they bind to nucleobase nitrogen's? Nucleic Acids Res 45(2):987–1004
77. Cruz-León S, Schwierz N (2022) Rna captures more cations than dna: insights from molecular dynamics simulations. J Phys Chem B 126(43):8646–8654 PMID: 36260822
78. McPherson A (2021) Structures of additional crystal forms of satellite tobacco mosaic virus grown from a variety of salts. Acta Crystallographica Sect F: Struct Biol Commun 77(12)
79. Lavery R, Maddocks JH, Pasi M, Zakrzewska K (2014) Analyzing ion distributions around DNA. Nucleic Acids Res 42(12):8138–8149
80. Pasi M, Maddocks JH, Lavery R (2015) Analyzing ion distributions around DNA: sequence-dependence of potassium ion distributions from microsecond molecular dynamics. Nucleic Acids Res 43(4):2412–2423
81. Manning GS (1969) Limiting laws and counterion condensation in polyelectrolyte solutions i. colligative properties. J Chem Phys 51(3):924–933
82. Deserno M, Holm C, May S (2000) Fraction of condensed counterions around a charged rod: comparison of poisson-boltzmann theory and computer simulations. Macromolecules 33(1):199–206
83. Remo R, Xiangshu J, West SM, Joshi R, Honig B, Mann RS (2010) Origins of specificity in protein-dna recognition. Ann Rev Biochem 79:233
84. Smith SB, Cui Y, Bustamante C (1996) Overstretching b-dna: the elastic response of individual double-stranded and single-stranded dna molecules 271(5250):795–799
85. Rief M, Clausen-Schaumann H, Gaub HE (1999) Sequence-dependent mechanics of single dna molecules. Nat Struct Biol 6(4):346–349
86. Murphy MC, Rasnik I, Cheng W, Lohman TM, Ha T (2004) Probing single-stranded dna conformational flexibility using fluorescence spectroscopy. Biophys J 86(4):2530–2537
87. Chen H, Meisburger SP, Pabit SA, Sutton JL, Webb WW, Pollack L (2012) Ionic strength-dependent persistence lengths of single-stranded rna and dna. Proc Nat Acad Sci 109(3):799–804
88. Sim AYL, Lipfert J, Herschlag D, Doniach S (2012) Salt dependence of the radius of gyration and flexibility of single-stranded dna in solution probed by small-angle x-ray scattering. Phys Rev E 86(2):021901
89. Seol Y, Skinner GM, Visscher K (2004) Elastic properties of a single-stranded charged homopolymeric ribonucleotide. Phys Rev Lett 93(11):118102
90. Bizarro CV, Alemany A, Ritort F (2012) Non-specific binding of na+ and mg 2+ to rna determined by force spectroscopy methods. Nucleic Acids Res 40(14):6922–6935
91. Jacobson DR, McIntosh DB, Saleh OA (2013) The snakelike chain character of unstructured rna. Biophys J 105(11):2569–2576
92. Camunas-Soler J, Ribezzi-Crivellari M, Ritort F (2016) Elastic properties of nucleic acids by single-molecule force spectroscopy. Annu Rev Biophys 45:65–84
93. Jacobson DR, McIntosh DB, Stevens MJ, Rubinstein M, Saleh OA (2017) Single-stranded nucleic acid elasticity arises from internal electrostatic tension. Proc Nat Acad Sci 114(20):5095–5100
94. Chakraborty D, Hori N, Thirumalai D (2018) Sequence-dependent three interaction site model for single-and double-stranded dna. J Chem Theory Comput 14(7):3763–3779
95. Engel MC, Romano F, Louis AA, Doye JPK (2020) Measuring internal forces in single-stranded dna: application to a dna force clamp. J Chem Theory Comput 16(12):7764–7775

96. Viader-Godoy X, Pulido CR, Ibarra B, Manosas M, Ritort F (2021) Cooperativity-dependent folding of single-stranded dna. Phys Rev X 11(3):031037
97. Cheatham TE, Kollman PA (1997) Molecular dynamics simulations highlight the structural differences among dna: Dna, rna: Rna, and dna: Rna hybrid duplexes. J Am Chem Soc 119(21):4805–4825
98. Noy A, Perez A, Lankas F, Luque FJ, Orozco M (2004) Relative flexibility of dna and rna: a molecular dynamics study. J Molec Biol 343(3):627–638
99. Faustino I, Pérez A, Orozco M (2010) Toward a consensus view of duplex rna flexibility. Biophys J 99(6):1876–1885
100. Gore J, Bryant Z, Nollmann M, Le MU, Cozzarelli NR, Bustamante C (2006) Mai U Le, Nicholas R Cozzarelli, and Carlos Bustamante. Dna overwinds when stretched. Nature 442(7104):836–839
101. Olson WK, Gorin AA, Lu X-J, Hock LM, Zhurkin VB (1998) Dna sequence-dependent deformability deduced from protein-dna crystal complexes. Proc Nat Acad Sci 95(19):11163–11168
102. Gross P, Laurens N, Oddershede LB, Bockelmann U, Peterman EJG, Wuite GJL (2011) Quantifying how dna stretches, melts and changes twist under tension. Nat Phys 7(9):731–736
103. Marin-Gonzalez A, Pastrana CL, Bocanegra R, Martín-González A, Vilhena JG, Pérez R, Ibarra B, Aicart-Ramos C, Moreno-Herrero F (2020) Understanding the paradoxical mechanical response of in-phase a-tracts at different force regimes. Nucleic Acids Res 48(9):5024–5036
104. Baumann CG, Smith SB, Bloomfield VA, Bustamante C (1997) Ionic effects on the elasticity of single dna molecules. Proc Nat Acad Sci 94(12):6185–6190
105. Wenner JR, Williams MC, Rouzina I, Bloomfield VA (2002) Salt dependence of the elasticity and overstretching transition of single dna molecules. Biophys J 82(6):3160–3169
106. Bryant Z, Stone MD, Gore J, Smith SB, Cozzarelli NR, Bustamante C (2003) Structural transitions and elasticity from torque measurements on dna. Nature 424(6946):338–341
107. Mosconi F, Allemand JF, Bensimon D, Croquette V (2009) Measurement of the torque on a single stretched and twisted dna using magnetic tweezers. Phys Rev Lett 102(7):078301
108. Moroz JD, Nelson P (1998) Entropic elasticity of twist-storing polymers. Macromolecules 31(18):6333–6347
109. Sheinin MY, Wang MD (2009) Twist-stretch coupling and phase transition during dna supercoiling. Phys Chem Chem Phys 11(24):4800–4803
110. Velasco-Berrelleza V, Burman M, Shepherd JW, Leake MC, Golestanian R, Noy A (2020) Serrana: a program to determine nucleic acids elasticity from simulation data. Phys Chem Chem Phys 22(34):19254–19266
111. Assenza S, Pérez R (2022) Accurate sequence-dependent coarse-grained model for conformational and elastic properties of double-stranded dna. J Chem Theo Comput 18(5):3239–3256
112. Freeman GS, Hinckley DM, Lequieu JP, Whitmer JK, De Pablo JJ (2014) Coarse-grained modeling of dna curvature. J Chem Phys 141(16):10B615_1
113. Snodin BEK, Randisi F, Mosayebi M, Šulc P, Schreck JS, Romano F, Ouldridge TE, Tsukanov R, Nir E, Louis AA et al (2015) Introducing improved structural properties and salt dependence into a coarse-grained model of dna. J Chem Phys 142(23):06B613_1
114. Geggier S, Vologodskii A (2010) Sequence dependence of dna bending rigidity. Proc Nat Acad Sci 107(35):15421–15426
115. Šulc P, Romano F, Ouldridge TE, Doye JPK, Louis AA (2014) A nucleotide-level coarse-grained model of rna. J Chem Phys 140(23):06B614_1, 2014
116. Assenza S, Mezzenga R (2019) Soft condensed matter physics of foods and macronutrients. Nat Rev Phys 1(9):551–566
117. Heenan PR, Perkins TT (2019) Imaging dna equilibrated onto mica in liquid using biochemically relevant deposition conditions. ACS Nano 13(4):4220–4229
118. Marko JF, Siggia ED (1995) Stretching dna. Macromolecules 28(26):8759–8770

119. Abels JA, Moreno-Herrero F, Van der Heijden T, Dekker C, Dekker NH (2005) Single-molecule measurements of the persistence length of double-stranded rna. Biophys J 88(4):2737–2744
120. Nomidis SK, Kriegel F, Vanderlinden W, Lipfert J, Carlon E (2017) Twist-bend coupling and the torsional response of double-stranded dna. Phys Rev Lett 118(21):217801
121. Gao X, Hong Y, Ye F, Inman JT, Wang MD (2021) Torsional stiffness of extended and plectonemic dna. Phys Rev Lett 127(2):028101
122. Noy A, Golestanian R (2012) Length scale dependence of dna mechanical properties. Phys Rev Lett 109(22):228101
123. Zhang Y, Yan M, Huang T, Wang X (2022) Understanding the structural elasticity of rna and dna: all-atom molecular dynamics. Adv Theo Simul 2200534
124. Savelyev A, Papoian GA (2010) Chemically accurate coarse graining of double-stranded dna. Proc Nat Acad Sci 107(47):20340–20345
125. Korolev N, Luo D, Lyubartsev AP, Nordenskiöld L (2014) A coarse-grained dna model parameterized from atomistic simulations by inverse monte carlo. Polymers 6(6):1655–1675
126. Mitchell JS, Glowacki J, Grandchamp AE, Manning RS, Maddocks JH (2017) Sequence-dependent persistence lengths of dna. J Chem Theo Comput 13(4):1539–1555
127. Skoruppa E, Laleman M, Nomidis SK, Carlon E (2017) Dna elasticity from coarse-grained simulations: the effect of groove asymmetry. J Chem Phys 146(21):214902
128. Uusitalo JJ, Ingólfsson HI, Marrink SJ, Faustino I (2017) Martini coarse-grained force field: extension to rna. Biophys J 113(2):246–256
129. Cruz-León S, Vázquez-Mayagoitia A, Melchionna S, Schwierz N, Fyta M (2018) Coarse-grained double-stranded rna model from quantum-mechanical calculations. J Phys Chem B 122(32):7915–7928
130. Luengo-Márquez J, Zalvide-Pombo J, Pérez R, Assenza S (2022) Force-dependent elasticity of nucleic acids. Nanoscale 15:6738
131. Marin-Gonzalez A, Vilhena JG, Moreno-Herrero F, Perez R (2019) Dna crookedness regulates dna mechanical properties at short length scales. Phys Rev Lett 122(4):048102
132. Marin-Gonzalez A, Vilhena JG, Moreno-Herrero F, Perez R (2019) Sequence-dependent mechanical properties of double-stranded rna. Nanoscale 44:21471–21478
133. Twarock R, Bingham RJ, Dykeman EC, Stockley PG (2018) A modelling paradigm for rna virus assembly. Current Opin Virol 31:74–81
134. Erdemci-Tandogan G, Orland H, Zandi R (2017) Rna base pairing determines the conformations of rna inside spherical viruses. Phys Rev Lett 119(18):188102
135. Erdemci-Tandogan G, Wagner J, van der Schoot P, Podgornik R, Zandi R (2016) Effects of rna branching on the electrostatic stabilization of viruses. Phys Rev E 94(2):022408
136. Li S, Erdemci-Tandogan G, van der Schoot P, Zandi R (2017) The effect of rna stiffness on the self-assembly of virus particles. J Phys: Condensed Matter 30(4):044002
137. Dong Y, Li S, Zandi R (2020) Effect of the charge distribution of virus coat proteins on the length of packaged rnas. Phys Rev E 102(6):062423
138. Zeng Y, Larson SB, Heitsch CE, McPherson A, Harvey SC (2012) A model for the structure of satellite tobacco mosaic virus. J Struct Biol 180(1):110–116
139. van den Worm SHE, Valegård K, Fridborg K, Liljas L, Stonehouse NJ, Murray JB, Walton C, Stockley PG (1998) Crystal structures of ms2 coat protein mutants in complex with wild-type rna operator fragments. Nucleic Acids Res 26(5):1345–1351
140. Speir JA, Munshi S, Wang G, Baker TS, Johnson JE (1995) Structures of the native and swollen forms of cowpea chlorotic mottle virus determined by x-ray crystallography and cryo-electron microscopy. Structure 3(1):63–78
141. Das R, Karanicolas J, Baker D (2010) Atomic accuracy in predicting and designing non-canonical rna structure. Nat Meth 7(4):291–294
142. Kerpedjiev P, Höner zu Siederdissen C, Hofacker IL (2015) Predicting RNA 3d structure using a coarse-grain helix-centered model. RNA 21:1110–1121
143. Popenda M, Szachniuk M, Antczak M, Purzycka KJ, Lukasiak P, Bartol N, Blazewicz J, Adamiak RW (2012) Automated 3d structure composition for large rnas. Nucleic Acids Res 40(14):e112–e112

144. Swati J, Tamar S (2017) F-rag: generating atomic coordinates from rna graphs by fragment assembly. J Molecular Biol 429(23):3587–3605
145. Parisien M, Major F (2008) The mc-fold and mc-sym pipeline infers rna structure from sequence data. Nature 452(7183):51–55
146. Lemieux S, Major F (2006) Automated extraction and classification of rna tertiary structure cyclic motifs. Nucleic Acids Res 34(8):2340–2346
147. Luwanski K, Hlushchenko V, Popenda M, Zok T, Sarzynska J, Martsich D, Szachniuk M, Antczak M (2022) Rnaspider: a webserver to analyze entanglements in rna 3d structures. Nucleic Acids Res
148. Thiel BC, Bussi G, Poblete S, Hofacker IL (2022) Sampling globally and locally correct rna 3d structures using ernwin, spqr and experimental saxs data. bioRxiv
149. Bottaro S, Di Palma F, Bussi B (2014) The role of nucleobase interactions in rna structure and dynamics. Nucleic Acids Res 42(21):13306–13314
150. Turoňová B, Sikora M, Schürmann C, Hagen WJH, Welsch S, Blanc FEC, von Bülow S, Gecht M, Bagola K, Hörner C et al (2020) In situ structural analysis of sars-cov-2 spike reveals flexibility mediated by three hinges. Science 370(6513):203–208
151. Schroeder SJ (2020) Perspectives on viral rna genomes and the rna folding problem. Viruses 12(10):1126
152. Spitale RC, Incarnato D (2022) Probing the dynamic rna structurome and its functions. Nat Rev Genet
153. Chojnowski G, Zaborowski R, Magnus M, Bujnicki JM (2021) Rna fragment assembly with experimental restraints. bioRxiv
154. Jumper J, Evans R, Pritzel A, Green T, Figurnov M, Ronneberger O, Tunyasuvunakool K, Bates R, Žídek A, Potapenko A et al (2021) Highly accurate protein structure prediction with alphafold. Nature 596(7873):583–589
155. Townshend RJL, Eismann S, Watkins AM, Rangan R, Karelina M, Das R, Dror RO (2021) Geometric deep learning of rna structure. Science 373(6558):1047–1051
156. Nguyen HT, Hori N, Thirumalai D (2019) Theory and simulations for rna folding in mixtures of monovalent and divalent cations. Proc Nat Acad Sci 116(42):21022–21030
157. Potestio R, Peter C, Kremer K (2014) Computer simulations of soft matter: linking the scales. Entropy 16(8):4199–4245

Chapter 3
The In Vitro Packaging of "Overlong" RNA by Spherical Virus-Like Particles

Ana Luisa Duran-Meza, Abigail G. Chapman, Cheylene R. Tanimoto, Charles M. Knobler, and William M. Gelbart

Abstract Of the myriad viruses, very few have been shown to be capable of self-assembly in vitro from purified components into infectious virus particles. One of these is Cowpea Chlorotic Mottle Virus (CCMV), an unenveloped spherical plant virus whose capsid self-assembles around its RNA genome without a packaging signal. While heterologous RNA, not just cognate viral RNA, can be packaged into individual CCMV virus-like particles (VLPs), the RNA needs to fall within a certain range of lengths. If it is too short, it is packaged into particles smaller than wild type, or with two or more RNAs per capsid. If the RNA is too long, multiple capsids assemble around one RNA, and the RNA associated with these multiplet structures is not as RNase resistant. Further, as shown in the present work, 4200 nt appears to be the limiting length of RNA that can be packaged into single RNase-resistant CCMV VLPs. We explore the extent to which "overlong" RNA can be packaged more efficiently upon the addition of spermine, a polyvalent cation whose increasing concentration has been shown to compactify RNA. Finally, we show that the capsid protein of Brome Mosaic Virus (BMV), a bromovirus closely related to CCMV, also gives rise to multiplets when it is self-assembled with the same "overlong" RNA constructs, but with different distributions of multiplets.

Keywords Virus-like particles · In vitro assembly · Capsid multiplets · RNase-resistant capsid packaging · Spermine compaction

Introduction

The capacity for in vitro self-assembly of viruses from their purified constituents—shown by the icosahedral plant viruses CCMV (cowpea chlorotic mottle virus) and BMV (brome mosaic virus), bacteriophage MS2, and the rod-like TMV (Tobacco Mosaic Virus)—makes them highly promising platforms for biomedical applications,

A. L. Duran-Meza · A. G. Chapman · C. R. Tanimoto · C. M. Knobler · W. M. Gelbart (✉)
Department of Chemistry and Biochemistry, University of California Los Angeles, 607 Charles E. Young Drive East, Los Angeles, CA 90095, USA
e-mail: gelbart@chem.ucla.edu

M. Comas-Garcia and S. Rosales-Mendoza (eds.), *Physical Virology*, Springer Series in Biophysics 24, https://doi.org/10.1007/978-3-031-36815-8_3

among them vaccines and gene delivery [1]. Their utility depends on the ability of the capsid proteins of these ssRNA viruses to assemble in vitro around heterologous RNAs, forming virus-like particles (VLPs) of high stability, which: protect the RNA against attack by RNase; are capable of delivering their contents to the ribosomal (translational) machinery; and are free of biological contaminants that might arise from assembly in bacterial or yeast cells. Once formed, the VLPs may be functionalized by chemically linking proteins to them that recognize specific cells and enhance uptake [2, 3].

While the packaging capacity of these VLPs is not a major issue in determining their suitability for delivering cargos of small molecules such as miRNAs or mRNAs that code for short protein sequences, the maximum length of RNA that can be packaged in spherical capsids is a limiting factor to their employment for delivering constructs such as self-replicating RNAs (replicons) that code for a replicase and a gene of interest [4], because the strongly preferred curvature (radius) of the capsid protein dictates the volume available to packaged RNA. This is not a limitation for TMV whose genomic RNA is 6400 nt long, and whose cylindrical VLPs can accommodate any length of heterologous RNA because the capsid length is simply proportional to RNA length and its curvature is independent of its length. But faithful and efficient TMV VLP assembly requires that the RNA includes an origin-of-assembly (OAS) sequence to initiate packaging [5]. While a terminal TLS-like sequence has been identified for the packaging of BMV RNAs [6] a packaging signal has not been found for CCMV assembly [7] and heterologous RNAs lacking a TLS have been packaged without difficulty [8]. The packaging of RNA in capsids formed in vitro from MS2 CP (in the absence of the maturation protein that breaks symmetry and ensures infectivity in the wt virus) depends on the presence of a *pac* site, a 19-base stem loop that initiates packaging [9], and in vitro packaging of heterologous RNA in MS2 as long as 3000 nt requires the presence of multiple pac sites [10].

CCMV and BMV have capsids with Caspar-Klug triangulation numbers $T = 3$, composed of 180 proteins. While all RNA viruses with this same capsid structure have essentially identical outer diameters of ~30 nm, their nucleotide content varies significantly. Wild-type CCMV and BMV capsids contain ~3000 nt of ssRNA. In contrast, Cucumber Mosaic Virus (CMV), Bacteriophage MS2, Turnip Crinkle Virus (TCV), Turnip Yellow Mosaic Virus (TYMV), and NoroVirus (NoV), all with $T = 3$ capsids and approximately 30-nm diameters, have genomes with lengths 3300, 3569, 4034, 6400 and 7600 nt, respectively. The remarkably long genomes for TYMV and NoV are attributable to the presence in their capsids of agents that act to condense the RNA, such as the polyamine spermidine in the case of TYMV [11] and small basic proteins in the case of NoV [12]. Their RNAs otherwise appear to lack the high degree of branching that results in compaction [13, 14].

In the case of CCMV and BMV the major driving force for the packaging of RNA is the electrostatic interaction between the negatively charged phosphate backbone and the positively-charged N-terminus of the CP. Direct evidence for this non-specific interaction is found in self-assembly studies with mutant CCMV CPs in which the number of N-terminal cationic residues is successively decreased and shown to lead to decreasing lengths of RNA being packaged [15]. Further evidence comes from the cryoelectron microscopy structural study by Beren et al. [16] of BMV virions in which the RNA is shown in high-resolution reconstructions to be highly disordered, in contrast to what is found, for example, in wt MS2 where the multiple packaging signals and interaction of the RNA ends with the maturation protein enforce a single well-defined structure [17]. The lack of any specific structure for the packaged RNA in BMV (and, presumably, CCMV) accounts for their ability to package (each individually) a broad range of RNAs—notably, the three very different molecules making up their genome, as well as heterologous RNAs.

It would appear then, from the standpoint solely of volume constraints, that CCMV and BMV should be capable of packaging RNAs with lengths well in excess of the 3000 nt found in the wild type. What then is their packaging limit? One might assume that this could be simply determined by a series of experiments in which assemblies with increasingly long RNAs are carried out until a limit is reached beyond which the RNAs are not packaged. However, it has been shown that attempts to package RNA longer than 3000 nt by CCMV CP leads to "multiplet" structures in which a single RNA is shared by two or more capsids [8]. Multiplets formed by CCMV CP and several heterologous "overlong" RNAs have also been observed in assemblies of homopolymeric polyU and CCMV CP [18], for a conjugated polyelectrolyte and CCMV CP [19], and for packaging by SV40 CP of RNA [20] and polystyrene sulfonate [21]. The existence of doublet structures has also been found by Elrad and Hagan [22] in molecular dynamics simulations of the packaging of long polymers by capsid protein in their coarse-grained modelling of virus-particle-like assembly [23].

To determine the packaging limit by experiment therefore requires analyses of the multiplet distributions as RNA length is increased. This is the strategy we employ here to determine the efficiency of packaging and maximum length of RNA that can be packaged by CCMV and BMV. In previous studies we examined the formation of multiplets in CCMV assemblies with CP purified from virus-infected plants, and in the present work we find the same multiplet distribution using CP expressed in E. coli (a protein source that is more readily scaled up). Further, by analyzing the RNA extracted from assembly mixes before and after RNase treatment, using RNA molecules with lengths up to 6400 nt, we observe that CCMV CP is unable to package RNA longer than 4200 nt in single capsids. The same in vitro-packaging-length limit is found for the CP from the closely related bromovirus BMV, but with a different distribution of singlets, doublets, triplets, and quadruplets.

If the amount of RNA that can be packaged is set by the volume and effective charge of the RNA, then the addition of compactifying (e.g., polyvalent cationic) agents would be expected to increase the packaging limit. Indeed, the compaction of TYMV RNA by polyamines has been demonstrated by their effect on the sedimentation rate, which is inversely proportional to the RNA hydrodynamic radius. The addition of 0.1 mM spermine resulted in a 40% decrease in the radius while addition of bis(3-aminopropyl amine) led to a 25% decrease [24]. More recently, fluorescence correlation spectroscopy determinations [25] of the effect of the addition of 1 mM spermine to MS2 RNA showed a 50% decrease in its hydrodynamic radius. Finally, recent dynamic light scattering measurements on a broad range of lengths and sequences of RNA molecules, as a function of increasing spermine concentrations, have established a continuous (saturating at 50%) decrease in hydrodynamic radius in all cases [26]. Consistent with these results we report here the extent to which RNA compaction by polyvalent cations affects the efficiency of packaging of "overlong" RNA by CCMV and BMV capsid protein. We find that both the number of singlet particles and the packaged-length limit can indeed be enhanced by spermine. But the effect is small, and increasing the polyamine concentration eventually results in RNA aggregation, which limits the extent to which spermine can enhance the relative number of singlets or the length of RNA packaged into RNase-resistant singlets.

Results

To generate a wide range of lengths and sequences of RNA molecules, BMV RNA1 (3234 nt), NOV-EYFP (4196 nt), NOV-R.Luc (4413 nt), NOV-STING* (4638 nt), and TMV (6395 nt) were prepared by in vitro transcription of T7 DNA plasmids, and SIN-19 (8985 nt) and FL-SIN (11,703 nt) from SP6 plasmids (Thermo Fisher, USA), and all of the RNAs were purified with an RNEasy Mini Kit (Qiagen, DEU). Recombinant capsid proteins were grown in ampicillin-chloramphenicol-resistant cells of E. coli strain Rosetta 2 BL21, and in vitro reconstitutions of VLPs were carried out using our published protocols [8]. RNA was extracted from VLPs with a QIAamp Viral Mini Kit (Qiagen DEU) following the manufacturer's specifications. To assess VLPs for RNA protection they were mixed with RNaseA at a ratio of 0.1 g RNase A:1 g RNA, and incubated at 4 °C for 1 h. Digestion of RNA was stopped by the addition of RNase inhibitor and the sample was washed through a 100 kDa MW-cutoff Amicon filter to purify the remaining RNase-resistant VLPs.

Multiplet VLP Products Are Observed for RNAs Longer Than 3000 nt

Using a Tecnai G2 TF20 electron microscope (FEI, USA), assembly mix samples were prepared and imaged with negative stain as reported elsewhere [8, 18]. Electron micrographs were manually analyzed to count the number of singlets, doublets, triplets, and higher-order multiplets in each of the images acquired; ImageJ (US National Institutes of Health) was used to measure the geometric mean of orthogonal measurements of the diameter of the particles.

First we confirmed that bacterial-expressed capsid proteins give rise to VLP multiplet distributions similar to those found earlier [8] for infected-plant-derived proteins, when self-assembled with RNAs of different lengths. When 3200 nt RNA is packaged by CCMV CP, mostly singlet VLPs are formed (see Fig. 3.1a). Assemblies with 4200 nt RNA (Fig. 3.1b) contain some multiplet particles—mostly doublets—where one RNA is shared by two or more capsids. Packaging of 6400 nt-RNA (Fig. 3.1c) leads to a mixture of singlet, doublet, triplet, and higher-order multiplet particles, but again mostly doublets. The histogram of sizes of singlets in Fig. 3.1d is consistent with that measured in the earlier experiments conducted with plant-derived CCMV

CP and heterologous RNA [8]. The size distributions of singlet particles—measured for each of the multiplet assemblies associated with the three different RNA lengths—have widths of several nm, centered around 29–31 nm. Slightly larger singlet sizes are observed (Fig. 3.1d) for the longer-RNA (4200 and 6400 nt, versus 3200 nt) assemblies.

This is consistent with the fact that the longer RNA samples include a significant number of shorter, prematurely-terminated, transcripts and degradation products, as seen in Fig. 3.2. More explicitly, the 4200 and 6400 nt samples include RNA lengths in the range 3200–4200 nt that are short enough to be self-assembled into single T = 3 capsids but that result in capsids that are less stable and less ordered—and hence somewhat *larger*—than those for 3200 nt-RNA, the length preferred by CCMV CP.

As suggested by studies that demonstrated no structural difference between recombinant CP expressed in E. coli and wt CP derived from plants [27], we observe that the measured multiplet distribution (see Fig. 3.3) is the same as reported earlier [8] for these RNA + CCMV CP assembly mixes, even though plant-derived CP is used in the latter case and recombinant CP in the former. For a 3200 nt RNA, 85% of the particles are singlets; for 4200 nt RNA, 65% of particles are singlets, and the rest are mostly doublets; and for 6400 nt RNA, 32% appear to be singlets while 56% percent are doublet particles. Upon serial dilution of the assembly mix, the multiplet

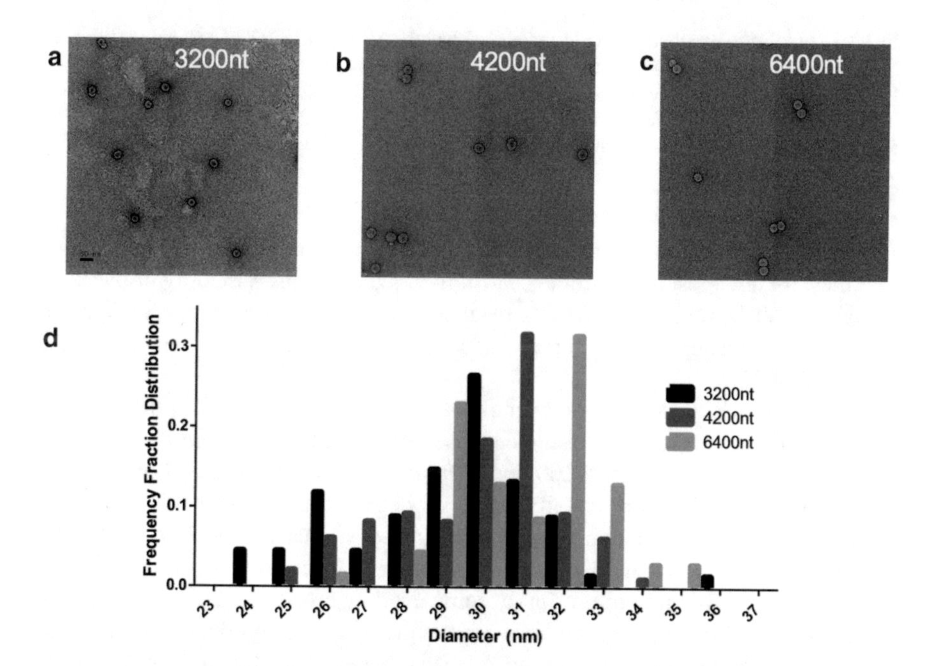

Fig. 3.1 Characterization of CCMV VLPs assembled with different lengths of RNA. Negative-stain electron micrographs of self-assembly mixes involving CCMV CP and: ~3200 nt-long RNA (**a**), ~4200 nt-long RNA (**b**), and ~6400 nt-long RNA (**c**). **d** shows the singlet capsid-size-distribution plots measured for each of these self-assemblies before RNase treatment

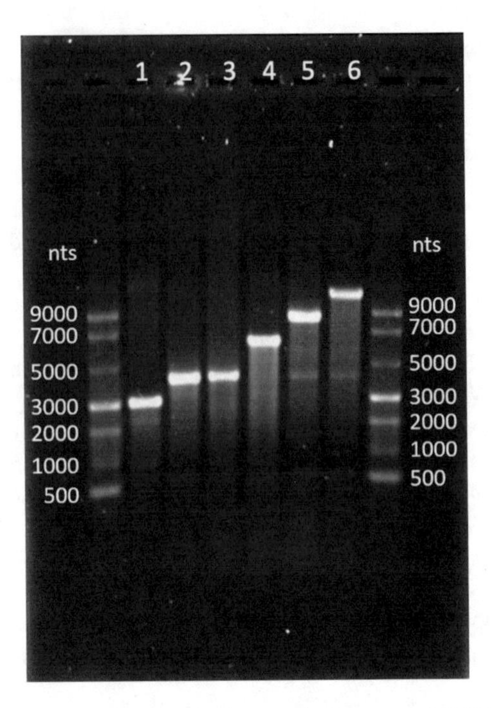

Fig. 3.2 In vitro transcribed RNAs of increasing length. Agarose gel (0.83%) showing different-length RNA transcript products. Lane 1: 3234 nt RNA, Lane 2: 4413 nt RNA, Lane 3: 4638 nt RNA, Lane 4: 6395 nt RNA, Lane 5: 8985 nt RNA, Lane 6: 11703 nt RNA. 1 μg of RNA was mixed with RNA loading dye and denatured by heating to 65 °C at a rate of 1 °C per sec, holding at 65 °C for 10 min, then cooling to 4 °C at a rate of 1 °C per sec. The gel was run at 100 V for 1.5 h, and visualized with GelRed

particles continue to be observed and the distribution remains unaltered, indicating that they are not a result of particle crowding or aggregation. Assembly of overlong RNAs into VLPs with *BMV* CP also results in multiplets, but gives rise to distributions that contain significantly larger fractions of higher-order multiplets than in the case of CCMV.

Importantly, as discussed below, most of the singlets that appear in the 4200 nt-RNA assemblies, and all of the singlets observed in the 6400 nt-RNA case, contain RNA molecules shorter than 4200 and 6400 nt, respectively. These shorter RNAs arise as premature transcripts and hydrolysis degradation products of the in vitro transcribed RNAs. As seen in Fig. 3.2, an agarose gel analysis of RNA molecules with lengths ranging from ~3200 to ~12000 nt, each lane shows a strong band (the corresponding full-length transcript) accompanied by broad streaks associated with faster-running, shorter, premature transcripts and degradation products. As discussed below in the context of Figs. 3.4 and 3.5, it is these shorter RNAs—up to lengths of 4200 nt—that are packaged into singlets when transcripts of RNAs longer than 3200 nt are mixed with CCMV CP under assembly conditions.

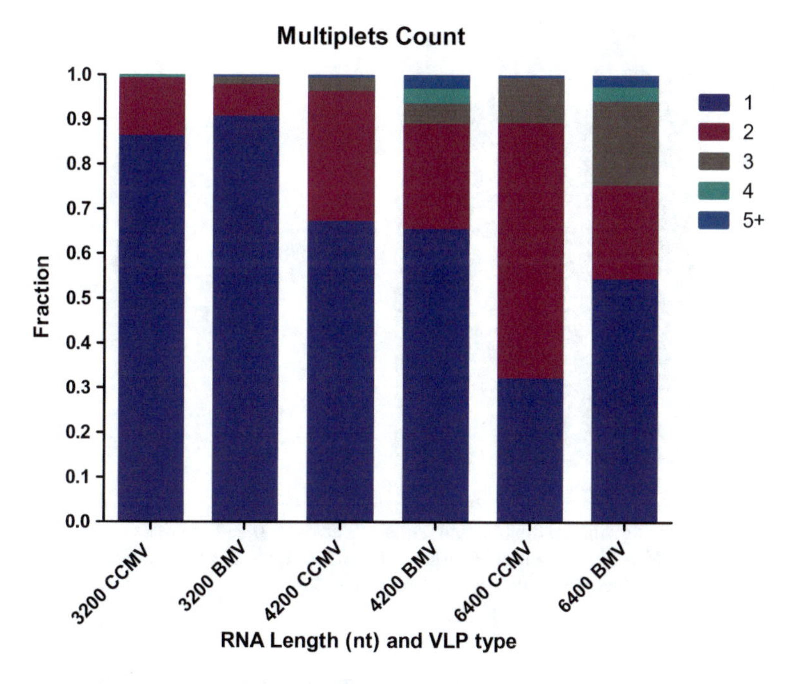

Fig. 3.3 Characterizing CCMV and BMV VLP particles by negative-stain EM. Counts of the frequency of singlets, doublets, triplets, quadruplets and larger particles in samples of CCMV and BMV VLP assemblies (before RNase treatment)

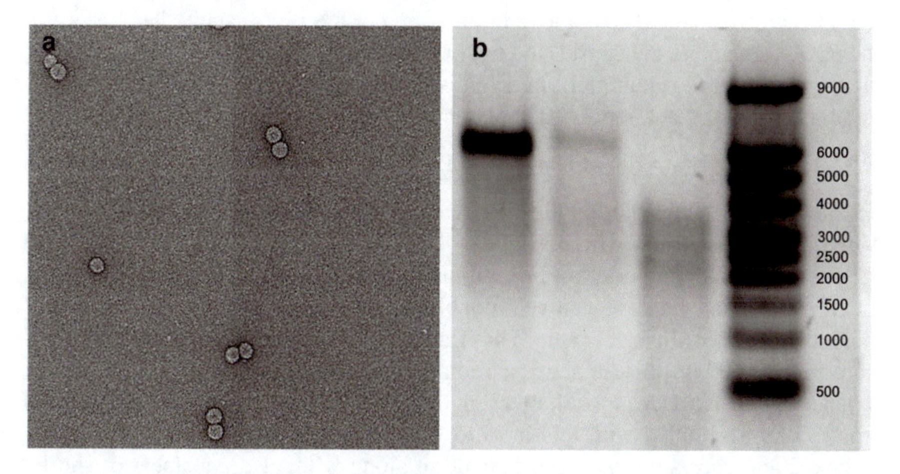

Fig. 3.4 6400 nt RNA in CCMV VLPs. **a** Negative-stain electron micrograph of CCMV VLPs assembled with in vitro transcribed TMV-(6400 nt-)RNA, showing singlet, doublet and triplet particles. **b** 0.83% agarose gel. Lane 1: in vitro transcribed 6400 nt RNA. Lane 2: RNA extracted from CCMV-VLPs assembled with 6400 nt RNA. Lane 3: RNA extracted from RNase-treated CCMV-VLPs assembled from 6400 nt RNA. Lane 4: RNA Ladder

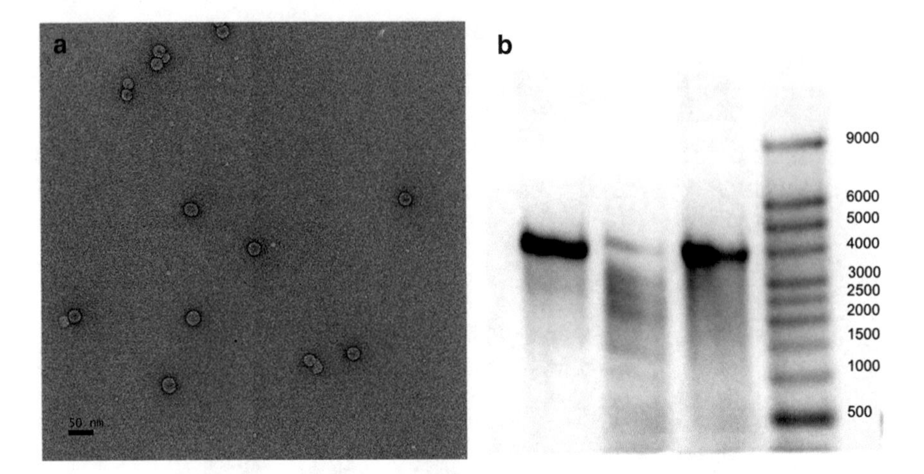

Fig. 3.5 4200 nt RNA in CCMV VLPs. **a** Negative-stain electron micrograph of 4200 nt-RNA CCMV-VLPs, showing singlet, doublet and triplet particles. **b** 0.83% agarose gel. Lane 1: RNA extracted from CCMV-VLPs assembled with 4200 nt-RNA. Lane 2: RNA extracted from RNase-treated CCMV-VLPs assembled with 4200 nt-RNA. Lane 3: in vitro transcribed 4200 nt RNA, Lane 4: RNA Ladder

After RNase A Treatment, Only Singlet VLPs Are Observed

When treated with RNase, all multiplet VLPs become sets of singlets, as evidenced by electron micrographs (not shown), i.e., no doublets, triplets, or quadruplets remain, from which we conclude that the multiplets are not RNase resistant: the portions of the RNA connecting the capsids are not protected. However, the singlets remaining after disassembly of multiplets are RNAase resistant (see Fig. 3.4b and discussion below), evidence that after disassembly of multiplets the capsids are intact.

RNAs Longer Than 4200 nt Cannot Be Packaged into RNase-Resistant Singlets

To show in particular that 6400 nt RNA cannot be packaged into singlet CCMV VLPs, we extracted RNA from samples of 6400 nt-RNA CCMV-VLPs before and after their treatment with RNase. Recall that assemblies of this length of RNA with CCMV CP result predominantly in doublets, but also in a significant number of singlet VLPs; see Figs. 3.1c and 3.4a. Lane 1 of Fig. 3.4b, loaded with in vitro-transcribed 6400 nt RNA, features a strong band where expected (see RNA ladder in lane 4), but also a smear of shorter lengths down to ~2000 nt. The RNA extracted from VLPs assembled with this in vitro transcribed RNA and CCMV CP is run in lane 2 and shows a *weak* band at 6400 nt along with, again, a smear of shorter lengths down to ~2000 nt. Lane 3 contains the RNA extracted from the assembly mix *after RNase treatment*. Most significantly, no band in this lane is present at 6400 nt and the shorter-length smear begins at ~4000 nt and runs down to ~2000 nt, from which it can be concluded that *no RNAs longer than ~4000 nt were packaged into singlet VLPs.* Rather, the predominant doublet VLPs involve the sharing by two capsids of > 4000 nt-long RNAs that are digested into smaller molecules upon RNase treatment. The strong signal in lane 3 from RNAs around 3000 nt in length (the average size of CCMV viral RNA) is consistent with these molecules being the ones that are preferentially packaged into singlet VLPs and hence RNase-resistant.

4200 nt RNA Can Be Packaged by CCMV CP, but Inefficiently

To establish whether it is possible to package 4200 nt RNA in CCMV VLP singlets, we assembled in vitro-transcribed 4200 nt RNA with CCMV CP and extracted the RNA from the resulting VLPs before and after treatment with RNase. The typical structures that are found when packaging the 4200 nt RNA are shown in Fig. 3.5a. The agarose gels in Fig. 3.5b show in vitro-transcribed 4200 nt RNA along with shorter transcript RNA (lane 3). Lane 1 in the agarose gel shown in Fig. 3.5b contains RNA extracted from 4200 nt-RNA CCMV-VLPs before RNase treatment, and lane 2 contains RNA extracted from RNase-treated 4200 nt-RNA CCMV-VLP samples. The in vitro-transcribed RNA used in these self-assemblies is run in lane 3, showing a strong band at 4200 nt corresponding to the full-length RNA transcript—but also showing a significant amount of faster-running shorter RNA that arises from hydrolysis degradation and premature transcripts. From the relatively weak 4200 nt band in lane 2 (involving the RNA extracted from RNase-treated VLP assemblies), along with the accompanying RNA intensity there associated with faster-running shorter RNA, we conclude that only a small percentage of the full-length RNA has been packaged into RNase-resistant singlets, i.e., most of the singlets observed for assemblies with in vitro-transcribed 4200 nt RNA correspond to the packaging of shorter RNA. By comparing the intensity profile in lane 1 (involving the RNA extracted from VLP assemblies that have *not* been treated with RNase) with the profile in lane 3 (in vitro-transcribed RNA), we conclude further that the full-length RNA packaged in multiplets has been cleaved by RNase to give the bands at shorter lengths in lane 2 which also include contributions from direct packaging of the shorter transcripts in the 4200 nt-RNA sample.

Size Differences in Doublet Pairs Suggest Preference for Packaging 3000–4000 nt RNA

It had been noted earlier [17] that while the diameters of capsids contained within doublets and triplets for assemblies of 9000 nt RNA are most likely to be closely similar, the distribution of size differences has a half-width of about 5 nm and differences as large as 10 nm are observed. As shown in Fig. 3.6, we have found this as well for the doublets formed in capsids assembled around 6400 nt RNA. Similarly, the largest size differences observed for the assemblies around 3200 and 4200 nt RNA are also as large as 10 nm, and the maxima in the distributions occur at around 5 nm. These results are consistent with the strong preference of CCMV for packaging 3000–4000 nt of RNA, which can be achieved in triplets or doublets for 9000 nt-RNA assemblies and in doublets for a 6400 nt RNA. For the 3200 and 4200 nt RNA assemblies, however, size differences in doublets must necessarily be larger in order to accommodate cargo sizes compatible with $T = 2$– and $T = 1$–sized capsids.

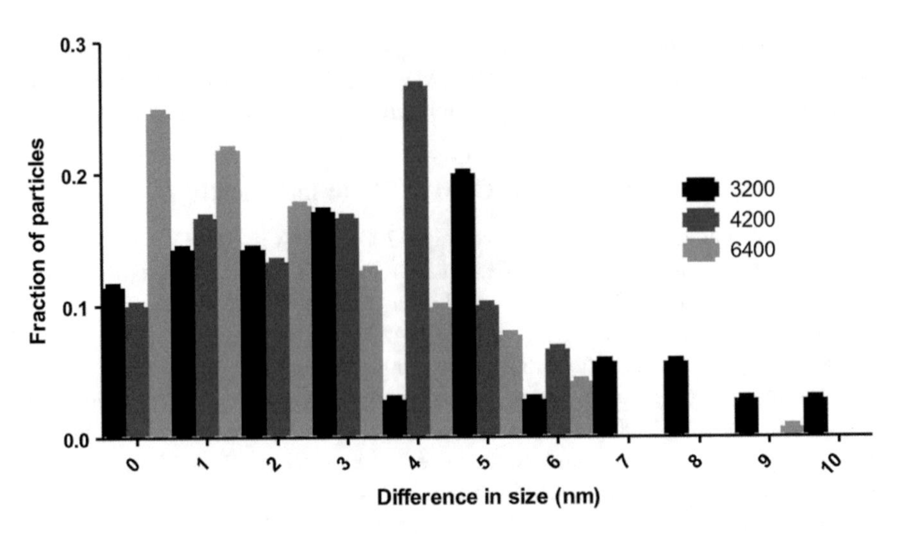

Fig. 3.6 Distribution of size differences between particles in doublets, for assemblies of CCMV CP with 3200, 4200, and 6400 nt RNA

Spermine Has a Small Enhancing Effect on Packaging

The ability of the polyamine spermine to compactify RNAs has been demonstrated for a variety of molecules [24, 25]. The control parameter is the ratio of total cationic (spermine) charge to total anionic (nucleotide/phosphate) charge, with each spermine having a +4 charge and each RNA nucleotide corresponding to a −1 charge from the phosphate backbone. In the case of the 4196 nt RNA, for example, the compaction by spermine is clear from the fact that RNA samples incubated (at room temperature for 30 min) at different spermine:RNA ratios are seen to run progressively faster in an agarose gel (not shown) as spermine concentration is increased, even though the effective charge on the RNA is decreased; for charge ratios above 1, however, the spermine also aggregates the RNA. Continuous compaction of the RNA with increasing cationic:anionic charge ratio is documented more directly and systematically by dynamic light scattering measurements [26], suggesting that there should be a corresponding effect of spermine on the distribution of multiplets in overlong-RNA assemblies and on the maximum length that can be packaged in RNase-resistant singlets.

Particle counts of VLPs made with spermine-compacted RNA show (Fig. 3.7) that for a 4200 nt RNA at a spermine:RNA charge ratio of 0.25:1 there is a 10% increase in the number of singlet particles observed. In assemblies at charge ratios of 0.5:1 and 0.75:1 there is no further increase in singlets; rather, at these higher charge ratios there is an increase in the number of aggregated RNAs and hence higher-order multiplets. For all of these experiments spermine was stored in degassed water overlaid with argon gas and frozen at −80 °C until used. Frozen RNA was thawed on ice and thermally denatured by heating to 90 °C at a rate of 1 °C s^{-1}, holding at

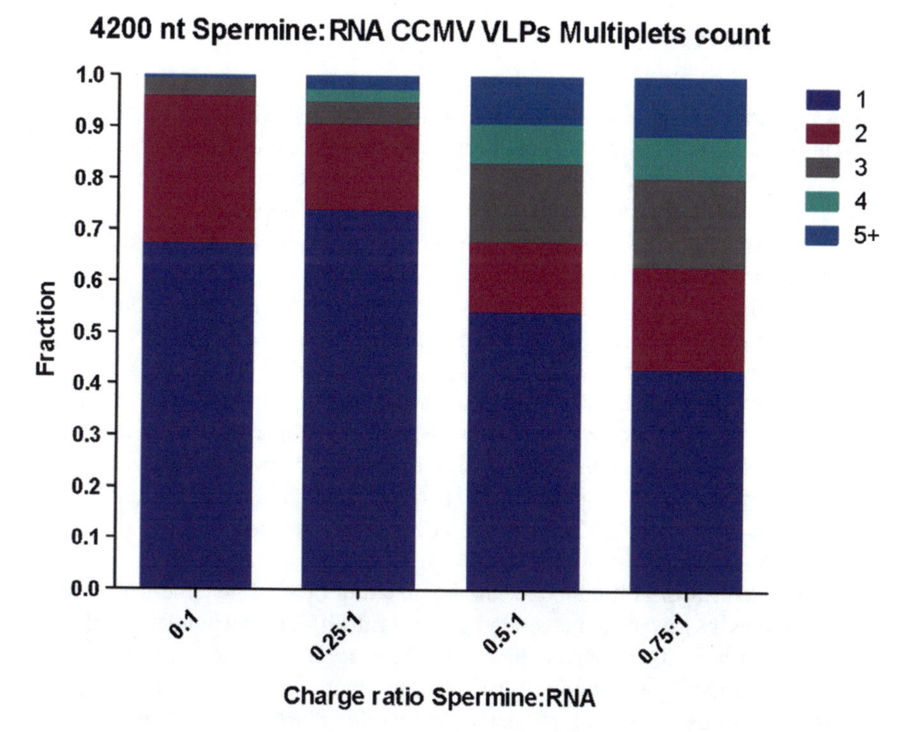

Fig. 3.7 Effect of spermine on RNA and its packaging by CCMV CP

90 °C for 1 s, then cooling to 4 °C at a rate of 1 °C s^{-1}. Thermally denaturing and slowly cooling the RNA allows for disruption of any duplexes formed between RNA molecules in the freezing process and for refolding of the RNA. The RNA was then mixed with spermine at a specified charge ratio and incubated at 4 °C for 30 min.

Fraction of singlets (blue), doublets (red), triplets (grey), quadruplets (green) and larger particles (light blue) in samples of CCMV VLP particles formed from 4200 nt RNA compacted with spermine at different charge ratios.

That the spermine does not prevent the full-length of 4200 nt RNA from being packaged by singlet CCMV VLPs is confirmed by extracting the RNA from a CP assembly carried out for a 0.25:1 spermine:RNA ratio and running it in a gel (not shown).

Conclusions/Perspectives

An examination of the distribution of multiplets formed when RNA longer than 3000 nt is mixed under assembly conditions with recombinant CCMV CP has shown that there is a limit to the length of RNA that can be packaged into a single, RNase-resistant, capsid. Consistent with results obtained previously for infected-plant-derived CCMV CP, the frequency of doublet and higher-order multiplets increases with increasing length of RNA.

More explicitly, with gel electrophoresis analyses of RNA extracted from RNase-treated assembly mixes, we have shown that CCMV CP cannot form VLPs that protect a significant fraction of RNA longer than about 4200 nt. In particular, gels of RNA extracted from CCMV VLPs indicate that in an assembly mix of CP and 6400 nt RNA there are no capsids containing RNAs longer than 4200 nt that are RNase resistant. This suggests that the majority of the singlets observed in EM contain only fragments shorter than 4200 nt, which are a natural result of in vitro RNA transcription of the 6400 nt-TMV-RNA. Multiplets also occur when recombinant CP of BMV is used to package long RNA.

As remarked in the Introduction, the compaction of viral-genome-length RNA by polyamines has been demonstrated by several different techniques, including sedimentation velocity measurements [24], fluorescence correlation spectroscopy [25], and dynamic light scattering [26]. We find that this effect of spermine can be used to improve the packaging efficiency of "overlong" RNA by CCMV and BMV capsid protein, e.g., the number of RNase-resistant singlet particles can be increased when RNA is incubated with spermine prior to assembly. However, this effect is small, and introducing spermine also results in RNA aggregation at spermine:RNA charge ratios greater than one, thereby limiting increase in the number of singlets.

Our work establishes the inability of CCMV or BMV CP to package RNA longer than 4200 nt into particles which are resistant to RNase. In doing so, we highlight the strongly-evolved preference of CCMV to form $T = 3$ particles. The formation of multiplets is evidence of the dominant role of the spontaneous curvature of the CP in the assembly of VLPs of flexible anionic polymers such as RNA in determining the diameter of the capsids. In contradistinction, as in the case of charged nanoemulsion particles [28] that are incompressible, multiplets cannot form and the size of the capsid therefore must increase to accommodate increasing size of the cargo. The larger fraction of multiplets in the assemblies with BMV as compared to CCMV is likely attributable to the lower positive charge on the N-termini of the capsid protein (+9 for BMV rather than +10 for CCMV) resulting in a greater degree of overcharging of the RNA compared to that of the CP and therefore a smaller amount of RNA that can be stabilized by a single capsid. Moreover, electrophoretic mobility measurements show that at the pH of assembly, 4.5, the magnitude of the charge on CCMV capsids is about twice that on BMV capsids [29], making assembly of capsids into multiplets more difficult.

These observations demonstrate both the promise and limitations of CCMV and BMV VLPs as gene-delivery platforms. The in vitro self assembly provides highly monodisperse and pure VLPs in good yields with easily obtained recombinant CP. However, these positive features come with the disadvantage of a major length restriction of 4200 nt in packaging capacity, which with the use of spermine may be only modestly increased. If, for example, the aim is to deliver self-amplifying RNA, the gene of interest must be genetically fused to an RNA-dependent-RNA-polymerase replicase, the shortest of which (that for the Nodamura virus) is 3000 nt long, thereby limiting the length of the gene of interest to only 1200 nt. If one wishes to retain the advantages of in vitro self assembly, then, the choice for packaging of longer RNA is TMV CP, despite the necessity of employing RNA constructs that encode an origin of assembly sequence (OAS) [30]. Here, in contrast to the situation with spherical viruses, the highly-evolved preference of capsid protein to form a hollow *cylinder* with fixed inner and outer radii can be satisfied by any length of RNA because the capsid curvature is independent of length.

Acknowledgements We are grateful to Jerrell Tisnado for expression and purification of recombinant capsid protein, and to all members of our research group for many helpful discussions about viral self-assembly over the past several years. This work has been supported financially by the National Science Foundation (Molecular and Cellular Biosciences Division, Genetic Mechanisms Program, Grants 1716975 and 2103700 to WMG).

References

1. Yildiz I, Tsvetkova I, Wen AM, Shukla S, Masarapu MH, Dragnea B, Steinmetz NF, Fu Y, Li J (2016) A novel delivery platform based on Bacteriophage MS2 virus-like particles. Virus Res 211:9–16
2. Vaidya AJ, Solomon KV (2022) Surface functionalization of rod-shaped viral particles for biomedical applications. ACS Appl Bio Mater 5(5):1980–1989
3. Smith MT, Hawes AK, Bundy BC (2013) Reengineering viruses and virus-like particles through chemical functionalization strategies. Curr Opin Biotech 24:620–626
4. Biddlecome A, Habte HH, McGrath KM, Sambanthamoorthy S, Wurm M et al (2019) RNA vaccines in in vitro reconstituted virus-like particles. PLoS One 14(6):e0215031
5. Saunders K, Thuenemann EC, Peyret H, Lomonossoff GT (2022) The Tobacco Mosaic virus origin of assembly sequence is dispensable for specific viral RNA encapsidation but necessary for initiating assembly at a single site. J Mol Biol 434:167873
6. Choi Y, Dreher TW, Rao ALN (2002) tRNA elements mediate the assembly of an icosahedral virus. Proc Natl Acad Sci USA 99:655–660
7. Annamalai P, Rao ALN (2005) Dispensability of 3' tRNA-like sequence for packaging cowpea chlorotic mottle virus genomic RNAs. Virol 332:650–658
8. Cadena-Nava RD, Comas-Garcia M, Garmann RF, Rao ALN, Knobler CM, Gelbart WM (2012) Self-assembly of viral capsid protein and RNA molecules of different sizes: requirement for a specific high protein. RNA mass ratio. J Virol 86:3318–3326
9. Stockley PG, Rolfsson O, Thompson GS, Basnak G, Francese S, Stonehouse NJ, Homans SW, Ashcroft AE (2007) A simple, RNA-mediated allosteric switch controls the pathway to formation of a T = 3 viral capsid. J Mol Biol 369:541–552
10. Zhan S, Li J, Xu R, Wang L, Zhang K, Zhang R (2009) Armored long RNA controls or standards for branched DNA assay for detection of human immunodeficiency virus type 1. J Clin Microbiol 47:2571–2576

11. Cohen SS, Greenberg ML (1981) Spermidine, an intrinsic component of turnip yellow mosaic virus. Proc Natl Acad Sci USA 78:54470–55474
12. Clarke IN, Lambden PR (2000) Organization and expression of calicivirus genes. J Infect Dis 181:S309–S316
13. Gopal A, Ececioglu D, Yoffe AM, Ben-Shaul A, Rao ALN, Knobler CM, Gelbart WM (2014) Viral RNAs are unusually compact. PLoS ONE 9:e105875
14. Erdemci-Tandogan G, Wagner J, van der Schoot P, Podgornik R, Zandi R (2014) RNA topology remolds electrostatic stabilization of viruses. Phys Rev E 89(3):032707
15. Garmann RF, Comas-Garcia M, Koay MST, Cornelissen JJLM, Knobler CM, Gelbart WM (2014) The role of electrostatics in the assembly pathway of a single-stranded RNA virus. J Virol 88:10472–10479
16. Beren C, Cui Y, Chakravarty A, Yang X, Rao ALN, Knobler CM, Zhou ZH, Gelbart WM (2020) Genome organization and interaction with capsid protein in a multipartite RNA virus. Proc Natl Acad Sci USA 117:10673–10680
17. Dai X, Li Z, Lai M, Shu S, Du Y, Zhou ZH, Sun R (2017) In situ structures of the genome and genome-delivery apparatus in an ssRNA virus. Nature 541(7635):112–116
18. Thurm AR, Beren C, Duran-Meza AL, Knobler CM, Gelbart WM (2019) RNA homopolymers form higher-curvature virus-like particles than do normal-composition RNAs. Biophys J 117(7):1331–1341
19. Brasch M, Cornelissen JJLM (2012) Relative size selection of a conjugated polyelectrolyte in virus-like protein structures. Chem Commun 48(10):1446–1448
20. Kler S, Wang JC, Dhason M, Oppenheim A, Zlotnick A (2013) Scaffold properties are a key determinant of the size and shape of self-assembled virus-derived particles. ACS Chem Biol 8(12):2753–2761
21. Li C, Kneller AR, Jacobson SC, Zlotnick A (2017) Single particle observation of SV40 VP1 polyanion-induced assembly shows that substrate size and structure modulate capsid geometry. ACS Chem Biol 12(5):1327–1334
22. Hagan MF, Elrad OM (2010) Understanding the concentration dependence of viral capsid assembly kinetics—the origin of the lag time and identifying the critical nucleus size. Biophys J 98(6):1065–1074
23. Perlmutter JD, Qiao C, Hagan MF (2013) Viral genome structures are optimal for capsid assembly. eLife 2:e00632
24. Mitra S, Kaeseberg P (1965) Biophysical properties of RNA from turnip yellow mosaic virus. J Mol Biol 14(2):558–571
25. Borodavka A, Dykeman EC, Schrimpf W, Lamb DC (2017) Protein-mediated RNA folding governs sequence-specific interactions between rotavirus genome segments. Elife 6:e27453
26. Duran-Meza AL, Oster L, Sportsman R, Phillips M, Knobler CM, Gelbart WM (2023) Long ssRNA Undergoes Continuous Compaction in the Presence of Polyvalent Cations. Biophys J. https://doi.org/10.1016/j.bpj.2023.07.022
27. Zhao X, Young MJ (1995) In vitro assembly of cowpea chlorotic mottle virus from coat protein expressed in *Escherichia Coli* and in vitro transcribed viral cDNA. Virology 207:486–494
28. Chang CB, Knobler CM, Gelbart WM, Mason TG (2008) Curvature dependence of viral protein structures on encapsidated nanoemulsion droplets. ACS Nano 2(2):281–286
29. Duran-Meza AL, Villagrana-Escareño MV, Ruiz-García J, Knobler CM, Gelbart WM (2021) Controlling the surface charge of simple viruses. PLoS ONE 16(9):e0255820
30. Smith ML, Corbo T, Bernales J, Lindbo JA, Pogue GP, Palmer KE, McCormick AA (2007) Assembly of trans-encapsidated viral vectors engineered from Tobacco mosaic virus and Semliki Forest virus and their evaluation as immunogens. Virology 358(2):321–333

Chapter 4
The Multiple Regulatory Roles of Single-Stranded RNA Viral Genomes in Virion Formation and Infection

Peter G. Stockley and Reidun Twarock

Abstract We describe our discovery, recent findings, and potential implications, of previously unsuspected roles of viral genomes in assembly regulation and infection. For a range of viral pathogens, including those that infect people, we have shown that assembly depends on multiple RNA-coat protein interactions. Spontaneous release of these constraints during assembly may also contribute to viral infectivity. These properties of viral genomes are conserved across strain variants and are therefore currently unexploited potential targets for directly-acting anti-viral drugs. Furthermore, grafting RNA assembly signals onto non-viral RNAs confers cognate packaging properties with appropriate coat proteins and is opening up an era of bespoke designer viral vectors.

Keywords RNA packaging signals (PSs) · Viral genome-directed assembly · Co-operativity and Hamiltonian paths · Molecular frustration · Conserved anti-viral drug targets

Introduction

Viral pathogens are astonishing in their ubiquity, using virtually all known cell types and organisms as hosts. Double-stranded (ds) DNA bacteriophages in the oceans are thought to be the most numerous biological objects on the planet [11, 36]. Their predation of planktonic organisms causes carbon to fall to the ocean depths, and may well regulate the natural carbon cycle, emphasizing their importance. Frequent viral infections have driven the development of the various immune systems in the

P. G. Stockley (✉)
Astbury Centre for Structural Molecular Biology, University of Leeds, Leeds LS2 9JT, UK
e-mail: p.g.stockley@leeds.ac.uk

R. Twarock (✉)
Departments of Mathematics and Biology, York Cross-Disciplinary Centre for Systems Analysis, University of York, York, UK
e-mail: reidun.twarock@york.ac.uk

M. Comas-Garcia and S. Rosales-Mendoza (eds.), *Physical Virology*, Springer Series in Biophysics 24, https://doi.org/10.1007/978-3-031-36815-8_4

biosphere, including the CRISPR gene editing system of bacteria [18]. Viral infections also play roles in gene transfers, e.g. the potential reuse of a gene(s) acquired from a retrovirus by placental mammals [64]. Indeed, large portions of the human genome seem to harbor inactivated retroviruses attesting to our long association with them and hinting at an eventual genetic solution to the challenges of infection.

Obviously, modern viruses can be deadly, as annual influenza epidemics [62] and the recent coronaviral (Covid-19) pandemic illustrate [6]. At the time of writing, the latter is thought to be responsible for >7 million human deaths. This figure is expected to climb sharply despite worldwide vaccination campaigns. It has however had the benefit of catapulting vaccine design and manufacture into the modern era [55]. It also reveals one of the challenges in modern medicine, namely the dearth of directly-acting anti-viral drugs (DAAs). This is worrying since vaccine campaigns for Covid-19 have been both exhaustive and expensive, yet they have exposed extensive vaccine hesitancy in populations across the world [49]. Zoonotic viral transfers into the ever-expanding human population, coupled to the consequences of climate change, are more likely to occur. DAAs that can be administered to people already infected with viral pathogens are therefore urgently needed. Two such DAAs that inhibit the actions of the surface glycoproteins from influenza virus, namely hemagglutinin and neuraminidase [58], are extremely inhibitory but are held in reserve from the clinic to avoid the evolution of resistance mutations that might limit their effectiveness during a flu pandemic. DAAs have, however, been assets in the worldwide HIV & HCV epidemics. A triple drug cocktail of inhibitors directed against the HIV viral polymerase, its protease and integrase, leads to effective "cures" for patients who tolerate this dosing regime [23]. For HCV elimination, it appears that protease inhibitors are sufficient for the same outcome [31]. However, for most viruses, effective DAAs are not yet available and novel anti-viral approaches are urgently required.

Given this history, the evolutionarily conserved mechanisms regulating virion assembly, and potentially infection, described here offer a range of currently unexploited targets for the development of DAAs against common human pathogens.

The Discovery of RNA Packaging Signal (PS)-Mediated Assembly

The earliest molecular biologists realised that viral particles represent simple systems in which to probe the relationship of nucleic acid base sequences to the amino acid sequences of virally-encoded proteins [15]. The simplest such viruses build capsids based on icosahedral geometry [8], extending this principle to allow larger structures to be built from more subunits, thus reducing the amount of genetic information needed to specify capsid building blocks—coat proteins (CPs). This idea of "quasi-equivalence" describes the architecture of many simple viruses, including human pathogens, predicting the surface layouts of many viral shells. It has been further refined [34, 59, 60], leading to a classification of virus architecture that also includes outliers to Caspar and Klug's theory.

One such quasi-equivalent viral protein shell is the $T = 3$ casing of bacteriophage MS2 [40] that in principle contains 180 CP subunits organized as 90 dimers surrounding 3569 nts of ssRNA (Fig. 4.1a). Assembly of this particle was believed to initiate via a sequence-specific interaction between the genome (gRNA) and a CP dimer that occurs at a 19 nt long stem-loop (TR) within the gRNA [39]. Uhlenbeck and his colleagues carried out ground-breaking structure–activity measurements of this RNA–protein recognition in vitro showing that the identity of only 3 nucleotides (the adenines at local positions –4, -10, and the uridine at -5) within this 19 nt long stem-loop were critical for complex formation [32]. We extended this work using direct chemical synthesis of variant TRs to probe the roles of the various nucleotide functional groups [56]. Subsequently, we showed that this complex triggers in vitro assembly of a $T = 3$ virus-like particle provided that the TR oligo is available in sub-stoichiometric amounts compared to the CP dimer [53]. The origin of this requirement is due to a TR-induced conformational change within a CP dimer (Fig. 4.1b), locking it into an asymmetric CP conformation able to occupy the 60 A/B quasi-conformer positions within the CP shell [63]. CP dimers without a bound TR remain symmetrical and occupy the 30 C/C positions completing the virion. Normal mode analysis of the CP dimer in isolation and in complex with TR are consistent with the sequence determinants required for this allosteric switch (from a symmetric C/C to an asymmetric A/B dimer) being minimal [19]. This result suggests that many other stem-loops (SLs) across the gRNA could also potentially trigger this switch.

When the technique of Selective Evolution of Ligands by Exponential Enrichment (SELEX) is used to identify RNA oligonucleotides that bind to the MS2 CP dimer, TR and several single nucleotide variants of it are found in the selected pool [30]. One of these has a higher affinity for the dimer than TR but does not appear in the natural gRNA sequence. These other variants are all of lower affinity than TR. This result is consistent with SLs other than TR in the phage gRNA being able to trigger CP dimer conformational switching. There are >100 SLs within the MS2 genome that encompass the sequence criteria identified by SELEX, i.e. that could in principle bind to the CP dimer. Since conformer switching results in formation of an A/B dimer only a maximum of 60 of these SLs can actually act as PSs, consistent with there being 60 A/B positions in the capsid shell. The relative positions of the RNA binding sites at the inner capsid surface impose geometric constraints on the potential SLs within the gRNA that could function as PSs. We introduced a novel data analysis method, Hamiltonian Paths Analysis (HPA) [20–22, 61] that combines these geometric constraints with a bioinformatics analysis of the viral genome in order to locate putative PSs in the genomic sequence. HPA is rooted in mathematical models of virus architecture that indicate the positions of the RNA-CP contacts [20, 35] (Fig. 4.2a). Connecting points corresponding to PS binding sites on neighbouring CP dimers by an edge, results in a polyhedral shell. Enumerating the vertices of this polyhedron in the order in which putative PSs, in the $5'$–$3'$ direction in the linear genomic sequence make contacts to these sites, defines a path on this polyhedron. If all these vertices are visited, i.e., if all 60 contacts are made, the

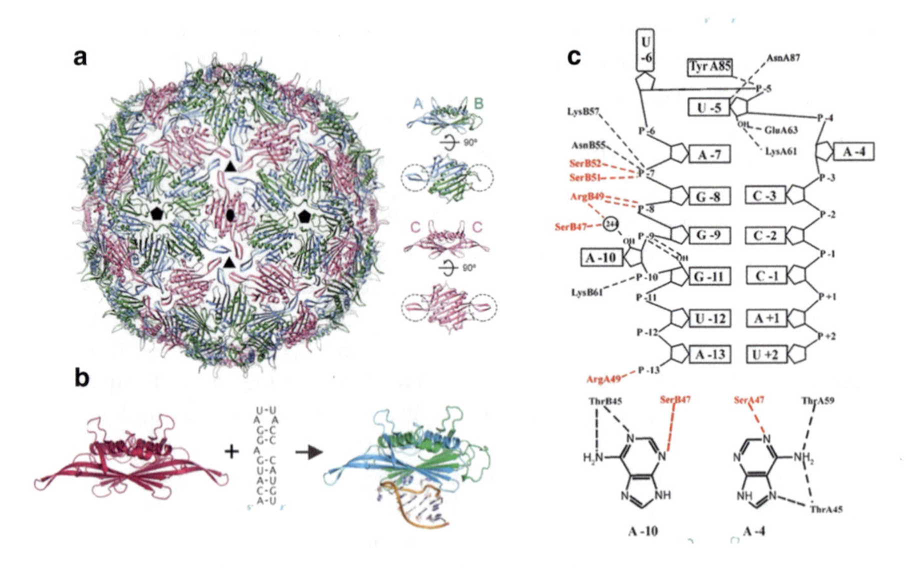

Fig. 4.1 The molecular architecture of bacteriophage MS2. **a** Structure of the bacteriophage particle determined by X-ray crystallography at atomic resolution (3.3 Å). The particle appears as a symmetrical T = 3 capsid composed entirely of 180 coat protein subunits, in the form of 90 non-covalent dimers, distinguished by their two distinct conformations. A/B & C/C dimers are identical except in the loop of polypeptide (38 amino acids out of 129 total) that connects the F & G β-strands. A-type loops are extended whereas B-type loops are folded back towards the main body of the subunit and are associated with a conserved amino acid (Pro72) adopting the rarer cis configuration. C/C dimers are identical in conformation and both FG-loops adopt an extended conformation that differs in detail with that of the A-subunit. See cartoons between **a** and **b** panels for more detail. A/B & C/C-dimers are orientated such that their extended FG-loops interdigitate at particle threefold axes, whereas B-type loops occupy the spatially more restricted 5-folds. There is no electron density in this structure for the MS2 gRNA (adapted from Chandler-Bostock et al. [10]). **b** The 19-nt hairpin TR from the gRNA, a packaging signal, triggers the conformational switch from a C/C to and A/B dimer (adapted from Dykeman et al. [19]). **c** Shows the chemical details of the non-covalent recognition of TR by the MS2 CP dimer. Dashed lines indicate the positions of hydrogen bonds

result is called a Hamiltonian path (Fig. 4.2b). However, it is not strictly required for all contacts to be formed as these constraints can also be applied locally. Note, the gRNA extends towards the particle interior in between PS binding sites. Using this idea as a constraint, we were able to predict the PSs of bacteriophage MS2 [20] (Fig. 4.2c). More recent, high-resolution, asymmetric cryo-EM reconstruction [16] or X-ray footprinting [10] identifies 15–23 PSs bound to the CP shell, which are all contained in our ensemble of predicted PSs. The reconstruction also identifies many SLs in the gRNA in proximity with the capsid shell beyond the 15 that are firmly bound according to the cryo-EM reconstruction (Fig. 4.2d). It is tempting to assume these acted as PSs during the assembly process, but subsequently dissociated from the CP shell.

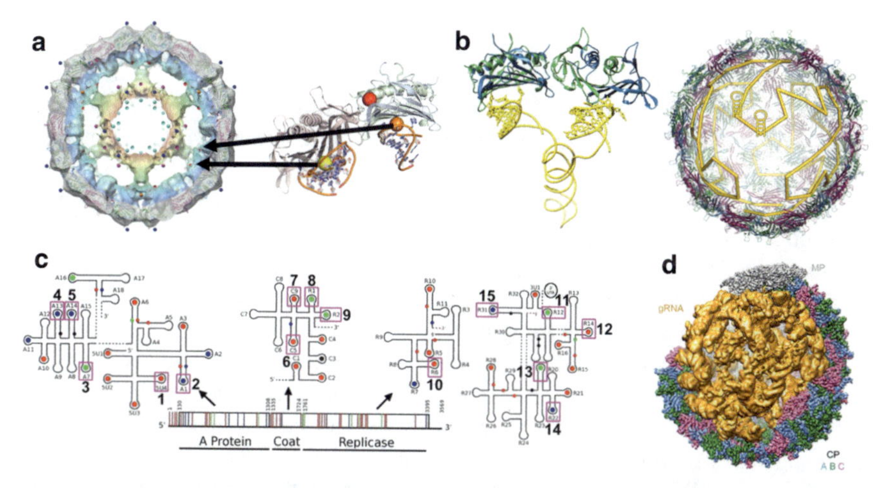

Fig. 4.2 Geometric constraints on genome organisation in viral capsids. **a** Affine extended symmetries model virus architecture at different radial levels simultaneously, here shown for bacteriophage MS2. Vertices map around capsid (grey) and packaged gRNA (radially coloured). There are vertices that mark the contacts between stem-loops in the gRNA (see also Fig. 4.2d) and CP, shown in magnification (adapted from Keef et al. [35]). **b** The distances between these vertices can be used as constraints on genome sequence analysis to identify packaging signals. Connecting vertices corresponding to neighbouring PSs in the gRNA defines a self-avoiding path on a polyhedral shell, called a Hamiltonian path, and this data analysis methods has therefore been called Hamiltonian Paths Analysis (HPA) (Adapted from Twarock et al. [61]). **c** HPA identified PSs in the MS2 gRNA, shown here as dots colour-coded according to their affinities and stabilities with reference to the gRNA secondary structure, and as lines with reference to a gene map (adapted from Twarock et al. [61]). **d** A high resolution asymmetric cryo-electron microscopy reconstruction identifies 15 PSs (boxed in **c**) and contains further density in proximity to the capsid shell that could correspond to stem-loops that transiently acted as PSs, validating HPA predictions. In contrast to the structure in Fig. 4.1a it contains the single copy Maturation Protein (MP, shown here as a grey space filling model), which replaces a CP dimer at a two-fold axis in the protein shell, and also reveals density for the 3569 nts long gRNA (shown here as a gold space filling model; other colours in the shell are the same as those in Fig. 4.1a) (Adapted from Dai et al. [16])

The Packaging Signal (PS)-Mediated Assembly Mechanism

In order to better understand how PSs function collectively in promoting virus assembly across a range of viral families, we are carrying out an interdisciplinary work program aimed at probing the different roles PSs can play in viral assembly and infectivity.

Specificity, Co-operativity and Compaction

Satellite Tobacco Necrosis Virus (STNV), so called because its gRNA codes only for the CP subunit that forms a protective $T = 1$ shell of 60 copies around it, has been used as a simple model for these studies [38]. SELEX experiments [7] identified an STNV CP-binding aptamer whose sequence matches a SL within the 5′ UTR of its 1239 nt long gRNA. Other SLs with similar loop sequences/motifs (-A.X.X.A-) also occur across the gRNA. In order to examine the specificity and co-operativity of CP-binding to RNA the MS2 and STNV viral gRNAs were fluorescently-end-labelled. This allowed us to monitor their assembly behavior in vitro using single-molecule fluorescence correlation spectroscopy (smFCS) [2, 3, 54]. This technique allows the time-dependent, hydrodynamic radius (R_h) of a fluorescently-labelled species in solution to be determined. At nanomolar concentrations of gRNA, multiple measurements are made for a focused volume (roughly a nanolitre) every 30 s, and the averaged R_h versus time plot calculated. At nanomolar concentrations this volume contains, on average 1 or 0 labelled species, i.e. it reports the properties of single molecules (sm). The resulting smFCS plots for STNV and MS2 gRNAs are shown in Fig. 4.3a. As the CP-free gRNAs are titrated by CPs purified from either of the two viruses, differing behavior occurs. When the gRNA binds to its cognate CPs there is a rapid (seconds) drop in its R_h value of ~30–40%. This compares to no change or a slight increase in R_h for the corresponding non-cognate interactions. Negative-stain electron micrographs (nsEM) of these samples show that, close to the point of maximal R_h decline, cognate interactions lead to formation of CP-gRNA complexes with the expected curvature of their final VLP. Eventually virus-like particles (VLPs) resembling the respective native virions are produced. In other words, in vitro reassembly at these low concentrations shows evidence of sequence-specificity with respect to gRNA encapsidation (Fig. 4.3b). The kinetics of such assembly is dependent on the sequences within the gRNAs (Fig. 4.3c, d). It appears that multiple TR-like sequences in MS2, and their equivalents in STNV, are binding to cognate CPs, facilitating their assembly into defined capsids. An MS2 CP variant that binds TR normally but is unable to make the CP-CP contacts required to form such a capsid fails to induce the collapse of the gRNA in these experiments. This confirms that collapse is a consequence of forming both RNA-CP and CP-CP interactions. These experiments demonstrate that PSs have specificity for their cognate protein partner, and that PS-CP contacts act together, i.e. collectively, in promoting compaction of the gRNAs. The need for such compaction can be seen in Fig. 4.3b where the dashed horizontal line indicates the R_h value of the final STNV virion. Clearly the CP-free gRNA is much larger than this in these solution conditions, so its collapse is an essential aspect of virion assembly.

PS Affinity Distribution is Important for Selective and Efficient Genome Packaging

MS2 gRNA PSs vary around the TR consensus sequence/structure motif, and therefore have a wide range of stabilities. In addition, their affinities for CP dimers are all lower than that of TR. Using stochastic modelling of virus assembly dynamics

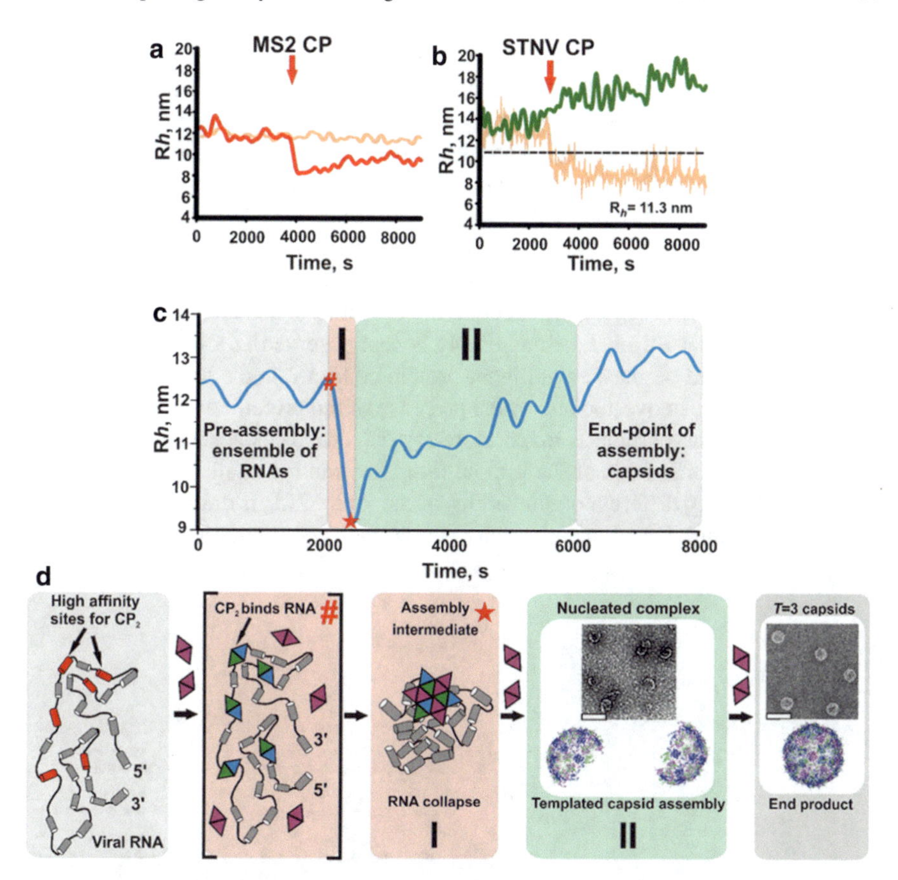

Fig. 4.3 Evidence of ssRNA genome packaging specificity from single-molecule studies. **a** CPs from bacteriophage MS2 bind to its own gRNA (fluorescently-labelled; red line) preferentially causing the R_h to collapse, whilst the gRNA of STNV (salmon coloured) is unaffected. **b** Conversely STNV CPs bind its gRNA (salmon coloured) causing it to collapse, whilst the MS2 RNA (green) binds CPs non-specifically creating an aggregate. Dashed black line represents the R_h value of the STNV virion revealing that its gRNA must collapse for assembly of the T = 1 capsid to occur. **c** Cartoon of panels showing the interpretation of these behaviours. High-affinity MS2 CP-binding sites in the gRNA bind CP in such a way that they favour additional subunits binding, leading to collapse in the size of the RNA. Negative stain EMs just following the collapse or after some time imply that the collapsed state is the direct precursor of the T = 3 capsid formed. Cartoons are colour-coded for the stages of the reaction above (adapted from Borodavka et al. [2])

[21], we demonstrated that this affinity variation is essential for the assembly mechanism. Using a dodecahedral model virus assembled from 12 pentagonal units that each have one RNA binding site (Fig. 4.4a), we computed capsid yield for different affinity distributions along the model genomic RNA. Whilst equal affinities across the genome resulted in the same outcome irrespective of whether the affinities were weak or strong, different ways of alternating strong and weak signals resulted in widely different outcomes (Fig. 4.4b). This suggests that PS affinities/stabilities have been

tuned by evolution to optimize genome packaging and virus assembly efficiency. However, even the best performing distribution amongst 30 k randomly generated sequences is only efficient if assembly takes place in the presence of a protein ramp, i.e. if protein subunit concentration builds as gradually as in a viral infection (Fig. 4.4c). Indeed, the protein ramp mimics the situation in the viral life-cycle, in which assembly takes place while viral CPs are still being synthesized. Addition of a full aliquot of CP subunits at the start of the simulation results in a reduced capsid yield, because the site of assembly nucleation on the genomic RNA is less well-defined in that case (Fig. 4.4d), reflecting the outcome of some experiments. These results suggest that PS-mediated assembly is best observed at low protein concentrations, and could be masked at higher protein concentrations, perhaps explaining why it has long been overlooked. In the presence of the protein ramp, genome organization within the particle is much more restricted than its absence. In particular under these circumstances of the protein ramp the number of distinct Hamiltonian paths, representing different organizations of the viral RNA, is dramatically reduced to only a small number of geometrically similar paths/configurations (Fig. 4.4d).

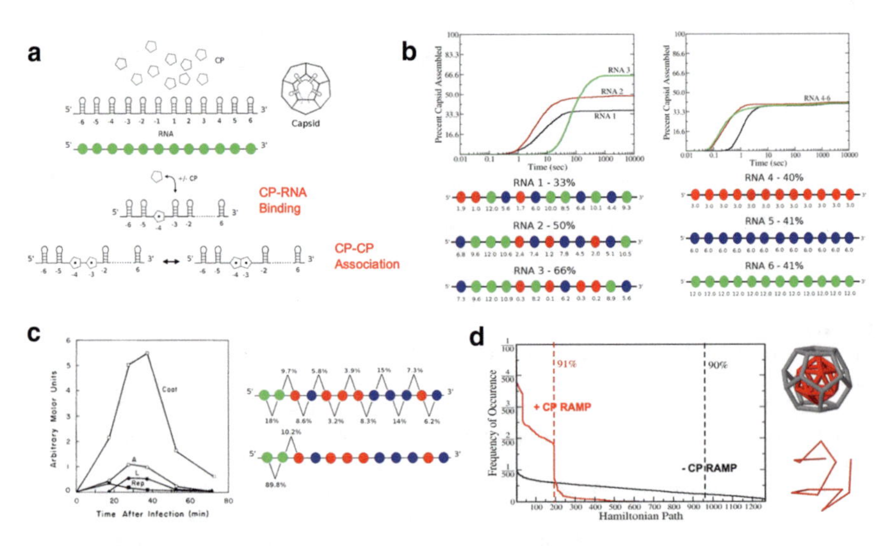

Fig. 4.4 A computational model of virus assembly reveals molecular features of the packaging signal-mediated assembly mechanism. **a** Assembly of 12 pentagons into a dodecahedral model capsid via a Gillespie algorithm based on the following reactions: each pentagonal unit can bind to and unbind from one of the 12 RNA PSs (shown as colour-coded beads according to their affinity; green high, blue intermediate, red low). **b** Simulations of virus assembly around 2000 copies of model gRNAs (i.e., strings of coloured beads) results in different assembly yields for distinct configurations (left). Strings with homogeneous affinities perform identically irrespective of the strengths of the interactions (right), demonstrating that variation in affinities across the gRNA is essential for the mechanism. **c** In the presence of the protein ramp (left), i.e., if protein concentration is increased gradually as in a real viral infection, nucleation is more localised, contributing to increased yields. **d** The protein ramp also dramatically reduces the spectrum of distinct gRNA configurations (or, Hamiltonian paths) within the particle, resulting in a more defined gRNA organisation (part **a**, **b**, and **d** are adapted from Dykeman et al. [21])

These features can also be observed experimentally, e.g. during in vitro reassembly of STNV (Fig. 4.5a). Use of RNA SELEX [28] identifies an aptamer sequence with a perfect match to a unique 14 nts long site within the >1 kb gRNA. This sequence can form a SL, which sequence variants suggest is the conformation bound by the cognate CP with high affinity. Presumably this is the highest affinity for CP within the cognate gRNA. This PS, termed PS3, is located toward the 5′-end of the gRNA and is flanked on either side by two additional PS-like sequences that can also form SLs. These five PS sites (PSs1-5) overlap part of the 5′ UTR of the gRNA and the start codon of the CP open reading frame. We have studied their collective properties as part of a 127 nts ssRNA fragment [46] (Fig. 4.5b). CP binding to this fragment is sequence-specific and highly co-operative, causing the fragment R_h value to collapse as CPs bind. This mimics the collapse seen, on a larger scale, when the entire gRNA is used as the assembly cargo.

Within this genome fragment co-operativity depends on the presence of a CP-recognition sequence (-A.X.X.A-) within each PS loop and the relative nucleotide spacing between them. Sequence changes that increase or decrease the SL folding propensity have similar impacts on reassembly with both the 127-mer and the full-length gRNA (Patel et al. [44]). The more stable PS folds within the 5′ gRNA fragment impart significantly faster assembly kinetics on full-length gRNAs, implying that assembly initiation is regulated by the PS SLs within the shorter fragment. There

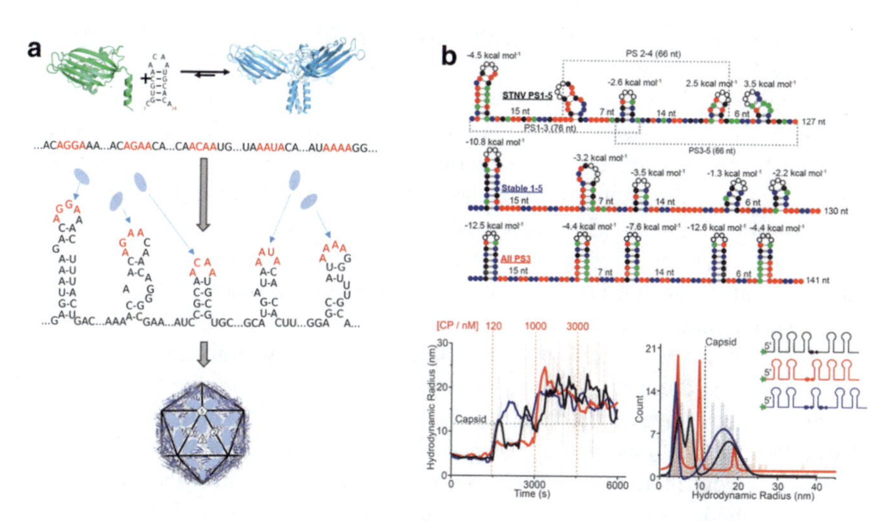

Fig. 4.5 Unpicking the genome sequences regulating assembly in a model virus. **a** STNV CP monomers bind packaging signals in the form of SLs within the gRNA creating CP trimers, 20 of which form the shell of the T = 1 capsid. The PSs are all SLs with loop sequences based around an—A.X.X.A-CP-recognition sequence (red) which occurs in their loops. There is co-operative binding to groups of PSs, at least towards the 5′ end of the gRNA where assembly is thought to initiate. Each PS differs in intrinsic affinity due to its folding propensity, CP-recognition motif homology and relative spacing in the gRNA. **b** Assembly substrates used to probe the sequence features governing assembly, including folding propensity. These were then used in smFCS plots revealing the consequences of their mutations (adapted from Patel et al. [44])

are 25 additional PS-like sites located across nucleotides 128–1239 in the gRNA. Experiments where full-length gRNA, mutated to lack PS sites (i.e. in which the CP-recognition sequence of each of these SLs is mutated to remove -A.X.X.A-), competes directly for in vitro reassembly (binding CP subunits) with the wild-type gRNA, or equivalent gRNA fragments lacking some or all of these non-initiation PSs, confirm that: (i) assembly is sequence-specific, (ii) the 5′ 5 × PSs are important for initiation, and (iii) that the additional PS-sites also contribute to assembly efficiency. Strikingly, moving the 5′ assembly initiation sequence (127-mer) from the 5′ to the 3′-end of the wild-type gRNA sequence creates a losing competitor against wild-type gRNA [66]. Thus, the natural sequence order of PS sites 5′ to 3′ defines the successful Hamiltonian path(s) in this case. Note, all of these results were obtained at low (nM) reagent concentrations. The observed packaging specificity in vitro disappears as these concentrations are raised, presumably because at higher concentrations electrostatic interactions predominate and overwhelm sequence-specificity [28]. This also appears to explain the outcomes when CP and gRNA are artificially over-expressed in vivo [37].

The Multiple Roles of PSs During Assembly

It is known from genome sequencing that many ssRNA virions contain the expected cognate gRNA cargo at close to 100% occupancy, implying a high degree of packaging specificity in vivo. As explained for STNV above, genomic RNAs displaying multiple PS-like sites across their length account for this specificity since few, if any, cellular RNAs will share a similar pattern of cognate CP-binding sites. The redundancy of multiple PSs moreover provides a resilience for this assembly mechanism against errors in replication created by the viral RNA polymerases, which are known to be very error-prone. This in turn is part of the biology of this class of virus since it allows them to exploit and adapt rapidly to new evolutionary niches. Note, the replication template (negative-sense strand) is the same length and net charge as the viral gRNA and is very likely to have a similar secondary structure. Complementary base pairing during replication, however, prevents presentation of cognate CP recognition sites, consistent with the non-appearance of this strand in virion samples. Many additional features encompassed within viral gRNAs, such as end to end base pairing, also contribute to their ability to be assembly substrates. Parsing these so that their individual effects can be seen in isolation is non-trivial.

Virion assembly in eukaryotic cells is thought to occur within "viral factories" where viral genomes are also being replicated [43]. Although such sites are not thought to be bounded by membranes, cognate gRNA-CP interactions might be the expected consequence of this sequestration. However, the viral factory idea does not explain the presence and functions of multiple PS sites within many viral genomes that clearly confer sequence-specific assembly to gRNAs and their fragments in vitro. Nor does it account for the many other roles gRNAs play in regulating virion assembly. As discussed above, PS affinities for CP along a gRNA vary in sequence and folding propensity, allowing evolution to direct assembly along a small subset of the geometrically possible assembly pathways, encoded by the limited number of Hamiltonian paths along which PS-CP contacts prefer to form, rapidly giving rise to the final virion. The fact that genome packaging specificity is not the only benefit from a PS-mediated assembly mechanism can also be seen from experiments with STNV where insights on the PS-mediated assembly mechanism were used to enhance packaging efficiency by changing the stability of PSs in a cassette at the 5′ end. Enhancing the stability of the 5 PSs in the first 127 nts increases packaging efficiency in vitro compared with the wild-type genome, clearly demonstrating that the benefits of the mechanism are more subtle than simply providing specificity.

The importance of PSs can also be seen by comparing virion assembly kinetics in the presence and absence of gRNA. For example, in vitro reassembly studies of MS2 show that both CP dimers and CP dimer-TR complexes are stable in the absence of each other. However, mixing them gives rise to a rapid assembly process in which it is possible to detect, using mass spectrometry (MS), potential intermediates in the formation of the $T = 3$ virion [53]. At room temperature in vitro phage assembly is a rapid process, consistent with the known biology of infection. Phage added to pili-bearing bacterial cells in culture at 37 °C, cause lysis due to infection in ~10 min. Given that this period includes initial internalization of the gRNA, phage gene expression and gRNA replication, followed by phage assembly, all before the phage-encoded lysis protein acts, it follows that all these processes must be rapid. This is consistent with in vitro studies of assembly which is rapid in mixtures of gRNA or its PS-containing fragments, and CPs. This contrasts with assembly of CP alone, which is very slow (>days) and mostly incomplete. These results reveal one aspect of the roles of PSs (regulation of assembly kinetics) within the gRNA. MS and NMR suggest that the non-assembling states of CP dimers alone or their complexes with TR are due to them primarily representing just one type of protein conformer/tile needed to make the MS2 capsid. TR-bound CP dimers are of the A/B type whilst RNA-free CP dimers are C/C type. Thus, we would expect that as well as TR, 59 additional TR-like stem-loops act as PSs throughout the MS2 gRNA, favouring formation of the necessary 60 A/B tiles in a $T = 3$ particle. We have seen similar increases in the assembly tendency of HBV core protein and the STNV CP in the presence of their cognate gRNAs, or sub-fragments [48]. In those cases, it is clear that PSs within the gRNA have at least one function, namely an electrostatic role that results in screening similarly charged regions of the respective CPs from each other. This creates assembly-competent capsomers of the respective virion/nucleocapsids.

PS-Mediated Assembly is Widespread, and Its Molecular Hallmarks Are Conserved Within Viral Families

RNA bacteriophages are model systems for studying assembly mechanisms that occur in more complex forms in wider families of viruses. By using interdisciplinary approaches, and expanding our repertoire of data analysis methods (e.g., Bernoulli plot analysis of SELEX data [52]), we have demonstrated that several viral families regulate their virion assembly mechanisms similarly. This includes the two distinct clades of *Picornaviridae* [9, 52], see Fig. 4.6a, b. One group that cleaves their VP0 subunits into VP2 & VP4 represent the mainstream viruses, such as poliovirus and human rhinovirus. We demonstrated that both the amino acids corresponding to the PS binding sites on the CP subunits and the molecular determinants of the PSs are highly conserved in the viral family, suggesting that PSs could serve as broad-spectrum anti-viral targets. The other group of *Picornaviridae,* include the Parechoviruses which are a major cause of infant sepsis, which is currently untreatable and often rapidly fatal. Other Picornaviral infections include life-threating infections of humans and livestock. Formation of the nucleocapsid (NCP) of the pseudo-retrovirus Hepatitis B Virus (HBV) is also regulated by PS-core protein contacts from its pre-genomic RNA (Fig. 4.6c), the first substrate packaged, that is copied into ssDNA and then dsDNA by an encapsidated polymerase. HBV is a major human pathogen and is the principal cause of liver cancer and cirrhosis worldwide, resulting in ~1 million deaths annually [65]. These occur despite the existence of an effective vaccine that is not universally deployed. There are thought to be >400 million people infected with HBV due to events that occurred before the production of the vaccine, and because of new infections that arise from the vertical transmission from mother to off-spring.

Other research groups have also identified similar mechanisms involved in virion assembly, e.g. in the alphaviruses [4, 5]. These pathogens are carried by biting insects (mosquitoes) and can cause partial paralysis, and even death of their human hosts. Their geographic spread is increasing because of climate change and greater human interaction. Beyond virology, the Hilvert group in Zurich showed that a bacterial enzyme cage (Fig. 4.7a), engineered to bind and encapsidate its own mRNA, encapsidates significantly more mRNA when subjected to rounds of artificial evolution [57] (Fig. 4.7b). In collaboration, we showed that this follows acquisition of more "PS-like" sites in its mRNA cargo (Fig. 4.7c), implying that there are significant "benefits" to RNAs carrying such signals. The frequency, scope, and advantages of the PS-mediated assembly mechanism in natural viruses is consistent with such an interpretation.

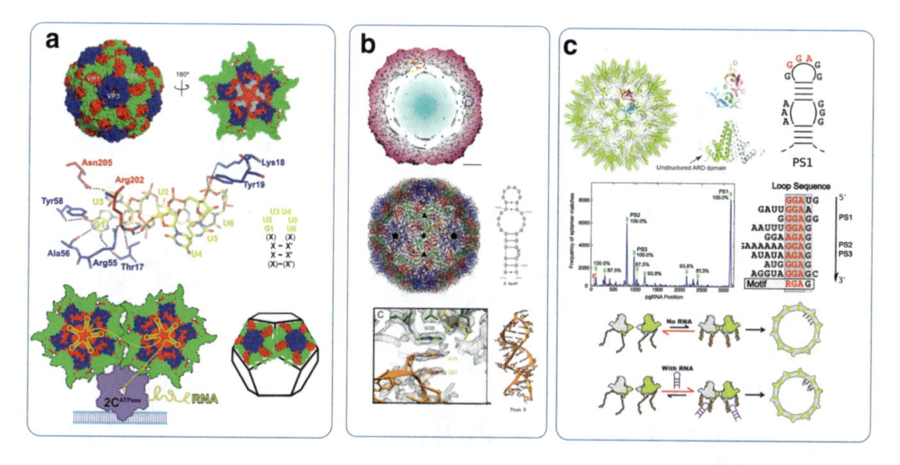

Fig. 4.6 Packaging signals occur widely in viral pathogens. **a** Ordered fragments of genomic RNA are present in the X-ray structure of Human Parecho virus and are sequence-specifically bound by CPs (top and middle panels); a model for formation of such structures via an assemblysome. [Adapted from [52] Nature Comms. 8:5]. **b** Differing parts of the picornaviral gRNA form sequence-specific contacts from SL PSs to the VP2 subunits in the more mainstream family members, EV-E in this case. (adapted from Chandler-Bostock et al. [9]). **c** Hepatitis B Virus forms an initial T = 4 nucleocapsid composed of a single core protein (Cp) around a single-stranded pgRNA copy of its genome. SELEX against the Cp identifies highly conserved sites in known HBV genomes, each of which can form an SL (top) with the highlighted red sequences as the Cp recognition motif (middle). Individual PS sites trigger sequence-specific association of HBV Cp dimers leading to formation of a $T = 4$ shell in vitro (bottom) (adapted from Patel et al. [48])

PS-Mediated Assembly as a Drug Target

Where we have analysed it, PSs and their recognition sites on cognate CPs are evolutionarily conserved across strain variants (e.g. Fig. 4.8b). This hints that they comprise a currently untapped potential anti-viral drug target, a directly-acting anti-viral (DAA). In order to assess the consequences of such targeting, we have mutated three of the most conserved PS sites within the HBV pre-genomic ssRNA [47]. Mutation of their Cp recognition sequences has a highly deleterious effect on in vitro reassembly at low concentrations. The largest effects occur with mutation of the recognition site in the dominant ("best") PS. All three highly conserved PS sites, however, make positive contributions to assembly efficiency. These mutations change very few nucleotides within a total of ~3200 but reduce reassembly efficiency by >90%. The fact that these sequence-specific effects are seen with a pgRNA of the same length as the wild-type genome are further confirmation that there is sequence-specificity in pgRNA encapsidation. Although favorable electrostatic interactions occur between the negatively charged pgRNA and the arginine-rich domains at the C-terminus of the Cp, these are not the primary driving force behind NCP assembly. Care needs to be used in interpreting such in vitro results, which can be misleading. For instance, the HBV pgRNA is effectively circularized by base

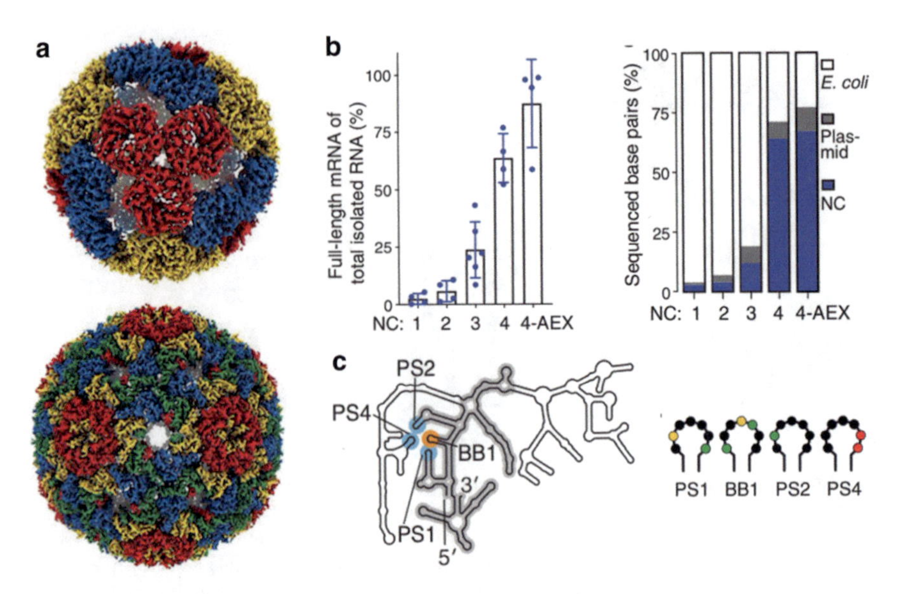

Fig. 4.7 Packaging signal-mediated assembly in a nonviral system. a Directed evolution of a bacterial enzyme that is engineered to package its own coding sequence (mRNA) results in the emergence of larger particle structures (e.g., a particle made of 180 proteins, top, and of 240 proteins, bottom). **b** During later rounds of directed evolution, particle size no longer increases, whilst packaging efficiency of full-length mRNA increases dramatically (in the step from NC3 to NC4). **c** This increase has been explained by the emergence of a cassette of four packaging signals, that have been identified using the X-ray Synchrotron Footprinting (XRF) method (all parts adapted from Tetter et al. [57])

pairing between complementary terminal sequences. Sequence variants ablating this interaction increase the pgRNA R_h value (~40%) and reduce encapsidation efficiency, showing that compaction is an important additional feature. Replacement of HBV sequences with "non-specific" non-viral sequences can also inadvertently add exogenous PS sites, since these comprise minimalistic secondary structures, e.g. SLs.

An oligonucleotide encompassing the highest affinity PS site within the HBV pgRNA was therefore used as a selection target for a library of small molecular weight, drug-like ligands (Fig. 4.8a). This screen identified a range of compounds having affinities in the nanomolar range for both the PS1 selection target and the other HBV PSs, confirming that they have partially conserved features. Addition of these high-affinity, small molecular weight ligands to in vitro reassembly reactions results in significant decreases in assembly efficiency around the wild-type pgRNA [45], confirming that RNA PS sites can act as targets of DAAs. The dispersed and co-operative nature of PS sites also confounds many simple mutational experiments, since within a gRNA, deletions and substitutions can disrupt dominant folding patterns creating PSs that were naturally absent. Additionally, mutation of individual or a few PS sites can be misleading about their importance in natural assembly, where co-operativity is essential. In several cases, the structures of CP-binding sites for PSs

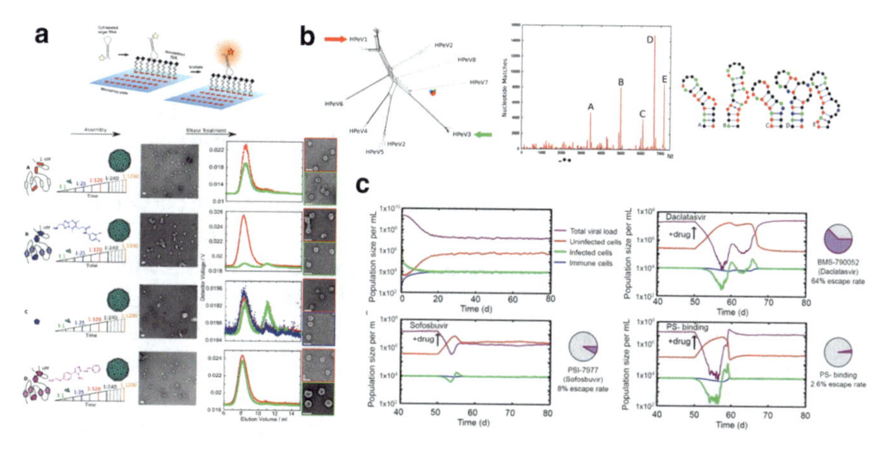

Fig. 4.8 Packaging signal-mediated assembly as a drug target. Small molecular weight compounds binding to PS1 of the Hepatitis B virus pgRNA have been identified (top) and shown to ablate virus assembly in vitro (bottom) (adapted from Patel et al. [45]). **a** The SELEX library used for the identification of packaging signals in the Parechovirus HPeV1 also identifies similar PSs in other strains, including the evolutionarily most distal HPeV3. Analysis of genomic fragments in HPeV3 with similarity to the selected library (Bernoulli peaks, middle) reveal stem-loops with the same molecular characteristics as the HPeV1 PSs. The PSs corresponding to the five highest peaks are shown colour-coded according to nucleotides (G = green; U = black; A = red; C = blue). **b** Mathematical infection models have been used to compare the impact of different treatment strategies on viral load (magenta), the numbers of uninfected (red) and infected (green) cells, and immune cells (blue). The impact of two known virus inhibitors (Daclatasvir and Sofosbuvir), and of a PS-binding drug are compared. The rate of drug escape (pie chart) is lowest for the PS-binding drug (adapted from Bingham et al. [1])

are known at high-resolution and these would also represent potential sites for the development of DAAs. These observations are important because modelling the consequences of antiviral therapy directed against PSs reveals a lower propensity for eliciting drug resistance than other forms of anti-viral therapy currently used in the clinic [1] consistent with this being a promising route for exploitation (Fig. 4.8c).

PSs in Viral Infection—Molecular Frustration Within Viral Protein Shells

PS-CP contacts in gRNAs can be thought of as a form of molecular Velcro, where each partner hook is only attracted to its cognate opposite eye. However, if there are up to 60 PS sites per gRNA, as we believe to be the case in MS2, the consequence of these many interactions between the gRNA and the protein shell might be such a stable virion that it may counteract genome release, preventing infection of another host cell. How then might gRNAs escape the clutches of their CP shells?

We have made insights into this problem for the specialized case of the RNA bacteriophage MS2. Given that many of the features of the PS-mediated assembly mechanism were first identified and characterised in that system, it is appropriate to observe its infectivity mechanism, and how its PSs may contribute to that process. It is important to note that in contrast to many viral CPs, MS2 together with its close family member bacteriophage Qβ, has a CP architecture that lacks flexible tails of polypeptide. Often such tails are rich in basic amino acids and help to neutralize the charges on the phosphates of encapsidated gRNAs. They are also flexible allowing contacts to occur to PS sites in their natural, non-identical contexts. This flexibility may also explain why to date the majority of packaged gRNA has been visualised using asymmetric cryo-EM reconstructions of the RNA phages only. Viruses with flexible structural elements in their CPs at most show ordered density for their genomes in close contact with the protein shell [9, 33, 52], i.e. at PS-binding sites. Even in those cases the resolution is often not good enough for modelling defined sequences.

For MS2, we have a range of crystal structures defining its $T = 3$ shell of identical CP subunits at atomic resolution. However, the act of crystallization results in averaging of the sample orientation, and processing of the X-ray diffraction data further assumes that the object has icosahedral symmetry. These assumptions turn out to be misleading with respect to the true architecture of the phage particle. Infection occurs in bacteria carrying an F pilus, hence the "male-specific, fraction 2" (MS2) name. Phage particles bind to these pili using a surface accessible maturation protein (MP), which we deduced previously must sit along one of the phage twofold axes [17]. Asymmetric cryo-EM structures of this particle have transformed our view of this structure, resolving its biology at the atomic level. On each phage particle one C/C lattice position in the protein shell is occupied by the single copy MP, which in turn is bound to a specific gRNA sequence adjacent to its 3′-end. The remaining 89 CP dimers are arranged as 60 A/B "tiles" and 29 C/C "tiles". In an asymmetric reconstruction, determined with a very large dataset of >300 k MS2 particles [16], multiple sequence-specific contacts with these CP dimers are seen as expected for a structure that assembles via PS-mediated assembly. This includes the TR stem-loop as one of 15 PS sites bound to CP dimers (Fig. 4.2d). In addition, other portions of the gRNA encompass folded stem-loops in proximity to the inner capsid surface. Their positions suggest that they could correspond to PSs that have released their contacts to the CP shell, i.e. PS that have been bound to CP only transiently during assembly.

This is a complex topic because it requires us to grapple with one of the fundamental assumptions in virology, namely that every viral particle is identical. In a structural sense many virions appear identical in crystallographic and cryo-EM structures. From these structures we get an image of virions as static, closed arrays composed of multiple, identical CPs. How then is infection achieved? In order to address this question, we have determined the secondary structure of the MS2 gRNA as transcript and within an infectious particle. This can currently only be achieved using RNA footprinting. Initially we used lead ions, which diffuse into the virion via pores at the fivefold axes, to investigate the gRNA structure adjacent to the TR PS. Complexation with Pb^{2+} ions leads to nucleotide cleavage which is secondary structure dependent, e.g. an SL, or intermolecular contact, e.g. to a CP dimer from a PS. These studies confirm that PS-like sequences adjacent to TR are bound by CP subunits [50, 51].

We recently, however, extended these studies to cover the entire gRNA secondary structure, both as transcript and as gRNA within infectious phage, using X-ray RNA footprinting (XRF) [10]. This approach produces modifications because the X-ray photons photolyse solvent water molecules even in frozen samples creating hydroxyl radicals in situ adjacent to nucleotide sugar residues. This allows modifications to be tightly controlled by using pre-frozen samples that cannot undergo conformational change post-modification. Frozen samples are then warmed in the absence of the cleavage moiety. Modified sugars undergo chemical rearrangement and hydrolysis of the phosphodiester chain leading to cleavage at the adjacent phosphodiester. The degree of modification at each phosphodiester is determined by capillary sequencing. We developed a purpose-designed suite of algorithms to process these data and imported them as constraints into the S-fold algorithm to predict gRNA secondary structures. Akin to other foot-printing approaches, benchmarking against a known set of secondary structure elements is required. We used TR and one of its neighbouring PSs for this purpose. Differential modification of the PS-CP recognition motif in the same PS-like stem-loops, present in both transcript and encapsidated gRNA, allowed us to identify PSs in contact with CP subunits in phage particles unambiguously. XRF identifies 31 PSs in contact with capsid in the infectious phage. These include the 15 PSs encompassing TR, that were previously determined via the highest resolution, asymmetric cryo-EM structure to contact the protein shell. An additional 34 unbound SLs, each with the molecular characteristics of a PS, are also predicted to be present, presumably corresponding to the SLs seen in proximity to the inner capsid surface via cryo-EM. Cryo-EM particle datasets are repeatedly winnowed by software to achieve the highest resolution structure. It is possible therefore that PSs that are bound only transiently are not detected in contact with the capsid shell. In a similar caveat, our XRF data were obtained on a sample of infectious phage. This sample could contain non-infectious particles. The resultant data are therefore of necessity an average of the entire sample, and we assume provide a realistic picture of a bacteriophage preparation.

Neither cryo-EM nor XRF can determine whether a particular phage particle is infectious. However, their combination leads to the conclusion that not all PSs remain bound by the CP shell. Interestingly, the positions of the bound sites in the capsid shell are not random. Many of the unbound PS sites occur towards the 3′-end of the gRNA, whilst many bound sites surround the CP dimer in contact with TR. This suggests that the CPs in contact with TR and its neighbours, together with the MP, form a highly stabilized region of the phage particle, consistent with a proposed role in assembly (Fig. 4.9a). Since the MP contacts occur to CP dimers that form part of differing pentameric facets of the $T = 3$ shell, it is possible that nucleation at these sites defines particle geometry, committing to the formation of a $T = 3$ shell during assembly. Defining capsid morphology is yet another regulatory role for the PSs. Subsequent phage infection then occurs via attachment of the MP protein to the F pilus of a bacterium which occasionally is internalized, drawing the bound phage and its genome towards the pilus extrusion machinery. The MP-gRNA contact seems to remain as the gRNA gets extruded, with the gRNA entering the cell in the 3′–5′ orientation, i.e. effectively backing into the cell.

This infection mechanism requires the capsid to release MP and its attached gRNA region at the start of an infection. Facilitating the first steps in this process CPs ascribed to the group of transient PSs, i.e. those that are not in contact with the CP shell in the mature phage, cluster at one side of gRNA next to the MP (magenta in Fig. 4.9b). This removes barriers to partial disruption of the capsid, i.e. the free energy of PS-CP contacts, facilitating its coordinated extrusion at the start of the infection (Fig. 4.9c). This is important for the downstream processing of the genome. At some point a cellular protease cleaves the MP into two fragments presumably freeing the 3′ end for cell entry and subsequent replication/gene expression. The first complete gene to enter the cell by this mechanism is the replicase, which cannot be translationally repressed because existing CP subunits are excluded from the infected cell. The replicase then has access to its 3′ gRNA binding site in the UTR and is free to copy the genome, potentially also facilitating extraction of the gRNA from its capsid and internalizing it into the cell.

In addition, dissociation of a PS from a CP dimer creates a conformational anomaly. The conformation of CP dimers in solution is normally determined by binding RNA SLs (PSs) leading to a preferred asymmetry in the conformations of the two loops of polypeptide connecting the F & G β-strands in the globular fold of each subunit (i.e. creating an A/B dimer). These loops are symmetrical in the absence of RNA, so PS dissociation in a capsid leads to a dimer having a conformational preference that is C/C, but in the phage particle adopting such a conformation would destabilize protein–protein interactions in the CP shell and presumably is resisted by other CP subunits. The coexistence of energetically similar, mutually exclusive states has been described as "molecular frustration" [24–27, 29], and it is thought to contribute to their biological functions. Here, gRNA conformations with PS-CP contacts present, and absent to enable gRNA reconfiguration within the capsid to rearrange, are such molecularly frustrated states. Reversion to the "preferred" A/B conformer in a pentameric facet to relieve local strain in the CP lattice would destabilize phage shells. This favors disassembly, and its immediate consequence, namely infection. This inference suggests that viral gRNAs play critical roles in virion biology from assembly to infection.

Conclusions

Many major human pathogens have ssRNA genomes and assemble isometric virions using instructions inherent within their gRNAs that require cognate RNA-CP contacts. The rapid artificial evolution of similar interfaces between an RNA cargo and its encapsidating protein shell implies that there are considerable advantages for such regulation. These CP-gRNA contacts play multiple roles in assembly and potentially also in infection. Where we have looked across strain variants, the gRNA components of this assembly regulation are conserved implying that they, as well

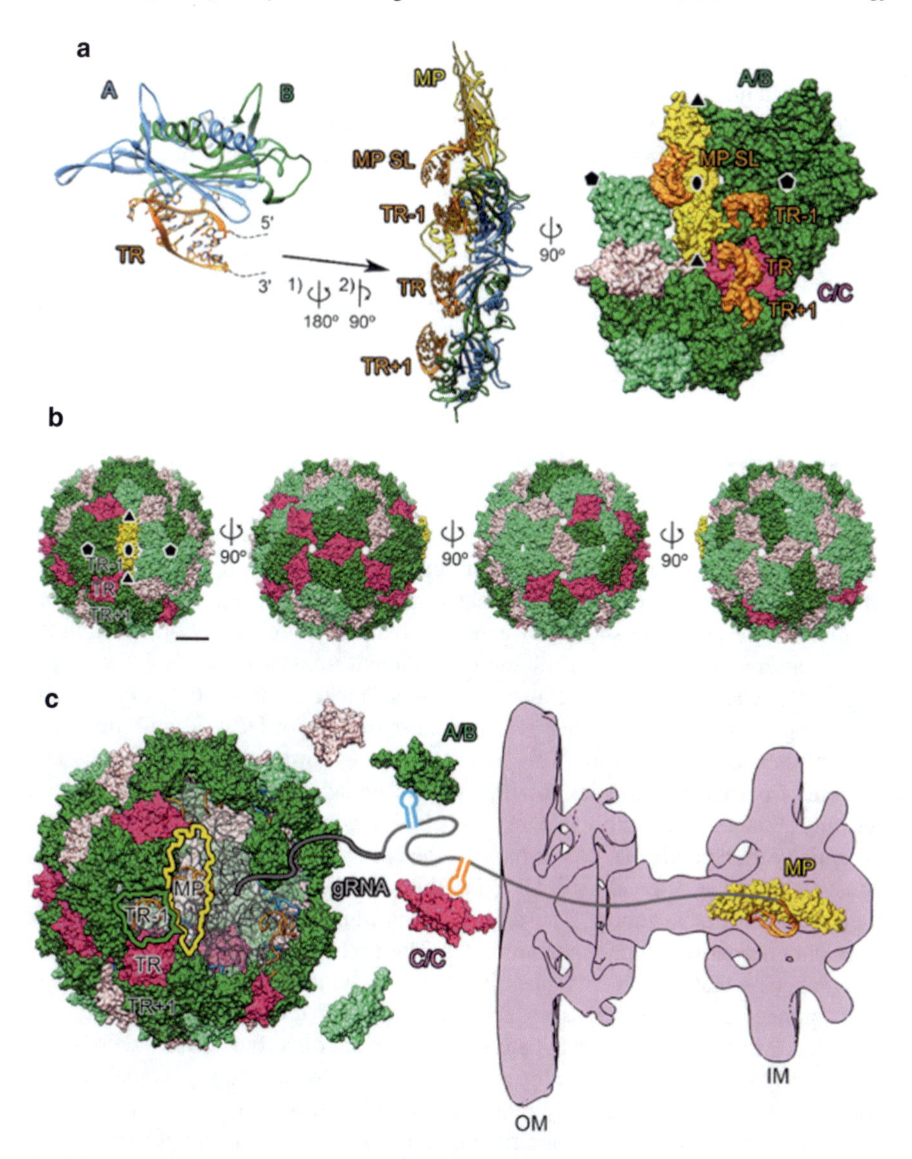

Fig. 4.9 Molecular frustration within a capsid protein layer could facilitate infection. **a** MS2 bacteriophage CP dimers bind to the highest affinity PS, the TR translation operator, and its two neighbouring PSs located within the gRNA. The single copy MP also binds a specific site close to the 3′ end of the gRNA, forming an extending assembly initiation complex. **b** Some of the PS-CP contacts are transient. CP dimers shown in magenta are not in contact with a PS, but RNA density in their vicinity and results from XRF suggest that these contacts could have played a role during virion assembly. **c** MP binds the cellular receptor a pilus extrusion pore and leads the gRNA into its inner channel allowing the 3′ end of the gRNA, which encompasses the PSs that dissociate in the mature phage, to enter the bacterial cell easily. A cluster of CP unbound from PSs close to MP suggests that the loss of PS contacts at specific CP positions could contribute to genome release (adapted from Chandler-Bostock et al. [10])

as their binding sites on viral CPs, represent untapped potential DAA targets. Some of the mechanistic assembly features enabled by PSs have been probed in detail in model and pathogenic viruses. In contrast to RNA phages where the CPs are entirely globular, many viral CPs have what have been termed "intrinsically unstructured" regions that contribute to gRNA contacts in a flexible way obscuring their presence. More complex virions, such as the Human Immunodeficiency Virus [12–14], and pleiotropic ones including Covid-19 [41], may also harbor similar RNA-CP contacts required to regulate their assembly. Similar PS-mediated contacts could also underlie the assembly and functioning of ssDNA viruses such as Adeno-associated Virus (AAV), which is a major gene therapy vector [67].

Perspectives

Since the earliest structural studies of relatively simple viruses in the middle of the last century, virions have been described as "a piece of bad news" in the form of their genomes "wrapped on a protective protein shell" [42]. It might be more accurate to describe the shell as a genome coating whose formation is regulated by the internal nucleic acid, facilitating its efficient transfer from one infected cell to another. Clearly, understanding the features that regulate virion assembly and uncoating is important to enable us to make use of a major facet of viral lifecycles, namely their cellular tropism. This would enable us to move genetic cargoes into and out of particular cell types at will, a goal of modern genetic medicine. We have recently shown that assembly features from natural genomes can be transplanted into potentially therapeutic mRNAs improving the ability of the latter to act as assembly substrates for the original viral CPs, i.e. this transplantation enhances aspects of their regulation of assembly. The fact that such transfer is possible reconfirms the important properties of these genomic sites. Their relative evolutionary conservation also suggests that despite advances in vaccine production, that are extensions of work pioneered in the west by Jesty & Jenner starting in the 1700s, DAAs need more attention. Treating an infected patient with a drug that directly interferes with highly conserved aspects of viral assembly and infection would have huge potential benefits for the patient, and significantly reduce their ability to spread infection. After all, infectious virions have had a long time to come to terms with, and get around, the human immune system whilst at least one of their cognate assembly mechanisms has remained hidden until now.

References

1. Bingham RJ, Dykeman EC, Twarock R (2017) RNA virus evolution via a quasispecies-based model reveals a drug target with a high barrier to resistance. Viruses. https://doi.org/10.3390/v9110347
2. Borodavka A, Tuma R, Stockley PG (2012) Evidence that viral RNAs have evolved for efficient, two-stage packaging. Proc Natl Acad Sci USA 109:15769–15774. https://doi.org/10.1073/pnas.1204357109
3. Borodavka A, Tuma R, Stockley PG (2013) A two-stage mechanism of viral RNA compaction revealed by single molecule fluorescence. RNA Biol 10:481–489. https://doi.org/10.4161/rna.23838
4. Brown RS, Kim L, Kielian M (2021) Specific recognition of a stem-loop RNA structure by the alphavirus capsid protein. Viruses. https://doi.org/10.3390/v13081517
5. Brown RS, Anastasakis DG, Hafner M, Kielian M (2020) Multiple capsid protein binding sites mediate selective packaging of the alphavirus genomic RNA. Nat Commun 11:4693. https://doi.org/10.1038/s41467-020-18447-z
6. Buchy P, Buisson Y, Cintra O, Dwyer DE, Nissen M, Ortiz de Lejarazu R, Petersen E (2021) COVID-19 pandemic: lessons learned from more than a century of pandemics and current vaccine development for pandemic control. Int J Infect Dis 112:300–317. https://doi.org/10.1016/j.ijid.2021.09.045
7. Bunka DH, Lane SW, Lane CL, Dykeman EC, Ford RJ, Barker AM, Twarock R, Phillips SE, Stockley PG (2011) Degenerate RNA packaging signals in the genome of Satellite Tobacco Necrosis Virus: implications for the assembly of a T=1 capsid. J Mol Biol 413:51–65. https://doi.org/10.1016/j.jmb.2011.07.063
8. Caspar DL, Klug A (1962) Physical principles in the construction of regular viruses. Cold Spring Harb Symp Quant Biol 27:1–24. https://doi.org/10.1101/sqb.1962.027.001.005
9. Chandler-Bostock R, Mata CP, Bingham RJ, Dykeman EC, Meng B, Tuthill TJ, Rowlands DJ, Ranson NA, Twarock R, Stockley PG (2020) Assembly of infectious enteroviruses depends on multiple, conserved genomic RNA-coat protein contacts. PLoS Pathog 16:e1009146. https://doi.org/10.1371/journal.ppat.1009146
10. Chandler-Bostock R, Bingham RJ, Clark S, Scott AJP, Wroblewski E, Barker A, White SJ, Dykeman EC, Mata CP, Bohon J, Farquhar E, Twarock R, Stockley PG (2022) Genome-regulated Assembly of a ssRNA Virus May Also Prepare It for Infection. J Mol Biol 434:167797. https://doi.org/10.1016/j.jmb.2022.167797
11. Clokie MR, Millard AD, Letarov AV, Heaphy S (2011) Phages in nature. Bacteriophage 1:31–45. https://doi.org/10.4161/bact.1.1.14942
12. Comas-Garcia M, Davis SR, Rein A (2016) On the selective packaging of genomic RNA by HIV-1. Viruses. https://doi.org/10.3390/v8090246
13. Comas-Garcia M, Datta SA, Baker L, Varma R, Gudla PR, Rein A (2017) Dissection of specific binding of HIV-1 Gag to the “packaging signal” in viral RNA. Elife. https://doi.org/10.7554/eLife.27055
14. Comas-Garcia M, Kroupa T, Datta SA, Harvin DP, Hu WS, Rein A (2018) Efficient support of virus-like particle assembly by the HIV-1 packaging signal. Elife. https://doi.org/10.7554/eLife.38438
15. Crick FH, Watson JD (1956) Structure of small viruses. Nature 177:473–475. https://doi.org/10.1038/177473a0
16. Dai X, Li Z, Lai M, Shu S, Du Y, Zhou ZH, Sun R (2017) In situ structures of the genome and genome-delivery apparatus in a single-stranded RNA virus. Nature 541:112–116. https://doi.org/10.1038/nature20589
17. Dent KC, Thompson R, Barker AM, Hiscox JA, Barr JN, Stockley PG, Ranson NA (2013) The asymmetric structure of an icosahedral virus bound to its receptor suggests a mechanism for genome release. Structure 21:1225–1234. https://doi.org/10.1016/j.str.2013.05.012
18. Doudna JA, Charpentier E (2014) Genome editing. The new frontier of genome engineering with CRISPR-Cas9. Science 346:1258096. https://doi.org/10.1126/science.1258096

19. Dykeman EC, Stockley PG, Twarock R (2010) Dynamic allostery controls coat protein conformer switching during MS2 phage assembly. J Mol Biol 395:916–923. https://doi.org/10.1016/j.jmb.2009.11.016
20. Dykeman EC, Stockley PG, Twarock R (2013) Packaging signals in two single-stranded RNA viruses imply a conserved assembly mechanism and geometry of the packaged genome. J Mol Biol 425:3235–3249. https://doi.org/10.1016/j.jmb.2013.06.005
21. Dykeman EC, Stockley PG, Twarock R (2014) Solving a Levinthal's paradox for virus assembly identifies a unique antiviral strategy. Proc Natl Acad Sci USA 111:5361–5366. https://doi.org/10.1073/pnas.1319479111
22. Dykeman EC, Grayson NE, Toropova K, Ranson NA, Stockley PG, Twarock R (2011) Simple rules for efficient assembly predict the layout of a packaged viral RNA. J Mol Biol 408:399–407. https://doi.org/10.1016/j.jmb.2011.02.039
23. Feder AF, Harper KN, Brumme CJ, Pennings PS (2021) Understanding patterns of HIV multi-drug resistance through models of temporal and spatial drug heterogeneity. Elife. https://doi.org/10.7554/eLife.69032
24. Ferreiro DU, Komives EA, Wolynes PG (2014) Frustration in biomolecules. Q Rev Biophys 47:285–363. https://doi.org/10.1017/S0033583514000092
25. Ferreiro DU, Komives EA, Wolynes PG (2018) Frustration, function and folding. Curr Opin Struct Biol 48:68–73. https://doi.org/10.1016/j.sbi.2017.09.006
26. Ferreiro DU, Hegler JA, Komives EA, Wolynes PG (2007) Localizing frustration in native proteins and protein assemblies. Proc Natl Acad Sci USA 104:19819–19824. https://doi.org/10.1073/pnas.0709915104
27. Ferreiro DU, Hegler JA, Komives EA, Wolynes PG (2011) On the role of frustration in the energy landscapes of allosteric proteins. Proc Natl Acad Sci USA 108:3499–3503. https://doi.org/10.1073/pnas.1018980108
28. Ford RJ, Barker AM, Bakker SE, Coutts RH, Ranson NA, Phillips SE, Pearson AR, Stockley PG (2013) Sequence-specific, RNA-protein interactions overcome electrostatic barriers preventing assembly of satellite tobacco necrosis virus coat protein. J Mol Biol 425:1050–1064. https://doi.org/10.1016/j.jmb.2013.01.004. PMCID: PMC3593212
29. Freiberger MI, Guzovsky AB, Wolynes PG, Parra RG, Ferreiro DU (2019) Local frustration around enzyme active sites. Proc Natl Acad Sci USA 116:4037–4043. https://doi.org/10.1073/pnas.1819859116
30. Hirao I, Spingola M, Peabody D, Ellington AD (1998) The limits of specificity: an experimental analysis with RNA aptamers to MS2 coat protein variants. Mol Divers 4:75–89. https://doi.org/10.1023/a:1026401917416
31. Jakobsen JC, Nielsen EE, Feinberg J, Katakam KK, Fobian K, Hauser G, Poropat G, Djurisic S, Weiss KH, Bjelakovic M, Bjelakovic G, Klingenberg SL, Liu JP, Nikolova D, Koretz RL, Gluud C (2017) Direct-acting antivirals for chronic hepatitis C. Cochrane Database Syst Rev 9:CD012143. https://doi.org/10.1002/14651858.CD012143.pub3
32. Johansson HE, Dertinger D, LeCuyer KA, Behlen LS, Greef CH, Uhlenbeck OC (1998) A thermodynamic analysis of the sequence-specific binding of RNA by bacteriophage MS2 coat protein. Proc Natl Acad Sci USA 95:9244–9249. https://doi.org/10.1073/pnas.95.16.9244
33. Kalynych S, Palkova L, Plevka P (2016) The structure of human Parechovirus 1 reveals an association of the RNA genome with the capsid. J Virol 90:1377–1386. https://doi.org/10.1128/JVI.02346-15
34. Keef T, Twarock R, Elsawy KM (2008) Blueprints for viral capsids in the family of polyomaviridae. J Theor Biol 253:808–816. https://doi.org/10.1016/j.jtbi.2008.04.029
35. Keef T, Wardman JP, Ranson NA, Stockley PG, Twarock R (2013) Structural constraints on the three-dimensional geometry of simple viruses: case studies of a new predictive tool. Acta Crystallogr A 69:140–150. https://doi.org/10.1107/S0108767312047150
36. Keen EC (2015) A century of phage research: bacteriophages and the shaping of modern biology. BioEssays 37:6–9. https://doi.org/10.1002/bies.201400152

37. Kotta-Loizou I, Peyret H, Saunders K, Coutts RH, Lomonossoff GP (2019) Investigating the biological relevance of in vitro-Identified putative packaging signals at the 5′ terminus of satellite tobacco necrosis virus 1 genomic RNA. J Virol 93(9):e02106-18. https://doi.org/10.1128/JVI.02106-18. PMCID: PMC6475782
38. Liljas L, Unge T, Jones TA, Fridborg K, Lovgren S, Skoglund U, Strandberg B (1982) Structure of satellite tobacco necrosis virus at 3.0 A resolution. J Mol Biol 159:93–108. https://doi.org/10.1016/0022-2836(82)90033-x
39. Lodish HF (1968) Independent translation of the genes of bacteriophage f2 RNA. J Mol Biol 32:681–685. https://doi.org/10.1016/0022-2836(68)90351-3
40. Loeb T, Zinder ND (1961) A bacteriophage containing RNA. Proc Natl Acad Sci USA 47:282–289. https://doi.org/10.1073/pnas.47.3.282
41. Masters PS (2019) Coronavirus genomic RNA packaging. Virology 537:198–207. https://doi.org/10.1016/j.virol.2019.08.031
42. Medawar PB et al (1983) Viruses. In: Aristotle to zoos: a philosophical dictionary of biology. Harvard University Press, Cambridge, pp 275
43. Novoa RR, Calderita G, Arranz R, Fontana J, Granzow H, Risco C (2005) Virus factories: associations of cell organelles for viral replication and morphogenesis. Biol Cell 97:147–172. https://doi.org/10.1042/BC20040058
44. Patel N, Wroblewski E, Leonov G, Phillips SEV, Tuma R, Twarock R, Stockley PG (2017) Rewriting nature's assembly manual for a ssRNA virus. Proc Natl Acad Sci USA 114:12255–12260. https://doi.org/10.1073/pnas.1706951114
45. Patel N, Abulwerdi F, Fatehi F, Manfield IW, Le Grice S, Schneekloth JS Jr, Twarock R, Stockley PG (2022) Dysregulation of hepatitis B virus nucleocapsid assembly in vitro by RNA-binding small ligands. J Mol Biol 434:167557. https://doi.org/10.1016/j.jmb.2022.167557
46. Patel N, Dykeman EC, Coutts RH, Lomonossoff GP, Rowlands DJ, Phillips SE, Ranson N, Twarock R, Tuma R, Stockley PG (2015) Revealing the density of encoded functions in a viral RNA. Proc Natl Acad Sci USA 112:2227–2232. https://doi.org/10.1073/pnas.1420812112
47. Patel N, Clark S, Weiss EU, Mata CP, Bohon J, Farquhar ER, Maskell DP, Ranson NA, Twarock R, Stockley PG (2021) In vitro functional analysis of gRNA sites regulating assembly of hepatitis B virus. Commun Biol 4:1407. https://doi.org/10.1038/s42003-021-02897-2
48. Patel N, White SJ, Thompson RF, Bingham R, Weiss EU, Maskell DP, Zlotnick A, Dykeman E, Tuma R, Twarock R, Ranson NA, Stockley PG (2017) HBV RNA pre-genome encodes specific motifs that mediate interactions with the viral core protein that promote nucleocapsid assembly. Nat Microbiol 2:17098. https://doi.org/10.1038/nmicrobiol.2017.98
49. Pires C (2022) Global predictors of COVID-19 vaccine hesitancy: a systematic review. Vaccines (Basel). https://doi.org/10.3390/vaccines10081349
50. Prevelige PE Jr (2016) Follow the yellow brick road: a paradigm shift in virus assembly. J Mol Biol 428:416–418. https://doi.org/10.1016/j.jmb.2015.12.009
51. Rolfsson O, Middleton S, Manfield IW, White SJ, Fan B, Vaughan R, Ranson NA, Dykeman E, Twarock R, Ford J, Kao CC, Stockley PG (2016) Direct evidence for packaging signal-mediated assembly of bacteriophage MS2. J Mol Biol 428:431–448. https://doi.org/10.1016/j.jmb.2015.11.014
52. Shakeel S, Dykeman EC, White SJ, Ora A, Cockburn JJB, Butcher SJ, Stockley PG, Twarock R (2017) Genomic RNA folding mediates assembly of human parechovirus. Nat Commun 8:5. https://doi.org/10.1038/s41467-016-0011-z
53. Stockley PG, Rolfsson O, Thompson GS, Basnak G, Francese S, Stonehouse NJ, Homans SW, Ashcroft AE (2007) A simple, RNA-mediated allosteric switch controls the pathway to formation of a T=3 viral capsid. J Mol Biol 369:541–552. https://doi.org/10.1016/j.jmb.2007.03.020
54. Stockley PG, Twarock R, Bakker SE, Barker AM, Borodavka A, Dykeman E, Ford RJ, Pearson AR, Phillips SE, Ranson NA, Tuma R (2013) Packaging signals in single-stranded RNA viruses: nature's alternative to a purely electrostatic assembly mechanism. J Biol Phys 39:277–287. https://doi.org/10.1007/s10867-013-9313-0

55. Szabo GT, Mahiny AJ, Vlatkovic I (2022) COVID-19 mRNA vaccines: Platforms and current developments. Mol Ther 30:1850–1868. https://doi.org/10.1016/j.ymthe.2022.02.016
56. Talbot SJ, Goodman S, Bates SR, Fishwick CW, Stockley PG (1990) Use of synthetic oligoribonucleotides to probe RNA-protein interactions in the MS2 translational operator complex. Nucleic Acids Res 18:3521–3528. https://doi.org/10.1093/nar/18.12.3521
57. Tetter S, Terasaka N, Steinauer A, Bingham RJ, Clark S, Scott AJP, Patel N, Leibundgut M, Wroblewski E, Ban N, Stockley PG, Twarock R, Hilvert D (2021) Evolution of a virus-like architecture and packaging mechanism in a repurposed bacterial protein. Science 372:1220–1224. https://doi.org/10.1126/science.abg2822
58. Toots M, Plemper RK (2020) Next-generation direct-acting influenza therapeutics. Transl Res 220:33–42. https://doi.org/10.1016/j.trsl.2020.01.005
59. Twarock R (2004) A tiling approach to virus capsid assembly explaining a structural puzzle in virology. J Theor Biol 226:477–482. https://doi.org/10.1016/j.jtbi.2003.10.006
60. Twarock R, Luque A (2019) Structural puzzles in virology solved with an overarching icosahedral design principle. Nat Commun 10:4414. https://doi.org/10.1038/s41467-019-12367-3
61. Twarock R, Leonov G, Stockley PG (2018) Hamiltonian path analysis of viral genomes. Nat Commun 9:2021. https://doi.org/10.1038/s41467-018-03713-y
62. Uyeki TM, Hui DS, Zambon M, Wentworth DE, Monto AS (2022) Influenza. Lancet 400:693–706. https://doi.org/10.1016/S0140-6736(22)00982-5
63. Valegard K, Murray JB, Stockley PG, Stonehouse NJ, Liljas L (1994) Crystal structure of an RNA bacteriophage coat protein-operator complex. Nature 371:623–626. https://doi.org/10.1038/371623a0
64. Van Blerkom LM (2003) Role of viruses in human evolution. Am J Phys Anthropol Suppl 37:14–46. https://doi.org/10.1002/ajpa.10384
65. WHO (2016) Combating Hepatitis B and C to Reach Elimination by 2030. World Health Organization, pp 1–16
66. Wroblewski E (2018) Defining RNA packaging signals for virus assembly. Leeds
67. Zeltner N, Kohlbrenner E, Clement N, Weber T, Linden RM (2010) Near-perfect infectivity of wild-type AAV as benchmark for infectivity of recombinant AAV vectors. Gene Ther 17:872–879. https://doi.org/10.1038/gt.2010.27

Chapter 5
Creating Artificial Viruses Using Self-assembled Proteins and Polypeptides

David Silverio Moreno-Gutierrez, Ximena del Toro Rios, and Armando Hernandez-Garcia

Abstract Viruses are extraordinary entities that have inspired biomolecular engineers to craft artificial versions of them. Recently, multiple efforts have been made to design a minimalistic, programmable, safer, and engineerable virus-like entity. This has opened the field of viromimetics, which aims to design or engineer macromolecules able to self-assemble into nanoparticles and to recreate viral properties. Using peptides, polypeptides, and proteins as building blocks, several strategies have been implemented to develop artificial viruses. Inspired in the way of how naturally occurring viral coat proteins work, peptides and polypeptides have been designed to mimic structural and functional properties of simple coat proteins. Other groups have repurposed protein cages or evolve them to display virus-like properties. These artificial virus-like entities are being applied for gene and drug delivery, antigen display, vaccine design, and also as a way to understand the fundamental nature of viruses.

Keywords Viromimetics · Protein polymers · Protein engineering · Viral capsids · Viral coat proteins · Viral assembly

Introduction

Viruses are remarkable entities that have triggered our curiosity and imagination during decades. Generally, they are particles smaller than microorganisms that are widely distributed in nature and can infect all domains of life: Bacteria, Archaea and Eukarya. The origin of viruses is in dispute [1]. Some people point out that they existed since the early stages of life, whereas others suggest that they are genetic mobile elements that appeared after life emerged on earth. In any case, viruses are a continuous source of inspiration for bionanotechnology [2].

Fundamental research about viruses has contributed to understand that the properties required for a functional viral particle are found in the structural components that

D. S. Moreno-Gutierrez · X. del Toro Rios · A. Hernandez-Garcia (✉)
Institute of Chemistry, National Autonomous University of Mexico, Ciudad Universitaria, Mexico City, Mexico
e-mail: armandohg@iquimica.unam.mx

M. Comas-Garcia and S. Rosales-Mendoza (eds.), *Physical Virology*, Springer Series in Biophysics 24, https://doi.org/10.1007/978-3-031-36815-8_5

form the virions [3–5]. This has led to ask if it is possible to create an artificial virus or a some kind of synthetic virus-like capsid [6–8]. In the recent years, several research groups have been working towards the aim of creating synthetic virus-like capsids. They have been designing peptides, proteins and other (macro)molecules that could mimic a virus to some extent. This has created the novel area of viromimetics [9–11].

In this chapter, we will discuss what viromimetics is and how research groups have implemented it through the design and engineering of peptide and protein-based materials. Particularly, we will focus on the strategies that have been applied to imitate key aspects of the viral coat proteins using non-viral peptides, polypeptides, and proteins. We will start with a historical perspective to visualize how basic virology has assisted to bring us closer to engineer a truly artificial counterpart of a natural virus.

Historical Perspective

Viruses have been studied during more than 100 years [12]. This has shown us that the virosphere, the pool of all viruses found in nature, is incredibly diverse Fig. 5.1. Until recently, we have started to understand in striking detail some of the fundamental molecular and physical mechanisms of a narrow portion of the virosphere. The Tobacco Mosaic Virus (TMV), like other plant viruses, is one of the simplest viruses existing in nature. TMV is one of the most studied viruses since the dawn of the virology [12]. TMV was firstly studied by Adolf Mayer in 1886 [13] and later by Dmitri Ivanovski in 1889 [14]. These initial studies lead to Martinus Beijerinck to propose in 1898 the existence of a biological entity smaller than bacteria, which he named as "*contagium vivum fluidum*" or "virus" [15, 16]. Later on, TMV played a crucial role to establish the foundations of virology by serving as a model system to address key questions about the biological and physical nature of viruses [17]. All this led to understand key aspects about viruses: viral coating, capsid architecture, assembly and disassembly mechanisms, replication of viruses, budding, release from cells, and others [4]. As new viruses were being discovered and studied, they enriched the initial knowledge about viruses.

The accumulation of knowledge in virology allowed the engineering of viruses and their technological application [18]. One great example of virus engineering are the virus-like particles (VLPs), which are created by the remotion of the genome from the interior of viral particles. These emptied virions can be filled up with foreign nucleic acids or any other molecule, or just used as templates. VLPs are commonly used as gene delivery vectors [19, 20] and carriers of small drugs, but can also encapsulate proteins, metals, polymers, nanoparticles, and others. This capacity to wrap and deliver a wide range of cargos has made VLPs workhorses for bionanotechnology [2, 21, 22]. With the arrival of novel molecular biology tools, it was possible to engineer capsid proteins and the inner and outer surfaces of viral particles. This quickly benefited the creation of the field of "synthetic virology" [21] which aims

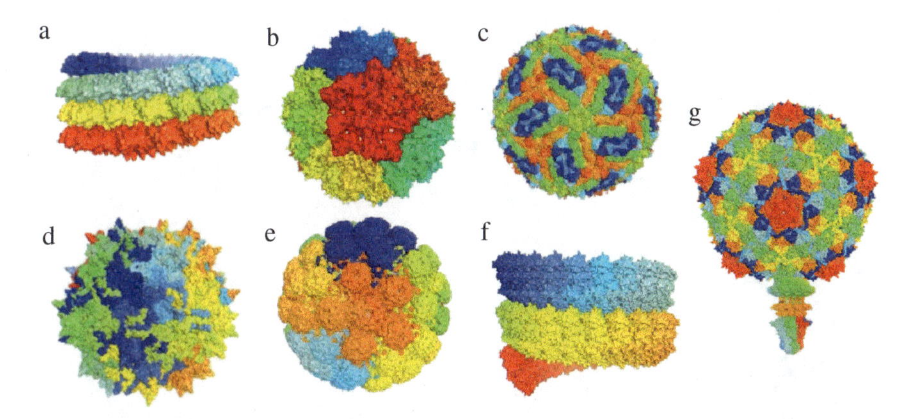

Fig. 5.1 Virosphere shows a large diversity in composition, size, shape, and molecular design. **a** Tobacco Mosaic Virus (TMV) (PDB: 2TMV). **b** Cowpea chlorotic mottle virus (CCMV) (PDB: 1ZA7). **c** ZikaVirus (PDB: 5IRE). **d** Adeno-associated virus (AAV) (PDB: 7RWL). **e** Polyoma virus (PDB: 5FUA). **f** Ebolavirus (PDB: 5Z9W). **g** T7 bacteriophage (PDB head: 2XVR & tail: 6R21)

to create more efficient versions of VLPs for various applications in biotechnology, biomedicine, material sciences and other disciplines [18].

The recognition of inherent limitations in the applicability of natural and engineered VLPs motivated the idea of creating an "artificial virus". The term "synthetic virus-like gene-transfer system" was first mentioned in the 90s [7, 23] and was proposed for achieving highly efficient non-viral gene therapy. In this report, a short peptide sequence derived from a virus coat protein was incorporated into polylysine-DNA complexes and helped to disrupt cell endosomes and promote the gene transfer. After this, other groups started to mimic functionalities of real viruses in their DNA-polymer complexes by incorporating functional viral sequences into non-viral gene transfer agents [24, 25]. With the time, and upon the observed success, research groups stepped away from using viral parts and tried to recreate viruses using non-viral macromolecules.

What is a Virus?

To understand how a virus can be physically imitated, first we need to specify what a virus is. Looking at the virosphere [26], we can conclude that all viruses are obligate intracellular parasites that have been fine-tuned by evolution to deliver their genomes (DNA or RNA) into host cells (infect) and replicate by hijacking the cellular machinery to make new copies of their genome and structural components. In order to replicate and create new fully functional viral particles, multiple steps of the so-called viral cycle are required to be accomplished (Fig. 5.2). The steps in the viral cycle include: (i) survival intra-organism circulation and reaching the target tissue

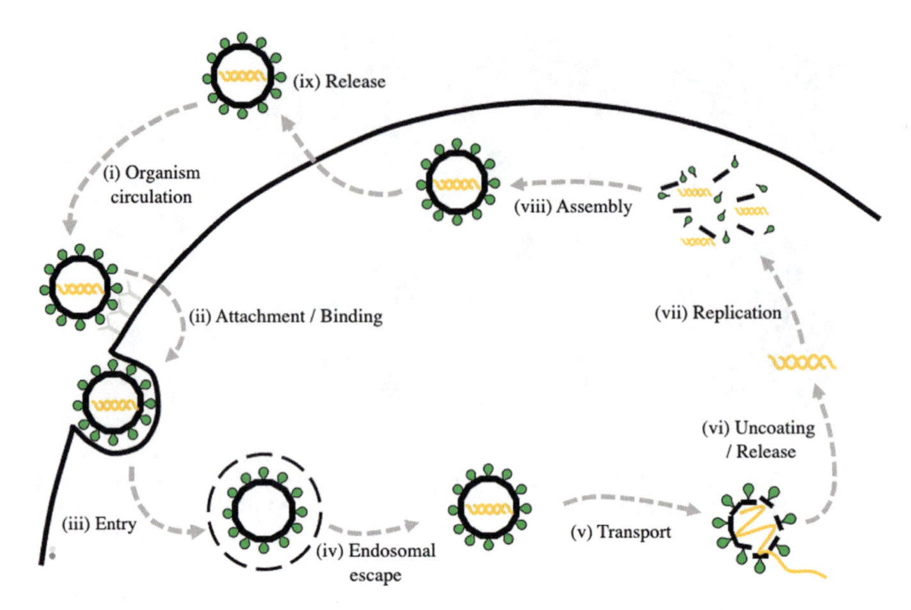

Fig. 5.2 Idealized viral cycle showing various of the required steps that a virus needs to infect a host cell and replicate

(for plant and animal viruses), (ii) attachment and binding to the target cell, (iii) permitting cell entry, (iv) endosomal escape, (v) transportation along the cell in the cytoplasm, might be directed to the nucleus, (vi) disassembly and genome releasing, (vii) expression of the non-structural and structural protein components and synthesis of new genome copies (molecular replication), (viii) genome packing and assembly of the new viral particles, and (ix) particle release from the cell.

The virosphere also shows how diverse and distinctive viruses can be in terms of their physical and chemical features (size, shape, architecture, composition, design, and functionality of the viral components) (Fig. 5.1). In physical terms, viruses are nanosized particles with a nucleic acid molecule (RNA or DNA) in their interior. In chemical terms, viruses are composed, at least, by a protein with self-assembly capabilities, which encapsulates a genetic cargo into a highly organized and regular coating lattice [3, 27–32] (Fig. 5.3). The viral coating forms nanoparticles that are, strikingly, very regular in shape, size, composition, and assembly mechanism and have determining consequences in the practical application of viruses or VLPs [29, 33]. Also, they served as a source of inspiration for designing and engineering an artificial virus-like counterpart.

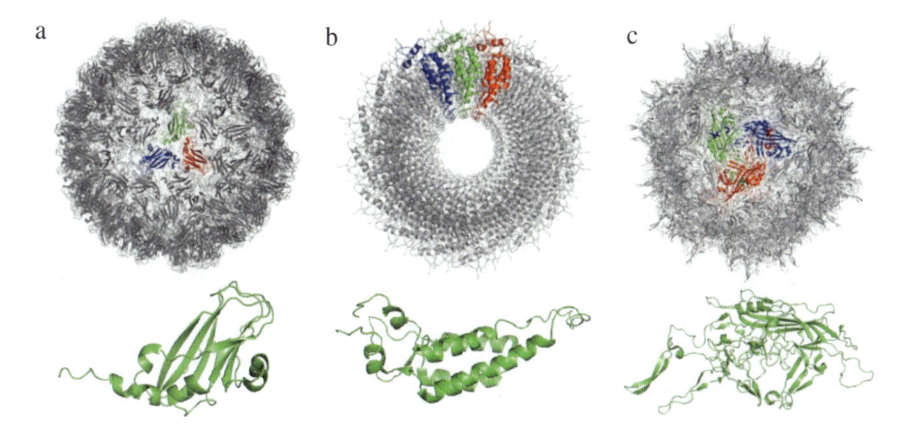

Fig. 5.3 RNA viruses are composed at least by a viral coat protein that self-assembles into nano-sized capsids which encapsulate a nucleic acid molecule. Viral nanoparticles (top) and coat proteins (bottom) of **a** CCMV (PDB: 1ZA7), **b** TMV (PDB: 2TMV) and **c** AAV-2 (PDB: 1LP3). Viral coat proteins determine many viral properties such as self-assembly, organization, shape, size, among others

Size

Viral particles span a wide range of sizes that can go from just a few to hundreds of nanometers (10–1200 nm). For example, icosahedral plant and mammalian viruses such as cowpea chlorotic mottle virus (CCMV), adeno associated virus (AAV) and adeno virus (AdV) have capsids with 24–28 [34, 35], 25 [36], and 70–100 nm in diameter [37], respectively. TMV and the animal-infective Ebola virus have an elongated rod-shaped capsid with dimensions of 15–18 $\times$ 280–300 [5, 38] and 80 $\times$ 600–1400 [39, 40] nm, for diameter and length, respectively.

Shape

Among all the variety of viral morphologies, the icosahedral and cylindrical (elongated rod-like particles) ones are the most representative and common. However, geometries different from the icosahedral and cylinders are also found. The poxvirus Vaccinia virus (VV) has bricked-shaped enveloped particles with a length, width and thickness of roughly 220–450, 140–260 and 140–260 nm, respectively [41]. The lentivirus human immunodeficiency virus (HIV) has a conical capsid enveloped into a spherical particle of about 100 nm in diameter [42, 43]. The herpesvirus has a unique structure made up of four-layered nanoparticles with a total diameter of $\sim$ 210 nm [44, 45]. Its core contains the dsDNA, followed by an icosahedral capsid of 125 nm, then a proteinaceous irregular matrix called tegument, and finally a lipid bilayer.

Composition

The capsids of the simplest viruses, such as TMV and other RNA plant viruses, only need of a single type of protein to encapsulate their genome. In contrast, viruses such as AAV and AdV have capsids composed by several types of proteins. Furthermore, other viruses have a more complex composition. Some wrap their capsids in a lipidic membrane constituted by components derived from the host cell where the virus is produced (e.g., Ebola and HIV).

Why to Design an Artificial Virus?

The interest to design an artificial version of a virus is mostly motivated by technological reasons, however designing an artificial virus also aims to answer fundamental scientific questions. Regarding the latter, the designing of an artificial version of a virus could be a way to discover novel and important properties or features of viruses. Also, artificial viruses could work as model systems to prove and test our current knowledge about viruses. The engineering of a molecule that could imitate a viral coat protein could be used to confirm how viral coat proteins work and how viruses, or at least simplified versions of them, are formed and have emerged, perhaps shedding some light into their origins. For example, nucleocapsids optimized by evolution in laboratory support the hypothesis that ancient viruses may have emerged from sub-cellular protein compartments and help to understand evolutionary assembly pathways [46, 47].

From a technological standpoint, the engineering of artificial viruses could overcome fundamental limitations that have been found in natural VLPs and lead to the development of more applicable artificial viruses. If we could artificially imitate the molecular mechanisms that make viruses to have a higher success rate of cargo delivery, then, we could develop efficient delivery systems. Also, an artificial VLP made of simpler building blocks could be easier to tune its physical, chemical, and biological properties. The generation of better and programmable genetic carriers can be achieved by creating a simplified artificial virus-like entity that incorporates the versatile and fine control mechanisms that viruses exert to enter the cells and replicate. The latter idea is very relevant considering that there is still an unmet demand of efficient carriers for attending genetic diseases. As we will see later, there are already several remarkable examples of how it is possible to design artificial virus-like coatings and how could they be developed into fully artificial viruses.

Viral Coat Proteins

Intensive virology research, especially on TMV and other simple plant RNA viruses, has made clear that the coat protein is the minimal component that carries the basic functionalities of a virus [3, 5, 28, 32, 48–51] (Fig. 5.3). The coat protein is responsible for the most commonly recognized property of viral capsids: to encapsulate the genome into a nanoparticle and protect it from environmental degradation. The capsid is a physical barrier between the viral genome and the environment and ultimately contributes to determine the fate, interactions, and stability of the viral particles. The physicochemical properties of the assembled viral particles such as shape, size, and surface charge are encoded in the coat protein. Eventually, the coat proteins participate in the release of the genome into the cytoplasm for its later expression and synthesis of new copies of the structural proteins and the genome itself.

Packaging and Condensing Its Genome

The viral coat proteins play the main role during the recognition and assembly into functional viral nanoparticles of the newly synthetized genome copies [3, 48, 52]. During this process, electrostatic interactions between the genome and capsid proteins are the main forces that drive the assembly [53–55]. Many viral capsids have cationic residues at their inner surface to efficiently condense their genomes inside them [54, 56]. The charge neutralization helps to reduce the repulsion between the negatively charges of the phosphate backbone of the nucleic acids during packaging. For some viruses, the interactions between capsid proteins and the genome can be sequence-specific and might be crucial at the time of assembly. Moreover, viruses have many ways to pack their nucleic acids into the capsids. While some of them might need the viral genome to template the assembly, other pack their genome in pre-formed capsids [57]. The latter involves the intervention of specialized enzymes that actively select the viral genomes and introduce them into the capsids [37].

Surface Zeta Potential

The surface charge or zeta potential of viral nanoparticles is related the coat protein in simple viruses [58]. This is an important factor that affects the colloidal stability, biodistribution, and the ability to evade immune system recognition by viruses [59]. The general dividing line between stable and unstable particles in suspension is generally taken at either +30 or −30 mV of zeta potential [60, 61]. Generally, particles positively charged externally are internalized faster by cells, however it is also correlated with a higher cytotoxicity [62]. Therefore, many viruses have evolved to have a negative external surface in physiological conditions [58]. In fact, enveloped viruses have a negatively charge surface because the lipidic envelope is derived from the negatively charged cellular membrane. Although the overall surface charge of a virus might be negative, cationic regions are still present in the ligands or glycoproteins used for cell attachment and entry. This localized cationic charge density have major implications in viral tropism [59, 63, 64].

What is Viromimetics?

The accumulation of fundamental knowledge about the molecular footprint of viruses and the successfully engineering of VLPs has led to ask if is possible to design an artificial virus. In this respect, basic research about viruses and VLPs has contributed to understand that the properties required for a minimally functional virus-like entity are molecularly encoded, fundamentally, in the viral coat. This has been supported by the observation that polyplexes deploying components taken from viral coats have enhanced delivery capabilities. Therefore, the answer to the question necessarily involves at least the design or engineering of a (macro)molecule able to imitate the basic properties of viral coat proteins [6]. Following this, VLPs cannot be considered as artificial versions of viruses because they are viruses that have been emptied of their genomes (hollow viral capsids). A truly viromimetic entity requires to be fabricated, completely or to a large extend, with non-viral components exhibiting properties associated to the virosphere [65].

Viromimetics can be defined as the scientific efforts aiming to design or engineer (macro)molecules mimicking physical and functional properties of viruses. To better understand this definition, we can refer to definitions of related concepts. A "viromimetic structure" has been defined as "a system that emulates very well the structural and functional properties of natural viruses from the aspects of capsid formation, genome package and gene transfection" [66]. Likewise, "virus-inspired nanosystem" has been defined as the one that "simulates the original composition and topological structure of viruses but have editable properties" [11].

The central idea of all the definitions about viromimetics is the capability to design components that mimic or emulate viral-like properties. A viromimetic entity needs to display shape, size, surface organization, self-assembly, particle stability among other properties, similar than viruses. Besides the physical properties, it also needs to have functional viral properties such as specific cell entry, disassembly, subcellular localization, and distribution, and, ultimately, make more copies of itself. Furthermore, other desirable feature of a viromimetic system is that it must be engineerable and allow its programming and tuning.

In a technological and practical context, artificial viruses could help to overcome the low transfection efficiencies inherent to the current non-viral gene delivery systems. In this sense, virus-like nanostructures built by viromimetics aim to maximize their transfection and delivery capabilities of gene-based drugs for achieving a desired therapeutic effect. So far, most of the efforts in viromimetics have been focused on attending limiting functional factors of available vectors for gene delivery. For example, increasing and promoting cell entry [67–69], targeted delivery [70–73], endosomal escape [7, 69, 74–76] and responsive delivery [66, 77]. However, viromimetics is also addressing physical and chemical properties [65, 78–80] such disassembly capabilities [77] and nuclear import of the cargo [73], with the aim of enhancing the efficiency of artificial virus-like vectors while also boosting their specificity and safety.

To date, there are just a few examples that greatly mimic virus-like assemblies, properties and structures [65, 81, 82] (mainly with spherical shapes in range of sizes in the sub-micron scale). However, most of them still lack of robust functional properties and thus, their transfection and delivery efficiencies are low or have not been tested yet. Even so, they are promising starting points for building systems with cumulative viral properties and for creating streamlined genetic vectors.

How Can We Fully Imitate a Virus?

The answer to this question passes through the design or engineering of a (macro)molecule that is able to (i) encapsulate DNA/RNA into nanoparticles, (ii) enter cells and, importantly, (iii) replicate their components and assemble inside of cells. Many groups have reported molecules forming virus-like capsids, however, none have been demonstrated that after uptake by a cell, they could generate new particles (replicate) inside the cell, and thus they cannot complete the whole viral cycle described above.

Some of the reported viral-inspired molecules have severe inherent restrictions to ultimately replicate and, thus, fully imitate a virus. Some examples of the type molecules are: dendrimers [83], "classic" polymers [84–86], lipids [87, 88], inorganic nanoparticles [89, 90], metal organic frameworks [91] and hybrids [7, 86, 90, 92–97]. Indeed, there are remarkable examples of lipids and polymers that form complexes with nucleic acids and efficiently enter cells and deliver their cargo. However, due to their chemical nature, they cannot be synthesized from the universal genetic code; thus, they cannot ultimately replicate inside a cell. Likewise for other inorganic and organic materials (e.g., dendrimers, carbon nanotubes, etcetera). Even DNA nanoparticles that have achieved strong cell transfection [98] cannot replicate inside a cell because they are made of single stranded DNA and require non-biological temperatures to be formed.

Since synthesis of amino acid chains can be encoded genetically, proteins and (poly)peptides are the most viable option for developing a fully artificial virus. This redraws the original question into how to imitate a virus using proteins and (poly)peptides. Indeed, already have been reported virus-like nanostructures based on proteins [69–71, 75, 76, 82, 99–104], polypeptides [65, 68, 70, 72, 73, 80, 105, 106] and peptides [66, 68, 107, 108]. Taking this into consideration, the question of how to imitate a virus can be simplified by restating the question as the following: how can we design a protein or (poly)peptide containing the minimal distinctive viral features?

Mimicking Strategies

Several approaches have been evaluated to design a peptide, polypeptide or protein that mimics physical features and functional properties of viruses. The most common strategy has been to put together functional peptide sequences [27, 65, 66, 80, 82, 100, 104]. This modular approach can be exploited to append all the required functionalities (encoded in various peptide sequences) needed to obtain a minimal virus-like coat [65, 66, 74]. By appending novel peptide sequences rationally designed, the final sequence increases its complexity. This approach has proved to be simple and effective. Inspired by the coat protein of the TMV, the most common molecular design incorporates at least three functionalities into the final sequence: (i) nucleic acid binding, (ii) particle self-assembly, and (iii) colloidal stability [65] (Fig. 5.4).

Besides using peptide sequences, a different approach uses proteins. A common strategy uses protein cages, which already form structures similar to a virus coat [82], and repurpose them to package and deliver nucleic acids [99, 104], via domain addition or mutagenesis. In contrast to the previous approach, proteins that display viral functionalities can be computationally designed from scratch [77, 99, 102, 104, 109–111]. In order to create virus-like structures, this approach has to deal with the hard problem of the protein folding. Perhaps advanced algorithms such as *AlphaFold* [112] will allow protein designers to move forward enough to be able to design novel virus-like proteins from scratch.

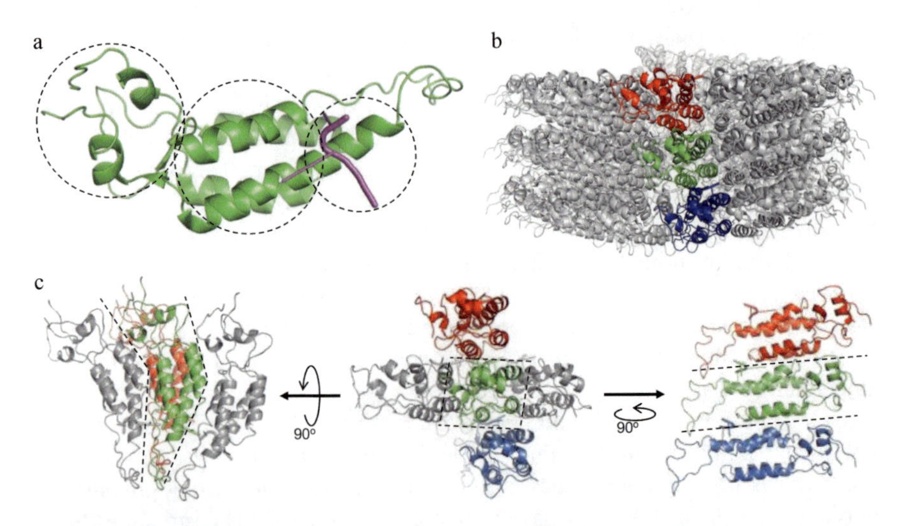

Fig. 5.4 Viral coat protein from TMV. **a** Protein sub-unit of TMV carry both structural and biological functionalities. Each circle indicates regions involved in colloidal stability (blue), self-assembly (purple) and RNA binding (red). **b** Lateral view from the TMV assembly highlighting three stacked protein sub-units. **c** Different views of the protein–protein interaction in the viral assembly

Nucleic Acid Binding

Nucleic acid binding represents the most basic function of a virus-like particle [3, 48, 51, 52, 113]. The most common way that viruses bind and pack their genome is through electrostatic interactions of cationic amino acids (arginine, lysine and histidine) [53, 55]. Many viromimetic strategies incorporate long stretches of cationic amino acids in the polypeptide chain to achieve DNA or RNA binding [65, 66, 68, 75, 80, 100, 114]. Although using electrostatic interactions is a very ubiquitous strategy, it presents several limitations. Firstly, they lack specificity for particular sequences of DNA/RNA or even for nucleic acids at all. This means that the artificial viral coat molecule won't be able to encapsulate efficiently the intended DNA or RNA in the presence of other polyelectrolytes, heavily limiting their capacity to replicate inside a cell [115, 116]. If the interest is just to use it as delivery system, this approach is enough (although large stretches of cationic amino acids could be toxic to cells [117]). However, using peptides with specific binding affinity for DNA/RNA sequences could offer the possibility to achieve a replicative virus-like particle [74, 118–120]. This is because after new copies of the genome and the structural components have been made, they can assemble together into new functional particles.

Particle Self-assembly

The viromimetic protein needs to encapsulate its genome in some kind of complex or particle in order to provide protection and enter the target cell [3, 30, 32, 121, 122]. The protein needs to establish interactions with other protein units to self-assemble into particles. The common way to achieve this has been by using hydrophobic sequences [30, 50, 122]. The utilization of sequences rich in nonpolar amino acids permit aggregation and self-assembly [65, 66, 68, 73, 75, 123–126], and depending on the kind of self-assembly process, the resultant nanoparticle could be highly regular and organized.

Colloidal Stability

A third functionality minimally required is the colloidal stability of the assembled particles between the protein and the nucleic acid [65, 73, 123, 127–129]. The formed virus-like nanoparticles need to be stable enough to remain soluble in aqueous solution and in presence of other components, otherwise they will precipitate and won't be functional.

Other Functionalities

It is clear that a fully viromimetic protein needs to achieve more functionalities to complete the whole viral cycle previously depicted [27, 64, 130–132] (Fig. 5.2). So far, there are very limited examples of proteins designed or engineered to carry advanced tasks. We believe that in the near future there will be more reports into mimicking further steps of the viral cycle.

Viromimetics with Peptides and Polypeptides

The peptides were one of the first building blocks used in viromimetics because of their short length and sequence simplicity (Fig. 5.5). The first viral-mimicking peptides were rationally designed based on the prediction of structure and function from scratch using bottom-up and modular approaches. Spurred by the need of achieving efficient gene delivery vectors, many designed peptides aimed to provide efficient nucleic acid complexation, cellular uptake, and the release of the genetic cargo inside targeted cells. However, other viral properties, such as replication, needed for a larger imitation of viral properties have not been explored yet. The physicochemical properties of the amino acid repertory have played a critical role in the attempt to mimic basic functionalities of the simplest viral coat proteins (nucleic acid binding, self-assembly and colloidal stability) (Fig. 5.5a). In this sense, cationic amino acids are chosen for nucleic acid binding, while amphipathic sequences are used for self-assembly and colloidal stability.

The "KALA" peptide (WEAKLAKALAKALAKHLAKALAKALKACEA) was the first synthetically designed sequence that enabled the condensation of DNA [134]. This peptide forms an α-helix secondary structure which exposes cationic lysines on one side and nonpolar leucines on the other. The amphipathic nature is the main driving force for self-assembly and colloidal stability in the KALA peptide. The solubility at physiological pH is enhanced by the presence of one glutamic acid residue at each end of the helix. Moreover, the lysine-rich regions are responsible for electrostatic interactions with the DNA. Further modifications have been made and have resulted in the RALA and GALA versions of the peptide. They have improved endosomal escape, biocompatibility, stimulus responsive disassembly and overall, higher transfection efficiency [68, 76]. There are other examples of peptides that have used other types of cationic-rich sequences for nucleic acid binding. Some examples include poly-lysine, arginine-rich peptides and even some natural binding domains from proteins such as histones [74, 101, 135, 136]. For self-assembly, α-helical coiled coils, silk-like, and elastin-like sequences have provided the amphipathic nature needed, and can also show responsive properties [73, 105, 106, 109]. Moreover, multifunctional peptides have been obtained by the addition of amino acid residues for cell targeting, nuclear localization, membrane fusion (cationic residues such as histidine, which promote the proton sponge effect for endosomal escape),

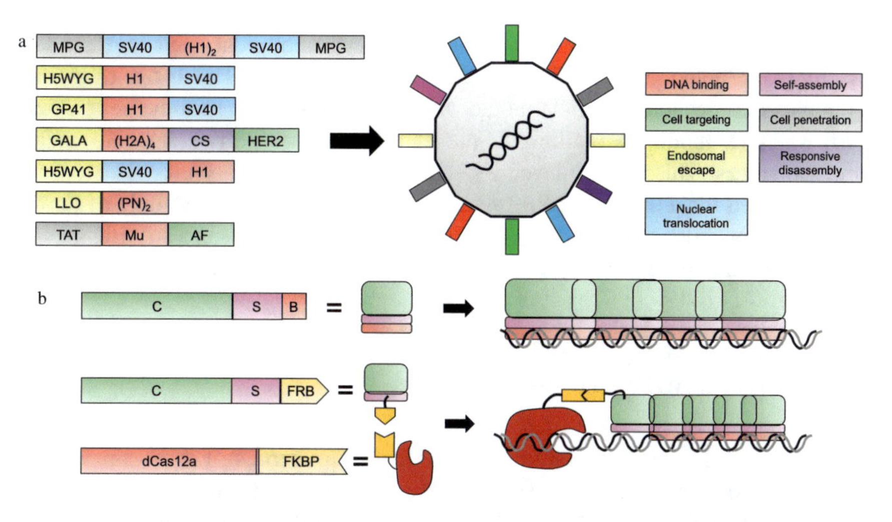

Fig. 5.5 Examples of modular peptides and polypeptides designed for gene delivery and viral mimicking. **a** Examples of peptides modularly designed using different blocks with specific viromimetic functions that self-assemble with nucleic acids into virus-like particles. The function of each block is color coded. MPG: cell penetrating peptide derived from HIV GP41 Protein. SV40: nuclear localization signal from Simian Virus 40. H1: histone 1 DNA binding domain. H5WYG: fusogenic peptide derived from influenza virus. GP41: HIV GP41 derived fusogenic peptide. GALA: fusogenic peptide. H2A: histone 2 DNA binding domain. CS: cleavage site of cathepsin D protease. HER2: synthetic peptide targeting HER2 receptor. LLO: endosomolytic pore-forming protein Listeriolysin O. PN: protamine DNA condensing peptide. TAT: HIV TAT derived peptide. Mu: DNA binding domain. AF: affibody ZHER2 towards HER2 receptor (references: [70, 74, 75, 101, 120, 133]). **b** Modular design of viromimetic polypeptide CSB. CSB polypeptide is a de novo designed TMV-inspired coat protein that assembles with DNA into rod-like nanoparticles [65] (top). Substitution of the DNA binding "B" block by a dimerization domain FRB heterodimerizes with a dCas12a-FBKP in presence of rapamycin. This complex recognizes specific sequences on DNA to initiate assembly [47] (bottom)

and responsivity to environmental stimuli, which allow the disassembly of peptide aggregates in specific physicochemical conditions such as potential redox and the presence of specific proteases [68, 70, 74, 75, 101, 120, 133].

Given the fact that the examples of peptides mentioned here were designed for nucleic acid delivery purposes rather than viral mimicry, little attention has been paid to other important viral properties. The design of peptides intended for use as truly viral mimetic vectors needs to assure that they recognize specific nucleic acid sequences and package them in an ordered structure. Until recently, viral mimicking based on polypeptides has adopted a more comprehensive design to address these aspects. One example is the control of the size and shape of the nanoparticle by modulating the strength and length of the hydrophilic vs hydrophobic blocks [136–138]. More recently, inspired by the functional design of the TMV coat protein, a modular triblock polypeptide called "CSB" was rationally designed by playing with the length of the colloidal, assembly and DNA binding modules [65] (Fig. 5.5b). The polypeptide CSB forms rod-shaped nanostructures with colloidal stability ("C"), self-assembly ("S") and nucleic acid interaction ("B") properties. This is one of the first examples of a designed artificial nucleocapsid containing one genome per particle. CSB was later modified to allow the nucleation of its assembly at specific nucleic acid sequences. This was achieved by substituting the dodecalysine "B" block for an inactivated form of the CRISPR-Cas system [47]. This is an example of how to program a versatile polypeptide system to encapsulate genomes in a specific manner and further engineer it into a more complete form of artificial virus.

Viromimetics with Proteins

Another fruitful approach to mimic the structural properties of viral capsids uses non-viral proteins as base. A straightforward way to obtain an artificial virus is to take available natural proteins that already self-assemble into nanometric compartments or cages and turn them into virus-like particles using protein engineering methodologies [139] (Fig. 5.6). Proteins have been conferred with nucleic acid binding by making their inner surface more cationic or by the conjugation with nucleic acid binding domains. Moreover, protein particles have been further engineered to protect and specifically encapsulate its own genome through in vitro directed evolution [46, 103, 139] (Fig. 5.6a). Examples of natural protein cages are ferritin, heat shock proteins, encapsulins and other bacterial enzymes that naturally form icosahedral nanoparticles similar to viruses [69, 77, 140]. However, some limitations of these protein nanocages includes their high level of porosity.

Another approach is to de novo design proteins that self-assemble into nanocages or compartments. In this approach, the icosahedral structure of viral capsids has been imitated using non-viral proteins, of any kind, that were designed from scratch to form dimeric, trimeric or pentameric building blocks [141, 142] (Fig. 5.6b). One can play with the symmetry of the building blocks and fuse them into a specific and oriented fashion, so they can spontaneously self-assemble into icosahedral nanostructures. Following this strategy, have been designed symmetrical macromolecular hollow icosahedral cages composed of 120 protein subunits with size of 24–40 nm [142]. In a similar fashion, it was designed a 30 nm particle by fusing three non-viral components (a pentamer, a trimer and a dimer) into a single chimeric protein to yield a 60-subunits icosahedral cage [141]. Other artificial nanocages have been engineered to display stimulus responsive disassembly to changes in the pH or as an allosteric effect [110, 111]. Moreover, they also have been subject to in vitro directed evolution to form particles that are able to encapsulate its own RNA genome [82], in spite, the protein nanoparticle interacts with nucleic acids in an electrostatic non-specific way.

Although there are a few extraordinary successful examples of virus-like protein nanocages designed from scratch, this approach requires high computational resources, and the rate of success is still modest. In the last decade, the report of designed nanostructures related to artificial viruses has increased. So far, many viral properties such as a highly ordered structure, self-assembly, nucleic acid binding,

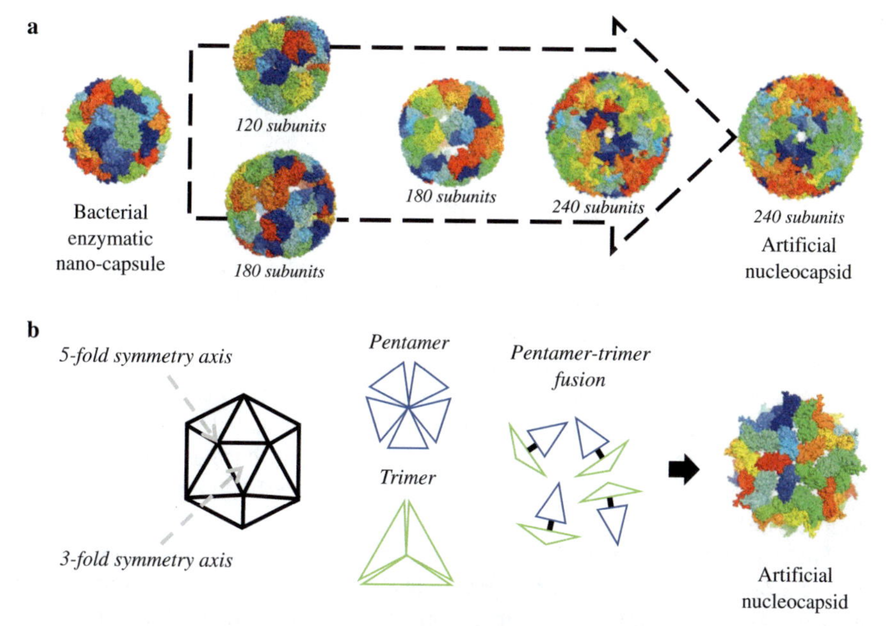

Fig. 5.6 Examples of viromimetic particles based on natural multimeric proteins. **a** In vitro evolution of a bacterial enzymatic nano-capsule into an artificial nucleocapsid able to encapsulate its own genome. Four rounds of in vitro evolution were done. Each structure moving to the head (left to right) of the black arrow represents the particles obtained after each round of evolution [103]. **b** Building of a nucleocapsid using pentameric and trimeric proteins as building blocks. The design is based on the icosahedral geometry considering the fivefold and three-fold symmetry axis. Pentameric and trimeric domains serve as building blocks that are fused into a single that its able to self-assembly into icosahedral particles. This example was also subject to in vitro evolution to be able to encapsulate its own genome [82]

specific genome recognition, cell targeting, endosomal escape, and "smart" disassembly have been successfully imitated, but combining all of them into a single structure is still challenging. To address this challenge effectively in the field of protein design, it is required to combine cutting edge technologies in protein engineering and structural biology. Tools like in silico structure prediction, high throughput cloning, expression, and purification, are needed to increase the rate of success of all these protein designs. In the years to come, more complex and sophisticated designs will emerge that will resemble more closely viral particles and that could be defined with certainty as artificial viruses.

Perspectives for Viromimetics

The creation of the frontier discipline of viromimetics has benefited from the combination of fundamental knowledge obtained from virology and the technological application of viruses. Important advances in the understanding of virus mechanisms to assemble, to deliver their genome and to interact with host cells, have given new perspectives of how to engineer or design de novo non-viral peptides, polypeptides, and proteins to form artificial VLPs. This has led viromimetics to achieve important milestones in the recent years. Viromimetics has succeeded in the creation of several nanoparticles displaying properties that closely reassemble viruses. However, to keep moving forward, viromimetics will have to keep looking closely into what viruses are doing and find ways to implement those natural strategies into their building blocks. Viromimetics will have to establish compelling strategies and clear designing rules inspired in viruses to keep advancing the field. Besides this, it is needed to find ways to combine multiple advanced viral features into a single building block or into a single carrier.

Viromimetics has the capability to generate uniform gene carriers with predictable surface properties and transfection capabilities, enhancing the delivery efficiency of therapeutics and imaging agents, among others. Thus, it is expected that the virus-like particles created following viromimetic principles will have a huge impact in the field of gene delivery and in other practical applications. Exploitation of sophisticate viral mechanisms could open new opportunities to maximize their efficiency and therapeutical effects. The latter is truth if advanced viral properties, such as dynamic and stimulus-responsive oligonucleotide release, programmable targeting and even replication, are carefully crafted into the artificial viral particles. Other strategies that can be exploited are the use of proteins to stabilize the initially formed nanoparticle, like the cementing protein IIIA of AdV linking several coat proteins [143].

It has become clear that engineering artificial viruses that incorporate molecular strategies taken from viruses has led to increase their delivery capabilities. However, in the future molecular evolution of peptide- or protein-based artificial viruses can be used to create novel properties. Darwinian evolution has driven viruses to transfect host cells with their DNA or RNA genome with high efficiency. Using this approach, it could be possible to evolve genetically encoded replicative artificial viral coat proteins to form particles that show high transfection efficiencies in specific tissues or environmental conditions, or to surpass physiological barriers such as the brain blood barrier. In this sense, viromimetics offers exciting possibilities to "maximize" desirable properties for technological applications. In the coming years, many important milestones will be achieved that will put us close to obtain truly artificial viruses that could be used for many biotechnological, biomedical, and material science applications.

Acknowledgements This work was funded by 2019 UC MEXUS-CONACyT Collaborative Research Grant (CN-19-118) and CONACyT Frontera de la Ciencia 2019 (160671). S. M.-G. is a doctoral student from Programa de Doctorado en Ciencias Biomédicas, Universidad Nacional Autónoma de México and received a doctoral fellowship (CVU: 662246) from CONACYT, México.

References

1. Krupovic M, Dolja VV, Koonin EV (2019) Origin of viruses: primordial replicators recruiting capsids from hosts. Nat Rev Microbiol 17(7):449–458. https://doi.org/10.1038/s41579-019-0205-6
2. Wen AM, Steinmetz NF (2016) Design of virus-based nanomaterials for medicine, biotechnology, and energy. Chem Soc Rev 45(15):4074–4126. https://doi.org/10.1039/C5CS00287G
3. Duggal R, Hall TC (1993) Identification of domains in brome mosaic virus RNA-1 and coat protein necessary for specific interaction and encapsidation. J Virol 67(11):6406–6412. https://doi.org/10.1128/JVI.67.11.6406-6412.1993
4. Savithri HS, Suryanarayana S, Murthy MRN (1989) Structure-function relationships of icosahedral plant viruses. Arch Virol 109(3):153–172. https://doi.org/10.1007/BF01311078
5. Harrison BD, Wilson TMA, Klug A (1999) The Tobacco Mosaic Virus particle: structure and assembly. Philos Trans R Soc London Ser B Biol Sci 354(1383):531–535. https://doi.org/10.1098/rstb.1999.0404
6. Mastrobattista E, van der Aa MAEM, Hennink WE et al (2006) Artificial viruses: a nanotechnological approach to gene delivery. Nat Rev Drug Discov 5(2):115–121. https://doi.org/10.1038/nrd1960
7. Wagner E, Plank C, Zatloukal K et al (1992) Influenza virus hemagglutinin HA-2 N-terminal fusogenic peptides augment gene transfer by transferrin-polylysine-DNA complexes: toward a synthetic virus-like gene-transfer vehicle. Proc Natl Acad Sci USA 89(17):7934–7938. https://doi.org/10.1073/pnas.89.17.7934
8. Wagner E (2004) Strategies to improve DNA polyplexes for in vivo gene transfer: will "artificial viruses" be the answer? Pharm Res 21(1):8–14. https://doi.org/10.1023/b:pham.0000012146.04068.56
9. Gao XH, Ding JQ, Long QQ et al (2021) Virus-mimetic systems for cancer diagnosis and therapy. WIREs Nanomed Nanobiotechnol 13(3). https://doi.org/10.1002/wnan.1692
10. de la Fuente IF, Sawant SS, Tolentino MQ et al (2021) Viral mimicry as a design template for nucleic acid nanocarriers. Front Chem 9. https://doi.org/10.3389/fchem.2021.613209
11. Liao Z, Tu L, Li X et al (2021) Virus-inspired nanosystems for drug delivery. Nanoscale 13(45):18912–18924. https://doi.org/10.1039/D1NR05872J
12. Harrison BD, Wilson TMA, Harrison BD et al (1999) Milestones in research on Tobacco Mosaic Virus. Philos Trans R Soc London Ser B Biol Sci 354(1383):521–529. https://doi.org/10.1098/rstb.1999.0403
13. Ueber MA, Tabaks DMD (1886) Landwirtsch Versuchs-stationen 32:451–467
14. Über ID, Tabakspflanze DMD (1903) Zeitschrift für Pflanzenkrankheiten 13(1):1–41
15. Beijerinck MW (1898) Über Ein Contagium Vivum Fluidum Als Ursache Der Fleckenkrankheit Der Tabaksblätter. Verh der K Akad van Wet te Amsterdam 65:1–22
16. Harrison BD, Wilson TMA, Bos L (1999) Beijerinck's work on Tobacco Mosaic virus: historical context and legacy. Philos Trans R Soc London Ser B Biol Sci 354(1383):675–685. https://doi.org/10.1098/rstb.1999.0420
17. Creager AN, Scholthof KB, Citovsky V et al (1999) Tobacco Mosaic Virus. Pioneering research for a century. Plant Cell 11(3):301–308. https://doi.org/10.1105/tpc.11.3.301
18. Steele JFC, Peyret H, Saunders K et al (2017) Synthetic plant virology for nanobiotechnology and nanomedicine. WIREs Nanomed Nanobiotechnol 9(4):e1447. https://doi.org/10.1002/wnan.1447
19. Sleat DE, Turner PC, Finch JT et al (1986) Packaging of recombinant RNA molecules into pseudovirus particles directed by the origin-of-assembly sequence from Tobacco Mosaic Virus RNA. Virology 155(2):299–308. https://doi.org/10.1016/0042-6822(86)90194-7
20. Takamatsu N, Ishikawa M, Meshi T et al (1987) Expression of bacterial chloramphenicol acetyltransferase gene in tobacco plants mediated by TMV-RNA. EMBO J 6(2):307–311. https://doi.org/10.1002/j.1460-2075.1987.tb04755.x

21. Guenther CM, Kuypers BE, Lam MT et al (2014) Synthetic virology: engineering viruses for gene delivery. Wiley Interdiscip Rev Nanomed Nanobiotech 6(6):548–558. https://doi.org/10.1002/wnan.1287
22. He JY, Yu LY, Lin XD et al (2022) Virus-like particles as nanocarriers for intracellular delivery of biomolecules and compounds. Viruses-Basel 14(9). https://doi.org/10.3390/v14091905
23. Plank C, Oberhauser B, Mechtler K et al (1994) The influence of endosome-disruptive peptides on gene transfer using synthetic virus-like gene transfer systems. J Biol Chem 269(17):12918–12924
24. Remy JS, Kichler A, Mordvinov V et al (1995) Targeted gene transfer into hepatoma cells with lipopolyamine-condensed DNA particles presenting galactose ligands: a stage toward artificial viruses. Proc Natl Acad Sci USA 92(5):1744–1748. https://doi.org/10.1073/pnas.92.5.1744
25. Touze A, Coursaget P (1998) In vitro gene transfer using human papillomavirus-like particles. Nucleic Acids Res 26(5):1317–1323. https://doi.org/10.1093/nar/26.5.1317
26. Ke V et al (2021) Viruses defined by the position of the virosphere within the replicator space. Microbiol Mol Biol Rev 85(4):e00193–20. https://doi.org/10.1128/MMBR.00193-20
27. Kežar A, Kavčič L, Polák M et al (2022) Structural basis for the multitasking nature of the potato virus y coat protein. Sci Adv 5(7):eaaw3808. https://doi.org/10.1126/sciadv.aaw3808
28. Jacob T, Usha R (2002) Expression of cardamom mosaic virus coat protein in *Escherichia Coli* and its assembly into filamentous aggregates. Virus Res 86(1):133–141. https://doi.org/10.1016/S0168-1702(02)00057-6
29. Le DT, Radukic MT, Müller KM (2019) Adeno-Associated virus capsid protein expression in *Escherichia Coli* and chemically defined capsid assembly. Sci Rep 9(1):18631. https://doi.org/10.1038/s41598-019-54928-y
30. Singh L, Hallan V, Zaidi AA (2011) Intermolecular Interactions of chrysanthemum Virus B coat protein: implications for capsid assembly. Indian J Virol 22(2):111. https://doi.org/10.1007/s13337-011-0049-9
31. Ludgate L, Liu K, Luckenbaugh L et al (2016) Cell-free hepatitis B virus capsid assembly dependent on the core protein C-terminal domain and regulated by phosphorylation. J Virol 90(12):5830–5844. https://doi.org/10.1128/JVI.00394-16
32. Tan TY, Fibriansah G, Kostyuchenko VA et al (2020) Capsid protein structure in zika virus reveals the flavivirus assembly process. Nat Commun 11(1):895. https://doi.org/10.1038/s41467-020-14647-9
33. de Ruiter MV, van der Hee RM, Driessen AJM et al (2019) Polymorphic assembly of virus-capsid proteins around DNA and the cellular uptake of the resulting particles. J Control Release 307:342–354. https://doi.org/10.1016/j.jconrel.2019.06.019
34. Speir JA, Munshi S, Wang G et al (1995) Structures of the native and swollen forms of cowpea chlorotic mottle virus determined by x-ray crystallography and cryo-electron microscopy. Structure 3(1):63–78. https://doi.org/10.1016/s0969-2126(01)00135-6
35. Garmann RF, Sportsman R, Beren C et al (2015) A simple RNA-DNA scaffold templates the assembly of monofunctional virus-like particles. J Am Chem Soc 137(24):7584–7587. https://doi.org/10.1021/jacs.5b03770
36. Horowitz ED, Rahman KS, Bower BD et al (2013) Biophysical and ultrastructural characterization of adeno-associated virus capsid uncoating and genome release. J Virol 87(6):2994–3002. https://doi.org/10.1128/JVI.03017-12
37. Kennedy MA, Parks RJ (2009) Adenovirus virion stability and the viral genome: size matters. Mol Ther 17(10):1664–1666. https://doi.org/10.1038/mt.2009.202
38. Caspar DL (1963) Assembly, and stability, of the tobacco mosaic virus particle. Adv Protein Chem 18:37–121. https://doi.org/10.1016/s0065-3233(08)60268-5
39. Aleksandrowicz P, Marzi A, Biedenkopf N et al (2011) Ebola virus enters host cells by macropinocytosis and clathrin-mediated endocytosis. J Infect Dis 204(Suppl 3):S957–67. https://doi.org/10.1093/infdis/jir326
40. Noda T, Sagara H, Suzuki E et al (2002) Ebola virus VP40 drives the formation of virus-like filamentous particles along with GP. J Virol 76(10):4855–4865. https://doi.org/10.1128/jvi.76.10.4855-4865.2002

41. Johnson L, Gupta AK, Ghafoor A et al (2006) Characterization of vaccinia virus particles using microscale silicon cantilever resonators and atomic force microscopy. Sens Actuat B Chem 115(1):189–197. https://doi.org/10.1016/j.snb.2005.08.047
42. Gelderblom HR (1991) Assembly and morphology of HIV: potential effect of structure on viral function. AIDS 5(6):617–637
43. Anonymous (2016) Human immunodeficiency virus (HIV). Transfus Med hemotherapy Off Organ der Dtsch Gesellschaft fur Transfusionsmedizin und Immunhamatologie 43(3):203–222. https://doi.org/10.1159/000445852
44. Brown JC, Newcomb WW (2011) Herpesvirus capsid assembly: insights from structural analysis. Curr Opin Virol 1(2):142–149. https://doi.org/10.1016/j.coviro.2011.06.003
45. Zhou ZH, Dougherty M, Jakana J et al (2000) Seeing the herpesvirus capsid at 8.5 Å. Science 288(5467):877–880. https://doi.org/10.1126/science.288.5467.877
46. Terasaka N, Azuma Y, Hilvert D (2018) Laboratory evolution of virus-like nucleocapsids from nonviral protein cages. Proc Natl Acad Sci 115(21):5432–5437. https://doi.org/10.1073/pnas.1800527115
47. Calcines-Cruz C, Finkelstein IJ, Hernandez-Garcia A (2021) CRISPR-guided programmable self-assembly of artificial virus-like nucleocapsids. Nano Lett 21(7):2752–2757. https://doi.org/10.1021/acs.nanolett.0c04640
48. Zamora M, Méndez-López E, Agirrezabala X et al (2017) Potyvirus virion structure shows conserved protein fold and RNA binding site in SsRNA viruses. Sci Adv 3(9):eaao2182. https://doi.org/10.1126/sciadv.aao2182
49. Borodavka A, Tuma R, Stockley PG (2013) A two-stage mechanism of viral RNA compaction revealed by single molecule fluorescence. RNA Biol 10(4):481–489. https://doi.org/10.4161/rna.23838
50. Capuano CM, Grzesik P, Kreitler D et al (2014) A hydrophobic domain within the small capsid protein of Kaposi's Sarcoma-associated herpesvirus is required for assembly. J Gen Virol 95(Pt 8):1755–1769. https://doi.org/10.1099/vir.0.064303-0
51. Baer ML, Houser F, Loesch-Fries LS et al (1994) Specific RNA binding by amino-terminal peptides of Alfalfa mosaic virus coat protein. EMBO J 13(3):727–735. https://doi.org/10.1002/j.1460-2075.1994.tb06312.x
52. Qian XY, Chien CY, Lu Y et al (1995) An amino-terminal polypeptide fragment of the influenza virus NS1 protein possesses specific RNA-binding activity and largely helical backbone structure. RNA 1(9):948–956
53. Belyi VA, Muthukumar M (2006) Electrostatic origin of the genome packing in viruses. Proc Natl Acad Sci 103(46):17174–17178. https://doi.org/10.1073/pnas.0608311103
54. Perlmutter JD, Hagan MF (2015) The role of packaging sites in efficient and specific virus assembly. J Mol Biol 427(15):2451–2467. https://doi.org/10.1016/j.jmb.2015.05.008
55. Ni P, Wang Z, Ma X et al (2012) An examination of the electrostatic interactions between the N-terminal tail of the brome mosaic virus coat protein and encapsidated RNAs. J Mol Biol 419(5):284–300. https://doi.org/10.1016/j.jmb.2012.03.023
56. Carrillo PJP, Hervás M, Rodríguez-Huete A et al (2018) Systematic analysis of biological roles of charged amino acid residues located throughout the structured inner wall of a virus capsid. Sci Rep 8(1):9543. https://doi.org/10.1038/s41598-018-27749-8
57. Garmann RF, Goldfain AM, Manoharan VN (2019) Measurements of the self-assembly kinetics of individual viral capsids around their RNA genome. Proc Natl Acad Sci USA 116(45):22485–22490. https://doi.org/10.1073/pnas.1909223116
58. Duran-Meza AL, Villagrana-Escareño MV, Ruiz-García J et al (2021) Controlling the surface charge of simple viruses. PLoS ONE 16(9):e0255820. https://doi.org/10.1371/journal.pone.0255820
59. Michen B, Graule T (2010) Isoelectric points of viruses. J Appl Microbiol 109(2):388–397. https://doi.org/10.1111/j.1365-2672.2010.04663.x
60. Sikora A, Bartczak D, Geißler D et al (2015) A systematic comparison of different techniques to determine the zeta potential of silica nanoparticles in biological medium. Anal Methods 7(23):9835–9843. https://doi.org/10.1039/C5AY02014J

61. Soheyla H, Foruhe Z (2013) Effect of zeta potential on the properties of nano-drug delivery systems—a review (part 1). Trop J Pharm Res 12(2):255–264
62. Fröhlich E (2012) The role of surface charge in cellular uptake and cytotoxicity of medical nanoparticles. Int J Nanomedicine 7:5577–5591. https://doi.org/10.2147/IJN.S36111
63. Fein DE, Limberis MP, Maloney SF et al (2009) Cationic lipid formulations alter the in vivo tropism of AAV2/9 vector in lung. Mol Ther 17(12):2078–2087. https://doi.org/10.1038/mt.2009.173
64. Méndez F, de Garay T, Rodríguez D et al (2015) Infectious bursal disease virus VP5 polypeptide: a phosphoinositide-binding protein required for efficient cell-to-cell virus dissemination. PLoS ONE 10(4):e0123470
65. Hernandez-Garcia A, Kraft DJ, Janssen AFJ et al (2014) Design and self-assembly of simple coat proteins for artificial viruses. Nat Nanotechnol 9:698
66. Cao M, Zhang Z, Zhang X et al (2022) Peptide self-assembly into stable capsid-like nanospheres and co-assembly with DNA to produce smart artificial viruses. J Colloid Interface Sci 615:395–407. https://doi.org/10.1016/j.jcis.2022.01.181
67. Bennett R, Yakkundi A, McKeen HD et al (2015) RALA-mediated delivery of FKBPL nucleic acid therapeutics. Nanomedicine 10(19):2989–3001. https://doi.org/10.2217/nnm.15.115
68. McCarthy HO, McCaffrey J, McCrudden CM et al (2014) Development and characterization of self-assembling nanoparticles using a bio-inspired amphipathic peptide for gene delivery. J Control Release 189:141–149. https://doi.org/10.1016/j.jconrel.2014.06.048
69. Wang H, Liu N, Yang FX et al (2022) Bioengineered protein nanocage by small heat shock proteins delivering MTERT SiRNA for enhanced colorectal cancer suppression. ACS Appl Bio Mater 5(3):1330–1340. https://doi.org/10.1021/acsabm.1c01221
70. Govindarajan S, Sivakumar J, Garimidi P et al (2012) Targeting human epidermal growth factor receptor 2 by a cell-penetrating peptide-affibody bioconjugate. Biomaterials 33(8):2570–2582. https://doi.org/10.1016/j.biomaterials.2011.12.003
71. Karjoo Z, McCarthy HO, Patel P et al (2013) Systematic engineering of uniform, highly efficient, targeted and shielded viral-mimetic nanoparticles. Small 9(16):2774–2783. https://doi.org/10.1002/smll.201300077
72. Mangipudi SS, Canine BF, Wang YH et al (2009) Development of a genetically engineered biomimetic vector for targeted gene transfer to breast cancer cells. Mol Pharm 6(4):1100–1109. https://doi.org/10.1021/mp800251x
73. Yigit S, Tokareva O, Varone A et al (2014) Bioengineered silk gene delivery system for nuclear targeting. Macromol Biosci 14(9):1291–1298. https://doi.org/10.1002/mabi.201400113
74. Alipour M, Hosseinkhani S, Sheikhnejad R et al (2017) Nano-biomimetic carriers are implicated in mechanistic evaluation of intracellular gene delivery. Sci Rep. https://doi.org/10.1038/srep41507
75. Kim NH, Provoda C, Lee KD (2015) Design and characterization of novel recombinant listeriolysin O-protamine fusion proteins for enhanced gene delivery. Mol Pharm 12(2):342–350. https://doi.org/10.1021/mp5004543
76. Nishimura Y, Takeda K, Ezawa R et al (2014) A display of PH-sensitive fusogenic GALA peptide facilitates endosomal escape from a bio-nanocapsule via an endocytic uptake pathway. J Nanobiotechnol 12. https://doi.org/10.1186/1477-3155-12-11
77. Jones JA, Cristie-David AS, Andreas MP et al (2021) Triggered reversible disassembly of an engineered protein nanocage**. Angew Chemie-International Ed 60(47):25034–25041. https://doi.org/10.1002/anie.202110318
78. Punter MTJJM, Hernandez-Garcia A, Kraft DJ et al (2016) Self-assembly dynamics of linear virus-like particles: theory and experiment. J Phys Chem B 120(26):6286–6297. https://doi.org/10.1021/acs.jpcb.6b02680
79. Marchetti M, Kamsma D, Cazares Vargas E et al (2019) Real-time assembly of viruslike nucleocapsids elucidated at the single-particle level. Nano Lett 19(8):5746–5753. https://doi.org/10.1021/acs.nanolett.9b02376
80. Armando H-G, Marc WW, Cohen SM et al (2012) Coating of single DNA molecules by genetically engineered protein Diblock copolymers. Small 8(22):3491–3501. https://doi.org/10.1002/smll.201200939

81. Edwardson TGW, Levasseur MD, Tetter S et al (2022) Protein cages: from fundamentals to advanced applications. Chem Rev 122(9):9145–9197. https://doi.org/10.1021/acs.chemrev.1c00877
82. Butterfield GL, Lajoie MJ, Gustafson HH et al (2017) Evolution of a designed protein assembly encapsulating its own RNA genome. Nature 552(7685):415–420. https://doi.org/10.1038/nature25157
83. Zhou W, Liu L, Huang J et al (2021) Supramolecular virus-like particles by co-assembly of triblock polypolypeptide and PAMAM dendrimers. Soft Matter 17(19):5044–5049. https://doi.org/10.1039/D1SM00290B
84. Sun XL, Liu CX, Liu DH et al (2012) Novel biomimetic vectors with endosomal-escape agent enhancing gene transfection efficiency. Int J Pharm 425(1–2):62–72. https://doi.org/10.1016/j.ijpharm.2012.01.010
85. Zhong R, Wang RP, Hou XM et al (2020) Polydopamine-doped virus-like structured nanoparticles for photoacoustic imaging guided synergistic chemo-/photothermal therapy. RSC Adv 10(31):18016–18024. https://doi.org/10.1039/d0ra02915g
86. Ding YF, Wang H, Wang YY et al (2022) Co-delivery of luteolin and TGF-beta 1 plasmids with ROS-responsive virus-inspired nanoparticles for microenvironment regulation and chemo-gene therapy of intervertebral disc degeneration. NANO Res 15(9):8214–8227. https://doi.org/10.1007/s12274-022-4285-7
87. Zhuang R, Chen MQ, Zhou YH et al (2021) Virus-mimicking liposomal system based on dendritic lipopeptides for efficient prevention ischemia/reperfusion injury in the mouse liver. ACS Macro Lett 10(2):215–222. https://doi.org/10.1021/acsmacrolett.0c00743
88. Shi Y, Feng XQ, Lin LM et al (2021) Virus-inspired surface-nanoengineered antimicrobial liposome: a potential system to simultaneously achieve high activity and selectivity. Bioact Mater 6(10):3207–3217. https://doi.org/10.1016/j.bioactmat.2021.02.038
89. Zhao X, Wang YY, Jiang WX, et al (2022) Herpesvirus-mimicking DNAzyme-loaded nanoparticles as a mitochondrial DNA stress inducer to activate innate immunity for tumor therapy. Adv Mater 34(37). https://doi.org/10.1002/adma.202204585
90. Liu Z, Wang P, Xie F et al (2022) Virus-inspired hollow mesoporous gadolinium-bismuth nanotheranostics for magnetic resonance imaging-guided synergistic photodynamic-radiotherapy. Adv Healthc Mater 11(6):2102060. https://doi.org/10.1002/adhm.202102060
91. Qiao C, Zhang R, Wang Y et al (2020) Rabies virus-inspired metal-organic frameworks (MOFs) for targeted imaging and chemotherapy of glioma. Angew Chemie Int Ed 59(39):16982–16988. https://doi.org/10.1002/anie.202007474
92. Ajithkumar KC, Pramod K (2018) Doxorubicin-DNA adduct entrenched and motif tethered artificial virus encased in PH-responsive polypeptide complex for targeted cancer therapy. Mater Sci Eng C 89:387–400. https://doi.org/10.1016/j.msec.2018.04.023
93. Wang YH, Zhang L, Guo ST et al (2013) Incorporation of histone derived recombinant protein for enhanced disassembly of core-membrane structured liposomal nanoparticles for efficient SiRNA delivery. J Control Release 172(1):179–189. https://doi.org/10.1016/j.jconrel.2013.08.015
94. Zhang Y, Feng SJ, Hu GT et al (2022) An adenovirus-mimicking photoactive nanomachine preferentially invades and destroys cancer cells through hijacking cellular glucose metabolism. Adv Funct Mater 32(13). https://doi.org/10.1002/adfm.202110092
95. Blum AP, Nelles DA, Hidalgo FJ et al (2019) Peptide brush polymers for efficient delivery of a gene editing protein to stem cells. Angew Chemie Int Ed 58(44):15646–15649. https://doi.org/10.1002/anie.201904894
96. Benli-Hoppe T, Ozturk SG, Ozturk O et al (2022) Transferrin receptor targeted polyplexes completely comprised of sequence-defined components. Macromol Rapid Commun 43(12). https://doi.org/10.1002/marc.202100602
97. Nie Y, Schaffert D, Rödl W et al (2011) Dual-targeted polyplexes: one step towards a synthetic virus for cancer gene therapy. J Control Release 152(1):127–134. https://doi.org/10.1016/j.jconrel.2011.02.028

98. Li J, Fan CH, Pei H et al (2013) Smart drug delivery nanocarriers with self-assembled DNA nanostructures. Adv Mater 25(32):4386–4396. https://doi.org/10.1002/adma.201300875
99. Edwardson TGW, Mori T, Hilvert D (2018) Rational engineering of a designed protein cage for SiRNA delivery. J Am Chem Soc 140(33):10439–10442. https://doi.org/10.1021/jacs.8b06442
100. Azuma Y, Edwardson TGW, Terasaka N et al (2018) Modular protein cages for size-selective RNA packaging in vivo. J Am Chem Soc 140(2):566–569. https://doi.org/10.1021/jacs.7b10798
101. Majidi A, Nikkhah M, Sadeghian F et al (2016) Development of novel recombinant biomimetic chimeric MPG-based peptide as nanocarriers for gene delivery: imitation of a real cargo. Eur J Pharm Biopharm 107:191–204. https://doi.org/10.1016/j.ejpb.2016.06.017
102. Noble JE, De Santis E, Ravi J et al (2016) A de novo virus-like topology for synthetic virions. J Am Chem Soc 138(37):12202–12210. https://doi.org/10.1021/jacs.6b05751
103. Tetter S, Terasaka N, Steinauer A et al (2021) Evolution of a virus-like architecture and packaging mechanism in a repurposed bacterial protein. Science 372(6547):1220. https://doi.org/10.1126/science.abg2822
104. Lilavivat S, Sardar D, Jana S et al (2012) In vivo encapsulation of nucleic acids using an engineered nonviral protein capsid. J Am Chem Soc 134(32):13152–13155. https://doi.org/10.1021/ja302743g
105. Bravo-Anaya LM, Garbay B, Nando-Rodriguez JLE et al (2019) Nucleic acids complexation with cationic elastin-like polypeptides: stoichiometry and stability of nano-assemblies. J Colloid Interface Sci 557:777–792. https://doi.org/10.1016/j.jcis.2019.09.054
106. Hatefi A, Karjoo Z, Nomani A (2017) Development of a recombinant multifunctional biomacromolecule for targeted gene transfer to prostate cancer cells. Biomacromol 18(9):2799–2807. https://doi.org/10.1021/acs.biomac.7b00739
107. Hernandez-Garcia A, Álvarez Z, Simkin D et al (2019) Peptide–SiRNA supramolecular particles for neural cell transfection. Adv Sci 6(3):1801458. https://doi.org/10.1002/advs.201801458
108. Matsuura K, Watanabe K, Matsuzaki T et al (2010) Self-assembled synthetic viral capsids from a 24-Mer viral peptide fragment. Angew Chemie Int Ed 49(50):9662–9665. https://doi.org/10.1002/anie.201004606
109. Villegas JA, Sinha NJ, Teramoto N et al (2022) Computational design of single-peptide nanocages with nanoparticle templating. Molecules 27(4). https://doi.org/10.3390/molecules27041237
110. Miller JE, Srinivasan Y, Dharmaraj NP et al. Designing protease-triggered protein cages. J Am Chem Soc. https://doi.org/10.1021/jacs.2c02165
111. Miaomiao X, Rongjin Z, Jun X et al (2020) Self-assembly of switchable protein nanocages via allosteric effect. CCS Chem 3(8):2223–2232. https://doi.org/10.31635/ccschem.020.202000437
112. Jumper J, Evans R, Pritzel A et al (2021) Highly accurate protein structure prediction with alphafold. Nature 596(7873):583–589. https://doi.org/10.1038/s41586-021-03819-2
113. Kao CC, Ni P, Hema M et al (2011) The coat protein leads the way: an update on basic and applied studies with the brome mosaic virus coat protein. Mol Plant Pathol 12(4):403–412. https://doi.org/10.1111/j.1364-3703.2010.00678.x
114. Hatefi A, Megeed Z, Ghandehari H (2006) Recombinant polymer-protein fusion: a promising approach towards efficient and targeted gene delivery. J Gene Med 8(4):468–476. https://doi.org/10.1002/jgm.872
115. Hernandez-Garcia A, Cohen Stuart MA, de Vries R (2018) Templated co-assembly into nanorods of polyanions and artificial virus capsid proteins. Soft Matter 14(1):132–139. https://doi.org/10.1039/C7SM02012K
116. Golinska MD, de Wolf F, Cohen Stuart MA et al (2013) Pearl-necklace complexes of flexible polyanions with neutral-cationic Diblock copolymers. Soft Matter 9(28):6406–6411. https://doi.org/10.1039/C3SM50536G

117. Lu H, Guo L, Kawazoe N et al (2009) Effects of poly(l-lysine), poly(acrylic acid) and poly(ethylene glycol) on the adhesion, proliferation and chondrogenic differentiation of human mesenchymal stem cells. J Biomater Sci Polym Ed 20(5–6):577–589. https://doi.org/10.1163/156856209X426402
118. Hernandez-Garcia A, Estrich NA, Werten MWT et al (2017) Precise coating of a wide range of DNA templates by a protein polymer with a DNA binding domain. ACS Nano 11(1):144–152. https://doi.org/10.1021/acsnano.6b05938
119. Balicki D, Putnam CD, Scaria PV et al (2002) Structure and function correlation in histone H2A peptide-mediated gene transfer. Proc Natl Acad Sci 99(11):7467–7471. https://doi.org/10.1073/pnas.102168299
120. Wang YH, Mangipudi SS, Canine BF et al (2009) A designer biomimetic vector with a chimeric architecture for targeted gene transfer. J Control Release 137(1):46–53. https://doi.org/10.1016/j.jconrel.2009.03.005
121. Kumar S, Karmakar R, Garg DK et al (2019) Elucidating the functional aspects of different domains of bean common mosaic virus coat protein. Virus Res 273:197755. https://doi.org/10.1016/j.virusres.2019.197755
122. Rat V, Pinson X, Seigneuret F et al (2020) Hepatitis B virus core protein domains essential for viral capsid assembly in a cellular context. J Mol Biol 432(13):3802–3819. https://doi.org/10.1016/j.jmb.2020.04.026
123. Wang Z, Hao D, Wang Y et al (2022) Peptidyl virus-like nanovesicles as reconfigurable "Trojan Horse" for targeted SiRNA delivery and synergistic inhibition of cancer cells. Small 2204959. https://doi.org/10.1002/smll.202204959
124. Cao M, Wang Y, Zhao W et al (2018) Peptide-induced DNA condensation into virus-mimicking nanostructures. ACS Appl Mater Interfaces 10(29):24349–24360. https://doi.org/10.1021/acsami.8b00246
125. Kong J, Wang Y, Zhang J et al (2018) Rationally designed peptidyl virus-like particles enable targeted delivery of genetic cargo. Angew Chemie Int Ed 57(43):14032–14036. https://doi.org/10.1002/anie.201805868
126. Vargas EC, Stuart MAC, de Vries R et al (2019) Template-free self-assembly of artificial de novo viral coat proteins into nanorods: effects of sequence, concentration, and temperature. Chem A Eur J 25(47):11058–11065. https://doi.org/10.1002/chem.201901486
127. Willems L, van Westerveld L, Roberts S et al (2019) Nature of amorphous hydrophilic block affects self-assembly of an artificial viral coat polypeptide. Biomacromol 20(10):3641–3647. https://doi.org/10.1021/acs.biomac.9b00512
128. Bekker S, Huismans H, van Staden V (2022) Generation of a soluble African horse sickness virus VP7 protein capable of forming core-like particles. Viruses 14(8). https://doi.org/10.3390/v14081624
129. Abidin RS, Lua LHL, Middelberg APJ et al (2015) Insert engineering and solubility screening improves recovery of virus-like particle subunits displaying hydrophobic epitopes. Protein Sci 24(11):1820–1828. https://doi.org/10.1002/pro.2775
130. Zhao X, Wang X, Dong K et al (2015) Phosphorylation of beet black scorch virus coat protein by PKA is required for assembly and stability of virus particles. Sci Rep 5(1):11585. https://doi.org/10.1038/srep11585
131. Staring J, Raaben M and Brummelkamp TR (2018) Viral escape from endosomes and host detection at a glance. J Cell Sci 131(15):jcs216259. https://doi.org/10.1242/jcs.216259
132. Sakuragi S, Goto T, Sano K et al (2002) HIV type 1 gag virus-like particle budding from spheroplasts of saccharomyces cerevisiae. Proc Natl Acad Sci 99(12):7956–7961. https://doi.org/10.1073/pnas.082281199
133. Sadeghian F, Hosseinkhani S, Alizadeh A et al (2012) Design, engineering and preparation of a multi-domain fusion vector for gene delivery. Int J Pharm 427(2):393–399. https://doi.org/10.1016/j.ijpharm.2012.01.062
134. Wyman TB, Nicol F, Zelphati O et al (1997) Design, synthesis, and characterization of a cationic peptide that binds to nucleic acids and permeabilizes bilayers. Biochemistry 36(10):3008–3017. https://doi.org/10.1021/bi9618474

135. Ni R, Chau Y (2020) Nanoassembly of oligopeptides and DNA mimics the sequential disassembly of a spherical virus. Angew Chem Int Ed Engl 59(9):3578–3584. https://doi.org/10.1002/anie.201913611
136. Ni R, Chau Y (2017) Tuning the inter-nanofibril interaction to regulate the morphology and function of peptide/DNA co-assembled viral mimics. Angew Chemie Int Ed 56(32):9356–9360. https://doi.org/10.1002/anie.201703596
137. Hernandez-Garcia A, Werten MWT, Stuart MC et al (2012) Coating of single DNA molecules by genetically engineered protein Diblock copolymers. Small 8(22):3491–3501. https://doi.org/10.1002/smll.201200939
138. Beun LH, Storm IM, Werten MWT et al (2014) From micelles to fibers: balancing self-assembling and random coiling domains in PH-responsive silk-collagen-like protein-based polymers. Biomacromol 15(9):3349–3357. https://doi.org/10.1021/bm500826y
139. Edwardson TGW, Hilvert D (2019) Virus-inspired function in engineered protein cages. J Am Chem Soc 141(24):9432–9443. https://doi.org/10.1021/jacs.9b03705
140. Khoshnejad M, Greineder CF, Pulsipher KW et al (2018) Ferritin nanocages with biologically orthogonal conjugation for vascular targeting and imaging. Bioconjug Chem 29(4):1209–1218. https://doi.org/10.1021/acs.bioconjchem.8b00004
141. Cannon KA, Nguyen VN, Morgan C et al (2020) Design and characterization of an icosahedral protein cage formed by a double-fusion protein containing three distinct symmetry elements. ACS Synth Biol 9(3):517–524. https://doi.org/10.1021/acssynbio.9b00392
142. Bale JB, Gonen S, Liu Y et al (2016) Accurate design of megadalton-scale two-component icosahedral protein complexes. Science 353(6297):389–394. https://doi.org/10.1126/science.aaf8818
143. Reddy VS, Nemerow GR (2014) Structures and organization of adenovirus cement proteins provide insights into the role of capsid maturation in virus entry and infection. Proc Natl Acad Sci USA 111(32):11715–11720. https://doi.org/10.1073/pnas.1408462111

Chapter 6
Construction of Higher-Order VLP-Based Materials and Their Effect on Diffusion and Partitioning

Nathasha D. Hewagama, Pawel Kraj, and Trevor Douglas

Abstract In spite of their role as harmful infectious entities, viruses have gained a significant research attention as they can be exploited as nanomaterials for biotechnology. The self-assembly of viral structural proteins give rise to diverse, robust, well-defined virus-like particle (VLP) architectures that can be manipulated for material synthesis. They have been explored to impart functionality creating biomimetic nanoreactors, as well as modular units in building three-dimensional (3D) hierarchical assemblies. These supramolecular structures exhibit collective properties and behaviors beyond single particles, including porosity and net surface charge, that can be exploited to tune the diffusion and partitioning of reaction components and thereby control biochemical processes. In this review, we focus on the use of VLPs as building blocks for hierarchical 3D materials, molecular diffusion and partitioning within the VLPs and VLP-derived 3D assemblies.

Keywords Virus-like particles (VLPs) · Higher-order assembly · Diffusion · Molecular partitioning · Enzyme catalysis

N. D. Hewagama · P. Kraj · T. Douglas (✉)
Department of Chemistry, Indiana University, 800 E Kirkwood Ave., Bloomington, IN 47405, USA
e-mail: trevdoug@indiana.edu

N. D. Hewagama
e-mail: nhewagam@iu.edu

P. Kraj
e-mail: pkraj@iu.edu

M. Comas-Garcia and S. Rosales-Mendoza (eds.), *Physical Virology*, Springer Series in Biophysics 24, https://doi.org/10.1007/978-3-031-36815-8_6

Overview

Viruses

Viruses and virus-like elements are by far the most abundant biological entities on the planet [47]. They are ubiquitous submicroscopic infectious agents that thrive and multiply on living host cells, across all domains of life. They comprise either (or both) a proteinaceous or lipid-based shell, a capsid, enclosing and protecting their genetic material, either DNA or RNA. These invisible invaders can cause deadly diseases [17], including human immunodeficiency virus-acquired immunodeficiency syndrome (HIV-AIDS), Ebola virus disease, smallpox, and severe acute respiratory syndrome (SARS) among many others, altering the environment we live in continuously. Even though viruses have been traditionally recognized as harmful entities due to their role as infectious agents [23], studies of viruses have historically provided valuable insights into understanding biochemical processes [30].

Viruses and Virus-Like Particles (VLPs) in Biotechnology

In a new paradigm viruses have emerged as useful synthetic nanoplatforms for a range of hard and soft materials applications within material science and medicine [12, 18, 20, 21, 61], rather than purely agents of disease. From a material science point of view, they are a remarkable class of building blocks with diverse shapes and sizes and an inspiration for creating a wide variety of functional biomimetic materials at a range of length scales when they are utilized as non-infectious shells [3, 8, 80, 91, 94]. Viral structural proteins can be utilized to self-assemble non-infectious shell structures, or virus-like particles (VLPs), that are nearly identical to the native virus structures but devoid of genetic material [3, 21, 29]. Protein cage architectures found in nature, including VLPs, inspire the design and synthesis of novel materials by taking advantage of their biocompatibility, conformational flexibility, homogeneity, symmetry, and our synthetic ability to fine-tune the structure and function [3, 36, 84, 85]. Because of these inherent properties of VLPs, they have found application in a range of fields including drug delivery, catalysis, immunology, biotechnology, and energy [8, 61, 77, 79, 96].

Virus capsids are macromolecular containers, where the interior, exterior, and the interface between the capsid subunits provide interfaces for a range of synthetic manipulations without losing the cage-like architecture [21]. The interior of the VLPs has been utilized to encapsulate a range of guest molecules via numerous encapsulation strategies [9, 16, 22, 27, 66, 69, 70]. One example is the use of VLPs as nanoreactor vessels by encapsulating enzymes, or even multiple enzymes catalyzing reaction cascades [9, 26, 35, 65]. The cage interior has been utilized not only to explore catalysis by enclosing the enzymes in confined environments but also as a recovery vessel for sequestering and solubilizing proteins, which can otherwise become aggregated and inactivated [71]. Cargo protection and stability are two other phenomena that are offered through encapsulation inside the VLP cages.

The exterior surfaces are often used by viruses to display certain functionalities required for the host cell-specific targeting and avoidance of host defense mechanisms [3, 21]. Inspired by these designs, the particles derived from bacteriophage P22, MS2, and Cowpea mosaic virus (CPMV) with decorated exterior surfaces have been used as potential tools for therapeutic delivery [11, 28, 45, 86]. Multivalent and highly symmetric presentations of fusion moieties can be obtained using a variety of methods, including genetic modifications and covalent or non-covalent conjugations. Modifying the exterior surfaces of the particle also plays an important role in constructing higher-order cage assemblies and the fabrication of bulk three-dimensional (3D) materials [2, 3, 10, 83, 88, 91].

Understanding the protein–protein interactions at the subunit interfaces provides insight for the design of materials [48, 62] taking advantage of the structural symmetry and dynamics of these cage-like structures including the assembly of subunits into particles and the pleomorphism of capsids. Baker and coworkers have demonstrated the construction of such particle systems by combining computational simulations and genetic constructions [4, 38, 39]. The accuracy and control of such designs in creating highly ordered and homogenous protein cage-like nanostructures could shape the future of custom-designed advanced materials with a broad-spectrum of applications.

The protein cages derived from VLPs serve as ideal robust macromolecular templates for crafting bioinspired and biomimetic materials across multiple length-scales. In this review, we focus on the 3D higher-order structures derived from the assembly of multiple VLPs, the diffusion of molecules across the pores of the particles and the channels within the VLP assemblies, and the partitioning of reaction components and their impact on the functionality of the VLP-derived materials.

Higher-Order Structures Derived from VLPs

Biological Composites

Self-assembly of molecular building blocks over multiple length scales gives rise to diverse hierarchically organized biomaterials with complex structural and functional properties found throughout living systems [76]. In determining bulk material properties, structural hierarchy serves as an important factor as the molecular building blocks can endow collective properties and behaviors in the assembled structure that are beyond the individual units [3, 90]. Nature provides many examples which exhibit this hierarchical phenomenon. Nacreous shells made of calcium carbonate and protein, bone and teeth made of hydroxyapatite and collagen, and bamboo made of cellulose fibers embedded in a lignin matrix, are some classic examples of high-performance biological composite structures with hierarchical organization and collective properties beyond those of the individual components. The assembled structures possess advanced mechanical properties like stiffness, strength, and toughness, which are poorly seen in their building block components [6, 54]. Many studies have been dedicated to understanding the underlying design principles of these hierarchical composite structures and implementing them to develop better biomimetic materials and devices.

VLPs as Building Blocks for Synthetic Hierarchical Assemblies

Viruses are a unique class of naturally occurring supramolecular structures, which are self-assembled from a limited number of subunit macromolecular building blocks generating highly organized architectures. VLPs, having cage-like structures, are an excellent choice as modular building blocks in constructing higher-order architectures as they offer unique advantages. They are chemically and mechanically robust structures. Genetic programmability of proteins, and therefore homogeneity of the expressed cages, is a critical feature of a building block component in constructing higher-order structures and facilitates the formation of materials with long-range order. Moreover, the ability to encapsulate a wide range of functional foreign cargoes in these cages can be implemented in creating functionalized bulk materials. The interparticle interactions can be tuned by genetic and chemical modifications on the exterior surface of these cages to control the assembly of individual cages into hierarchically ordered structures [3, 88–90]. Building synthetic hierarchical structures from VLPs is not exclusively an artificial process but can be viewed as an inspiration from the natural viral particle assemblies seen in viral factories; a step in the virus life cycle where virus components are densely accumulated in a certain area inside host cells with high local concentrations, sometimes creating paracrystalline arrays of virions, during viral replication and assembly [68, 75].

When designing higher-order superlattices from nanoparticle building blocks, it is important to balance and optimize both attractive and repulsive interactions between the system components; interparticle interactions that are too strong could lead to disordered gel-like aggregates while interactions that are too weak might fail to produce any assemblies [43, 97]. Different types of driving forces in the presence of various co-crystallization agents have been used to construct supramolecular assemblies from VLPs [40]. Using physical interactions such as complementary electrostatic interactions between the building block components is one of the well-studied driving forces. Covalent or non-covalent linkers also can facilitate the formation of higher-order assemblies by bridging the particles with tunable interaction strengths, geometries, linker lengths, and flexibilities to design superlattice structures with desired properties [90].

Spherical VLPs derived from cowpea chlorotic mottle virus (CCMV), Qβ, MS2, and P22 as well as anisotropic rod-shaped particles, including tobacco mosaic virus

(TMV), and M13 bacteriophage have been used as modular units to construct higher-order structures. Two-dimensional (2D) assemblies of VLPs have been demonstrated with several types of particles, including CPMV, turnip yellow mosaic virus (TYMV) through enhancing electrostatically induced interfacial adsorption onto oppositely charged lipid monolayers [25, 37], and Qβ, MS2 through convective assembly generating highly ordered close-packed 2D lattice monolayers onto solid substrates [1]. Moreover, the crystallization of P22 VLPs into 2D arrays was obtained on a positively charged lipid monolayer at the water–air interface, where the assembly of the particles was primarily dependent on the interparticle interactions at the exterior surfaces regardless the encapsulated cargo or the morphology of the capsid [98].

Fabricating 3D higher-order structures from VLP cages is an active area of research. Particles derived from CCMV VLPs having negatively charged exterior surfaces have been arranged into crystalline 3D structures using modified oppositely charged gold nanoparticles where the interparticle interactions were controlled by changing the ionic strength and the pH of the medium [43]. In addition, electrostatic complexation of CCMV particles into 3D structures has been reported with positively charged polypeptides [41], polymers [42], dendrons [44], amphiphilic dendrimers [64], and proteins [56] often resulting in ordered packing of the assembly components (Fig. 6.1a, b). Crystallization of wild-type Simian virus 40 nanoparticles with divalent Mg^{2+} cations into bcc crystal structures is another example of using electrostatic interactions to govern the higher-order VLP assembly process. Temperature-dependent, complementary oligonucleotide-directed assembly has been demonstrated with CPMV [87]. The sequence-dependent nucleotide base-pairing allows an effective strategy to control the interparticle interactions, which has also been utilized in assembling Qβ cages into 3D lattices using gold nanoparticles by decorating their surfaces with DNA [15]. Significant work has been done on exploring the 3D assembly of P22 particles which will be discussed in detail in section "Hierarchical 3D Structures Derived from P22 VLPs".

Other than spherical VLPs, rodlike particles, including TMV and M13 have been used to construct both 2D and 3D complex structures via several strategies such as depletion forces [49, 52, 53], electrostatics [55, 57, 58], and metal ion coordination [50, 51, 67, 78]. Altogether, this illustrates the diversity of the VLP cages and the availability of various assembling techniques in tailoring complex higher-order materials with preferred structural and functional properties.

Hierarchical 3D Structures Derived from P22 VLPs

The P22 VLP is a versatile nanocontainer for encapsulating cargoes including enzymes that catalyze single or multiple chemical transformations, thus creating a synthetic nanoreactor. Several studies have been committed to combining the role of P22 VLP as a nanoreactor and a modular unit to build higher-order 3D superlattices via several different techniques. These techniques direct the assembly of particles into ordered or disordered 3D materials, which can be exploited to understand the impact of the complex structure on properties like molecular diffusion, partitioning,

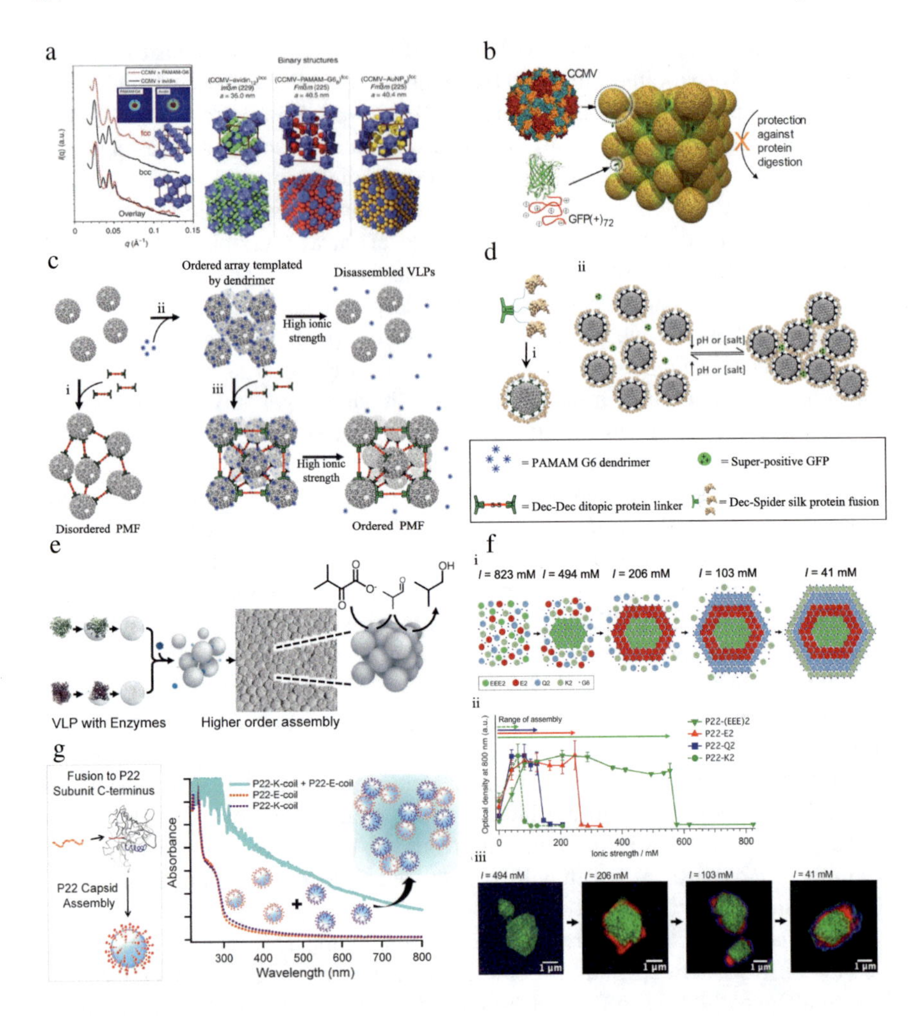

and enzyme activity. Electrostatic linkers, including charged dendrimers and polymers, or protein linkers, such as modified ditopic linkers derived from capsid exterior decoration protein (Dec), and coiled-coil peptide motifs have been successfully used to fabricate 3D materials from P22 particles. Dec-Dec ditopic protein linkers bind to the symmetry-specific sites on the exterior of the expanded and wiffleball morphologies of P22, and the addition of Dec-Dec linkers to P22 has resulted in the formation of unstructured disordered P22 assemblies in the absence of templating agents like dendrimers (Fig. 6.1c) [63, 90]. A stimuli-responsive 3D material has been constructed by decorating the P22 surface with spider silk (Ss) proteins using Dec-Ss protein fusions, where the hierarchical assembly of capsids was reversible and dependent on the pH of the solution due to the dimerization of Ss proteins

◀**Fig. 6.1** Higher-order 3D assemblies derived from VLPs. **a** VLPs derived from CCMV form 3D crystal structures with avidin, dendrimers, and gold nanoparticles (AuNP). Left: 2D scattering patterns of binary assemblies, CCMV-avidin and CCMV-PAMAM G6 dendrimers forming bcc and fcc ordered structures, respectively. Right: Crystal structure models for the binary CCMV-avidin, CCMV-PAMAM G6, and CCMV-AuNP assemblies. Figure modified from [56] with permission from Springer Nature. Copyright 2014. **b** Complexation of negatively charged CCMV VLPs with supercharged cationic polypeptide fused with GFP forming ordered crystal structures. Figure reproduced from [41]. Copyright 2018 American Chemical Society. **c** Assembly of P22 VLPs into disordered PMF with Dec-Dec ditopic linkers (i) or ordered arrays templated with dendrimers (ii). Increase of ionic strength causes the dissociation of VLPs and dendrimers due to charge screening. The addition of Dec-Dec linkers to the P22 ordered arrays locks the particles in place (iii), from which the dendrimers dissociate as the ionic strength increases, leaving the ordered PMF structure remaining intact. Figure adapted with permission from [63]. Copyright 2018 American Chemical Society. **d** (i) Decoration of P22 VLP exterior surface with Dec-Spider silk protein fusions (ii) to assemble a higher-order VLP material with response to pH. The assembly can be controlled in response to the ionic strength of the solution with superpositive GFP. Figure adapted with permission from [2]. Copyright 2018 American Chemical Society. **e** The roles of P22 particle as a nanoreactor and an assembly unit to build higher-order structures are combined to create a catalytic superlattice. Two different populations of VLPs, separately encapsulated with ketoisovalerate decarboxylase (KivD) and alcohol dehydrogenase A (AdhA) enzymes are assembled into an fcc ordered lattice, catalyzing the two-step reaction for isobutanol synthesis. Figure reprinted with permission from [91]. Copyright 2018 American Chemical Society. **f** (i) Starting with a mixed population of P22 VLP variants, each population carrying a different surface charge, they can be assembled into ordered core–shell structures in the presence of G6 dendrimers by gradual modulation of the ionic strength of the solution, as shown in this schematic. (ii) Each variant shows a different assembly behavior with dendrimers at various ionic conditions monitored by the optical density at 800 nm. Each P22 variant has a threshold ionic strength (I_t) above which they do not assemble (iii) Super-resolution fluorescence micrographs of the VLP arrays with increasing number of shell layers after each dialysis step performed to lower the ionic strength below the I_t of each variant when started with a mixture of P22 variants. P22-EEE2 and P22-K2 variants are labeled with Alexa-488 fluorescent dye; P22-E2 and P22-Q2 variants are labeled with Texas Red and CF405M fluorescent dyes, respectively. Figure adapted with permission from [88]. Copyright 2022 American Chemical Society. **g** Two P22 capsid variants with surface exposed three heptad complementary coiled-coil peptide motifs assemble into 3D structures due to the formation of antiparallel coiled-coil heterodimers which can be examined by the light scattering to monitor the assembly. Figure reprinted with permission from [83]. Copyright 2013 American Chemical Society

below a certain pH causing the capsids to assemble (Fig. 6.1d). The addition of positively supercharged GFP to this system facilitated the control of the assembly and disassembly in response to the ionic strength of the medium [2].

When mixed with oppositely charged polyamidoamine (PAMAM) G6 dendrimers, the negatively charged wild-type P22 particles assemble into a kinetically trapped mostly amorphous aggregate, possibly due to the strong electrostatic attractions. Exterior modification of P22 capsids with certain peptides can tune the capsid surface charge in a refined approach to building VLP-dependent ordered 3D structures. Thus, a P22 variant decorated with small peptide repeats on the exterior surface leads to formation of an ordered superlattice structure when mixed with G6 dendrimers; the fused peptides could provide sterically repulsive interactions between particles refining the assembly to achieve an energetically more favored ordered structure [91]. Fusing the C-terminus of the P22 coat protein (CP) with specific peptide sequences generates P22 VLP variants with different surface charges where the magnitude of the charge can be modulated by the peptide sequence fused to the CP. Using this approach, negatively charged P22 particles have been assembled into 3D ordered arrays (with face-centered cubic packing) in the presence of oppositely charged G6 dendrimers, where the assembly exhibits a sharp dependence on the ionic strength of the solution; this behavior has been effectively modeled [10]. The higher-order assembly of particles is independent of the encapsulated cargo but rather determined by the nature of the capsid exterior. P22 particles are proven to be a versatile platform for encapsulating a library of functional cargoes via self-assembly processes. Therefore, this modular assembly approach of particles, with encapsulated cargoes of choice, into higher-order structures provides the opportunity to design functional 3D materials. To demonstrate functionality in these P22-derived supramolecular structures, enzymes performing a two-step reaction for the synthesis of isobutanol have been incorporated (Fig. 6.1e) [91]. When mixed with G6 dendrimers, P22 variants with different surface charges exhibit different assembly behaviors in response to changes in the ionic strength of the solution. Starting from a mixed population of capsid variants, this ionic strength-dependent assembly behavior of capsids has been used to create ordered core–shell structures of VLPs (up to four layers) by tuning the electrostatic interactions between the capsids and linkers by a gradual lowering of the ionic strength of the solution (Fig. 6.1f) [88]. This method allows for control over the spatial arrangement of individual P22 particles (therefore the encapsulated cargo) within the lattice, which is a great challenge in many other synthetic systems.

P22 VLPs have been assembled into ordered protein macromolecular frameworks (PMFs) by combining both PAMAM dendrimer- and ditopic Dec-Dec linker-mediated assembly concepts [63]. Dendrimers template the ordered lattice and Dec-Dec linkers lock the ordered lattice in place, whereupon the dendrimers can be removed leaving highly charged PMF (Fig. 6.1c). P22 VLP nanoreactors have been used to fabricate these ordered functional PMFs with enhanced catalytic properties [82], which will be discussed in detail in section "Partitioning Within Hierarchically Assembled 3D VLP-Based Materials".

The use of complementary electrostatics to create P22 hierarchical structures has also been demonstrated with cationic polyallylamine polymers to make functional P22 clusters [46]. The attachment of complementary coiled-coil peptide motifs to the capsid exterior serves as another technique to bridge between P22 capsids creating higher-order structures [83]. Fusion of the C-terminus of the P22 CP with three repeats of either E-coil (VAALEKE) or K-coil (VAALKEK) peptide sequences has facilitated the 3D assembly of the two P22 variants where the assembly is held together by the heterodimeric E/K-coiled-coil motifs on the capsid surface (Fig. 6.1g). This serves as another method to control the position of the particles (thus, potentially the encapsulated cargo of choice) relative to each other in the 3D space.

The availability of multiple methods to assemble P22 particles into 3D hierarchical structures with desired structural order and functionality holds the potential of using P22 nanoparticle building blocks for constructing advanced catalytic bulk materials designed for specific applications. 3D structures derived from P22 VLPs offer a versatile platform to gain useful insights into how molecules partition, retain, diffuse, and undergo chemical transformations in regard to the complex structure, as both the spatial position of the particles and their packing can be controlled to form ordered or disordered assemblies.

Diffusion in Hierarchically Ordered Protein Materials

Diffusion is the thermally induced motion of particles in a fluid and originates from fluctuations present in chemical systems. Diffusion is the ensemble result of Brownian motion, which is the consequence of collisions between molecules, the shape and size of those molecules, and the effects of intermolecular interactions with their environment. Because all chemical systems involve some movement of molecules, diffusion is a major driving force for chemical transformations.

Because the proximity of reaction components is a critical part of any chemical reaction, diffusion is an important factor in chemical processes. Models of single enzyme reactions depict substrate binding to an enzyme, its chemical transformation, and the release of a product from the enzyme active site. In multistep enzyme reaction models, such as the majority of biochemical processes, the release of the product from the first enzyme is followed its diffusion to the active site of the next enzyme in the pathway. As diffusion is a rapid process and enzymatic pathways only need to be as fast as required for the host organism, there is no evolutionary pressure for post-turnover sequestration of most enzyme products [5]. Exceptions arise in pathways where an intermediate product is unstable, or where its escape would lead to unfavorable chemical consequences [32]. In these pathways enzymes have evolved substrate channeling, the transfer of the product of the first enzyme to the active site of the second without the reentry of the substrate into bulk solution.

Co-localization of Enzymes Does Not Inherently Offer Catalytic Enhancement

Recent advancements in enzyme immobilization have led to the inducement of substrate channeling in enzyme pathways which do not normally exhibit this behavior [13, 24]. In these cases, the rate of reaction can be enhanced through the enforcement of close proximity between the enzyme active sites in the pathway. Because VLPs can be designed to co-encapsulate multiple enzymes in a sequential multistep pathway and these enzymes are constrained within a small space, it has been suggested that the close proximity of the encapsulated enzymes may increase the rate of the encapsulated pathway in comparison to the same enzymes free in solution, though this has not been conclusively demonstrated. Substrate channeling between enzymes requires that the enzyme active sites are both in close proximity and oriented toward each other [7], factors which have not yet been incorporated into VLP nanoreactor systems.

Several multi-step reaction pathways have been encapsulated within VLPs. The cleavage of lactose into galactose and glucose, catalyzed by the β-galactosidase CelB, and subsequent phosphorylation of the glucose product by glucokinase (GLUK) is one example (Fig. 6.2a) [74]. Through fusion of one or both enzymes to the scaffolding protein (SP) of P22, VLPs with co-encapsulated CelB and GLUK (CelB-GLUK-P22) or separate VLPs each with only one of these enzymes encapsulated (CelB-P22 and GLUK-P22) can be synthesized. Enzyme kinetics experiments show no difference in production of glucose-6-phosphate between the CelB-GLUK-P22 and a mixture of CelB-P22 and GluK-P22, indicating that there is no enhancement of the rate of the reaction pathway conferred purely by colocalization of the two catalysts. A model of the system suggests that for catalytic enhancement to occur, the K_M should be much higher and the V_{max} much lower for the first enzyme in the pathway than for the second [7]. As diffusion is a rapid process, escape of the product of the initial reaction from the capsid is therefore highly likely. The escape of intermediate species has been confirmed in studies of a bifunctional glutathione biosynthesis enzyme encapsulated within P22 VLPs [95] and a reaction pathway encapsulated within CCMV VLPs [9].

Intermediate molecules escape from the interior of VLPs to the surrounding environment. Individual VLPs are surrounded by solvent molecules, with low concentrations of enzyme near each VLP. As a result, molecules diffusing out of the VLP immediately enter the bulk solution. In VLP systems assembled into higher-order assemblies the distance between VLPs is much shorter (in P22 PMFs, about 8 nm between particles), and molecules diffusing out of a VLP are more likely to encounter a neighboring VLP. The packing of VLPs in a 3D assembly generates a region of high enzyme concentration which occupies a volume equivalent to hundreds or thousands of VLPs. The release of a substrate into an area of high enzyme concentration could potentially lead to catalytic enhancement due to the increased likelihood of collision with an enzyme.

Higher-order assemblies of VLPs provide regions of high enzyme concentration by constraining many copies of an enzyme within one material. In studies of enzyme activity, materials co-assembled from individual P22 VLPs with separately encapsulated ketoisovalerate dehydrogenase (KivD-P22) and alcohol dehydrogenase A (AdhA-P22) show the same kinetic parameters as the enzyme encapsulated P22 particles free in solution (Figs. 6.1e and 6.2b) [91]. As a result, even the formation of regions of high catalyst concentration through concentration of the enzymes into assembled materials does not induce substrate channeling and acceleration of the reaction rate.

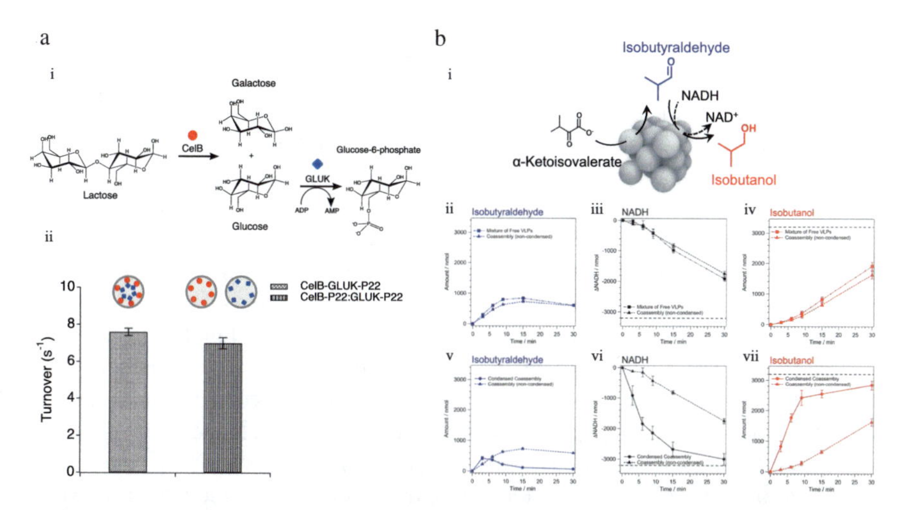

Fig. 6.2 Small molecule diffusion between reaction components. **a** The use of P22 VLPs to create a synthetic metabolon by encapsulating a fusion of two enzymes involved in lactose metabolism. (i) Reaction scheme: CelB enzyme catalyzes the hydrolysis of lactose, generating galactose and glucose. Glucose acts as a substrate for the second enzyme, GLUK, which phosphorylates glucose into glucose-6-phosphate (G6P). (ii) Activity evaluation of P22 capsids co-encapsulated with CelB and GLUK (CelB-GLUK-P22) for the conversion of lactose to G6P in comparison to singly encapsulated capsids (CelB-P22 and GLUK-P22 at a 1:1 ratio between CelB and GLUK). Observed overall turnover rates for the co-encapsulated and singly encapsulated enzyme systems suggests that there was no channeling advantage in the co-encapsulated system. Figure adapted with permission from [74]. Copyright 2014 American Chemical Society. **b** A catalytic superlattice co-assembled from two populations of P22 VLPs separately encapsulated with KivD and AdhA enzymes catalyzing sequential reactions. (i) KivD catalyzes the conversion of α-Ketoisovalerate to isobutyraldehyde, and AdhA catalyzes the reduction of isobutyraldehyde to isobutanol using NADH. Reaction progress in terms of isobutyraldehyde and isobutanol production (measured by gas chromatography-mass spectrometry) and NADH consumption (measured by UV–Vis spectroscopy) with a mixture of free nanoreactors, co-assembled nanoreactor superlattices (ii–iv), and condensed superlattices (v–vii). Figure adapted with permission from [91]. Copyright 2018 American Chemical Society

Catalytic Advantages from Protein Framework Materials

While the assembly of VLPs with encapsulated enzymes does not confer direct enhancement of enzymatic catalytic activity, the properties of the assembled material can be used to enhance overall catalytic activity. The interior volume of VLPs, and thus the local concentration and solvation of cargo enzymes within the VLPs, does not change when the heterogeneous VLP phase is dispersed or condensed. Also, single VLPs or soluble proteins require large amounts of solvation, limiting the practical amount that can be dispersed in solution. As a result, assembled VLPs can be concentrated into a much smaller volume than free VLPs (or unencapsulated enzyme) [63, 91]. The condensed material allows for usage of the same amounts of catalyst and substrate in much lower volumes, allowing for acceleration of the reaction rate through high concentration (Fig. 6.2b). This condensed form additionally avoids problems with enzyme solubility and recoverability common in enzyme catalysis and is an example of an indirect way of increasing catalytic efficiency upon incorporation into a higher-order material.

The organization of VLPs into a 3D structure introduces an interstitial space entered by substrates as they transit through the material. While macromolecular concentration inside VLPs and in the assembly as a whole is very high, the interstitial space between VLPs is less occupied. This space has been exploited in assemblies prepared from lumazine synthase, a non-viral protein cage derived from *Aquifex aeolicus* (AaLS) [14]. AaLS cages can be assembled into clusters through the surface presentation of proteins which dimerize in presence of a small molecule. The cage surface can be used to also present an active enzyme (β-lactamase), which is incorporated into the space between assembled cages. The clusters can be grown in layers deposited on a surface with high or low loading of β-lactamase into the space between cages. Interestingly, the reaction rate per enzyme is higher when the enzyme loading is low. In the high loading condition, the authors suggest that the interstitial space is blocked by the enzymes, inhibiting access to enzymes deep in the material. This result shows the potential of not only assembling VLPs into a material, but also of using the material to gain access to another way of modifying collective VLP behavior.

While diffusion itself is largely uninhibited within VLP-based materials, the emergent properties of these materials allow for the restriction of molecules through other means. For example, the co-localization of many VLPs causes the accumulation of negative charge, leading to the transit of molecules such as super positively charged GFP into but not out of the material [2]. Molecules exhibiting this behavior can be said to "partition" into the material. This partitioning is the subject of intense interest in the field.

Partitioning

The effective localization of molecules requires forces that can overcome molecular diffusion, which tends to drive the system towards a mixed homogeneous state. Local concentration of a species can be maintained and differentiated from the bulk medium by concentrating the species but also by excluding certain species. Molecular partitioning is thus a mechanism for discrimination that relies on selective exclusion (depletion) or the localization (concentration) of certain molecules.

Control over the movement and localization of molecules within porous materials structures is a key component to their functionality. Thus membranes, with pores of defined size, can selectively allow the passage of some molecules across the membrane while blocking the passage of others. This is the basis of a dialysis membrane that allows the passage to small molecules while selectively blocking molecules larger than the pore size. Porous gel filtration chromatography resins allow small molecules to access pores and voids within the material thus providing a more tortuous and longer path through the material than molecules that are too large to enter the pores; this difference in pathlength provides the basis for an effective separation based on size. Porous materials, both natural and synthetically derived, have found application in petroleum cracking [31], gas storage [60], and separations [19, 93].

Direct intermolecular interactions play a key role in the process of partitioning but other factors including shape, size and charge also play extremely important roles in effective molecular discrimination resulting in spatial segregation of molecules. While intermolecular interactions have a large impact on the diffusion of small molecules, these effects are magnified when we consider the interactions of small molecules with extended 3D materials. Individual intermolecular interactions based on electrostatics become significantly enhanced in the interaction of charged molecules with charged surfaces or when confined between charged surfaces where they experience strong electrostatic fields. The behavior of ions in the vicinity of charged surfaces can be described through the double layer theory [33]; where an accumulation of ions at the surface, due to electrostatic interactions, falls off exponentially over a few nm from the surface (Debye length) until bulk concentration of the ions is achieved. The electrostatic interaction can also be effectively screened through increased ionic strength effectively decreasing the Debye length and minimizing the concentration difference of ions at the surface layer and the bulk.

Molecular Partitioning in Individual VLPs

The localization of species within individual VLPs has been of interest for a few decades. Early efforts for the selective entrapment of cargo species within a VLP included the selective nucleation and growth of low solubility nanoparticle crystals within a preformed empty CCMV VLP architecture. Similarly, polymer entrapment using complementary electrostatic interactions was also demonstrated using either a preformed empty capsid or by a self-assembly mechanism resulting in polymer entrapped within the VLP [20, 34].

Electrostatic interactions between the coat protein and a scaffold protein or SP-cargo fusions during VLP self-assembly of bacteriophage P22 also drives encapsulation or a wide range of gene products [35, 69, 72, 73]. This electrostatic based encapsulation has been shown effective in the packaging or partitioning of many cargo species within a wide range of VLP systems. Other approaches have relied on stronger interactions between the capsid and the cargo. Thus, polymers have been grown selectively on the interior of the P22 capsid through covalent attachment of an ATRP polymer initiation site [59]. Small molecules are routinely attached via covalent interactions to the interior of VLP capsids effectively localizing them as cargo on the VLP interior [3]. Also, attachment of metal binding sites to the interior has allowed the controlled growth of metal coordination polymers within the P22 capsid [92]. All these strategies effectively create a 'ship in a bottle' where the cargo is held within the assembled capsid, either through strong interactions with the capsid or because the pores in the assembled capsid are too small to allow passage, and the cargo therefore cannot equilibrate with the bulk solution.

Partitioning Within Hierarchically Assembled 3D VLP-Based Materials

Spatial partitioning is important in biology, which can be seen reflected in the range of subcellular compartments, often with highly specialized function, within an otherwise chaotic cellular environment. VLPs have emerged as useful biomimetic materials for the synthetic recapitulation of confined, specialized, nanoscale compartments. VLPs which sequester enzyme catalysts, have also importantly been used as building blocks in the construction of highly selective 3D framework materials [82]. The VLP-based 3D materials are a separate phase where the interstitial spaces between the VLP particles are filled with surrounding solution and provide molecular pathways for the diffusion of small molecules into and out of the framework.

Hierarchically assembled VLP-based materials can be formed through surface modification of individual VLPs to create individuals with complementary interactions. Thus, early work on attachment of complementary ssDNA to the surface of CPMV resulted in the mass assembly of particles through complementary dsDNA formation [87]. Similarly, decoration of the exterior surface of P22 with a spider silk protein that dimerized upon changes in pH resulted in reversible mass particle assembly upon acidification (Fig. 6.3a) [2]. The resulting VLP-based material was highly negatively charged and when incubated with a superpositive GFP showed a significant accumulation (partitioning) of the GFP within the P22-Ss array. In contrast, there was no discernible interaction between individual P22-Ss and the superpositive GFP in solution; the collective behavior was different than individual behavior and could potentially be used to engineer new functionality into these materials.

VLPs can be formed into ordered crystalline arrays based on charge complementary templating with a charged macromolecule/nanoparticle and are formed and held together through electrostatic interactions. At high ionic strength these interactions can be screened/disrupted and the arrays disassemble [10, 88, 91]. However, when the initially formed arrays are locked in place, as demonstrated with the addition of ditopic Dec-Dec molecules, the arrays are stable to high ionic strength and the templating macromolecule can be removed to leave the stable PMFs [63]. These ordered PMF arrays carry a high (net negative) surface charge making them ideal candidates for controlled molecular partitioning.

Using molecular partitioning within PMFs a significant enhancement of catalytic properties could be demonstrated which were a result of the collective properties of VLPs within the material and were quite different from the behavior of individual VLPs free in solution. Due to the relatively large inter-particle distances (i.e. porosity)

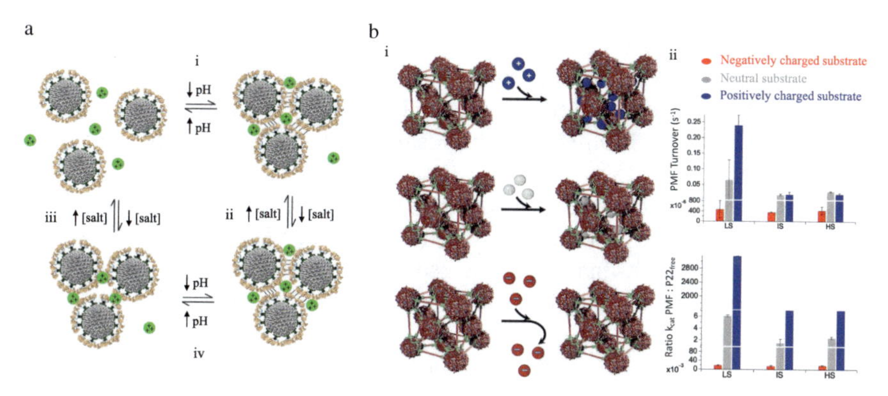

Fig. 6.3 Partitioning in VLP-derived higher-order 3D assemblies. **a** (i) P22 VLPs decorated with Dec-Spider silk protein fusions assemble into higher-order structures mediated by pH. (ii) Lowering the salt concentration results in the interaction between superpositive GFP and the particles, thereby allowing the Dec-Ss and GFP mediated 3D assembly. (iii) Alternatively, when the salt concentration is decreased first, superpositive GFP guides the electrostatically driven 3D assembly of particles, followed by (iv) the Ss protein mediated interparticle interactions in response to pH. Supercharged GFP shows no significant interaction with the individual particles at high salt conditions, yet they are colocalized into the Ss protein mediated 3D assemblies at the same ionic strength, possibly due to the increased local negative charge in the assembly compared to the free particles. Figure adapted with permission from [2]. Copyright 2018 American Chemical Society. **b** Enhancement of catalytic activity due to charged substrate partitioning into PMF derived from P22 VLP nanoreactors. (i) Positive, neutral, and negatively charged substrates interact differently with the highly negatively charged PMF. (ii) The catalytic turnover is highest with the positively charged substrates as they accumulate in the PFM. In contrast, the negatively charged substrates are excluded from the PMF, resulting in a negligible activity. This electrostatically driven partitioning of substrates, thus the observed catalytic activities are dependent on the ionic strength of the medium; it is more pronounced under low salt (LS) conditions and less significant under high salt (HS) and intermediate salt (IS) conditions. Figure adapted with permission from [82]. Copyright 2021 American Chemical Society

and high charge density of the porous framework, multiply-charged cationic macromolecules were selectively partitioned into the negatively charged PMF material. In contrast, negatively charged macromolecules were excluded from the PMF lattice [82]. This behavior is different from individual VLPs, which do not possess high enough charge density to accumulate or exclude these charged macromolecules from their vicinity. This collective effect is similar to that described above for the accumulation of supercharged GFP^{++} within a P22 assembly but no discernible interaction with the individual P22s.

Using the porosity and high charge density of the framework, assembled from active enzyme encapsulated P22 building blocks, charged substrates could be partitioned within the PMF at concentrations that far exceeded the bulk concentration (Fig. 6.3b). When these charged molecules were small enough to traverse the pore of individual P22 building blocks, and could access the encapsulated enzyme, this resulted in a significant enhancement of overall catalytic activity. The effect was even more pronounced when comparing the positively charged and negatively charged substrates because while the positively charge molecules were partitioned into the

PMF and subsequently turned over, the negatively charged substrates were effectively excluded from the PMF and could never access the encapsulated enzyme. Thus, the molecular discrimination based on electrostatic partitioning was further enhanced by the selective catalytic turnover of the positively charged small-molecule enzyme substrate [81, 82].

The PMF versatility lies in the ability to incorporate functionality through the encapsulation of selective enzymes within individual P22 VLPs, while their assembly into 3D materials results in an interstitial space, porosity, and surface charge in the PMF that allows diffusion and provides selective partitioning based on molecular size and charge.

Conclusions

The use of VLP nanocontainers to design and develop bioinspired materials over a wide range of length scales is an emerging area of research. The vast diversity and synthetic potential of VLPs have made them useful platforms in many fields, including catalysis, immunology, and medicine. They are naturally occurring hierarchical structures, self-assembled from a limited number of macromolecular building blocks. VLPs serve as a modular unit to build higher-order 3D VLP assemblies via various techniques where the packing order and the spatial position of these particles in the 3D material can be controlled. VLPs can be repurposed as catalytically active nanoreactors, and this allows for the design and construction of a wide range of functional materials. Hierarchically complex structures assembled with VLPs exhibit emergent properties, including porosity and surface charge effects, that are different from individual particles. These properties have the potential to alter the localization, diffusion rates and paths of molecules, which can be exploited to alter chemical processes and design advanced functional biomaterials.

Perspectives

As small molecule diffusion is a rapid process, and controlling their movement and localization benefits functionality. 3D assembly of VLPs gives rise to hierarchically organized structures in which the order and the spatial arrangement of capsids can be controlled. Surface charge and porosity are two emergent properties that appear in the VLP-based hierarchical systems that provide means to tune chemical transformations by imposing forces on molecular diffusion and partitioning. Even though there are a limited number of studies, these properties can be explored strategically to develop novel biomaterials with advanced features such as tunable catalytic efficiency, selectivity, and heterogeneous catalysis.

VLPs are building blocks for modular material assembly across multiple length scales. These building blocks in the 3D material are held together by tunable interactions between the system components, thus allows for orchestrating controlled assembly and disassembly of the particles. We envision that this concept can be extended to replace parts of the 3D material with different types of VLP building blocks embedded with different functionalities, by design, which would reshape the VLP-based biomaterial designs.

References

1. Ashley CE, Dunphy DR, Jiang Z, Carnes EC, Yuan Z, Petsev DN, Atanassov PB, Velev OD, Sprung M, Wang J (2011) Convective assembly of 2D lattices of virus-like particles visualized by in-situ grazing-incidence small-angle X-ray scattering. Small 7:1043–1050
2. Aumiller Jr WM, Uchida M, Biner DW, Miettinen HM, Lee B, Douglas T (2018) Stimuli responsive hierarchical assembly of p22 virus-like particles. Chem Mater 30:2262–2273
3. Aumiller WM, Uchida M, Douglas T (2018) Protein cage assembly across multiple length scales. Chem Soc Rev 47:3433–3469
4. Bale JB, Gonen S, Liu Y, Sheffler W, Ellis D, Thomas C, Cascio D, Yeates TO, Gonen T, King NP (2016) Accurate design of megadalton-scale two-component icosahedral protein complexes. Science 353:389–394
5. Bar-Even A, Noor E, Savir Y, Liebermeister W, Davidi D, Tawfik DS, Milo R (2011) The moderately efficient enzyme: evolutionary and physicochemical trends shaping enzyme parameters. Biochemistry 50:4402–4410
6. Barthelat F (2007) Biomimetics for next generation materials. Philos Trans R Soc A: Math Phys Eng Sci 365:2907–2919
7. Bauler P, Huber G, Leyh T, McCammon JA (2010) Channeling by proximity: the catalytic advantages of active site colocalization using Brownian dynamics. J Phys Chem Lett 1:1332–1335
8. Bhaskar S, Lim S (2017) Engineering protein nanocages as carriers for biomedical applications. NPG Asia Mater 9:e371–e371
9. Brasch M, Putri RM, de Ruiter MV, Luque D, Koay MS, Castón JR, Cornelissen JJ (2017) Assembling enzymatic cascade pathways inside virus-based nanocages using dual-tasking nucleic acid tags. J Am Chem Soc 139:1512–1519
10. Brunk NE, Uchida M, Lee B, Fukuto M, Yang L, Douglas T, Jadhao V (2019) Linker-mediated assembly of virus-like particles into ordered arrays via electrostatic control. ACS Appl Bio Mater 2:2192–2201
11. Chatterji A, Ochoa W, Shamieh L, Salakian SP, Wong SM, Clinton G, Ghosh P, Lin T, Johnson JE (2004) Chemical conjugation of heterologous proteins on the surface of cowpea mosaic virus. Bioconjug Chem 15:807–813
12. Chen C, Daniel M-C, Quinkert ZT, De M, Stein B, Bowman VD, Chipman PR, Rotello VM, Kao CC, Dragnea B (2006) Nanoparticle-templated assembly of viral protein cages. Nano Lett 6:611–615
13. Chen Y, Ke G, Ma Y, Zhu Z, Liu M, Liu Y, Yan H, Yang CJ (2018) A synthetic light-driven substrate channeling system for precise regulation of enzyme cascade activity based on DNA origami. J Am Chem Soc 140:8990–8996
14. Choi H, Choi B, Kim GJ, Kim HU, Kim H, Jung HS, Kang S (2018) Fabrication of nanoreaction clusters with dual-functionalized protein cage nanobuilding blocks. Small 14:1801488
15. Cigler P, Lytton-Jean AK, Anderson DG, Finn M, Park SY (2010) DNA-controlled assembly of a NaTl lattice structure from gold nanoparticles and protein nanoparticles. Nat Mater 9:918–922

16. Comellas-Aragonès M, Engelkamp H, Claessen VI, Sommerdijk NA, Rowan AE, Christianen P, Maan JC, Verduin BJ, Cornelissen JJ, Nolte RJ (2007) A virus-based single-enzyme nanoreactor. Nat Nanotechnol 2:635–639
17. Crawford D (2002) The invisible enemy: a natural history of viruses. OUP Oxford
18. de la Rica R, Matsui H (2010) Applications of peptide and protein-based materials in bionanotechnology. Chem Soc Rev 39:3499–3509
19. Denny MS, Moreton JC, Benz L, Cohen SM (2016) Metal–organic frameworks for membrane-based separations. Nat Rev Mater 1:1–17
20. Douglas T, Young M (1998) Host–guest encapsulation of materials by assembled virus protein cages. Nature 393:152–155
21. Douglas T, Young M (2006) Viruses: making friends with old foes. Science 312:873–875
22. Fiedler JD, Brown SD, Lau JL, Finn M (2010) RNA-directed packaging of enzymes within virus-like particles. Angew Chem 122:9842–9845
23. Fields BN (2007) Fields' virology. Lippincott Williams & Wilkins
24. Fu J, Yang YR, Johnson-Buck A, Liu M, Liu Y, Walter NG, Woodbury NW, Yan H (2014) Multi-enzyme complexes on DNA scaffolds capable of substrate channelling with an artificial swinging arm. Nat Nanotechnol 9:531–536
25. Fukuto M, Nguyen QL, Vasilyev O, Mank N, Washington-Hughes CL, Kuzmenko I, Checco A, Mao Y, Wang Q, Yang L (2013) Crystallization, structural diversity and anisotropy effects in 2D arrays of icosahedral viruses. Soft Matter 9:9633–9642
26. Giessen TW, Silver PA (2016) A catalytic nanoreactor based on in vivo encapsulation of multiple enzymes in an engineered protein nanocompartment. ChemBioChem 17:1931–1935
27. Glasgow JE, Capehart SL, Francis MB, Tullman-Ercek D (2012) Osmolyte-mediated encapsulation of proteins inside MS2 viral capsids. ACS Nano 6:8658–8664
28. Goodall CP, Schwarz B, Selivanovitch E, Avera J, Wang J, Miettinen H, Douglas T (2021) Controlled modular multivalent presentation of the CD40 ligand on P22 virus-like particles leads to tunable amplification of CD40 signaling. ACS Appl Bio Mater 4:8205–8214
29. Heddle JG, Chakraborti S, Iwasaki K (2017) Natural and artificial protein cages: design, structure and therapeutic applications. Curr Opin Struct Biol 43:148–155
30. Hershey AD, Chase M (2017) Independent functions of viral protein and nucleic acid in growth of bacteriophage. In: Nickelsen K (ed) Die Entdeckung der Doppelhelix: Die grundlegenden Arbeiten von Watson, Crick und anderen. Springer, Berlin
31. Htay MM, Oo MM (2008) Preparation of Zeolite Y catalyst for petroleum cracking. World Acad Sci Eng Technol 48:114–120
32. Huang X, Holden HM, Raushel FM (2001) Channeling of substrates and intermediates in enzyme-catalyzed reactions. Annu Rev Biochem 70:149–180
33. Jing Y, Jadhao V, Zwanikken JW, Olvera De La Cruz M (2015) Ionic structure in liquids confined by dielectric interfaces. J Chem Phys 143:194508
34. Jolley C, Klem M, Harrington R, Parise J, Douglas T (2011) Structure and photoelectrochemistry of a virus capsid–TiO_2 nanocomposite. Nanoscale 3:1004–1007
35. Jordan PC, Patterson DP, Saboda KN, Edwards EJ, Miettinen HM, Basu G, Thielges MC, Douglas T (2016) Self-assembling biomolecular catalysts for hydrogen production. Nat Chem 8:179–185
36. Kang S, Douglas T (2010) Some enzymes just need a space of their own. Science 327:42–43
37. Kewalramani S, Wang S, Lin Y, Nguyen HG, Wang Q, Fukuto M, Yang L (2011) Systematic approach to electrostatically induced 2D crystallization of nanoparticles at liquid interfaces. Soft Matter 7:939–945
38. King NP, Bale JB, Sheffler W, McNamara DE, Gonen S, Gonen T, Yeates TO, Baker D (2014) Accurate design of co-assembling multi-component protein nanomaterials. Nature 510:103–108
39. King NP, Sheffler W, Sawaya MR, Vollmar BS, Sumida JP, André I, Gonen T, Yeates TO, Baker D (2012) Computational design of self-assembling protein nanomaterials with atomic level accuracy. Science 336:1171–1174

40. Korpi A, Anaya-Plaza E, Välimäki S, Kostiainen M (2020) Highly ordered protein cage assemblies: a toolkit for new materials. Wiley Interdisc Rev Nanomed Nanobiotechnol 12:e1578
41. Korpi A, Ma C, Liu K, Nonappa Herrmann A, Ikkala O, Kostiainen MA (2018) Self-assembly of electrostatic cocrystals from supercharged fusion peptides and protein cages. ACS Macro Lett 7:318–323
42. Kostiainen MA, Hiekkataipale P, Jose Á, Nolte RJ, Cornelissen JJ (2011) Electrostatic self-assembly of virus–polymer complexes. J Mater Chem 21:2112–2117
43. Kostiainen MA, Hiekkataipale P, Laiho A, Lemieux V, Seitsonen J, Ruokolainen J, Ceci P (2013) Electrostatic assembly of binary nanoparticle superlattices using protein cages. Nat Nanotechnol 8:52
44. Kostiainen MA, Kasyutich O, Cornelissen JJ, Nolte RJ (2010) Self-assembly and optically triggered disassembly of hierarchical dendron–virus complexes. Nat Chem 2:394–399
45. Kovacs EW, Hooker JM, Romanini DW, Holder PG, Berry KE, Francis MB (2007) Dual-surface-modified bacteriophage MS2 as an ideal scaffold for a viral capsid-based drug delivery system. Bioconjug Chem 18:1140–1147
46. Kraj P, Selivanovitch E, Lee B, Douglas T (2021) Polymer coatings on virus-like particle nanoreactors at low ionic strength—charge reversal and substrate access. Biomacromol 22:2107–2118
47. Kristensen DM, Mushegian AR, Dolja VV, Koonin EV (2010) New dimensions of the virus world discovered through metagenomics. Trends Microbiol 18:11–19
48. Kumar M, Markiewicz-Mizera J, Olmos JDJ, Wilk P, Grudnik P, Biela AP, Jemioła-Rzemińska M, Górecki A, Chakraborti S, Heddle JG (2021) A single residue can modulate nanocage assembly in salt dependent ferritin. Nanoscale 13:11932–11942
49. Lee JH, Fan B, Samdin TD, Monteiro DA, Desai MS, Scheideler O, Jin H-E, Kim S, Lee S-W (2017) Phage-based structural color sensors and their pattern recognition sensing system. ACS Nano 11:3632–3641
50. Lee S-K, Yun DS, Belcher AM (2006) Cobalt ion mediated self-assembly of genetically engineered bacteriophage for biomimetic Co−Pt hybrid material. Biomacromol 7:14–17
51. Li T, Winans RE, Lee B (2011) Superlattice of rodlike virus particles formed in aqueous solution through like-charge attraction. Langmuir 27:10929–10937
52. Li T, Zan X, Sun Y, Zuo X, Li X, Senesi A, Winans RE, Wang Q, Lee B (2013) Self-assembly of rodlike virus to superlattices. Langmuir 29:12777–12784
53. Li T, Zan X, Winans RE, Wang Q, Lee B (2013) Biomolecular assembly of thermoresponsive superlattices of the tobacco mosaic virus with large tunable interparticle distances. Angew Chem Int Ed 52:6638–6642
54. Libonati F, Buehler MJ (2017) Advanced structural materials by bioinspiration. Adv Eng Mater 19:1600787
55. Liljeström V (2017) Cooperative colloidal self-assembly of metal-protein superlattice wires. Nat Commun 8:671
56. Liljeström V, Mikkilä J, Kostiainen MA (2014) Self-assembly and modular functionalization of three-dimensional crystals from oppositely charged proteins. Nat Commun 5:1–9
57. Liu K, Chen D, Marcozzi A, Zheng L, Su J, Pesce D, Zajaczkowski W, Kolbe A, Pisula W, Müllen K (2014) Thermotropic liquid crystals from biomacromolecules. Proc Natl Acad Sci 111:18596–18600
58. Liu S, Zan T, Chen S, Pei X, Li H, Zhang Z (2015) Thermoresponsive chiral to nonchiral ordering transformation in the Nematic liquid-crystal phase of rodlike viruses: Turning the survival strategy of a virus into valuable material properties. Langmuir 31:6995–7005
59. Lucon J, Qazi S, Uchida M, Bedwell GJ, Lafrance B, Prevelige PE, Douglas T (2012) Use of the interior cavity of the P22 capsid for site-specific initiation of atom-transfer radical polymerization with high-density cargo loading. Nat Chem 4:781–788
60. Ma S, Zhou H-C (2010) Gas storage in porous metal–organic frameworks for clean energy applications. Chem Commun 46:44–53

61. Ma Y, Nolte RJ, Cornelissen JJ (2012) Virus-based nanocarriers for drug delivery. Adv Drug Deliv Rev 64:811–825
62. Majsterkiewicz K, Biela AP, Maity S, Sharma M, Piette BM, Kowalczyk A, Gaweł S, Chakraborti S, Roos WH, Heddle JG (2022) Artificial protein cage with unusual geometry and regularly embedded gold nanoparticles. Nano Lett 22:3187–3195
63. McCoy K, Uchida M, Lee B, Douglas T (2018) Templated assembly of a functional ordered protein macromolecular framework from P22 virus-like particles. ACS Nano 12:3541–3550
64. Mikkila J, Rosilo H, Nummelin S, Seitsonen J, Ruokolainen J, Kostiainen MA (2013) Janus-dendrimer-mediated formation of crystalline virus assemblies. ACS Macro Lett 2:720–724
65. Minten IJ, Claessen VI, Blank K, Rowan AE, Nolte RJ, Cornelissen JJ (2011) Catalytic capsids: the art of confinement. Chem Sci 2:358–362
66. Minten IJ, Hendriks LJ, Nolte RJ, Cornelissen JJ (2009) Controlled encapsulation of multiple proteins in virus capsids. J Am Chem Soc 131:17771–17773
67. Nedoluzhko A, Douglas T (2001) Ordered association of tobacco mosaic virus in the presence of divalent metal ions. J Inorg Biochem 84:233–240
68. Novoa RR, Calderita G, Arranz R, Fontana J, Granzow H, Risco C (2005) Virus factories: associations of cell organelles for viral replication and morphogenesis. Biol Cell 97:147–172
69. O'Neil A, Reichhardt C, Johnson B, Prevelige PE, Douglas T (2011) Genetically programmed in vivo packaging of protein cargo and its controlled release from bacteriophage P22. Angew Chem Int Ed 50:7425–7428
70. O'Neil A, Prevelige PE, Basu G, Douglas T (2012) Coconfinement of fluorescent proteins: spatially enforced communication of GFP and mCherry encapsulated within the P22 capsid. Biomacromol 13:3902–3907
71. Patterson DP, Lafrance B, Douglas T (2013) Rescuing recombinant proteins by sequestration into the P22 VLP. Chem Commun 49:10412–10414
72. Patterson DP, Prevelige PE, Douglas T (2012) Nanoreactors by programmed enzyme encapsulation inside the capsid of the bacteriophage P22. ACS Nano 6:5000–5009
73. Patterson DP, Schwarz B, El-Boubbou K, van der Oost J, Prevelige PE, Douglas T (2012) Virus-like particle nanoreactors: programmed encapsulation of the thermostable CelB glycosidase inside the P22 capsid. Soft Matter 8:10158–10166
74. Patterson DP, Schwarz B, Waters RS, Gedeon T, Douglas T (2014) Encapsulation of an enzyme cascade within the bacteriophage P22 virus-like particle. ACS Chem Biol 9:359–365
75. Reddy VR, Campbell EA, Wells J, Simpson J, Nazki S, Hawes PC, Broadbent AJ (2022) Birnaviridae virus factories show features of liquid-liquid phase separation and are distinct from paracrystalline arrays of virions observed by electron microscopy. J Virol 96:e02024-e2121
76. Restuccia A, Seroski DT, Kelley KL, O'Bryan CS, Kurian JJ, Knox KR, Farhadi SA, Angelini TE, Hudalla GA (2019) Hierarchical self-assembly and emergent function of densely glycosylated peptide nanofibers. Commun Chem 2:1–12
77. Roldão A, Silva A, Mellado M, Alves P, Carrondo M (2011) Viruses and virus-like particles in biotechnology: fundamentals and applications. Compr Biotechnol 625
78. Schenk AS, Eiben S, Goll M, Reith L, Kulak AN, Meldrum FC, Jeske H, Wege C, Ludwigs S (2017) Virus-directed formation of electrocatalytically active nanoparticle-based Co_3O_4 tubes. Nanoscale 9:6334–6345
79. Schwarz B, Uchida M, Douglas T (2017) Biomedical and catalytic opportunities of virus-like particles in nanotechnology. Adv Virus Res 97:1–60
80. Selivanovitch E, Douglas T (2019) Virus capsid assembly across different length scales inspire the development of virus-based biomaterials. Curr Opin Virol 36:38–46
81. Selivanovitch E, Lafrance B, Douglas T (2021) Molecular exclusion limits for diffusion across a porous capsid. Nat Commun 12:1–12
82. Selivanovitch E, Uchida M, Lee B, Douglas T (2021) Substrate partitioning into protein macromolecular frameworks for enhanced catalytic turnover. ACS Nano 15:15687–15699
83. Servid A, Jordan P, O'Neil A, Prevelige P, Douglas T (2013) Location of the bacteriophage P22 coat protein C-terminus provides opportunities for the design of capsid-based materials. Biomacromol 14:2989–2995

84. Sharma J, Uchida M, Miettinen HM, Douglas T (2017) Modular interior loading and exterior decoration of a virus-like particle. Nanoscale 9:10420–10430
85. Steinmetz NF, Lim S, Sainsbury F (2020) Protein cages and virus-like particles: from fundamental insight to biomimetic therapeutics. Biomater Sci 8:2771–2777
86. Steinmetz NF, Manchester M (2009) PEGylated viral nanoparticles for biomedicine: the impact of PEG chain length on VNP cell interactions in vitro and ex vivo. Biomacromol 10:784–792
87. Strable E, Johnson JE, Finn M (2004) Natural nanochemical building blocks: icosahedral virus particles organized by attached oligonucleotides. Nano Lett 4:1385–1389
88. Uchida M, Brunk NE, Hewagama ND, Lee B, Prevelige JR, Jadhao V, Douglas T (2022) Multilayered ordered protein arrays self-assembled from a mixed population of virus-like particles. ACS Nano
89. Uchida M, Klem MT, Allen M, Suci P, Flenniken M, Gillitzer E, Varpness Z, Liepold LO, Young M, Douglas T (2007) Biological containers: protein cages as multifunctional nanoplatforms. Adv Mater 19:1025–1042
90. Uchida M, Lafrance B, Broomell CC, Prevelige Jr PE, Douglas T (2015) Higher order assembly of virus-like particles (VLPs) mediated by multi-valent protein linkers. Small 11:1562–1570
91. Uchida M, McCoy K, Fukuto M, Yang L, Yoshimura H, Miettinen HM, Lafrance B, Patterson DP, Schwarz B, Karty JA (2018) Modular self-assembly of protein cage lattices for multistep catalysis. ACS Nano 12:942–953
92. Uchida M, Morris DS, Kang S, Jolley CC, Lucon J, Liepold LO, Lafrance B, Prevelige jr PE, Douglas T (2012) Site-directed coordination chemistry with P22 virus-like particles. Langmuir 28:1998–2006
93. van de Voorde B, Bueken B, Denayer J, de Vos D (2014) Adsorptive separation on metal–organic frameworks in the liquid phase. Chem Soc Rev 43:5766–5788
94. Wang Y, Douglas T (2022) Bioinspired approaches to self-assembly of virus-like particles: from molecules to materials. Acc Chem Res 2903–3310
95. Wang Y, Uchida M, Waghwani HK, Douglas T (2020) Synthetic virus-like particles for glutathione biosynthesis. ACS Synth Biol 9:3298–3310
96. Wen AM, Steinmetz NF (2016) Design of virus-based nanomaterials for medicine, biotechnology, and energy. Chem Soc Rev 45:4074–4126
97. Whitesides GM, Boncheva M (2002) Beyond molecules: self-assembly of mesoscopic and macroscopic components. Proc Natl Acad Sci 99:4769–4774
98. Yoshimura H, Edwards E, Uchida M, McCoy K, Roychoudhury R, Schwarz B, Patterson D, Douglas T (2016) Two-dimensional crystallization of P22 virus-like particles. J Phys Chem B 120:5938–5944

Chapter 7
Assembly of Coronaviruses and CoV-Like-Particles

Denisse Cadena-López, Maria Villalba-Nieto, Fernanda Campos-Melendez, Sergio Rosales-Mendoza, and Mauricio Comas-Garcia

Abstract Developing SARS-CoV-2 virus-like particles assembly (VLPs) systems is fundamental to understanding the biophysical principles of assembly and applying them in the biomedical and biotechnology fields. However, given that SARS-CoV-2 virions are enveloped, and the four structural proteins interact with each other at the viral membrane, the in vitro assembly of these VLPs from recombinant and purified proteins is extremely challenging. The current assembly systems are based on the co-transfection of cultured cells with multiple plasmids. One problem with this approach is that not all cells produce all the structural proteins. Furthermore, the yield of assembly and the morphology of the VLPs seems to depend on the type of cells used. Also, the fusion of tag proteins with structural proteins can affect the biological activity of the VLPs. Despite these problems, the production of Coronavirus VLPs has been crucial to understanding the minimal requirements for assembly and RNA packaging. Nonetheless, further studies are required to optimize these systems to increase the yield of assembly and the homogeneity of the VLPs. This optimization would allow an understanding of the specific interaction that gives rise to the assembly and selective packaging of the genomic RNA.

Keywords Virus-like-particle assembly · Coronaviruses · SARS-CoV-2 · Structural virology · Physical virology

---

Denisse Cadena-López, and Maria Villalba-Nieto—These authors contributed equally.

D. Cadena-López · M. Villalba-Nieto · F. Campos-Melendez · M. Comas-Garcia (✉)
High-Resolution Microscopy Section, Research Center for Health Sciences and Biomedicine, Autonomous University of San Luis Potosi, San Luis, S.L.P, México
e-mail: mauricio.comas@uaslp.mx

S. Rosales-Mendoza
Biotecnology Section, Research Center for Health Sciences and Biomedicine, Autonomous University of San Luis Potosi, San Luis, S.L.P, México

Department of Chemical Sciences, Autonomous University of San Luis Potosi, San Luis, S.L.P, México

M. Comas-Garcia
Department of Sciences, Autonomous University of San Luis Potosi, San Luis, S.L.P, México

M. Comas-Garcia and S. Rosales-Mendoza (eds.), *Physical Virology*, Springer Series in Biophysics 24, https://doi.org/10.1007/978-3-031-36815-8_7

Introduction

Coronaviruses (CoVs) are a group of positive-sense single-stranded RNA viruses with the largest genome known for an RNA virus. These viruses belong to the *Nidovirales* order, *Coronaviridae* family, and *Orthocoronavirinae* subfamily, which is divided into four genera (alpha, beta, gamma, and deltacoronavirus). The genera *Betacoronavirus* includes a series of mild to non-pathogenic (e.g., OC43 and HKU1) and highly pathogenic human viruses (e.g., SARS-CoV, MERS-CoV, and SARS-CoV-2), as well as animal viruses (e.g., Bovine coronavirus, mouse hepatitis virus, and several bat coronaviruses [55].

The study of the assembly of CoVs virus-like particles (VLPs) is fundamental to understanding the physical principles governing virion assembly and multiple aspects of the viral cycle. Furthermore, understanding these principles is critical to developing efficient systems for producing VLPs and making viable their exploitation in biomedical and/or biotechnological applications.

In general, CoVs virions contain at least four structural proteins: spike (S), envelope (E), membrane (M), and nucleocapsid (N). This is the case for most Betacoronaviruses (e.g., SARS-CoV-2); however, there are virions from other CoVs (e.g., Bovine coronavirus (BCV), infectious bronchitis virus (IBV), and murine hepatitis virus (MHV)) that can contain other proteins like hemagglutinin-esterases [28, 57, 63]. CoV assembly of CoVs occurs at the Golgi apparatus and the ER-Golgi intermediate compartment (ERGIC) [51], where the virion′s cell-derived membrane is acquired. It is important to point out that given that these virions are enveloped and that this lipid bilayer is key for the interaction between the structural proteins, the in vitro assembly of CoVs VLPs from reconstituted and purified components is extremely challenging. One of the reasons is that, except for the N protein, the other structural components are highly hydrophobic membrane proteins, whose purification is difficult [51]. Furthermore, the in vitro assembly of this type of VLPs using purified recombinant proteins would require using liposomes containing all the structural proteins in the appropriate proportions. Nonetheless, the study of the assembly process for CoVs VLPs can be performed by either understanding in detail the physicochemical properties of the structural proteins or following the assembly process in cultured cells. These are one of the few reasons why this chapter focuses on the efforts to assemble SARS-CoV and SARS-CoV-2 VLPs by transfecting cultured cells rather than on the in vitro assembly of VLPs with pure recombinant proteins.

Virion Structure and the Interactions Among Structural Proteins

In this section, we describe the virion content of a typical CoV and summarize some of the interactions among the structural proteins. Unlike most virions that package a (+)ssRNA genome, the viral particle of CoV does not assemble into symmetrical capsids (i.e., neither icosahedral nor helical). Instead, the CoV virion is pleomorphic and contains a variable number of viral proteins: on average, there are about 1000 trimers of the S protein; 2000, 20, and 1000 copies of the M, E, and E proteins, respectively; and one copy of the full-length genomic RNA [46].

The spike is a trimeric, transmembranal, and highly glycosylated protein responsible for virion entry and tropism [22, 58]. This protein interacts with the receptor (e.g., ACE-2) and co-receptors (e.g., TMRPSS2) and gives the "corona" shape characteristic of these viruses [25]. In infected cells, the S protein can be found in the plasma membrane, the Golgi apparatus, and the ERGIC [29, 30, 35]. This protein is highly immunogenic; thus, it has been used as the main target antigenic for vaccine development [6] and isolate therapeutic antibodies [1, 2, 65]. The S protein can be divided into several structural features: the receptor binding domain (RBD), the transmembranal and cytoplasmic domains, the S1 and S2 subunits, and the fusion peptide [12, 27]. These subunits and peptide are the product of posttranslational modifications by cellular proteases that can occur either during the secretory pathway of the S protein and/or the egress of the virion from the producer cell or during viral entry [21]. Figure 7.1 shows a Cryo-EM reconstruction of a SARS-CoV-2 virion. This data shows that the RBD can be either in an "Up" or "Down" conformation [60]. Furthermore, unlike many enveloped (+)ssRNA viruses, the number and distribution of the S protein at the surface of the virion are highly heterogenous [27, 60]. It also shows that within in trimeric spike, only on Receptor-binding domain (RBD) is in the up conformation and that not all timers have an RBD up conformation. This might reflect a weak interaction between the spike and membrane proteins.

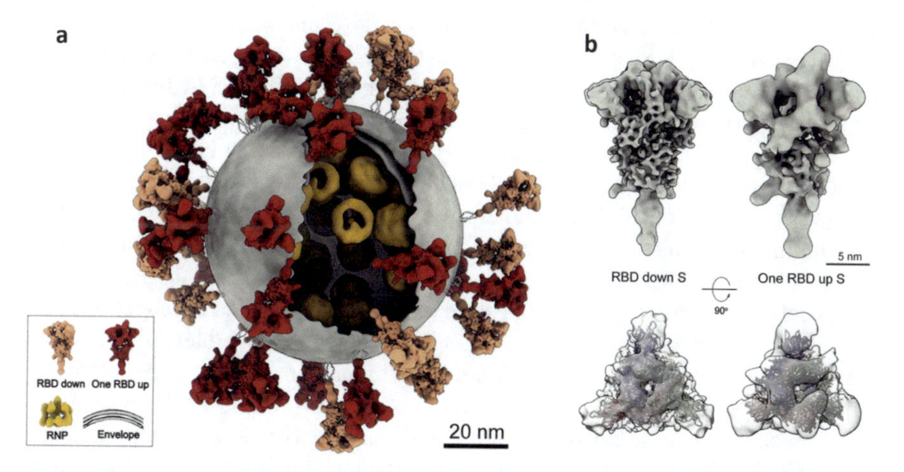

Fig. 7.1 **a** Cryo-EM reconstruction of a SARS-CoV-2 virion. The trimeric S protein that has the three Receptor-binding domains (RBD) is shown in pink, the trimers with one RBD in the up conformation are shown in red. This reconstruction shows that not all the trimeric spikes have an RBD in the up conformation. The ribonucleoprotein complexes are shown in yellow. These reconstruction shows that the spike proteins are randomly distributed. **b** Comparison of the trimeric spikes when the RBD is either in the down or up conformation. This figure was modified from [60]

The nucleocapsid (N) protein is key for genome packaging because it interacts with the genomic RNA (gRNA) and the M protein [38, 40]. Unlike most (+)ssRNA viruses, the CoVs RNA-binding protein (N) does not have a disordered and highly basic N-terminal domain that interacts with the genomic RNA [14, 26]. Furthermore, in most icosahedral (+)ssRNA viruses, the regions of the capsid protein responsible for RNA–protein and protein–protein interactions are located in different domains and interfaces; however, in CoVs the N protein can use the same domain for N-RNA and N–N interactions [14].

Figure 7.2 shows a schematic representation of the common features found in a Betacoronavirus N protein. The RNA binding region (N-RBD) is highly ordered and is located at the N-terminal domain (N-NTD), whereas the nucleocapsid protein dimerization domain is located at the C-terminus (N-CTD), and both domains are linked to each other by a highly disordered serine-rich region [19, 47]. This serine-rich region can also interact with nucleic acids. The interaction between N proteins and binding to the viral RNA is complex; it involves electrostatic and non-electrostatic interactions between the RNA and N-NTD, and N-CTD. For example, it has been shown that the C-CTD can also bind to nucleic acids in vitro with a similar affinity as the N-NTD [5, 31, 62]. The structural data suggest that there could be multiple RNA binding sites (i.e., RNA binding sites 1 and 2 (N-RBD1 and 2)) [4, 46]. The tyrosine 109 located at the N-NTD is key for RNA binding specificity [24], although the contribution of this amino acid to packaging selectivity during assembly is yet to be determined. The interactions that result in packaging and assembly are also complex because the N protein is highly dynamic. For example, the N-RBD has at least three conformations that coexist with each other [14]. This domain establishes attractive interactions with the N-NTD and repulsive interactions with the N-CTD [14]. Also, the N-NTD can form transient α-helices that contribute to both N–N and N-RNA interactions [14]. Altogether, this data indicates that RNA-N binding and multimerization of N on the gRNA is a complex process. As will be explained later, these interactions result in interesting physical phenomena that could contribute to a decrease in the overall size of the ribonucleic complex so that it can be packaged.

Molecular Dynamic (MD) simulations of SARS-CoV-2 have shown that RNA packaging may require multimerization of the N protein, which results in "condensation" of the viral RNA [33] (see Fig. 7.3). By condensation, we do not mean the "classical condensation" phenomena observed with DNA [44], but a reduction of the overall size of the RNA [10]). Also, it has been shown that the interaction between the RNA and N protein leads to a liquid–liquid phase separation (LLPS) [24, 46], which is responsible for this "condensation" phenomenon. The RNA-driven LLPS phenomenon is specific for the 5′ and 3′ ends and double-stranded regions of the SARS-CoV-2 gRNA. In particular, the first 500 nucleotides of the 5′-end are essential for LLPS. Roden and co-workers proposed that this condensation might not be related to a direct interaction between the N protein and the Packaging Signal (PS), but it might contribute to the decrease of the overall size of the RNA [46]. This observation is consistent with the MD from Liu and Zandi, who reported that condensation of the gRNA is essential for the budding of the nucleocapsid complex into the vesicles that contain the M protein [33]. A surprising result is that the N-RBD2 stabilizes high-oligomeric order states during this phase separation by selective binding to dsRNA regions [46]. In other words, the N-RBD2 was thought to mediate binding to the nucleic acid, but it also contributes to the assembly of the ribonucleoprotein complex.

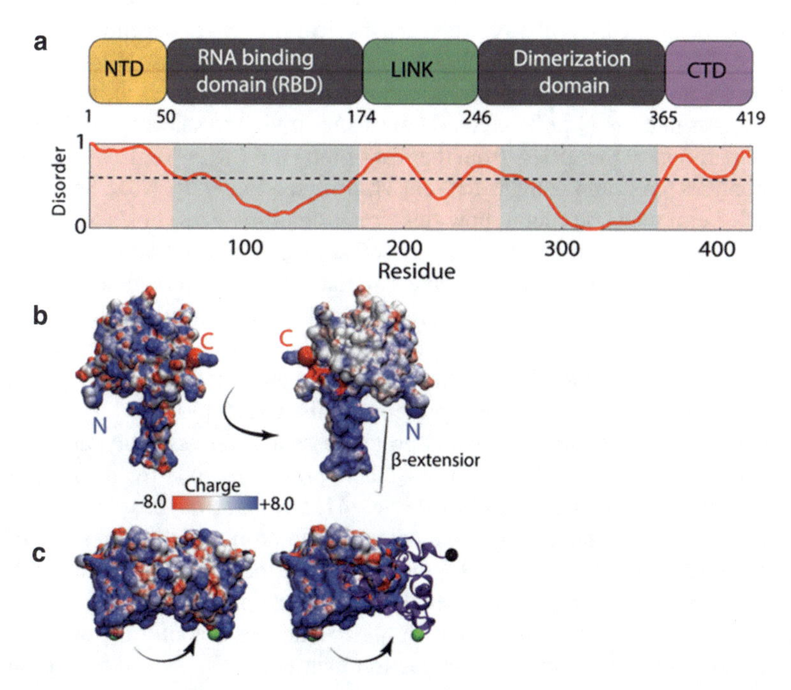

Fig. 7.2 **a** Architecture of the functional and structural domains of the SARS-CoV-2 N protein. The N-RBD and dimerization domains are well ordered structures, while the NTD, CTD, and serine-rich linker (LINK) are disordered. **b** Structure of the RNA binding domain (PDB: 6yi3) colored according to its electrostatic surface potential. **c** Structure of the N-dimerization domain (PDB: 6yun) colored according to its electrostatic surface potential. This figure was modified from Cubuk et al. [14]

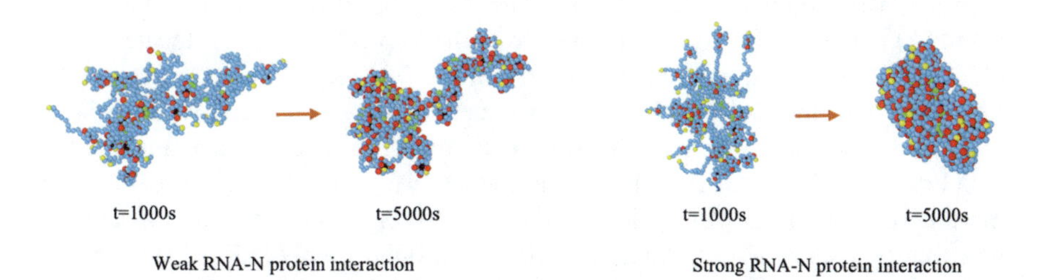

Fig. 7.3 Molecular dynamic simulation of a branched RNA interacting with the N protein. Under a regime of weak RNA-N protein interactions the overall shape and size of the viral barely changes. However, when RNA-N protein interactions are strong the degree of branching decreases and thus the RNA is condensed. This figure was modified from Li and Zandi [33]

On the one hand, as mentioned before, the selective interaction between N and RNA has been associated with Tyrosine 109. On the other hand, while the LLPS phenomena might not directly contribute to selective packaging, mutations at the tyrosine 109 result in ds-RNA containing-droplets with altered morphology and complete inhibition of condensation of a 5′-end containing RNA. Furthermore, Iserman and collaborators showed that the N droplets' physical properties depend on the RNA sequence [24]. This is because the oligomeric state of N within these N-RNA droplets depends on the RNA nature. Overall, the results from Roden et al. and Iserman et al. suggest that the N protein could contribute in vivo to selective packaging by phase-separation of the gRNA from the rest of the viral (i.e., (−)ssRNA and nested subgenomic RNAs) and cellular RNAs [24, 46] (see Fig. 7.4). Nonetheless, there are still several questions about how this protein multimerizes, binds to RNA, and contributes to viral assembly. Most importantly, it is not known how the ribonucleoprotein complexes and the N liquid droplets interact with the M protein so that they are selectively packaged during assembly. Interestingly, unlike most (+)ssRNA viruses, the assembly of SARS-CoV-2 VLPs in cultured cells can occur in the absence of the N protein and, thus, of RNA. Nonetheless, as explained here, assembly efficiency increases with the presence of the N protein and the gRNA [50].

The M protein is a transmembranal protein essential for assembly, and it is the most abundant protein in the virion [64]. This protein recruits (or interacts with) the other three structural proteins [16–18, 40, 41, 50] (see Fig. 7.5). In most CoVs, M interacts with the ribonucleoprotein complex and promotes the packaging of the genomic RNA; for example, in MHV, the strength of the interaction between the M protein and ribonucleoprotein complex depends on the presence of the PS [38–41]. This evidence has led to a model where the PS induces a conformational change in N that results in stronger M–N interaction than when N is bound to other RNA sequences.

Zhang et al. showed that the intravirion surface of the SARS-CoV-2M protein is highly basic and that mutations in this region result in impaired binding to the N protein [64]. On the one hand, the incorporation of the E protein into de virion is mediated by the interaction of the C-termini of the M and E proteins [11, 34]. On the other hand, the M-S interaction occurs between the amphipathic domain of M and the cytoplasmic domain of S [54]. For SARS-CoV, the specific interaction between S and M is mediated by the tyrosine 195 from the M protein, which is located at M-CTD [54]. Although it has been proposed that the ORF3a protein evolved from the M protein, the latter does not have the characteristic ion channel activity associated with the former [43]. The main hypothesis is that M changes the curvature of the membranes in the ERGIC to allow virion assembly [33]. Also, it is possible that multimerization of the M protein is one of the driving forces for virion assembly; however, by itself is not sufficient for efficient virion assembly [33].

The E protein is thought to be essential for particle infectivity. It regulates the localization and maturation of the S protein in the Golgi apparatus [3]; however, its role during assembly is not clear. Also, it has been proposed that this protein can be an ion channel [48, 49] and thus it might mediate a pH-induced virion maturation during the egress through the ERGIC, exosomes, and lysosomes. Nonetheless, as it

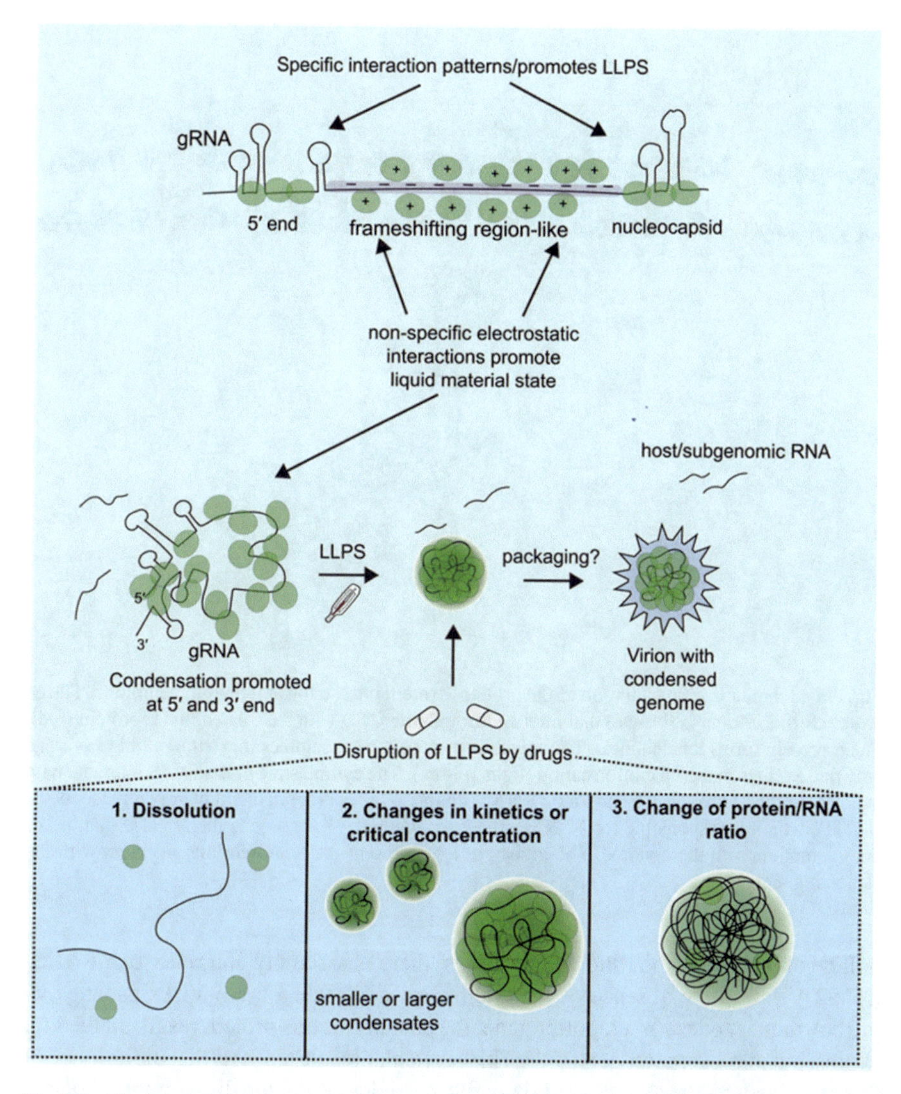

Fig. 7.4 The interaction between the RNA and the N protein results in a liquid–liquid phase transition. The interaction between the N protein and the RNA depends on specific sequence that are primarily located at the 5′ and 3′ ends and in double-stranded RNA regions. The selective LLPS of the genomic RNA results in the condensation of this molecule and it could segregate it from the subgenomic viral RNAs and host RNAs. The condensation of the genomic RNA could lead to selective packaging of the SARS-CoV-2 genome. This figure obtained from [24]

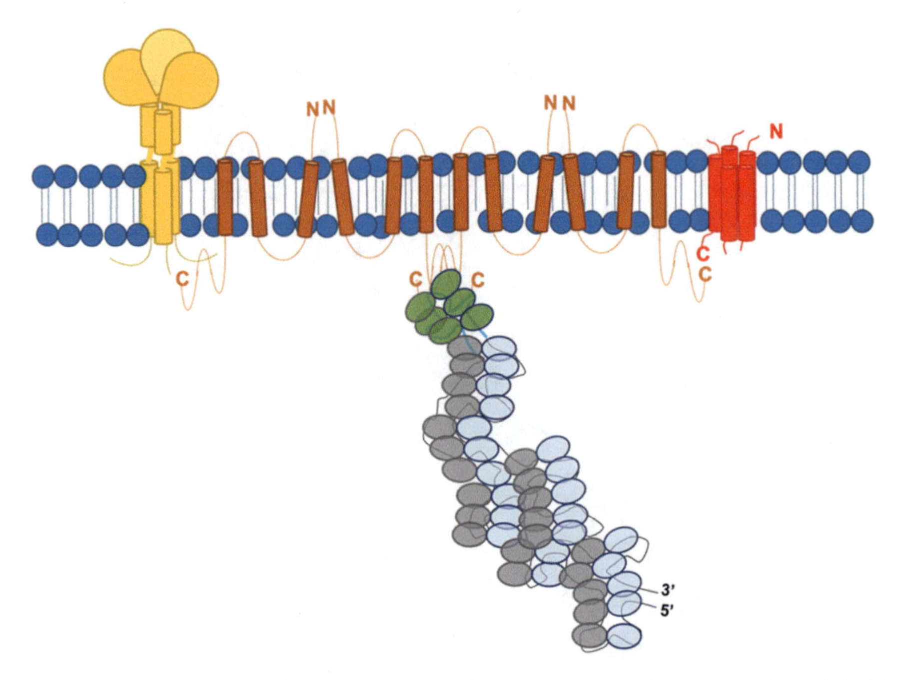

Fig. 7.5 Schematic representation of the protein–protein interaction at the viral membrane. The M protein forms tetramers (brown) that interact through the CTD with the CTD of the E protein (red). The E protein forms ion channels. The spike protein (yellow) assembles into trimers and in average only one RBD is in the up conformation (light yellow). The cytoplasmic domain of S interacts with the CTD of the M protein. The genomic RNA is bound to the N protein; the light gray represents the N-CTD, which is responsible for N–N interactions; while the light blue is the N-NTD that drives the interaction with the genome. The green circles represent the N protein that interacts with the packaging signal

will be explained in the following sections, there is assembly and release of SARS-CoV-2 VLPs in the absence of E protein. However, when these VLPs package an mRNA that encodes for a reporter gene, the absence of this protein results in the null expression of the reporter gene [53]. This particular result suggests that the absence of this protein alters the assembly of defective particles and/or the disassembly process.

The Genomic RNA and Its Role in Assembly

The genomic RNA (gRNA) is not required for VLP assembly; as it will be explained in the following section the assembly of CoV VLPs can proceed in the absence of any RNA by not co-expressing N along with S and M. In other words, the co-expression of the M and S proteins in cultured cells is sufficient for the assembly of VLPs. This characteristic makes CoVs unique among (+)ssRNA viruses; when assembled in vitro or cultured cells, the overwhelming majority of (+)ssRNA viruses require RNA [8]. Nonetheless, this requirement can be omitted by replacing the RNA-binding domain of the proteins that interact with the gRNA [13] or by in vitro assembly under conditions that screen the electrostatic repulsion of the N-terminal domain of the CP [32]. However, as will be explained in the following section, the assembly of "empty" VLPs can be achieved for CoV by omitting co-expressing the N protein.

In the case of SARS-CoV-2, Syed and collaborators showed that a sequence between nucleotides 20,080 and 22,222 (T20) is selectively packaged when compared to equal-length fragments of the gRNA [53]. Interestingly, when this sequence was further delimited to nucleotides 20,080 and 21,171 (PS9), the packaging efficiency increased by twofold with respect to T20 [53] (see Fig. 7.6). It is not clear why PS9 has a higher packaging efficiency than T20, but it is possible that the deletion of the last 1051 nucleotides stabilizes a structure that specifically interacts with the N protein.

On the one hand, the study from Syed and co-workers is consistent with findings on SARS-CoV, MERS-CoV, MHV, and Bovine coronavirus (BCV), where the PSs are also located at the 3′ end of the ORF1ab [7, 20, 23, 36, 41]. On the other hand, the PS from these four viruses has been delimited to shorter sequences (less than 200 nts) than PS9. This suggests that the minimal SARS-CoV-2 PS could be located somewhere within PS9. It is worth mentioning that unlike in the case of SARS-CoV-2, the determination of the MHV PS reported by Molenkamp and Spaan was performed using head-to-head competition experiments [36]. This approach has been shown to be extremely useful in other viruses, such as cowpea chlorotic mottle virus [9] and HIV-1 [10]; thus, it eliminates artifacts produced by non-specific interactions. Therefore, in the case of SARS-CoV-2, it is possible that if the packaging efficiencies of T20 or PS9 are compared against other segments by co-transfections, the selectivity of these two sequences could be much higher than the twofold difference that was originally estimated. However, carrying out head-to-head competition in cultured cells is not trivial; one needs to be sure that all the cells that produce VLPs are transcribing both competitor RNAs. This problem could be solved by having plasmids with multiple Pol-II transcription and termination signals.

The mechanism by which the interaction between the PS-N complex and the M protein results in selective packaging of the SARS-CoV-2 has not been solved. Nonetheless, studies with other CoVs shed some light on this process. Narayanan and Makino showed for the case of MHV that there is a significant difference when N binds to RNAs not containing the packaging signal compared to those bearing it [41]. It turns out that MHV N protein can bind to viral and non-viral RNAs that do not contain the PS and these interactions result in the assembly of ribonucleoprotein complexes. However, M selectively interacts with N when the RNA contains the PS. This specific interaction results in efficient RNA packaging. These observations suggest that the strength of the vRNA-N interaction depends on the presence of the PS. Furthermore, it was also shown that interaction between MHV N and M proteins is mediated by RNA [40]. These data suggest, at least of MHV, that the interaction between the RNA and N could result in either an allosteric change that enhanced binding of the ribonucleoprotein complex to M or that the direct interaction between

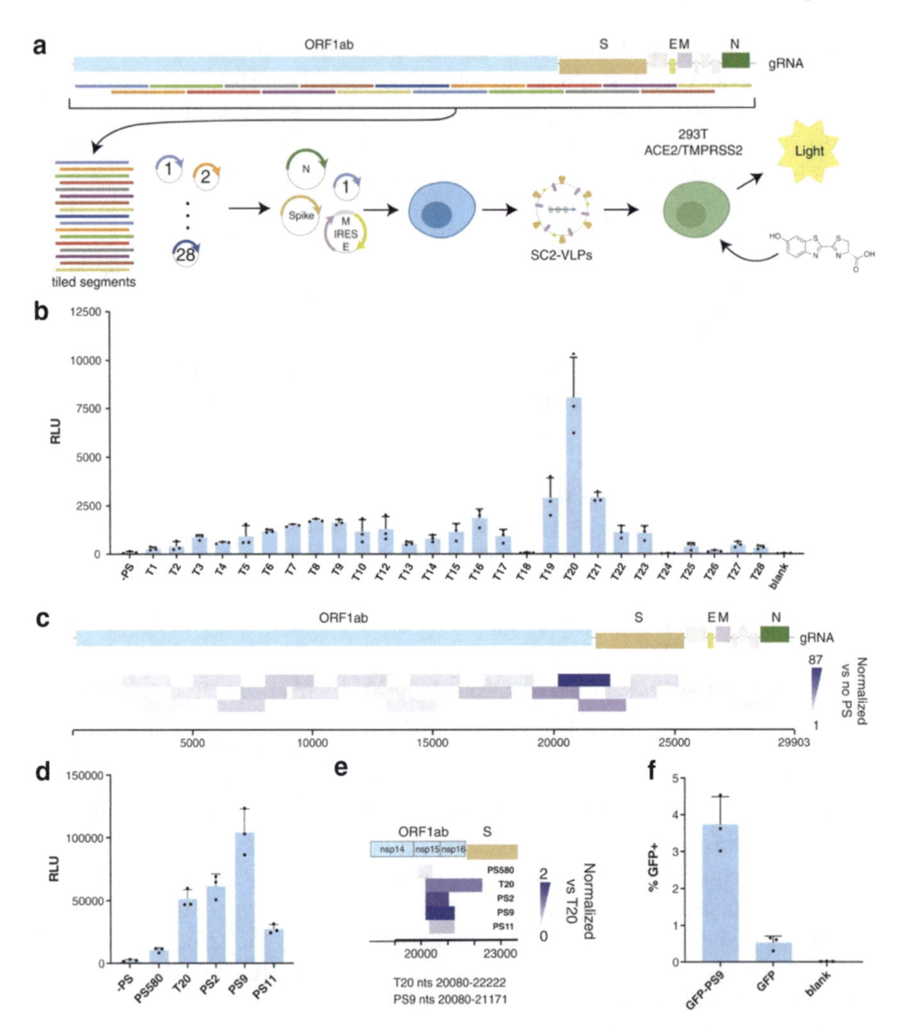

Fig. 7.6 RNA packaging by SARS-CoV-2 VLPs. **a** Screening of the SARS-CoV-2 of a *cis*-acting sequence that is selectively packaged by the VLPs (packaging signal or PS) by co-transfecting three plasmids that code for the structural proteins with a plasmid where the luciferase sequence is fused at the 3′ end with a segment of the viral genome. **b** Luciferase expression allows to determine the segment of the viral genome that is preferentially packaged by the VLPs. Fragment 20 (T20) had the highest expression among the initial 28 candidates. **c** Heatmap visualization of the data from B showing the location of each segment within the genome. **d** The PS was further delimited by generating shorter RNA segments. **e** Heatmap of the data from **d**. **f** Flow cytometry data of the percentage of positive cells in entry assays using the VLPs that packaged an mRNA of GFP that is 3′ end fused to PS9. Figure modified from [53]

M and N is mediated by RNA (the inner surface of the M protein is basic). It is important to mention that the MD simulations for the assembly of SARS-CoV-2 from Li and Roya support both models [33]. However, for SARS-CoV-2, it is still not clear how the PS promotes selective packaging in CoVs.

There are two main possible scenarios. In the first case, the binding of PS to N results in an allosteric change in the N protein that enhances the binding of the entire ribonucleoprotein complex to M. In the second scenario, the PS could lower the activation energy of the assembly process and thus the advantage could be governed by kinetics and not thermodynamics. A kinetic solution to this problem has been recently addressed by de Bruijn and collaborators for the assembly of a "linear" virus [15]. Given the fact that the assembly ribonucleoprotein SARS-CoV-2 and the role of the PS in the formation of this complex could be explained by the kinetic approach from de Bruijn and collaborators.

In the following section, we will discuss different experimental approaches that have been developed to understand the assembly of SARS-CoV and SARS-CoV-2 VLPs in cultured cells.

The Assembly of SARS-CoV and SARS-CoV-2 VLPs in Cultured Cells

SARS-CoV and SARS-CoV-2 VLPs can be assembled by co-transfecting Vero E6 and HEK-2943T cells. Siu and collaborators transfected cells with plasmids coding for SARS-CoV E, M, and N proteins [50]. The assembly of VLPs was monitored by Western blot and negative staining transmission electron microscopy. The presence of M, E, and N was analyzed by Western blots using specific antibodies against M and E, whereas antibodies identified N against a fused flag tag. Importantly, to increase the assembly yield, the M and E proteins were subcloned into a single plasmid (M − E). Siu et al. determined which proteins are required to assemble and release VLPs by co-transfected cells with N, M, M − E, M + N, or M − E + N. The cell culture medium was harvested at 24 and 48 h post-transfection (h.p.t), and the VLPs were purified by ultracentrifugation in a 20% sucrose cushion. However, even though the E protein is in the same plasmid as M, its expression is only noticeable only at 48 h.p.t.

On the one hand, at 24 h.p.t., the VLPs were only detected when all three proteins were co-expressed. On the other hand, at 48 h.p.t., all transfection combinations yield the release of some sort of extracellular vesicles. Nonetheless, based on the Western blot, the highest (qualitative) yield of the assembly was achieved by co-transfecting cells with M − E + N. Unfortunately, because these VLPs are missing the S protein, it was not possible to determine the structure of the secreted vesicles. To solve this problem, they co-transfected cells with M-E, N, and S. From this data is clear that the VLPs are not homogenous in composition and that only fractions 9–11 contained the highest amount of the four proteins. As expected, the purified VLPs from this co-transfection have the characteristic morphology of CoVs.

Vannema and co-workers studied the release of MHV-A59 S and M in OST7-1 cells [56]. They found that when these cells are co-trasfected with either M or S, neither protein is secreted into the media and that the release of uncleaved S and M requires the co-transfection with E, M, and S. The inhibition of N-glycosylation of S results in the aggregation of this protein in the ER; thus this protein is not incorporated into the VLPs. These findings are consistent with previous studies that showed that the M/S complexes are assembled in the pre-Golgi compartment when M is still unglycosylated [42]. On the one hand, in the absence of the membrane, protein S accumulates at the plasma membrane rather than at the Golgi apparatus. On the other hand, the intracellular localization of M is independent of S [42]. Finally, they found that M and S are incorporated into the virion at different rates; M is immediately incorporated, while S associates slowly.

Nakauchi and collaborators produced SARS-CoV VLPs by transfecting HEK-293T cells [37]. The genes for S, M, E, and N were subcloned into four different mammalian expression vectors. These cells were transfected with different combinations of these plasmids (i.e., M + E + S + N, M + E + N, M + S + N, M + N, M, and N) and the assembly of VLPs was confirmed by thin-section electron micrographs, immunoprecipitation, and ELISAs (with antibodies either against N or S). In addition, the culture media was analyzed by immunoprecipitation and ELISA to determine if the structural proteins were secreted. Co-transfections with a plasmid that produced an RNA containing the SARS-CoV packaging signal (PS) were also performed. Figure 7.7a shows that co-transfection of the plasmids coding for M, N, and a PS-containing RNA is sufficient to release VLPs that contain N. In other words, the presence of E and S does not affect the presence of N of M-induced vesicles, suggesting that Spike is not required for vesicle release. Interestingly, Fig. 7.7b demonstrates, based on ELISAs with antibodies that recognize S rather than M, that the co-expression of M and S is not sufficient to efficiently release VLPs or vesicles containing both proteins. This data shows that the presence of N is required for the efficient incorporation of S into VLPs and that E is dispensable for assembly.

To demonstrate a direct interaction between M and N, Nakauchi et al. carried out immunoprecipitation assays [37]. The co-expression of M and N in any of the combinations led to the immunoprecipitation of both proteins. It is important to point out that they showed that the antibodies used to precipitate M and N do not cross-react with each other. Interestingly, the ratio of extracellular N was 2.3–3 times higher when co-expressed with M than when expressed alone. This suggests that N can be secreted in some sort of secretory vesicles in the absence of M. Altogether, their results indicate that the co-expression of the SARS-CoV structural proteins could result in a heterogenous mixture of VLPs and secretory vesicles containing the viral proteins.

Plescia et al. and Swann et al. reported the assembly of SARS CoV-2 VLPs by co-expressing M, N, and E [45, 52]. In both cases, they transfected HEK-293T with plasmids driven by the CMV promoter. However, Plescia and co-workers fused the E protein with a 3xFlag, while Swann and co-workers did not fuse the viral proteins with any tags. While the presence of only three of the four structural proteins of SARS-CoV-2 seems to be the minimal system for the assembly of VLPs, Plescia

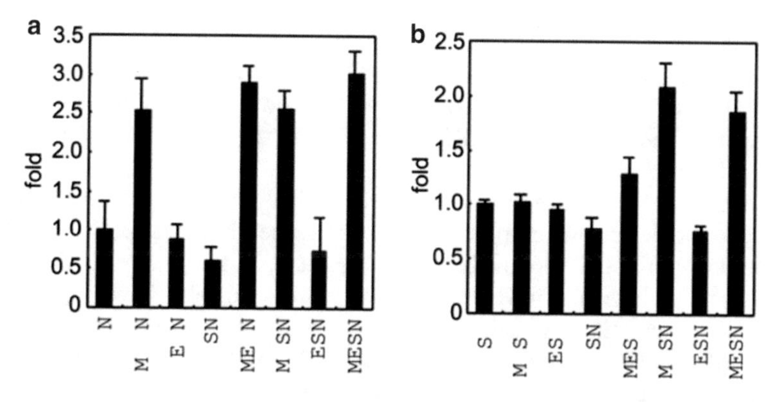

Fig. 7.7 Interaction between N and M proteins in HEK-293T cells. **a** ELISA to detect the N protein in the supernatant of cell transfected with N, M + N, E + N, S + N, M + E + N, M + S + N, E + S + N, or M + E + S + N. The ratio of extracellular N to the total N was calculated and normalized to cells only transfected with N. This data shows that the presence of the M is fundamental to enrich the supernatant with VLPs that contain the N protein. Also, it shows that the presence of E a S and dispensable for the assembly of particles that contain M and N. **b** ELISA to detect the S protein in the supernatant of cells transfected with the same plasmid combination as in **a**. The ratio of extracellular S to the total S was calculated and normalized to cells only transfected with S. This data shows that M and N are required for the efficient incorporation of S into VLPs. Figure modified from Nakauchi et al. [37]

et al. suggested that the expression of M + N + E + S is the most efficient SARS-CoV-2 system for VLP production. This is in accordance with the data for SARS-CoV [37, 50].

The main problem for studying of SARS-CoV-2 VLPs assembly is the requirement of co-transfecting cultured cells with four or more plasmids. Hence, there has been an effort to generate polycistronic plasmids [53, 61]. The specific characteristics of the plasmids depend on the applications intended for the VLPs. For example, VLPs used for neutralization assays require the co-expression of a reporter gene that can be packaged by the VLPs [53]. Also, VLPs used for vaccines may require modifications in the S protein to increase its stability [61].

Once the plasmids encoding the structural proteins are obtained, these are expressed by transient co-transfection. The most used host for expressing VLPs are HEK-293T cells; however, Vero E6 and Expi293 cell assembly has also been reported [59, 61]. Transfection reagent/DNA and a molar ratio of the plasmids are key variables that must be considered and optimized for each system. Xu et al. found that an 8:6:8:3 molar ratio for S, M, E, and N plasmids were optimal for assembling SARS CoV-2 VLPs. Sometimes these proteins are fused to flags. Syed and co-workers reported that mass ratios of CoV-2-N (0.67), CoV-2-M-IRES-E (0.33), CoV-2-S (0.0016), and Luc-T20 (1.0) ensured SARS CoV-2 VLP formation. First, they showed that while S-containing VLPs can be produced in the absence of at least of the other three structural proteins (see Fig. 7.8a), only expression of the four structural proteins results in VLPs that package a reporter gene that is fused to the PS (see Fig. 7.8b). Furthermore, they showed that the presence of a strep tag in the

M protein affects the functionality but not the assembly of the VLPs (see Fig. 7.8c). In other words, they found that strep-tagged M protein results in VLPs that cannot deliver an mRNA coding for a reporter gene. This observation adds a new level of complexity to the assembly of SARS-CoV-2 since the successful assembly of VLPs in cell cultures requires specific conditions.

As expected, several conditions modify the attributes of the obtained VLPs, such as the plasmid design and proportions used during the transfections, the expression system, post-translation modification, etc. For example, the yield of SARS-CoV-2 VLPs assembly is greater in HEK-293T than in Vero E6 cells; however, the VLPs expressed in Vero E6 exhibited the typical corona-like structure characteristic of SARS-CoV-2, while those expressed in HEK-293T are heterogenous in shape and size [59]. This suggests that transfection of HEK-293T cells could result in vesicles containing some structural proteins but not in the right conformation. This could result from different membrane compositions of the organelles involved in the assembly.

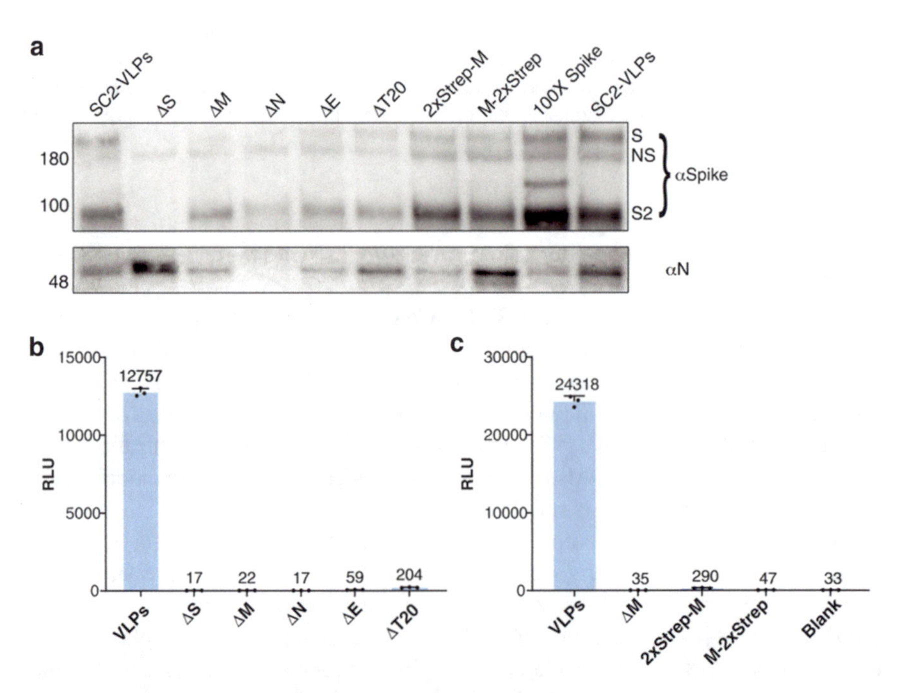

Fig. 7.8 Assembly of functional SARS-CoV-2 VLPs. **a** Western blots using antibodies that recognize S or N proteins in the supernatants of transfected cells under different conditions. The absence of M, N, E or an RNA that contains the PS (T20) results in the efficient release of VLPs bearing the S protein. Also, the presence of Strep tagged M protein does not affect the release of VLPs. **b** Entry assays with the VLPs from **a**. Only the expression of the four structural protein results in VLPs that package T20 and express the gene of interest. **c** Entry assays with Strep-tagged M-containing VLPs. While Strep-tagged M proteins does result in the assembly of the VLPs, entry assays with these VLPs do not result in the expression of the reporter gene. Figure modified from Syed et al. [53]

Conclusions

The assembly of SARS-CoV and SARS-CoV-2 VLPs requires at least the co-expression, in cultured cells, of either the matrix and spike or the matrix and nucleocapsid proteins. However, the co-expression of M and N results in extracellular vesicles that do not have the "classic corona-like" morphology. Also, the choice of the producer cell line plays a role in the morphology of the particles. However, it is not clear why the cell type influences VLP morphology and what are the cellular factors (e.g., the composition of the endomembranes involved during assembly) that influence the morphology of VLPs.

The assembly of functional SARS-CoV-2 VLPs (i.e., that can deliver a gene of interest to a target cell) requires the co-expression of all four structural proteins and the presence of an mRNA containing at least, a PS that is located between nucleotides 20,080–21,171 of the viral genome. However, it is not known if selective packaging requires the full-length PS sequence or a smaller portion, as is the case for other betacoronaviruses (e.g., MHV). The presence of protein tags (e.g., Strep-tag) does not interfere with assembly but affects the ability of the particles to deliver a gene of interest. These tags may alter the interaction between M and N and the stability of the particle, thus inhibiting the release of RNA. Interestingly, the presence of the envelope protein is completely dispensable for assembly but is required for the release of the RNA and expression of the gene of interest.

The fact that VLPs are assembled by co-expressing N and M, or N and S, can become a problem when co-transfecting cultured cells with multiple plasmids; it is almost impossible to transfect all cells with an equal copy number of each plasmid. Therefore, it is likely that the VLPs described in this chapter are a heterogenous mixture of particles. This heterogenicity could result in a population of particles with diverse biophysical and biological properties.

The studies discussed here show that the M protein is a key element that interacts with the ribonucleoprotein complex, S, and E. However, it is not known how these interactions result in the selective packaging of the gRNA. In fact, this mechanism could depend on several factors like the ability of the RNA and N protein to go through a liquid–liquid phase transition, selective binding of M to PS-containing RNPs, and the lipid composition of the assembly sites.

The results shown here are extremely promising as they set up the basis for generating new insights and ultimately a detailed comprehension of the SARS-CoV-2 assembly, which is required for the systematic production of VLPs to study fundamental physical aspects of this process. Moreover, the understanding on the assembly mechanism for SARS-CoV-2 and the interactions that control this process are particularly interesting because this knowledge will allow for the development of optimized platforms to produce VLPs that could be applied in the development of vaccines, diagnosis methods and therapeutic strategies; and serve as virological or biophysical tools.

Perspectives

The fact that the assembly of SARS-CoV-2 VLPs requires co-transfection of mammalian cell lines with multiple plasmids showcases the need to generate novel plasmids having either a multicistronic arrangement or a gene coding for a structural polyprotein that is further processed into individual proteins. In other words, the efficient assembly of homogenous SARS-CoV-2 VLPs is currently hampered by the lack of systems that allow for every single transfected cell to express all four structural proteins at the same rate.

The use of SARS-CoV-2 VLPs to deliver genes, as has been done for retroviruses or vesicular stomatitis virus, could become an attractive application. The CoV VLP system has several advantages over the aforementioned: (i) the length of the packaged RNA can be up to 30,000 nts, which is not possible with the other viral systems, and (ii) the wide tropism of this virus allows to generate therapeutic VLPs that could deliver a gene of interest to many target tissues and cells. Furthermore, these VLPs could be used for high-throughput assays to find novel antiviral drugs that inhibit assembly, selective packaging of the genomic RNA, disassembly, or release of the viral genome.

The study of the biophysical and physicochemical properties of the SARS-CoV-2 structural proteins has been done with recombinant proteins expressed in bacteria. While the expression of recombinant proteins in bacteria is an extremely powerful approach, it has its own limitations. For example, in such system it is extremely challenging to produce proteins with eukaryotic post-translation modifications or assemble VLPs that require the presence of an endomembrane. In this context, there is a need for expression systems of SARS-CoV-2 VLPs in eukaryotic systems. Therefore, developing a single plasmid coding all the structural proteins would allow for determining the protein–protein and protein-RNA interactions that control VLP assembly and selective RNA packaging during assembly in cultured cells. This scenario would be fundamental to understanding the assembly process of SARS-CoV-2.

References

1. Barnes CO, Jette CA, Abernathy ME, Dam KA, Esswein SR, Gristick HB, Malyutin AG, Sharaf NG, Huey-Tubman KE, Lee YE, Robbiani DF, Nussenzweig MC, West AP Jr, Bjorkman PJ (2020) SARS-CoV-2 neutralizing antibody structures inform therapeutic strategies. Nature 588:682–687
2. Barnes CO, West AP Jr, Huey-Tubman KE, Hoffmann MAG, Sharaf NG, Hoffman PR, Koranda N, Gristick HB, Gaebler C, Muecksch F, Lorenzi JCC, Finkin S, Hagglof T, Hurley A, Millard KG, Weisblum Y, Schmidt F, Hatziioannou T, Bieniasz PD, Caskey M, Robbiani DF, Nussenzweig MC, Bjorkman PJ (2020) Structures of human antibodies bound to SARS-CoV-2 spike reveal common epitopes and recurrent features of antibodies. Cell 182:828-42e16

3. Boson B, Legros V, Zhou B, Siret E, Mathieu C, Cosset FL, Lavillette D, Denolly S (2021) The SARS-CoV-2 envelope and membrane proteins modulate maturation and retention of the spike protein, allowing assembly of virus-like particles. J Biol Chem 296:100111
4. Chang CK, Hou MH, Chang CF, Hsiao CD, Huang TH (2014) The SARS coronavirus nucleocapsid protein–forms and functions. Antiviral Res 103:39–50
5. Chen CY, Chang CK, Chang YW, Sue SC, Bai HI, Riang L, Hsiao CD, Huang TH (2007) Structure of the SARS coronavirus nucleocapsid protein RNA-binding dimerization domain suggests a mechanism for helical packaging of viral RNA. J Mol Biol 368:1075–1086
6. Chen WH, Strych U, Hotez PJ, Bottazzi ME (2020) The SARS-CoV-2 vaccine pipeline: an overview. Curr Trop Med Rep 7:61–64
7. Cologna R, Hogue BG (2000) Identification of a bovine coronavirus packaging signal. J Virol 74:580–583
8. Comas-Garcia M (2019) Packaging of genomic RNA in positive-sense single-stranded RNA viruses: a complex story. Viruses 11
9. Comas-Garcia M, Cadena-Nava RD, Rao AL, Knobler CM, Gelbart WM (2012) In vitro quantification of the relative packaging efficiencies of single-stranded RNA molecules by viral capsid protein. J Virol 86:12271–12282
10. Comas-Garcia M, Datta SA, Baker L, Varma R, Gudla PR, Rein A (2017) Dissection of specific binding of HIV-1 Gag to the 'packaging signal' in viral RNA'. Elife 6
11. Corse E, Machamer CE (2003) The cytoplasmic tails of infectious bronchitis virus E and M proteins mediate their interaction. Virology 312:25–34
12. Coutard B, Valle C, de Lamballerie X, Canard B, Seidah NG, Decroly E (2020) The spike glycoprotein of the new coronavirus 2019-nCoV contains a furin-like cleavage site absent in CoV of the same clade. Antiviral Res 176:104742
13. Crist RM, Datta SA, Stephen AG, Soheilian F, Mirro J, Fisher RJ, Nagashima K, Rein A (2009) Assembly properties of human immunodeficiency virus type 1 Gag-leucine zipper chimeras: implications for retrovirus assembly. J Virol 83:2216–2225
14. Cubuk J, Alston JJ, Incicco JJ, Singh S, Stuchell-Brereton MD, Ward MD, Zimmerman MI, Vithani N, Griffith D, Wagoner JA, Bowman GR, Hall KB, Soranno A, Holehouse AS (2021) The SARS-CoV-2 nucleocapsid protein is dynamic, disordered, and phase separates with RNA. Nat Commun 12:1936. https://doi.org/10.1038/s41467-021-21953-3
15. de Bruijn R, Wielstra PCM, Calcines-Cruz C, van Waveren T, Hernandez-Garcia A, van der Schoot P (2022) A kinetic model for the impact of packaging signal mimics on genome encapsulation. Biophys J 121:2583–2599
16. de Haan CA, Smeets M, Vernooij F, Vennema H, Rottier PJ (1999) Mapping of the coronavirus membrane protein domains involved in interaction with the spike protein. J Virol 73:7441–7452
17. de Haan CA, Vennema H, Rottier PJ (1998) Coronavirus envelope assembly is sensitive to changes in the terminal regions of the viral M protein. Adv Exp Med Biol 440:367–375
18. de Haan CA, Vennema H, Rottier PJ (2000) Assembly of the coronavirus envelope: homotypic interactions between the M proteins. J Virol 74:4967–4978
19. Dinesh DC, Chalupska D, Silhan J, Koutna E, Nencka R, Veverka V, Boura E (2020) Structural basis of RNA recognition by the SARS-CoV-2 nucleocapsid phosphoprotein. PLoS Pathog 16:e1009100
20. Fosmire JA, Hwang K, Makino S (1992) Identification and characterization of a coronavirus packaging signal. J Virol 66:3522–3530
21. Heinz FX, Stiasny K (2021) Distinguishing features of current COVID-19 vaccines: knowns and unknowns of antigen presentation and modes of action. NPJ Vaccines 6:104
22. Hoffmann M, Kleine-Weber H, Schroeder S, Kruger N, Herrler T, Erichsen S, Schiergens TS, Herrler G, Wu NH, Nitsche A, Muller MA, Drosten C, Pohlmann S (2020) SARS-CoV-2 cell entry depends on ACE2 and TMPRSS2 and is blocked by a clinically proven protease inhibitor. Cell 181:271-80e8
23. Hsin WC, Chang CH, Chang CY, Peng WH, Chien CL, Chang MF, Chang SC (2018) Nucleocapsid protein-dependent assembly of the RNA packaging signal of Middle East respiratory syndrome coronavirus. J Biomed Sci 25:47

24. Iserman C, Roden CA, Boerneke MA, Sealfon RSG, McLaughlin GA, Jungreis I, Fritch EJ, Hou YJ, Ekena J, Weidmann CA, Theesfeld CL, Kellis M, Troyanskaya OG, Baric RS, Sheahan TP, Weeks KM, Gladfelter AS (2020) Genomic RNA elements drive phase separation of the SARS-CoV-2 nucleocapsid. Mol Cell 80:1078–1091
25. Jackson CB, Farzan M, Chen B, Choe H (2021) Mechanisms of SARS-CoV-2 entry into cells. Nat Rev Mol Cell Biol 23(1):3–20
26. Kang S, Yang M, Hong Z, Zhang L, Huang Z, Chen X, He S, Zhou Z, Zhou Z, Chen Q, Yan Y, Zhang C, Shan H, Chen S (2020) Crystal structure of SARS-CoV-2 nucleocapsid protein RNA binding domain reveals potential unique drug targeting sites. Acta Pharm Sin B 10:1228–1238
27. Ke Z, Oton J, Qu K, Cortese M, Zila V, McKeane L, Nakane T, Zivanov J, Neufeldt CJ, Cerikan B, Lu JM, Peukes J, Xiong X, Krausslich HG, Scheres SHW, Bartenschlager R, Briggs JAG (2020) Structures and distributions of SARS-CoV-2 spike proteins on intact virions. Nature 588:498–502
28. King B, Potts BJ, Brian DA (1985) Bovine coronavirus hemagglutinin protein. Virus Res 2:53–59
29. Klein S, Cortese M, Winter SL, Wachsmuth-Melm M, Neufeldt CJ, Cerikan B, Stanifer ML, Boulant S, Bartenschlager R, Chlanda P (2020) SARS-CoV-2 structure and replication characterized by in situ cryo-electron tomography. Nat Commun 11:5885
30. Klumperman J, Locker JK, Meijer A, Horzinek MC, Geuze HJ, Rottier PJ (1994) Coronavirus M proteins accumulate in the Golgi complex beyond the site of virion budding. J Virol 68:6523–6534
31. Kuo L, Koetzner CA, Hurst KR, Masters PS (2014) Recognition of the murine coronavirus genomic RNA packaging signal depends on the second RNA-binding domain of the nucleocapsid protein. J Virol 88:4451–4465
32. Lavelle L, Gingery M, Phillips M, Gelbart WM, Knobler CM, Cadena-Nava RD, Vega-Acosta JR, Pinedo-Torres LA, Ruiz-Garcia J (2009) Phase diagram of self-assembled viral capsid protein polymorphs. J Phys Chem B 113:3813–3819
33. Li S, Zandi R (2022) Biophysical modeling of SARS-CoV-2 assembly: genome condensation and budding. Viruses 14:2089. https://doi.org/10.3390/v14102089
34. Lim KP, Liu DX (2001) The missing link in coronavirus assembly. Retention of the avian coronavirus infectious bronchitis virus envelope protein in the pre-Golgi compartments and physical interaction between the envelope and membrane proteins. J Biol Chem 276:17515–17523
35. Martinez-Menarguez JA, Geuze HJ, Slot JW, Klumperman J (1999) Vesicular tubular clusters between the ER and Golgi mediate concentration of soluble secretory proteins by exclusion from COPI-coated vesicles. Cell 98:81–90
36. Molenkamp R, Spaan WJ (1997) Identification of a specific interaction between the coronavirus mouse hepatitis virus A59 nucleocapsid protein and packaging signal. Virology 239:78–86
37. Nakauchi M, Kariwa H, Kon Y, Yoshii K, Maeda A, Takashima I (2008) Analysis of severe acute respiratory syndrome coronavirus structural proteins in virus-like particle assembly. Microbiol Immunol 52:625–630
38. Narayanan K, Chen CJ, Maeda J, Makino S (2003) Nucleocapsid-independent specific viral RNA packaging via viral envelope protein and viral RNA signal. J Virol 77:2922–2927
39. Narayanan K, Kim KH, Makino S (2003) Characterization of N protein self-association in coronavirus ribonucleoprotein complexes. Virus Res 98:131–140
40. Narayanan K, Maeda A, Maeda J, Makino S (2000) Characterization of the coronavirus M protein and nucleocapsid interaction in infected cells. J Virol 74:8127–8134
41. Narayanan K, Makino S (2001) Cooperation of an RNA packaging signal and a viral envelope protein in coronavirus RNA packaging. J Virol 75:9059–9067
42. Opstelten DJ, Raamsman MJ, Wolfs K, Horzinek MC, Rottier PJ (1995) Envelope glycoprotein interactions in coronavirus assembly. J Cell Biol 131:339–349
43. Ouzounis CA (2020) A recent origin of Orf3a from M protein across the coronavirus lineage arising by sharp divergence. Comput Struct Biotechnol J 18:4093–4102

44. Park SY, Harries D, Gelbart WM (1998) Topological defects and the optimum size of DNA condensates. Biophys J 75:714–720
45. Plescia CB, David EA, Patra D, Sengupta R, Amiar S, Su Y, Stahelin RV (2021) SARS-CoV-2 viral budding and entry can be modeled using BSL-2 level virus-like particles. J Biol Chem 296:100103
46. Roden CA, Dai Y, Giannetti CA, Seim I, Lee M, Sealfon R, McLaughlin GA, Boerneke MA, Iserman C, Wey SA, Ekena JL, Troyanskaya OG, Weeks KM, You L, Chilkoti A, Gladfelter AS (2022) Double-stranded RNA drives SARS-CoV-2 nucleocapsid protein to undergo phase separation at specific temperatures. Nucleic Acids Res 50:8168–8192
47. Rozycki B, Boura E (2022) Conformational ensemble of the full-length SARS-CoV-2 nucleocapsid (N) protein based on molecular simulations and SAXS data. Biophys Chem 288:106843
48. Saurabh K, Solovchuk M, Sheu TW (2022) A detailed study of ion transport through the SARS-CoV-2 E protein ion channel. Nanoscale 14:8291–8305
49. Singh Tomar PP, Arkin IT (2020) SARS-CoV-2 E protein is a potential ion channel that can be inhibited by Gliclazide and Memantine. Biochem Biophys Res Commun 530:10–14
50. Siu YL, Teoh KT, Lo J, Chan CM, Kien F, Escriou N, Tsao SW, Nicholls JM, Altmeyer R, Peiris JS, Bruzzone R, Nal B (2008) The M, E, and N structural proteins of the severe acute respiratory syndrome coronavirus are required for efficient assembly, trafficking, and release of virus-like particles. J Virol 82:11318–11330
51. Stertz S, Reichelt M, Spiegel M, Kuri T, Martinez-Sobrido L, Garcia-Sastre A, Weber F, Kochs G (2007) The intracellular sites of early replication and budding of SARS-coronavirus. Virology 361:304–315
52. Swann H, Sharma A, Preece B, Peterson A, Eldredge C, Belnap DM, Vershinin M, Saffarian S (2020) Minimal system for assembly of SARS-CoV-2 virus like particles. Sci Rep 10:21877
53. Syed AM, Taha TY, Tabata T, Chen IP, Ciling A, Khalid MM, Sreekumar B, Chen PY, Hayashi JM, Soczek KM, Ott M, Doudna JA (2021) Rapid assessment of SARS-CoV-2-evolved variants using virus-like particles. Science 374:1626–1632
54. Ujike M, Taguchi F (2015) Incorporation of spike and membrane glycoproteins into coronavirus virions. Viruses 7:1700–1725
55. V'Kovski P, Kratzel A, Steiner S, Stalder H, Thiel V (2021) Coronavirus biology and replication: implications for SARS-CoV-2. Nat Rev Microbiol 19:155–170
56. Vennema H, Godeke GJ, Rossen JW, Voorhout WF, Horzinek MC, Opstelten DJ, Rottier PJ (1996) Nucleocapsid-independent assembly of coronavirus-like particles by co-expression of viral envelope protein genes. EMBO J 15:2020–2028
57. Winter C, Schwegmann-Wessels C, Cavanagh D, Neumann U, Herrler G (2006) Sialic acid is a receptor determinant for infection of cells by avian Infectious bronchitis virus. J Gen Virol 87:1209–1216
58. Wrapp D, Wang N, Corbett KS, Goldsmith JA, Hsieh CL, Abiona O, Graham BS, McLellan JS (2020) Cryo-EM structure of the 2019-nCoV spike in the Prefusion conformation. bioRxiv, 367:1260–1263
59. Xu R, Shi M, Li J, Song P, Li N (2020) Construction of SARS-CoV-2 virus-like particles by mammalian expression system. Front Bioeng Biotechnol 8:862
60. Yao H, Song Y, Chen Y, Wu N, Xu J, Sun C, Zhang J, Weng T, Zhang Z, Wu Z, Cheng L, Shi D, Lu X, Lei J, Crispin M, Shi Y, Li L, Li S (2020) Molecular architecture of the SARS-CoV-2 virus. Cell 183:730-38e13
61. Yilmaz IC, Ipekoglu EM, Bulbul A, Turay N, Yildirim M, Evcili I, Yilmaz NS, Guvencli N, Aydin Y, Gungor B, Saraydar B, Bartan AG, Ibibik B, Bildik T, Baydemir I, Sanli HA, Kayaoglu B, Ceylan Y, Yildirim T, Abras I, Ayanoglu IC, Cam SB, Ciftci Dede E, Gizer M, Erganis O, Sarac F, Uzar S, Enul H, Adiay C, Aykut G, Polat H, Yildirim IS, Tekin S, Korukluoglu G, Zeytin HE, Korkusuz P, Gursel I, Gursel M (2022) Development and preclinical evaluation of virus-like particle vaccine against COVID-19 infection. Allergy 77:258–270
62. Zeng W, Liu G, Ma H, Zhao D, Yang Y, Liu M, Mohammed A, Zhao C, Yang Y, Xie J, Ding C, Ma X, Weng J, Gao Y, He H, Jin T (2020) Biochemical characterization of SARS-CoV-2 nucleocapsid protein. Biochem Biophys Res Commun 527:618–623

63. Zhang XM, Kousoulas KG, Storz J (1991) The hemagglutinin esterase glycoprotein of bovine coronaviruses—sequence and functional comparisons between virulent and avirulent strains. Virology 185:847–852
64. Zhang Z, Nomura N, Muramoto Y, Ekimoto T, Uemura T, Liu K, Yui M, Kono N, Aoki J, Ikeguchi M, Noda T, Iwata S, Ohto U, Shimizu T (2022) Structure of SARS-CoV-2 membrane protein essential for virus assembly. Nat Commun 13:4399
65. Zheng M, Song L (2020) Novel antibody epitopes dominate the antigenicity of spike glycoprotein in SARS-CoV-2 compared to SARS-CoV. Cell Mol Immunol 17:536–538

Chapter 8
Norovirus—A Viral Capsid in Perpetual Flux

Lars Thiede, Ronja Pogan, and Charlotte Uetrecht

Abstract Noroviruses are the prime causative agents for viral gastroenteritis globally. The icosahedral capsid of noroviruses is usually comprised of 180 copies of VP1, which contains the shell (S) and protruding (P) domain. The S-domain forms the protective shell surrounding the genome, while the P-domain sits on top of the S-domain and facilitates binding processes. Research on different model systems has revealed dynamic processes governing the life of noroviral capsids. Assembly of the capsid occurs in an environmental and strain dependent manner. The fully assembled capsid can further alter its shape when interacting with divalent metal ions and other binding partners, contracting the P-domains to enable receptor binding and immune evasion. These findings aid in understanding the complexity of noroviruses and caliciviruses as a whole. Even though there are still several hurdles to overcome, pertaining to human virus production and receptor identification.

Keywords Norovirus · *Caliciviridae* · Capsid assembly · Structure dynamics · Virus-like particles · Bile acid

Introduction

The first viral agent identified as the cause for gastroenteritis was the human norovirus (hNoV), described in 1972 as Norwalk virus according to the location of discovery, Norwalk, Ohio, USA [28]. Noroviruses belong to the family of *Caliciviridae* and can be divided into 10 genogroups (GI-GX), of which GI, GII, GIV, GVIII and GIX infect humans [7]. Noroviruses frequently cause sporadic seasonal outbreaks [2], the majority of those have been traced back to GII.4 strains in the last decade [53]. Symptoms include nausea, vomiting, diarrhea and fever. The agent may pose severe

L. Thiede · R. Pogan · C. Uetrecht
CSSB Centre for Structural Systems Biology, Deutsches Elektronen-Synchrotron (DESY) and Leibniz Institute of Virology (LIV), Notkestraße 85, Hamburg, Germany

L. Thiede · R. Pogan · C. Uetrecht (✉)
Faculty V: School of Life Sciences, University of Siegen, Siegen, Germany
e-mail: charlotte.uetrecht@cssb-hamburg.de

M. Comas-Garcia and S. Rosales-Mendoza (eds.), *Physical Virology*, Springer Series in Biophysics 24, https://doi.org/10.1007/978-3-031-36815-8_8

health risks for children and immunocompromised patients as well as presenting an immense economic burden [1, 37] (Fig. 8.1).

Noroviruses have a non-segmented positive-strand RNA genome with an approximate size of 7.5 kb, containing three open-reading frames (ORFs). ORF1 encodes a non-structural polyprotein, which includes for example the RNA dependent RNA polymerase (RdRp). ORF2 encodes the major structural protein VP1 and ORF3 the minor structural protein VP2 [24, 25]. The norovirus capsid is composed of 90 VP1 dimers, forming a shell with $T = 3$ symmetry. Empty capsids also assemble in other

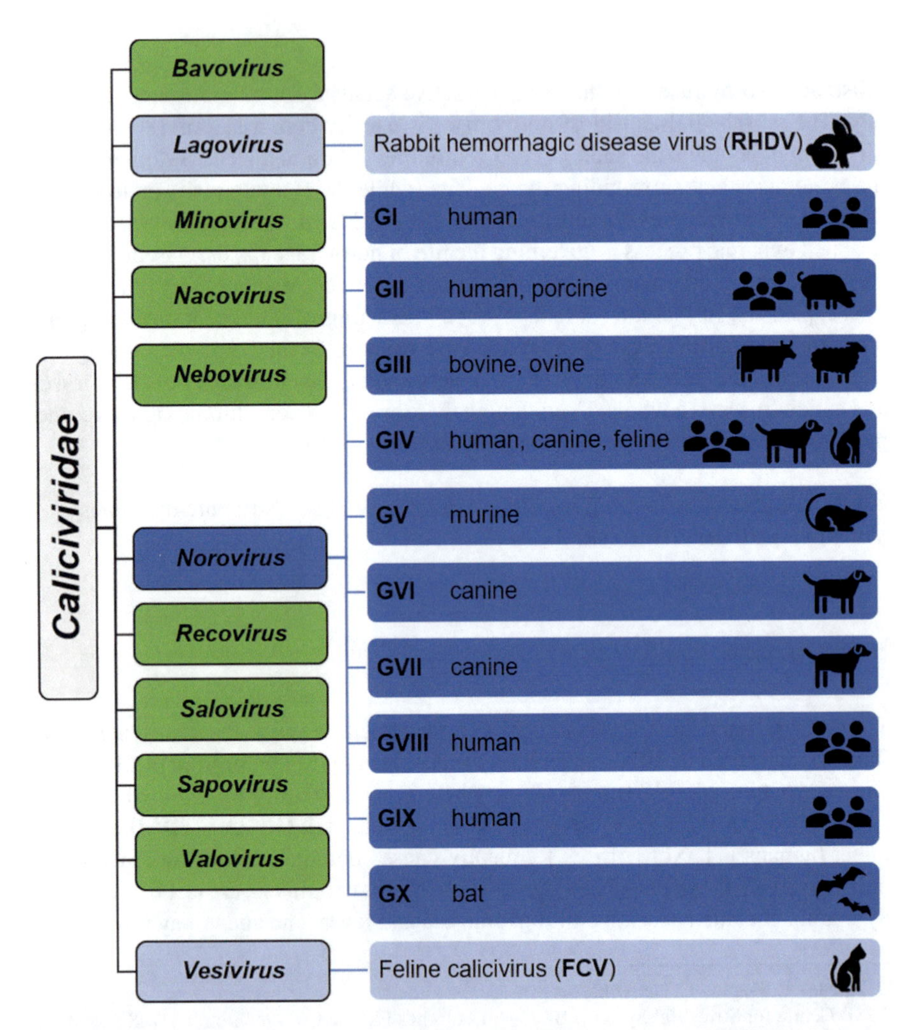

Fig. 8.1 Schematic diagram of the *Norovirus* genus within the *Caliciviridae* family. The 11 genera of calicivirus, as defined by [75] and the ten established genogroups in the genus norovirus (GI—GX), as defined by [7] are shown, as well as the species wherein each genogroup has been detected

symmetries [14, 26]. In the capsid, VP1 can be found in three quasi-equivalent conformational states. The general architecture of the capsid can be subdivided into three types of VP1: A-types forming the fivefold icosahedral axis, B-types forming dimers with the A-types, and C-types forming dimers with themselves at the twofold-axis [29, 30]. VP1 itself is divided into the distinct subdomains of shell (S) domain, the C-terminal protruding (P) domain, and the N-terminus. The S-domain forms the bulk of the capsid with the P-domain as protrusions on the capsid surface [59, 60].

Currently available hNoV cell culture systems are based on stem-cell derived enteroids [17, 18]. Recent advances include the progression of existing enteroid systems [48] as well as colon-derived cell lines [58]. Virus production in salivary glands has been proposed as a potential alternative [19]. Though promising, all of these are still quite novel and produce insufficient amounts of virions for structural studies. Additionally, zebra fish larvae have emerged as an animal-based system for cultivation [74]. While murine noroviruses (MNV) have been the most common surrogate, other mammalian caliciviruses have been investigated for potential similarities to their human counterpart, like rabbit hemorrhagic disease virus (RHDV) and feline calicivirus (FCV), [10, 29]. Virus-like particles (VLPs) are solely comprised of the protein components of interest, thus devoid of any genome, rendering them non-infectious. This makes them a convenient tool for virus studies [81]. Norovirus-like particles (NVLP), made from recombinant VP1, assembling into empty capsids constitute another widespread proxy for hNoV research.

With this virus production problem in mind, it is no surprise that vaccine development for hNoV has been unsuccessful and that antivirals are not available yet [51, 70]. Our current understanding of norovirus structure is created by an array of research based on different hNoV surrogates, protein assemblies and even conformations, presenting a difficult puzzle. This chapter attempts to shed light on the norovirus capsid and its ability to structurally adjust itself [66]. Substantial progress has been made in recent years due to integrating various techniques and technological advancements [15, 39, 47, 47].

The P-Domain Facilitates Virus-Host Interaction

The P-domain, located atop the S-domain, is comprised of a P1 and a P2 subdomain, with P2 being inserted in P1. These subdomains form a functional dimer that serves as an interaction site for several binding partners. This flexible P-dimer is responsible for interaction with the host cell via a receptor and several attachment factors. Histo-blood group antigens (HBGAs) have been proposed as an attachment factor for many hNoVs [8, 23, 44]. However, the actual receptor is still elusive and distinct from those found in MNV and FCV.

HGBAs are abundant on epithelial cell surfaces, particularly in the gastrointestinal mucosa as glycolipids or glycoproteins, and similar structures are found on intestinal bacteria. The central element of every HGBA is the α-L-fucose-(1, 2)-D-galactose H-disaccharide equivalent to blood group 0. Extension of this disaccharide with either galactose or N-acetyl-galactosamine residues yields the corresponding antigens of either the A or B blood group. Multiple biophysical methods have been employed to probe the poor mM binding affinities of the P-dimer to HGBAs. Several works based on native mass spectrometry (MS), saturation transfer difference and chemical shift perturbation NMR (STD and CSP NMR, respectively) have investigated K_Ds [20, 21, 45, 77]. However, recent investigations by NMR and HDX-MS revealed much lower affinities than previously thought [12, 16, 44]. Overestimation of binding is technique specific and this has been recently reviewed [54]. In native MS, overestimation has been attributed to enhanced clustering of glycans to β-sheet structures [80]. It has now been established that much higher K_Ds ranging from 0.5 to 30 mM are observed for glycans [12, 44]. The structural effects of HBGA binding on P-dimers are mostly subtle [16].

The first cell receptor discovered for noroviruses comes from the murine system. CD300lf was initially found by genome-wide CRISPR screens [52] and confirmed by crystal structure [50]. As demonstrated by Fig. 8.2, the P2 subdomains directly interact with CD300lf in a multivalent manner [50]. Additionally, both subdomains are shown to interact with divalent metal ions and bile acids, acting as cofactors, these have been demonstrated to aid binding; and in MNV they stabilize the P-dimers in the necessary conformation to allow for receptor attachment [12]. In some other calicivirus families, protein receptors have been identified, the feline junctional adhesion molecule A (fJAM-A) for FCV for example [43]. This suggests that hNoV also requires a proteinaceous receptor for entry besides HBGAs.

Bile acids bind, when compared to HGBAs, with high affinities, in the single μM range. Binding them has far reaching structural consequences on the P-domain and the capsid itself. Glycochenodeoxycholic acid (GCDCA) was shown to enhance binding in this manner [12, 79]. It is noteworthy, that GCDCA apparently binds the P-dimer of murine and human noroviruses in distinct modes. MNV accommodates bile deep within the P-dimer [50]. The P-dimer of hNoV on the other hand has conserved bile binding pockets on top of the P2 subdomain [32]. This interaction causes either contraction or extension of the P domains. The ramifications of this bile induced conformational change for the norovirus capsid will be explored later on in this chapter and demonstrates clearly the dynamic nature of binding processes through the P-domain as well as the capsid.

Assembly of the Norovirus Capsid Protein

While the above described $T = 3$ is the most common geometry among norovirus capsids, others have been observed, such as the far smaller $T = 1$ comprised of 60 VP1 monomers [78]. Moreover, $T = 4$ capsids containing 240 VP1 have been observed [14, 22, 26]. Molecular mechanisms of NVLP $T = 4$ assembly are still obscure, though not specific to human GII.4 strains as they are known in other caliciviruses [40]. A noteworthy structural modifier is the N-terminal domain of VP1. A 34 and 98 amino acid truncation completely abolishes capsid formation in

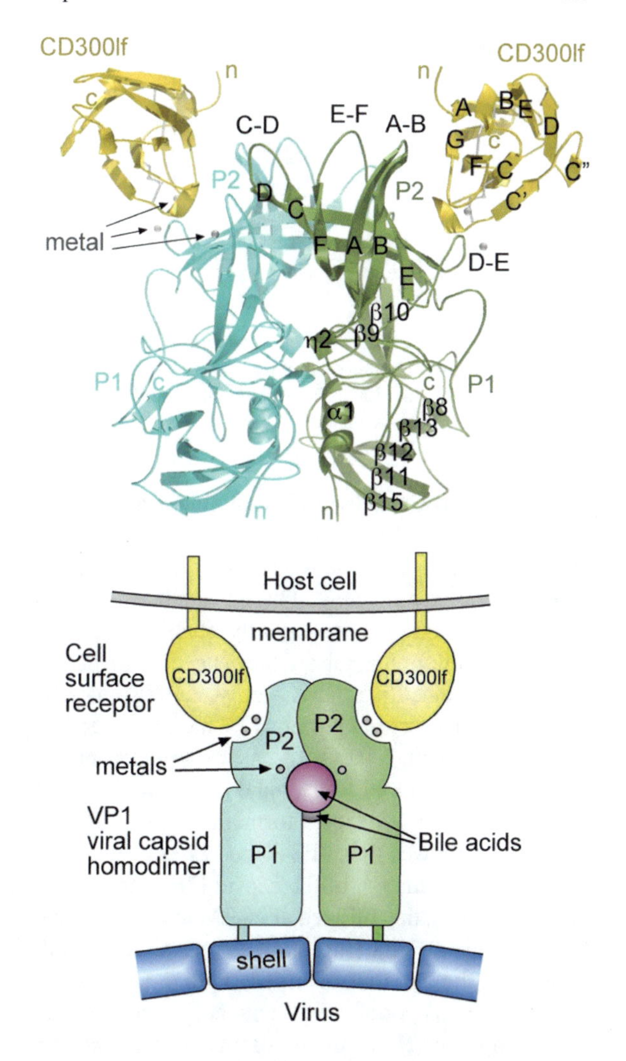

Fig. 8.2 CD300lf binds to the P2 subdomain of MNV VP1. Top: Ribbon model of CD300lf-P-domain complex, with two P-domains (cyan and green) and the receptor (yellow). Secondary structures are labeled, metal ions are depicted as silver spheres. Bottom: Schematic of the complex in context of the host cell and virus particle. Adapted from [50] under CC BY-NC-ND 4.0 license

GI.1 Norwalk, while a 20-residue deletion retains assembly [6]. Again, experiments on RHDV indicate that this is an overarching mechanism in caliciviruses [5]. The role of the N-terminal domain became more intriguing, when highly homogenous $T = 1$ batches of NVLPs were discovered (Fig. 8.3). Several N-terminal truncations up to 45 missing amino acids were revealed [56]. Importantly, it remains unclear whether N-termini are complete in $T = 1$ particles reconstructed from cryo-EM as density is usually missing, which could be attributed to flexibility or absence of the N-terminus. A more recent study investigated whether fusing VP2-derived peptides to the N- and C-terminus of VP1 affected capsid assembly. Images from EM showed intact capsids of various sizes [42]. Though sufficiently observed, the mechanisms that tie the N-terminal domain to the capsid assembly remain unknown.

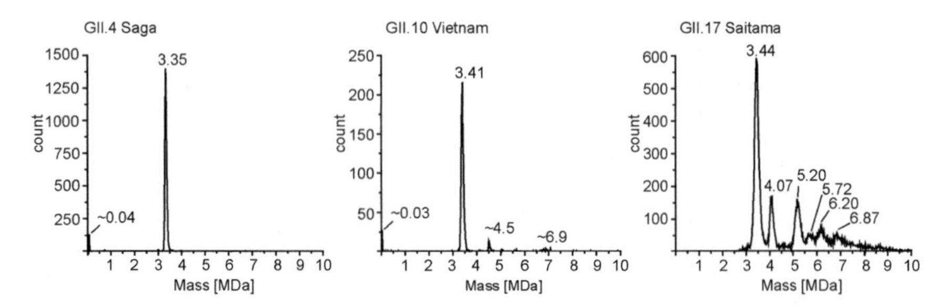

Fig. 8.3 Charge detection mass spectrometry (CDMS) on $T = 1$ dominant NVLP preparations. Strains and masses are annotated with ~3.4 MDa for $T = 1$. GII.10 Vietnam and GII.17 Saitama especially display additional, less abundant assemblies with unknown symmetry. Adapted from [56] under CC BY 4.0 license

VLP stability, usually probed by investigating disassembly, is impacted by environmental cues like temperature, ionic strength and pH [4, 64]. As with other attributes, disassembly is a strain-dependent process. Where GI.1 Norwalk abolishes assembly completely at high pH and low ionic strength [64], GII.17 Kawasaki is barely affected [57]. Importantly, stability is also modified by N-terminal truncation [56]. Therefore, much attention needs to be put towards the strains and attributes in question, something that has only recently entered norovirus research [55].

The minor structural protein VP2 has proven to be a difficult study subject. When co-expressed with VP1, VP2 can be detected but does rarely co-assemble into a capsid [22]. Bertolotti-Ciarlet and colleagues suggested early on that VP2 was not strictly necessary for capsid formation using VP1 VLPs [6]. Later studies from the same group showed that VP2 indeed associated with VP1 and thus contributed to its stability and formation of the capsid [76]. Structurally more informative are results on FCV virions, that utilized cryo-EM to demonstrate a portal-like assembly formed by 12 copies of VP2 (Fig. 8.4). A rearrangement of VP1 at the site of this portal allows for the opening of a pore with 1 nm diameter in the capsid. The authors hypothesize this portal to be a channel for genome delivery into the host cell, where the hydrophobic N-termini of VP2 enter the endosomal membrane. However, RNA egress via this route could not yet be demonstrated [11]. This may be due to the pore lacking size or missing additional structural requirements. Dynamic rearrangements like these are an integral part of the norovirus capsid and therefore need to be explored in detail.

Dynamics of the Assembled Capsid

The capsid structures, much like the assembly process itself, are subjected to environmental cues. Cryo-EM experiments first revealed raised P-domains on both an MNV and RHDV [29]. Indeed, association with bile acids such as GCDCA leads to widespread, though reversible, conformational changes affecting this extension.

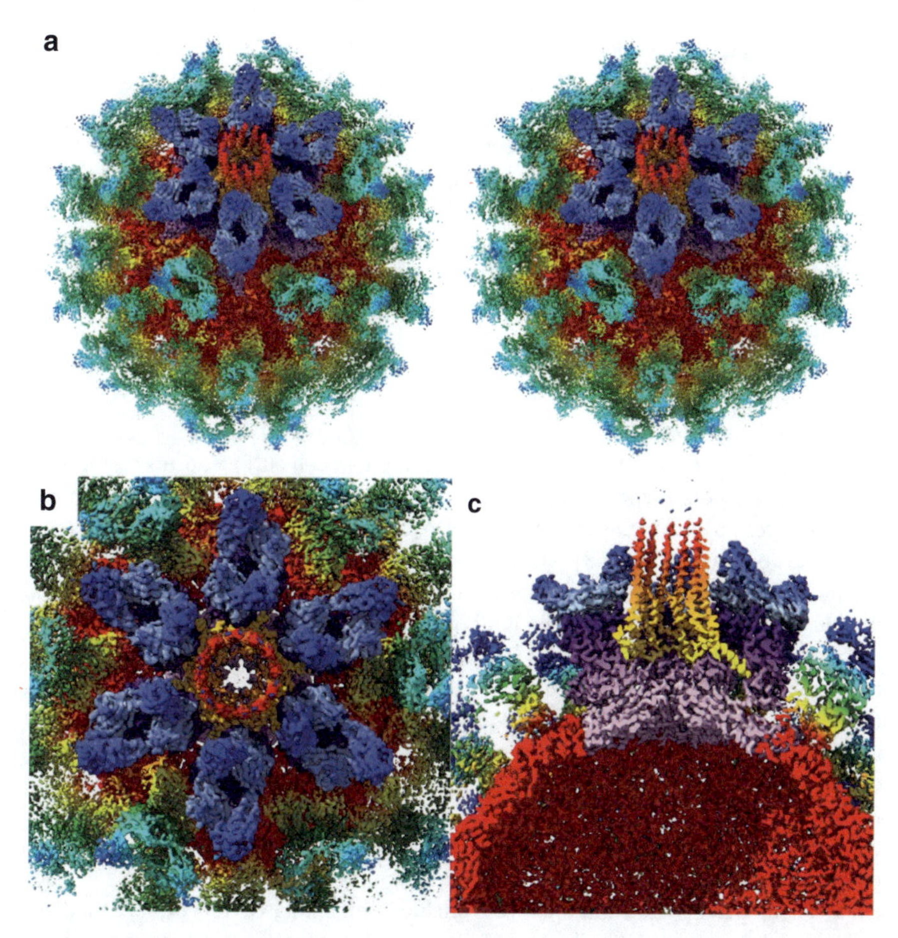

Fig. 8.4 VP2 portal structure in FCV. **a** A stereo-pair view of the reconstruction along the icosahedral two-fold axis is shown to highlight the portal vertex. **b** A close-up view of the portal assembly along the unique three-fold axis highlights the presence of a pore in the capsid shell at the center of this symmetry axis. **c** A cut-away view perpendicular to the portal axis. VP2 is colored orange and red, VP1 dimers arranged about the portal vertex are colored purple, and associated fJAM-A molecules are colored blue. Reprinted by permission from Springer: Springer Nature Limited, Nature, [11] Copyright © 2019

Upon binding GCDCA, the murine capsid contracts, tightly associating the P-domains to the shell surface. When not in this contracted state, these domains hover 10–15 Å above the S-domains [62]. The extension is achieved by twisting the loop that connects both domains of VP1. Furthermore, a high pH encourages the extended conformation, while low pH causes contraction. Notably, this feature is strain dependent: MNV-1 moved at pH 7 to an extended conformation, while MNV-S7 moved at pH 8. Metal ions such as Ca^{2+} and Mg^{2+} stabilized the resting conformation. The same conformations were found in the hNoV capsid of a GII.3 variant [68]. Other human variants displayed similar conformations, as earlier work demonstrated with a GII.10 variant [21] and later with a GII.4 capsid [13]. A study from the same year also showed a contracted MNV, both with and without the presence of bile acids [67]. Additionally, a hNoV GII.4 Minerva-derived capsid assumed a contracted conformation when bound to Cd^{2+} and an extended one, where the P-domains were raised up to 24 Å from the shell, when a chelating agent was applied. The removal of the metal ion creates a gap in the dimer interface of the P-domain, where the ion was originally bound. Several of the key residues in the ion binding cleft are conserved throughout multiple hNoVs, including the aforementioned GII.3 and GII.10. Thus, transition from one state to another may be caused by allosteric changes in the hinge region of the P-domain induced by binding or removal of metal ions [22]. The preservation of such an intricate mechanism points towards an important function that may be critical for caliciviruses as a whole.

The conformational change as a response to environmental cues, may indicate that the virus prefers one of the states when attaching to a cell surface, priming the viral shell after arrival in the appropriate location in the intestines [66]. Though binding does not seem to favor the expanded MNV [50] nor expanded hNoV capsids, but rather the contracted state [68]. The most likely reason for the transitory nature of the capsid is the evasion of the host immune response. For MNV, it was demonstrated that the immune response included antibodies for buried parts of the S-domain [33]. Antibodies binding portions of the P-domain, which are buried in the rested conformation were found for several human strains such as GII.10, GII.4 and GI.1 [21, 36, 61], suggesting that the contracted state served as a protective function by hiding the antibody epitopes. This was demonstrated in a GII.4 capsid, where resting P-domains occlude the binding site of the well-characterized human antibody NORO-320 [3]. Raised P-domains on the other hand leave the binding site fully exposed [22]. The difficulties in determining cell surface receptors for humans could be tributary to this dynamic quality of the capsid and the interaction process. The investigated capsids existed in a 'wrong' state and thus were unable to bind putative receptors. Since this process is driven by changes in the P-domain, a closer look into the local dynamics is warranted (Fig. 8.5).

The P-domain undergoes severe shape shifting, allowing for binding or the lack thereof, highlighting the necessity of a flexible P-domain. As explained, metal ion binding causes the P-subdomains to take the contracted conformation in human and murine noroviruses. This transition is caused by bile and pH as well [62, 79] and mediated by several loops within the subdomains moving to accommodate the interaction partner [50]. While moving onto the S-domain, the P-domains rotate by ~90° and create a complementary contact surface, increasing the number of interacting residues with the shell domain. A detailed step-by-step guide for this can be found elsewhere [63]. Furthermore, binding of an attachment factor causes the P-domains of two VP1 monomers to come together, both in MNV [50, 71] and hNoV [32]. In hNoV, the convening P-domains form two highly conserved glycan binding pockets, located on top of the P-domain at the cleft of the monomer/dimer interface [9]. This is where HGBAs or other glycans bind [38, 65]. A notable exception is a GII.10 Vietnam strain, that has demonstrated two more binding pockets in the P2 cleft at least for the minimal fucose ligand [34]. In hydrogen–deuterium-exchange

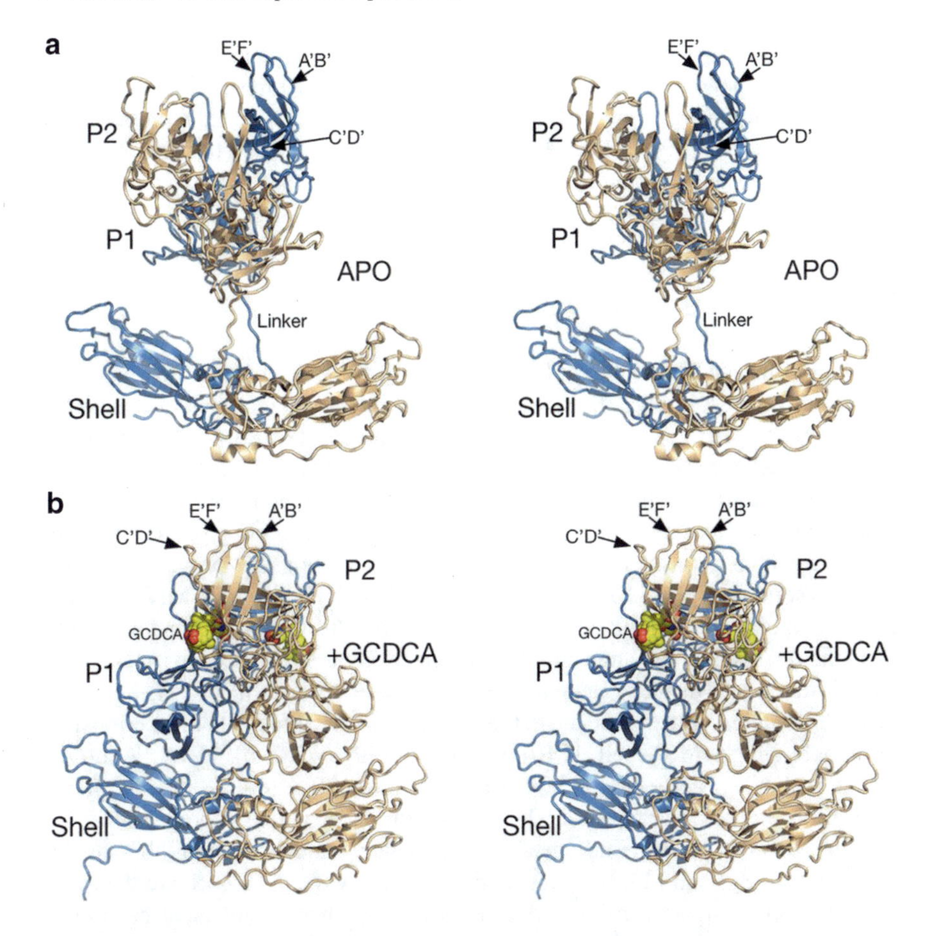

Fig. 8.5 Atomic models of the A/B subunit pairs for the apo and GCDCA MNV complexes. In these images, the A and B subunits are colored blue and tan, respectively. **a** Structure of the apo form of MNV. **b** The P domain rotates by nearly 90° and drops down onto the shell in the presence of bile acid. The bound GCDCA is highlighted in yellow and red spheres. The A, B, and C subunits are shown in blue, green, and red, respectively. The electron density of bound GCDCA is shown in gray. Adapted from [62] under CC BY 4.0 license. Modified for clarity

(HDX)-MS, this was the only strain so far showing long range effects upon glycan binding for the wildtype sequence, which is likely due to a small helix forming upon glycan binding [16]. Interaction between the P-domain and HGBAs is modulated by deamidation of critical residues [44]. Deamidation is a post-translational modification (PTM) that transforms an asparagine into aspartic acid in VP1 with a half-life of 1.5 days at 37 °C. This was observed in GII.4 Saga and MI001 P-dimers. HDX MS yielded that indeed partial and full deamidation resulted in a more exposed P2 domain, indicating an increased flexibility. This became even more pronounced upon glycan binding. Additionally, deamidation altered the monomer–dimer equilibrium towards monomeric P-domains [16]. Similar changes were observed in MNV upon bile or metal ion binding [12]. It seems that hovering and contraction, rotation as well as changes in monomer–dimer equilibria are conserved features across noroviruses or even *Caliciviridae*. Since the transitions occur under distinct circumstances, intrinsic features of the VP1 have been evolutionary repurposed to meet the needs of infectious cycles in different host species.

Conclusion

The norovirus capsid is subjected to a myriad of different, multifaceted processes during and after assembly. VP1 mostly assembles into $T = 3$ icosahedral capsids but other sizes are formed as well in a strain-dependent process [14, 26, 47]. Susceptibility to environmental cues and subsequent disassembly also falls in this category [57, 64]. The N-terminus of VP1 has been identified as an important modifier for the assembly with varying degrees of impact [42, 56]. Investigating the binding properties of the P-domain, the part of the capsid responsible for interaction, has proven difficult at times, as some data proved erroneous [54]. Even though receptors for MNV [50] and FCV [43] are known, the human receptor still remains elusive. The reason for this may lie in the surprisingly dynamic structural changes on the surface of the capsid itself. Upon binding bile, divalent metal ions or being subject to specific pH, the otherwise raised P-domains tightly associate to the S-domains. This turned out to be critical for receptor binding in MNV and HGBA interaction in hNoV [22, 50, 62, 68]. Additionally, it likely plays a role in immune response evasion [3]. However, the interplay is unclear. All of these intricacies are hence vital for understanding noroviruses.

Fully grasping these minute details requires using and integrating all available technologies. EM and MS, both conventional native MS and CDMS, are the prime methods to investigate capsids and their assembly. They complement each other especially well, as MS provides the time domain EM is lacking [15, 35]. CDMS is especially applicable to the often heterogenous samples, like norovirus [47]. HDX-MS and CSP NMR have been implemented in order to study the complex structure variations found in the P-dimer [54]. These are, together with STD NMR, also well suited to investigate binding to different ligands [16, 46]. Crystallography is one of the 'classical' tools that has provided plenty of structural data on the P-dimer [50], though it struggles with the megadalton-sized capsids. This is where cryo-EM and cryo-electron tomography have stepped in, progressing research on viral assemblies [22, 39]. Novel advancements in biophysical and imaging methodologies could push these boundaries even further [27, 31].

Perspectives

The most pressing issue within norovirus research, is the absence of proper hNoV cell culture systems to generate sufficient amounts of virions for structural studies [18, 19, 48, 58]. As this hurdle forces research to focus on model systems, it is important to keep in mind that studies on MNV, FCV, any other animal model or NVLPs do not necessarily reflect the infection cycle of the human strains. Rodents for example have a higher synthesis of bile acids [69], which could heavily impact the infection mechanisms and could explain the different modes of interaction with bile. To comprehend these mechanisms, it will be necessary to determine the exact roles of all known interaction partners. While bile, pH and metal ions can all cause critical structural rearrangements, it is yet unclear what the exact interplay between these factors is, does the presence of all of them differ to that of one? Are there more environmental cues or binding partners that cause the rise and fall of P-domains? Bile acid enhances HGBA binding in hNoV [32], so there is definitely some host-dependent attachment factor interplay at work in some cases. The answers to these questions will ultimately contribute towards finding human receptors and vaccine development. Moreover, capsid assembly still poses many questions. We know some factors that determine the size of capsids, but clearly not all, and even then, the comprehension is still incomplete. Furthermore, the critical intermediates of assembly, i.e. the assembly nucleus, are still a matter of debate [72, 73]. Their nature may shed light on, how the different sizes can be formed, how the size can be tweaked and assembly be poised by small molecules offering treatment options. The symmetries in non-icosahedral spherical assemblies are yet unknown [56]. For example, icosahedral capsid polymorphism can be seen in the hepatitis B virus (HBV) as well, forming $T = 3$ and $T = 4$ particles. Simulation work suggests that this is rooted in the energetics of different conformations [49]. CDMS-based studies on HBV demonstrated an 'overshooting' in the capsid assembly [41]. Do similar mechanisms exist in noroviruses and if they do, is there a biological impact, or can this be exploited for nanotechnology approaches or vaccine development? The norovirus capsid is yet an incomplete puzzle, for which solving will include the answers to these questions and likely more.

References

1. Adler JL, Zickl R (1969) Winter vomiting disease. J Infect Dis 119:668–673. https://doi.org/10.1093/infdis/119.6.668
2. Ahmed SM, Hall AJ, Robinson AE, Verhoef L, Premkumar P, Parashar UD, Koopmans M, Lopman BA (2014) Global prevalence of norovirus in cases of gastroenteritis: a systematic review and meta-analysis. Lancet Infect Dis 14:725–730. https://doi.org/10.1016/S1473-3099(14)70767-4
3. Alvarado G, Salmen W, Ettayebi K, Hu L, Sankaran B, Estes MK, Venkataram Prasad BV, Crowe JE (2021) Broadly cross-reactive human antibodies that inhibit genogroup I and II noroviruses. Nat Commun 12:4320. https://doi.org/10.1038/s41467-021-24649-w

4. Ausar SF, Foubert TR, Hudson MH, Vedvick TS, Middaugh CR (2006) Conformational stability and disassembly of norwalk virus-like particles. J Biol Chem 281:19478–19488. https://doi.org/10.1074/jbc.M603313200
5. Bárcena J, Verdaguer N, Roca R, Morales M, Angulo I, Risco C, Carrascosa JL, Torres JM, Castón JR (2004) The coat protein of Rabbit hemorrhagic disease virus contains a molecular switch at the N-terminal region facing the inner surface of the capsid. Virology 322:118–134. https://doi.org/10.1016/j.virol.2004.01.021
6. Bertolotti-Ciarlet A, White LJ, Chen R, Prasad BVV, Estes MK (2002) Structural requirements for the assembly of Norwalk virus-like particles. J Virol 76:4044–4055. https://doi.org/10.1128/JVI.76.8.4044-4055.2002
7. Chhabra P, de Graaf M, Parra GI, Chan MC-W, Green K, Martella V, Wang Q, White PA, Katayama K, Vennema H, Koopmans MPG, Vinjé J (2019) Updated classification of norovirus genogroups and genotypes. J Gen Virol 100:1393–1406. https://doi.org/10.1099/jgv.0.001318
8. Choi J-M, Hutson AM, Estes MK, Prasad BVV (2008) Atomic resolution structural characterization of recognition of histo-blood group antigens by Norwalk virus. Proc Natl Acad Sci USA 105:9175–9180. https://doi.org/10.1073/pnas.0803275105
9. Cong X, Li H, Sun X, Qi J, Zhang Q, Duan Z, Xu Y, Liu W (2022) Functional and Structural Characterization of Norovirus GII.6 in Recognizing Histo-blood Group Antigens. Virologica Sinica S1995820X22001584. https://doi.org/10.1016/j.virs.2022.09.010
10. Conley MJ, Bhella D (2019) Asymmetric analysis reveals novel virus capsid features. Biophys Rev 11:603–609. https://doi.org/10.1007/s12551-019-00572-9
11. Conley MJ, McElwee M, Azmi L, Gabrielsen M, Byron O, Goodfellow IG, Bhella D (2019) Calicivirus VP2 forms a portal-like assembly following receptor engagement. Nature 565:377–381. https://doi.org/10.1038/s41586-018-0852-1
12. Creutznacher R, Maass T, Dülfer J, Feldmann C, Hartmann V, Lane MS, Knickmann J, Westermann LT, Thiede L, Smith TJ, Uetrecht C, Mallagaray A, Waudby CA, Taube S, Peters T (2022) Distinct dissociation rates of murine and human norovirus P-domain dimers suggest a role of dimer stability in virus-host interactions. Commun Biol 5:563. https://doi.org/10.1038/s42003-022-03497-4
13. Devant JM, Hansman GS (2021) Structural heterogeneity of a human norovirus vaccine candidate. Virology 553:23–34. https://doi.org/10.1016/j.virol.2020.10.005
14. Devant JM, Hofhaus G, Bhella D, Hansman GS (2019) Heterologous expression of human norovirus GII.4 VP1 leads to assembly of T=4 virus-like particles. Antiviral Res 168:175–182. https://doi.org/10.1016/j.antiviral.2019.05.010
15. Dülfer J, Kadek A, Kopicki J-D, Krichel B, Uetrecht C (2019) Structural mass spectrometry goes viral. In: Advances in virus research. Elsevier, pp 189–238. https://doi.org/10.1016/bs.aivir.2019.07.003
16. Dülfer J, Yan H, Brodmerkel MN, Creutznacher R, Mallagaray A, Peters T, Caleman C, Marklund EG, Uetrecht C (2021) Glycan-induced protein dynamics in human norovirus P dimers depend on virus strain and deamidation status. Molecules 26:2125. https://doi.org/10.3390/molecules26082125
17. Ettayebi K, Crawford SE, Murakami K, Broughman JR, Karandikar U, Tenge VR, Neill FH, Blutt SE, Zeng X-L, Qu L, Kou B, Opekun AR, Burrin D, Graham DY, Ramani S, Atmar RL, Estes MK (2016) Replication of human noroviruses in stem cell-derived human enteroids. Science 353:1387–1393. https://doi.org/10.1126/science.aaf5211
18. Ettayebi K, Tenge VR, Cortes-Penfield NW, Crawford SE, Neill FH, Zeng X-L, Yu X, Ayyar BV, Burrin D, Ramani S, Atmar RL, Estes MK (2021) New insights and enhanced human norovirus cultivation in human intestinal enteroids. mSphere 6:e01136–20. https://doi.org/10.1128/mSphere.01136-20
19. Ghosh S, Kumar M, Santiana M, Mishra A, Zhang M, Labayo H, Chibly AM, Nakamura H, Tanaka T, Henderson W, Lewis E, Voss O, Su Y, Belkaid Y, Chiorini JA, Hoffman MP, Altan-Bonnet N (2022) Enteric viruses replicate in salivary glands and infect through saliva. Nature 607:345–350. https://doi.org/10.1038/s41586-022-04895-8

20. Han L, Kitov PI, Kitova EN, Tan M, Wang L, Xia M, Jiang X, Klassen JS (2013) Affinities of recombinant norovirus P dimers for human blood group antigens. Glycobiology 23:276–285. https://doi.org/10.1093/glycob/cws141
21. Hansman GS, Shahzad-ul-Hussan S, McLellan JS, Chuang G-Y, Georgiev I, Shimoike T, Katayama K, Bewley CA, Kwong PD (2012) Structural basis for norovirus inhibition and Fucose Mimicry by citrate. J Virol 86:284–292. https://doi.org/10.1128/JVI.05909-11
22. Hu L, Salmen W, Chen R, Zhou Y, Neill F, Crowe JE, Atmar RL, Estes MK, Prasad BVV (2022) Atomic structure of the predominant GII.4 human norovirus capsid reveals novel stability and plasticity. Nat Commun 13:1241. https://doi.org/10.1038/s41467-022-28757-z
23. Huang P, Farkas T, Marionneau S, Zhong W, Ruvoën-Clouet N, Morrow AL, Altaye M, Pickering LK, Newburg DS, LePendu J, Jiang X (2003) Noroviruses bind to human ABO, lewis, and secretor histo-blood group antigens: identification of 4 distinct strain-specific patterns. J INFECT DIS 188:19–31. https://doi.org/10.1086/375742
24. Jiang X, Graham DY, Wang K, Estes MK (1990) Norwalk virus genome cloning and characterization. Science 250:1580–1583. https://doi.org/10.1126/science.2177224
25. Jiang X, Wang M, Wang K, Estes MK (1993) Sequence and genomic organization of norwalk virus. Virology 195:51–61. https://doi.org/10.1006/viro.1993.1345
26. Jung J, Grant T, Thomas DR, Diehnelt CW, Grigorieff N, Joshua-Tor L (2019) High-resolution cryo-EM structures of outbreak strain human norovirus shells reveal size variations. Proc Natl Acad Sci USA 116:12828–12832. https://doi.org/10.1073/pnas.1903562116
27. Kadek A, Lorenzen K, Uetrecht C (2021) In a flash of light: X-ray free electron lasers meet native mass spectrometry. Drug Discov Today Technol 39:89–99. https://doi.org/10.1016/j.ddtec.2021.07.001
28. Kapikian AZ, Wyatt RG, Dolin R, Thornhill TS, Kalica AR, Chanock RM (1972) Visualization by immune electron microscopy of a 27-nm particle associated with acute infectious nonbacterial gastroenteritis. J Virol 10:1075–1081. https://doi.org/10.1128/jvi.10.5.1075-1081.1972
29. Katpally U, Voss NR, Cavazza T, Taube S, Rubin JR, Young VL, Stuckey J, Ward VK, Virgin HW, Wobus CE, Smith TJ (2010) High-resolution cryo-electron microscopy structures of murine norovirus 1 and rabbit hemorrhagic disease virus reveal marked flexibility in the receptor binding domains. J Virol 84:5836–5841. https://doi.org/10.1128/JVI.00314-10
30. Katpally U, Wobus CE, Dryden K, Virgin HW, Smith TJ (2008) Structure of antibody-neutralized murine norovirus and unexpected differences from viruslike particles. J Virol 82:2079–2088. https://doi.org/10.1128/JVI.02200-07
31. Kierspel T, Kadek A, Barran P, Bellina B, Bijedic A, Brodmerkel MN, Commandeur J, Caleman C, Damjanović T, Dawod I, De Santis E, Lekkas A, Lorenzen K, Morillo LL, Mandl T, Marklund EG, Papanastasiou D, Ramakers LAI, Schweikhard L, Simke F, Sinelnikova A, Smyrnakis A, Timneanu N, Uetrecht C (2023) Coherent diffractive imaging of proteins and viral capsids: simulating MS SPIDOC. Anal Bioanal Chem 415(18):4209–4220. https://doi.org/10.1007/s00216-023-04658-y. Epub 2023 Apr 4. PMID: 37014373
32. Kilic T, Koromyslova A, Hansman GS (2019) Structural basis for human norovirus capsid binding to bile acids. J Virol 93:e01581-e1618. https://doi.org/10.1128/JVI.01581-18
33. Kolawole AO, Xia C, Li M, Gamez M, Yu C, Rippinger CM, Yucha RE, Smith TJ, Wobus CE (2014) Newly isolated mAbs broaden the neutralizing epitope in murine norovirus. J Gen Virol 95:1958–1968. https://doi.org/10.1099/vir.0.066753-0
34. Koromyslova AD, Leuthold MM, Bowler MW, Hansman GS (2015) The sweet quartet: binding of fucose to the norovirus capsid. Virology 483:203–208. https://doi.org/10.1016/j.virol.2015.04.006
35. Laue M (2010) Electron microscopy of viruses. In: Methods in cell biology. Elsevier, pp 1–20. https://doi.org/10.1016/S0091-679X(10)96001-9
36. Lindesmith LC, Mallory ML, Debbink K, Donaldson EF, Brewer-Jensen PD, Swann EW, Sheahan TP, Graham RL, Beltramello M, Corti D, Lanzavecchia A, Baric RS (2018) Conformational occlusion of blockade antibody epitopes, a novel mechanism of GII.4 human norovirus immune evasion. mSphere 3:e00518–17. https://doi.org/10.1128/mSphere.00518-17

37. Lindsay L, Wolter J, De Coster I, Van Damme P, Verstraeten T (2015) A decade of norovirus disease risk among older adults in upper-middle and high income countries: a systematic review. BMC Infect Dis 15:425. https://doi.org/10.1186/s12879-015-1168-5
38. Lingemann M, Taube S (2018) Open sesame: new keys to unlocking the gate to norovirus infection. Cell Host Microbe 24:463–465. https://doi.org/10.1016/j.chom.2018.09.018
39. Luque D, Castón JR (2020) Cryo-electron microscopy for the study of virus assembly. Nat Chem Biol 16:231–239. https://doi.org/10.1038/s41589-020-0477-1
40. Luque D, González JM, Gómez-Blanco J, Marabini R, Chichón J, Mena I, Angulo I, Carrascosa JL, Verdaguer N, Trus BL, Bárcena J, Castón JR (2012) Epitope insertion at the N-terminal molecular switch of the rabbit hemorrhagic disease virus T=3 capsid protein leads to larger T=4 capsids. J Virol 86:6470–6480. https://doi.org/10.1128/JVI.07050-11
41. Lutomski CA, Lyktey NA, Zhao Z, Pierson EE, Zlotnick A, Jarrold MF (2017) Hepatitis B virus capsid completion occurs through error correction. J Am Chem Soc 139:16932–16938. https://doi.org/10.1021/jacs.7b09932
42. Ma J, Liu J, Zheng L, Wang C, Zhao Q, Huo Y (2022) Sequence addition to the N- or C-terminus of the major capsid protein VP1 of norovirus affects its cleavage and assembly into virus-like particles. Microb Pathog 169:105633. https://doi.org/10.1016/j.micpath.2022.105633
43. Makino A, Shimojima M, Miyazawa T, Kato K, Tohya Y, Akashi H (2006) Junctional adhesion molecule 1 is a functional receptor for feline calicivirus. J Virol 80:4482–4490. https://doi.org/10.1128/JVI.80.9.4482-4490.2006
44. Mallagaray A, Creutznacher R, Dülfer J, Mayer PHO, Grimm LL, Orduña JM, Trabjerg E, Stehle T, Rand KD, Blaum BS, Uetrecht C, Peters T (2019) A post-translational modification of human Norovirus capsid protein attenuates glycan binding. Nat Commun 10:1320. https://doi.org/10.1038/s41467-019-09251-5
45. Mallagaray A, Lockhauserbäumer J, Hansman G, Uetrecht C, Peters T (2015) Attachment of norovirus to Histo blood group antigens: a cooperative multistep process. Angew Chem Int Ed 54:12014–12019. https://doi.org/10.1002/anie.201505672
46. Mallagaray A, Rademacher C, Parra F, Hansman G, Peters T (2017) Saturation transfer difference nuclear magnetic resonance titrations reveal complex multistep-binding of l-fucose to norovirus particles. Glycobiology 27:80–86. https://doi.org/10.1093/glycob/cww070
47. Miller LM, Jarrold MF (2022) Charge detection mass spectrometry for the analysis of viruses and virus-like particles. Essays Biochem EBC20220101. https://doi.org/10.1042/EBC20220101
48. Mirabelli C, Santos-Ferreira N, Gillilland MG, Cieza RJ, Colacino JA, Sexton JZ, Neyts J, Taube S, Rocha-Pereira J, Wobus CE (2022) Human norovirus efficiently replicates in differentiated 3D-human intestinal enteroids. J Virol e00855–22. https://doi.org/10.1128/jvi.00855-22
49. Mohajerani F, Tyukodi B, Schlicksup CJ, Hadden-Perilla JA, Zlotnick A, Hagan MF (2022) Multiscale modeling of hepatitis B virus capsid assembly and its dimorphism. ACS Nano 16:13845–13859. https://doi.org/10.1021/acsnano.2c02119
50. Nelson CA, Wilen CB, Dai Y-N, Orchard RC, Kim AS, Stegeman RA, Hsieh LL, Smith TJ, Virgin HW, Fremont DH (2018) Structural basis for murine norovirus engagement of bile acids and the CD300lf receptor. Proc Natl AcadSci USA 115. https://doi.org/10.1073/pnas.1805797115
51. Netzler NE, Enosi D, White PA (2019) Norovirus antivirals: where are we now? Med Res Rev 39:860–886. https://doi.org/10.1002/med.21545
52. Orchard RC, Wilen CB, Doench JG, Baldridge MT, McCune BT, Lee Y-CJ, Lee S, Pruett-Miller SM, Nelson CA, Fremont DH, Virgin HW (2016) Discovery of a proteinaceous cellular receptor for a norovirus. Science 353:933–936. https://doi.org/10.1126/science.aaf1220
53. Parikh MP, Vandekar S, Moore C, Thomas L, Britt N, Piya B, Stewart LS, Batarseh E, Hamdan L, Cavallo SJ, Swing AM, Garman KN, Constantine-Renna L, Chappell J, Payne DC, Vinjé J, Hall AJ, Dunn JR, Halasa N (2020) Temporal and genotypic associations of sporadic norovirus gastroenteritis and reported norovirus outbreaks in middle tennessee, 2012–2016. Clin Infect Dis 71:2398–2404. https://doi.org/10.1093/cid/ciz1106

54. Peters T, Creutznacher R, Maass T, Mallagaray A, Ogrissek P, Taube S, Thiede L, Uetrecht C (2022) Norovirus–glycan interactions—how strong are they really? Biochem Soc Trans 50:347–359. https://doi.org/10.1042/BST20210526
55. Pogan R, Dülfer J, Uetrecht C (2018) Norovirus assembly and stability. Curr Opin Virol 31:59–65. https://doi.org/10.1016/j.coviro.2018.05.003
56. Pogan R, Weiss VU, Bond K, Dülfer J, Krisp C, Lyktey N, Müller-Guhl J, Zoratto S, Allmaier G, Jarrold MF, Muñoz-Fontela C, Schlüter H, Uetrecht C (2020) N-terminal VP1 truncations favor T = 1 norovirus-like particles. Vaccines 9:8. https://doi.org/10.3390/vaccines9010008
57. Pogan R, Schneider C, Reimer R, Hansman G, Uetrecht C (2018b) Norovirus-like VP1 particles exhibit isolate dependent stability profiles. J Phys: Condens Matter 30:064006. https://doi.org/10.1088/1361-648X/aaa43b
58. Pohl C, Szczepankiewicz G, Liebert UG (2022) Analysis and optimization of a Caco-2 cell culture model for infection with human norovirus. Arch Virol 167:1421–1431. https://doi.org/10.1007/s00705-022-05437-3
59. Prasad BVV, Hardy ME, Dokland T, Bella J, Rossmann MG, Estes MK (1999) X-ray crystallographic structure of the norwalk virus capsid. Science 286:287–290. https://doi.org/10.1126/science.286.5438.287
60. Prasad BV, Rothnagel R, Jiang X, Estes MK (1994) Three-dimensional structure of baculovirus-expressed Norwalk virus capsids. J Virol 68:5117–5125. https://doi.org/10.1128/jvi.68.8.5117-5125.1994
61. Ruoff K, Kilic T, Devant J, Koromyslova A, Ringel A, Hempelmann A, Geiss C, Graf J, Haas M, Roggenbach I, Hansman G (2019) Structural basis of nanobodies targeting the prototype norovirus. J Virol 93:e02005-e2018. https://doi.org/10.1128/JVI.02005-18
62. Sherman MB, Williams AN, Smith HQ, Nelson C, Wilen CB, Fremont DH, Virgin HW, Smith TJ (2019) Bile salts alter the mouse norovirus capsid conformation: possible implications for cell attachment and immune evasion. J Virol 93:e00970-e1019. https://doi.org/10.1128/JVI.00970-19
63. Sherman MB, Williams AN, Smith HQ, Pettitt BM, Wobus CE, Smith TJ (2021) Structural Studies on the Shapeshifting Murine Norovirus. Viruses 13:2162. https://doi.org/10.3390/v13112162
64. Shoemaker GK, van Duijn E, Crawford SE, Uetrecht C, Baclayon M, Roos WH, Wuite GJL, Estes MK, Prasad BVV, Heck AJR (2010) Norwalk virus assembly and stability monitored by mass spectrometry. Mol Cell Proteomics 9:1742–1751. https://doi.org/10.1074/mcp.M900620-MCP200
65. Singh BK, Leuthold MM, Hansman GS (2015) Human noroviruses' fondness for histo-blood group antigens. J Virol 89:2024–2040. https://doi.org/10.1128/JVI.02968-14
66. Smith H, Smith T (2019) The dynamic capsid structures of the noroviruses. Viruses 11:235. https://doi.org/10.3390/v11030235
67. Snowden JS, Hurdiss DL, Adeyemi OO, Ranson NA, Herod MR, Stonehouse NJ (2020) Dynamics in the murine norovirus capsid revealed by high-resolution cryo-EM. PLoS Biol 18:e3000649. https://doi.org/10.1371/journal.pbio.3000649
68. Song C, Takai-Todaka R, Miki M, Haga K, Fujimoto A, Ishiyama R, Oikawa K, Yokoyama M, Miyazaki N, Iwasaki K, Murakami K, Katayama K, Murata K (2020) Dynamic rotation of the protruding domain enhances the infectivity of norovirus. PLoS Pathog 16:e1008619. https://doi.org/10.1371/journal.ppat.1008619
69. Straniero S, Laskar A, Savva C, Härdfeldt J, Angelin B, Rudling M (2020) Of mice and men: murine bile acids explain species differences in the regulation of bile acid and cholesterol metabolism. J Lipid Res 61:480–491. https://doi.org/10.1194/jlr.RA119000307
70. Tan M (2021) Norovirus vaccines: current clinical development and challenges. Pathogens 10:1641. https://doi.org/10.3390/pathogens10121641
71. Taube S, Rubin JR, Katpally U, Smith TJ, Kendall A, Stuckey JA, Wobus CE (2010) High-resolution X-ray structure and functional analysis of the murine norovirus 1 capsid protein protruding domain. J Virol 84:5695–5705. https://doi.org/10.1128/JVI.00316-10

72. Tresset G, Le Coeur C, Bryche J-F, Tatou M, Zeghal M, Charpilienne A, Poncet D, Constantin D, Bressanelli S (2013) Norovirus capsid proteins self-assemble through biphasic kinetics via long-lived stave-like intermediates. J Am Chem Soc 135:15373–15381. https://doi.org/10.1021/ja403550f
73. Uetrecht C, Heck AJR (2011) Modern biomolecular mass spectrometry and its role in studying virus structure, dynamics, and assembly. Angew Chem Int Ed 50:8248–8262. https://doi.org/10.1002/anie.201008120
74. Van Dycke J, Ny A, Conceição-Neto N, Maes J, Hosmillo M, Cuvry A, Goodfellow I, Nogueira TC, Verbeken E, Matthijnssens J, de Witte P, Neyts J, Rocha-Pereira J (2019) A robust human norovirus replication model in zebrafish larvae. PLoS Pathog 15:e1008009. https://doi.org/10.1371/journal.ppat.1008009
75. Vinjé J, Estes MK, Esteves P, Green KY, Katayama K, Knowles NJ, L'Homme Y, Martella V, Vennema H, White PA (2019) ICTV report consortium, 2019. ICTV virus taxonomy profile: caliciviridae. J General Virol 100:1469–1470. https://doi.org/10.1099/jgv.0.001332
76. Vongpunsawad S, Venkataram Prasad BV, Estes MK (2013) Norwalk virus minor capsid protein VP2 associates within the VP1 shell domain. J Virol 87:4818–4825. https://doi.org/10.1128/JVI.03508-12
77. Wegener H, Mallagaray Á, Schöne T, Peters T, Lockhauserbäumer J, Yan H, Uetrecht C, Hansman GS, Taube S (2017) Human norovirus GII.4(MI001) P dimer binds fucosylated and sialylated carbohydrates. Glycobiology 27:1027–1037. https://doi.org/10.1093/glycob/cwx078
78. White LJ, Hardy ME, Estes MK (1997) Biochemical characterization of a smaller form of recombinant Norwalk virus capsids assembled in insect cells. J Virol 71:8066–8072. https://doi.org/10.1128/jvi.71.10.8066-8072.1997
79. Williams AN, Sherman MB, Smith HQ, Taube S, Pettitt BM, Wobus CE, Smith TJ (2021) A norovirus uses bile salts to escape antibody recognition while enhancing receptor binding. J Virol 95:e00176-e221. https://doi.org/10.1128/JVI.00176-21
80. Yan H, Lockhauserbäumer J, Szekeres GP, Mallagaray A, Creutznacher R, Taube S, Peters T, Pagel K, Uetrecht C (2021) Protein secondary structure affects glycan clustering in native mass spectrometry. Life 11:554. https://doi.org/10.3390/life11060554
81. Zeltins A (2013) Construction and characterization of virus-like particles: a review. Mol Biotechnol 53:92–107. https://doi.org/10.1007/s12033-012-9598-4

Chapter 9
Structural Alterations in Non-enveloped Viruses During Disassembly

Kimi Azad, Debajit Dey, and Manidipa Banerjee

Abstract The capsid shells of non-enveloped viruses can be categorized as large macromolecules constituted from protein building blocks, with the primary purpose of protecting and delivering the viral genome inside living cells during an infection. Non-enveloped virus capsids incorporate stability as well as dynamicity, which allows them to survive inhospitable conditions in the environment, and disassemble within host cells to ensure transfer of the genome for downstream replication and translation. In this chapter, we discuss the metastability of the non-enveloped capsids, ensured by molecular switches, that allow the highly stable assemblies to be destabilized under mild but specific physiological conditions. The structural alterations induced in non-enveloped capsids as well as in nucleocapsids of enveloped viruses during disassembly of the capsid shell, culminating in the release of genome, is extensively discussed. Disassembly is a key step in the establishment of infection and any spatiotemporal alteration in this step may diminish infectivity. Thus, a detailed understanding of the molecular pathways of disassembly may result in the development of more effective intervention methods and antiviral compounds.

Keywords Capsid · Non-enveloped · Nucleocapsid cores · Disassembly · Cryoelectron microscopy · X-ray crystallography · Cellular cues

---

Kimi Azad and Debajit Dey—Equal contribution.

K. Azad
Institut Pasteur, 25-28 Rue du Dr Roux, 75015 Paris, France
e-mail: kimi.azad@pasteur.fr

D. Dey
School of Medicine, University of Maryland, Rockville, MD, USA
e-mail: ddey@som.umaryland.edu

M. Banerjee (✉)
Kusuma School of Biological Sciences, Indian Institute of Technology, Delhi, New Delhi 110016, India
e-mail: mbanerjee@bioschool.iitd.ac.in

M. Comas-Garcia and S. Rosales-Mendoza (eds.), *Physical Virology*, Springer Series in Biophysics 24, https://doi.org/10.1007/978-3-031-36815-8_9

Overview of Non-enveloped Capsid Disassembly

Non-enveloped virus capsids contain highly stable, homogeneous, protein shells organized in either icosahedral or helical fashion. The stability of the capsid is designed to ensure protection and transportation of the infectious genome packaged within the shell; however, establishment of infection in a host cell requires partial or complete disassembly of the shell to allow release of genome [1]. The optimal functionality of a non-enveloped capsid thus requires the incorporation of stability and dynamicity in the same structural unit. These capsids usually undergo a post-assembly maturation process, typically involving proteolytic cleavage, which imparts metastability to the structure [2, 3]. The initial stages in host cell entry such as receptor binding, low endosomal pH, compartment-specific presence of proteases, reduced cytoplasmic Ca^{2+} concentration etc. induce conformational alterations in the particles which first result in the exposure of sequestered, relatively flexible capsid components [4], followed by partial or complete dissociation of the capsid structure and externalization of genome [5]. These structural alterations may be concurrent or sequential, and are primarily observed by triggering disassembly in capsids in vitro.

The majority of non-enveloped animal viruses enter cells through the endosomal route (Fig. 9.1); thus an essential function of the initially exposed, flexible components is the penetration of the membrane compartments in the late endosomal stage [4]. These components are usually membrane-active peptides, or membrane-lytic enzymes such as phospholipases, that utilize diverse mechanisms for cellular membrane disruption [4, 5]. In addition to the endosomal route, non-enveloped animal viruses also undergo partial or complete disassembly in cellular compartments such as the Endoplasmic Reticulum (SV40), Nucleus (Adenovirus) and Lysosomes (Hepatitis A Virus) [6–8] (Fig. 9.1). Partial disassembly that makes particles competent for genome release, has also been reported at the plasma membrane through receptor interaction (Poliovirus) [9–11]. Non-enveloped plant viruses with helical or icosahedral geometry like Tobacco Mosaic Virus or Brome Mosaic Virus also appear to have molecular switches that allow them to disassemble upon exposure to altered conditions such as basic pH, alteration in divalent cation concentration etc. [12–14]; while the release of genome from bacteriophages is a concerted and complex process requiring injection of the genome in the bacterial cytoplasm [15]. The essential factor in these transitions is the dynamicity of the particles that allow the requisite conformational changes. While the cellular triggers or molecular switches for disassembly are by no means similar across the board; specific features like proteolytic cleavage during maturation, exposure of flexible capsid components in the early stages of disassembly, capsid stability-enhancing mutations decreasing uncoating efficiency etc. appear to be common in certain groups of non-enveloped viruses (Table 9.1).

Capturing structural details of non-enveloped particles in the process of disassembly is challenging, as isolating disassembly intermediates in a relatively homogeneous and stable form is not an easy task. In the absence of a homogeneous and concentrated population of "altered" particles, X-ray crystallography cannot be utilized [16]. The recent advances in cryo-electron microscopy and single particle

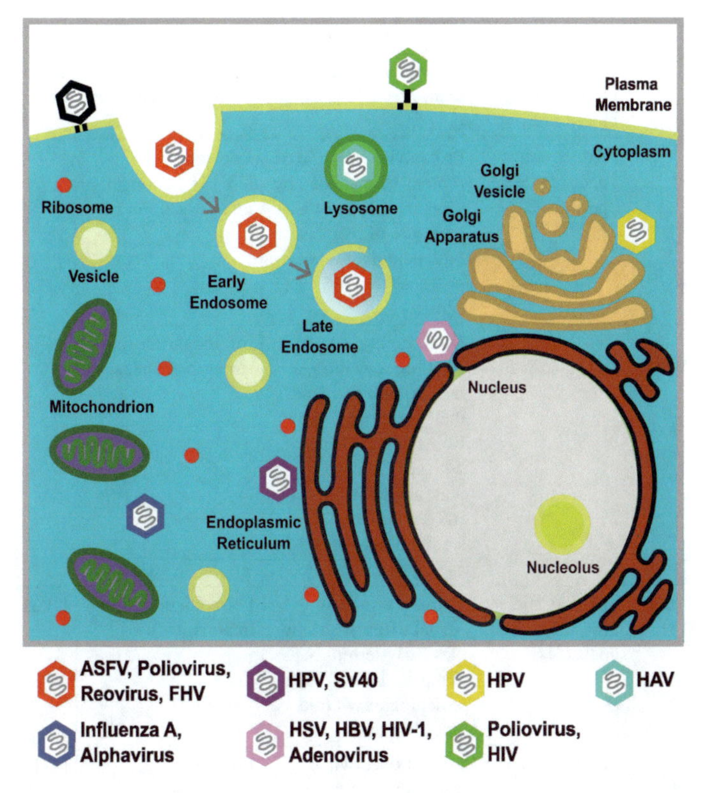

Fig. 9.1 A Schematic highlighting cellular compartments utilized for capsid and nucleocapsid disassembly. Representation of a cell with examples of viruses (bottom; marked with color) that undergo partial or complete disassembly at different cellular compartments. Endosome: red; cytoplasm: dark blue; endoplasmic reticulum: magenta; light pink: nucleus; golden: Golgi apparatus; dark green: plasma membrane; cyan: lysosome. ASFV: African swine fever virus; FHV: Flock House virus; HPV: Human papillomavirus; SV40: Simian virus 40; HSV: Herpes simplex virus; HBV: Hepatitis B virus; HIV: Human immunodeficiency virus; HAV: Hepatitis A virus

reconstruction [17] have made it possible to computationally assess and sort out multiple disassembly states in the same population, and to build a molecular route towards disassembly in case of some viruses [18–20]. Cryo-electron tomography is another technique that is potentially applicable towards resolving structural details of disassembly intermediates in situ, however, this method is currently limited by resolution [21]. Computational methods like Molecular Dynamics simulations have also been applied to understand conformational alterations during virus disassembly with varied success [22].

In this Chapter, we have attempted to provide a comprehensive discussion of the structural alterations in non-enveloped virus capsids, with animal, plant or bacterial hosts, during disassembly or uncoating. We begin the discussion with a description of the "molecular switches" built in particles during maturation which are required for structural alterations in response to cellular triggers. These switches impart stability

Table 9.1 Disassembly intermediates of non-enveloped viruses

Virus	Disassembly intermediates identified	Structural alterations identified	Mode of generation of intermediates	Reference(s)
Adenovirus	Adenovirus disassembling particles	Proteolytic cleavage of protein VI mediated by viral capsid protease, exposure of membrane lytic peptide, loss of pentons from the capsid	Receptor binding, low pH, heating, mechanical fatigue	[44, 125, 167]
Brome Mosaic Virus (BMV)	BMV intermediate particles	Disassembly via a swelling transition, a near complete capsid that has released the RNA, empty shells with small missing part, and the released RNA complexed to a small number of the capsid proteins	pH and buffer ionic strength changes	[13]
Coxsackievirus A16 (CAV16)	120S particles (or 135S-like particles)	Capsid expansion, partial externalization of VP1 N-terminus, complete loss of VP4, widening of twofold axis channel	Formaldehyde treatment	[103]
Echovirus 18	Open particles	Loss of one, two, or three adjacent capsid-protein pentamers, capsid opening of more than 120 Å in diameter enabling the release of double-stranded RNA genome	Low pH	[19]
Enterovirus 71 (EV71)	82S particles (or 80S-like particles)	Capsid expansion, partial externalization of VP1 N-terminus, complete loss of VP4, widening of twofold axis channel	Formaldehyde treatment	[104]
Equine rhinitis A virus (ERAV)	Transient 80S particles	No capsid expansion, complete loss of RNA from intact intermediate before quick dissociation to pentameric subunits	Low pH	[86]

(continued)

Table 9.1 (continued)

Virus	Disassembly intermediates identified	Structural alterations identified	Mode of generation of intermediates	Reference(s)
Flock House Virus (FHV)	Eluted and Puffed Particles	Capsid contraction, global loss of γ peptide, localized loss of capsid protein components, compaction of genome into rod-shaped structure, genome release via twofold axes, capsid disassembly following RNA release	Incremental heating	[20]
Israeli Acute Bee Paralysis Virus (IAPV)	Empty IAPV particles	No capsid expansion, no pore formation in capsid, detachment of VP4, partial loss of structure of VP2 N-terminus, loss of RNA	Incremental heating	[100]
Orthoreovirus	ISVP (Infectious subviral particle) and ISVP*	Autocleavage of outer capsid protein μ1, release of myristoylated μ1N peptide	Protease treatment, heating and incubation with monovalent cations	[40]
Poliovirus	135S particles, empty 80S particles	Capsid expansion, externalization of VP1 N-terminus, loss of VP4, opening of large holes at twofold and quasi-threefold symmetry axes, loss of RNA	Receptor binding, heating	[85, 96, 97]
Reovirus	Infectious subviral particles (ISVPs); core particles	Release of σ3, structural rearrangements in σ1; release of σ1 and μ1, structural rearrangements in λ2	Protease treatment, exposure to monovalent cations	[88]
Rhinovirus (HRV2)	A particles, empty B particles	Capsid expansion, structural rearrangements in VP1, VP2 and VP3, externalization of VP1 N-terminus, loss of VP4, opening of channels at canyon base and twofold axis, altered association with RNA with eventual loss of RNA from capsid	Heating, low pH	[99]

(continued)

Table 9.1 (continued)

Virus	Disassembly intermediates identified	Structural alterations identified	Mode of generation of intermediates	Reference(s)
Rotavirus	Rotavirus intermediate particle	Spike protein, VP4 cleavage into VP5* and VP8* by cellular proteases, VP5* rearrangement and membrane binding	Protease treatment, reduced Ca^{2+} concentration	[45, 46, 48–50]
Triatoma virus	Transient empty TrV particles	No capsid expansion or shell thinning, complete loss of RNA from intact intermediate before quick dissociation into small symmetrical subunits	Heating	[98]

and dynamicity to the capsids and are typically introduced post-assembly, and sometimes during virus entry as well. We then discuss the techniques that have been widely utilized for understanding conformational alterations in detail; followed by sections on preparation or isolation of disassembly intermediates, initial stages in structural alterations, progress of disassembly and global alterations, and molecular pathways of genome release. In addition to non-enveloped virus dynamicity and disassembly, mechanism of nucleocapsid disassembly in context of enveloped viruses has also been discussed. We conclude with the importance of studying disassembly for understanding non-enveloped virus infectivity and the prospects for developing intervention methods. It is important to note that while the sections have been devised to provide a broad overview of the steps involved in disassembly, not all non-enveloped viruses go through the same sequence of events for genome release. Also, while we have attempted to include a cross-section of significant work on non-enveloped viruses in this Chapter, the vastness of the topic might have prevented the inclusion of all relevant research. We apologize to all colleagues whose work could not be included.

Virus Maturation and Structural Changes: A Prelude to Disassembly

Virus maturation is a process that converts a non-infectious capsid into an infectious virion. The primary purpose of this event is to ensure that the particles are stable as well as dynamic, and capable of irreversible conformational changes upon reaching the correct energy landscape in the host cell. Maturation typically involves at least one proteolytic cleavage event, and is either executed by a viral protein in response to internal or external cues; or by a cellular protease [2, 3]. The process in many non-enveloped mammalian and insect viruses, as well as bacteriophages, involve covalent separation of membrane penetrating peptides and strengthening of inter-subunit association [2–4].

Maturation typically occurs post-assembly, and during or after the release of the particles in the extracellular milieu [2, 3]. Such spatiotemporal specification is to ensure favourable placement of inner capsid components such as membrane active peptides [4], which should ideally be exposed after release from infected host cells, and not within. The maturation cleavage in nodaviruses and tetraviruses have been studied in biochemical and structural detail, and appear to follow similar molecular pathways (Fig. 9.2). The capsids of nodaviruses have icosahedral $T = 3$ geometry and are generated from 180 copies of a chemically identical capsid protein, α [23]. Post-assembly, each copy of α undergoes autoproteolytic cleavage into β and γ subunits [24] (Fig. 9.2a). While β forms the inner and outer capsid shell, γ is a membrane active peptide that remains associated with the inner surface of the capsid [25]. The cleavage of α into β and γ is orchestrated by an aspartate residue at position 75 [26]. Post-assembly, Asp75 comes in proximity with the cleavage site, between AsN363 and Gly364, and functions as a base to polarize the side chain of AsN363, resulting in a nucleophilic attack on the main chain carbonyl by the latter (Fig. 9.2a). The resultant cyclic imide intermediate is hydrolysed by a water molecule, resulting in the cleavage of the peptide bond between residues 363 and 364 [26]. The cleavage mechanism depends on the local hydrophobic environment around Asp75, which causes its nucleophilic behaviour. This highlights the importance of correct assembly for maturation cleavage. Indeed, the proteolytic maturation for nodaviruses appears to be entirely assembly-dependent. The cleavage of subunits proceeds spontaneously upon assembly and cannot be controlled or regulated by other external factors like pH.

Tetraviruses, on the other hand, undergo a pH controlled autoproteolytic maturation process which is mechanistically similar to the nodavirus cleavage [27]. In case of tetraviruses like NωV (*Nudaurelia capensis* omega virus), maturation is associated with the compaction of the assembled particle from a diameter of 480–410 Å [28]. Maturation is triggered by apoptotic conditions in the parent cell, which reduces the environmental pH to ~5.0, causing charge neutralization of acidic residues in neighbouring subunits, consequent reduction of electrostatic repulsion and closer association between subunits [3, 29, 30]. This is followed by autoproteolytic cleavage of all capsid proteins (α) into β and γ, which proceeds with varied speed for the four different subunits in an icosahedral asymmetric unit (iASU) in the $T = 4$ capsid [31] (Fig. 9.2b). The cleavage for A and B subunits proceeds relatively faster and reaches completion within the first 10 min [31]. C and D subunits undergo a slower cleavage, which eventually results in the wedging of the covalently separated γ peptides wedging into the subunit contacts to form “flat” surfaces [31] (Fig. 9.2b).

The temporal difference in the covalent separation of the γ peptides has been shown to have biological significance in case of NωV. The γ peptides from both nodaviruses and tetraviruses, either in synthetic form or in context of the capsid, are capable of liposomal membrane disruption and are involved in endosomal membrane

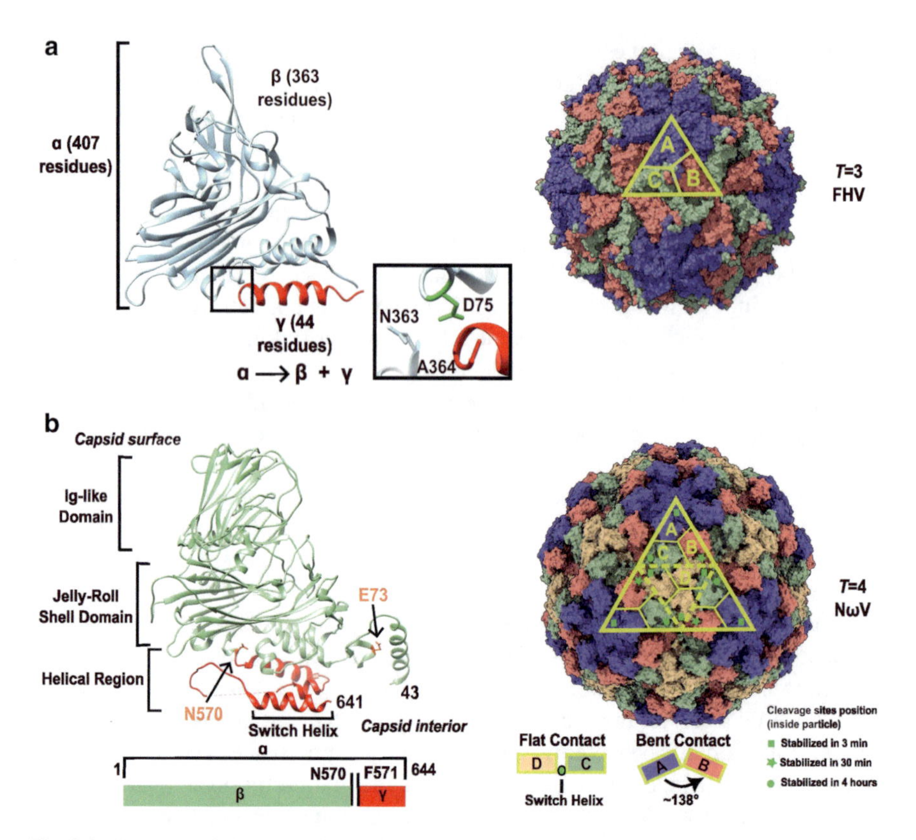

Fig. 9.2 Autoproteolytic maturation cleavage in Flock House virus (FHV) and Nudaurelia capensis omega virus (NωV). **a** Ribbon representation (left) of the α, β, and γ (red) regions of the capsid protein. The catalytic residue D75 (green) is involved in the autocatalytic maturation cleavage between N363 and A364 residues. Crystal structure (right) of the mature T = 3 icosahedral FHV capsid (PDB ID: 4FTB) with an icosahedral asymmetric unit (iASU), marked with a yellow triangle, containing A (blue), B (salmon) and C (green) subunits. **b** Ribbon representation (left) of one subunit of the icosahedral T = 4 NωV particle (PDB ID: 1OHF). One arrow points to the autocatalytic maturation cleavage site at the N570 residue. Another arrow highlights the E73 residue, where mutations result in maturation defects. Crystal structure of the mature T = 4 NωV icosahedral capsid is shown on the right with subunits A (blue), B (salmon), C (green) and D (tan). Bent contacts (~ 138°) and flat contacts in the capsid are represented by yellow solid and dashed lines, respectively. Cleavage sites stabilizing at different time points in the wild-type particle are also highlighted by different green symbols on the capsid protein subunits

penetration during virus entry [32, 33]. The peptides at the fivefold axes of symmetry of the capsids form a pentameric, amphipathic, helical bundle which is structurally indicative of a membrane channel or pore; while the peptides at twofold or threefold axes of symmetry are oriented differently [34]. In NωV, complete maturation cleavage of only the A subunit results in the particle reaching its maximum membrane penetrating ability, indicating that the peptides at the A subunit are primarily responsible for cellular membrane penetration during endosomal entry of the virus, while those from the other subunits likely have different functions such as association with genomic RNA and providing conformational stability to the particle [31, 32]. While a similar distinction in the functionalities of the γ peptide based on its icosahedral positioning has also been proposed in the case of Flock House Virus (FHV) [34], the inability to control maturation cleavage in this case makes the experimental proof untenable.

Membrane penetrating peptides like γ, after being covalently separated from non-enveloped capsids during maturation, are transiently exposed from the capsid interior, in a process termed “breathing” [35, 36]. This phenomenon indicates the dynamic character of a mature virion capable of establishing cellular infection and distinguishes infectious particles from immature, non-infectious ones. Several lines of evidence involving biophysical studies such as calorimetry and mass spectroscopy has established that the inhibition of maturation cleavage in capsids, by mutating the relevant catalytic or substrate residues, results in particles that are relatively more stable than mature, infectious particles, and display significantly reduced dynamicity of the capsid-sequestered membrane active components [20, 37]. Thus, maturation cleavage alters the physical properties of the capsid, imparts dynamicity and puts it in an energy landscape conducive to eventual dismantling and genome release.

Picornaviruses, Reoviruses and Adenoviruses also undergo autoproteolytic maturation cleavage processes that result in covalent separation of the membrane penetrating component. Picornaviruses contain a pseudo $T = 3$ capsid initially generated from 60 copies each of the capsid proteins VP0, VP1 and VP3. Autocatalytic cleavage of VP0 into VP2 and the membrane active peptide VP4 catalysed by His195 of VP2, and results in the breakage of the peptide bond between AsN68 of VP2 and Ser1 of VP4 [37]. The VP4 peptide is myristoylated at the N-terminus, and remains associated with the interior surface of the capsid till host cell interaction, when it is externalized [11]. Hepatitis A Virus (HAV), a picornavirus, contains an unusual non-myristoylated VP4 component, which is nonetheless capable of membrane penetration [38]. A similar processing occurs in mammalian orthoreovirus, a large $T = 13\ l$ icosahedral virus, where the $\mu 1$ capsid component autocatalytically cleaves to produce a membrane active, 42 residue, myristoylated peptide $\mu 1N$ [39]. The cleavage occurs between residues AsN42 and Pro43 of $\mu 1$, which is embedded within the trimeric form of $\mu 1$ in a $\mu 1$-$\sigma 3$ complex [39, 40]. The $\mu 1$-$\sigma 3$ complex on the outer surface of Orthoreovirus is thought to protect the entry-specific components in the extracellular milieu. The virus particle goes through multiple steps of capsid protein cleavage and release during cellular entry, which results in the formation of discrete disassembly states such as ISVP and ISVP* [39, 40]. It is thought that, given the position of the $\mu 1$ cleavage site, the maturation cleavage step is partially associated with and triggered by step-wise disassembly [40–43] (Table 9.1).

Apart from auto-cleavage, viral capsid associated proteases as well as cellular proteases are involved in the maturation of animal viruses. In case of adenovirus, the cellular membrane disrupting component constitutes residues 34–54 of protein VI which forms an amphipathic helix [44]. The proteolytic cleavage of protein VI, allowing the exposure of this region under specific cellular conditions, is mediated by the viral protease which is also part of the capsid [44]. In case of the triple-layered rotavirus particles, the outer layer consists of capsid proteins VP7 and VP4. The VP4 protein, which forms spikes on the virion surface, is cleaved into VP5* and VP8* by cellular proteases during rotavirus entry. This cleavage rigidifies the spikes from a relatively flexible conformation and increases the infectivity of the particles significantly [45–48]. While this maturation cleavage is primarily dependent on trypsin in cell culture, it has been shown that host proteases TMPRSS2 and TMPRSS11D mediate the trypsin-independent cleavage in host cells [49, 50]. In all such cases, the role of maturation cleavage is to ensure unidirectional progress of conformational alterations in the capsid during host cell entry. It has been shown in case of tetraviruses that in the absence of maturation cleavage, some of the low pH triggered alterations in the capsid can also be reversed [50].

In the case of several non-enveloped viruses discussed above, the importance of proteolytic maturation cleavage in disassembly is highlighted by the inability of maturation defective, uncleaved particles to infect host cells. Many such particles were unable to release the packaged genome in a spatiotemporally correct fashion to initiate infection. In FHV, mutations at the AsN residue at the cleavage site or the catalytic Asp can result in the formation of maturation defective particles that do not undergo auto-cleavage of the capsid protein α into β and γ [24, 26]. The particles generated are structurally similar to wild-type, infectious FHV, however, they do not display the phenotypic alterations typical of disassembly, when such conditions are triggered by incremental heating in vitro [20]. The increased stability of the capsid, and inability to follow the same conformational pathway to disassembly as wildtype particles, appears to lead to the non-infectious phenotype [20]. Similar maturation defective variants, that are unable to release capsid components to mediate the early stages of infection, have been studied in Adenovirus, Poliovirus and Orthoreovirus [39, 51, 52].

Bacteriophages undergo multi-step maturation processes requiring proteolytic cleavage, formation of covalent linkages and association with accessory proteins to ensure the stability of the capsid head [2]. The majority of the bacteriophages belong to the order *Caudovirales* and contain a head–tail structure, where an icosahedral head packaging the dsDNA genome is associated with an elongated tail. Based on the length and contractile nature of the tail, the order is divided into three families—*Myoviridae*, represented by bacteriophage T4; *Podoviridae*, represented by phage P22 and *Siphoviridae*, of which HK97 and λ bacteriophage are prominent members [2]. Maturation of the capsids follow an elaborate assembly of the icosahedral head, packaging of the genome to "head-full" capacity and association of the tail to complete the structure. Several transition states in the maturation pathway of bacteriophage HK97, which has a T-7*l* icosahedral head, have been structurally characterized [53]. Following the assembly of the head from constituent proteins, the phage protease is activated and cleaves the scaffolding domain of the major capsid protein as well as itself, resulting in their removal through the large pores in the metastable capsid. This is followed by packaging of the genome, expansion and thinning of the capsid shell and formation of isopeptide bonds, reminiscent of chain-mail, between Lys169 of one subunit and Asn356 of an adjacent subunit, throughout the icosahedral head [53]. During this transition, significant structural movements in the head results in the symmetrisation and flattening of the hexamers from a skewed conformation [54]. The final mature form, termed Head-II, is a highly stable structure due to the interlocking of protein subunits throughout the capsid [53]. In this context, it should be noted that the release of packaged genome from bacteriophages is through an injection syringe like structure, and differs considerably from the corresponding process in animal viruses. It is quite possible that in bacteriophage maturation, stability and not dynamicity, is the primary consideration. Bacteriophages like the λ phage, incorporate cementing proteins during maturation, which fastens adjacent capsid proteins together, thus ensuring maximum stability of the structure and preventing premature exposure or release of the genome [55].

An extraordinary example of evolutionary similarity has been noted in the context of assembly/maturation of tailed bacteriophages and Herpes Simplex Virus (HSV), a mammalian virus with a complex architecture. Post-assembly maturation events including protease activation, proteolytic cleavage, removal of the scaffolding protein and subsequent symmetrisation of the hexamers and strengthening of lateral contacts between adjacent capsomeres are similar features between HSV and *Caudovirales*, indicating that they may have been descended from a common ancestor [56, 57].

Methods for Capturing Structural Alterations During Disassembly: X-Ray Crystallography, Cryo-electron Microscopy, Cryo-tomography and Dynamic Simulation

Crystallography and cryo-electron microscopy are two powerful techniques that have shaped the field of virus structure determination. These methods have provided robust platforms to study the three-dimensional (3D) structure of viruses at atomic level, which is crucial for understanding the mechanisms of virus disassembly, genome release and neutralization by antiviral drugs.

Protein X-ray crystallography is the approach for determining 3D structure of proteins at atomic resolution. This technique relies on the ability of proteins to form crystals with well-defined symmetry and that of X-rays to diffract off the regular arrangement of atoms in a crystal. The methodology involves growing large, high-quality crystals of a protein of interest, and then bombarding these crystals with X-rays to produce a diffraction pattern. This pattern can then be computationally analyzed to determine the positions of all the atoms in the protein, as well as any other associated molecules or ligands [58]. Advances in X-ray sources, detectors, and computational methods have tremendously enhanced the speed and accuracy of protein crystallography in the last few decades, making it possible to study an increasing number of protein complexes, including those that were previously considered difficult to crystallize. Crystallography has played a pivotal role in understanding the structure of viruses [59]. Tomato Bushy Stunt Virus (TBSV), a member of the

Tombusviridae family, was the first ever virus structure solved using x-ray crystallography in 1984 [60]. This was closely followed by the structures of Southern Bean Mosaic Virus (SBMV) (PDB ID: 4SBV) and Human Rhinovirus (2,856,083). X-ray crystallography has contributed to a total of 437 PDB depositions of high resolution virus structures till date.

Compared to X-ray crystallography, cryo-electron microscopy (cryo-EM) and single particle reconstruction (SPR) is a relatively new addition in the structural biologist's toolbox [61]. Cryo-EM and SPR has gained tremendous momentum in the last 10 years as the method of choice for determining virus structures. For cryo-EM imaging, viruses are flash-frozen at ultra-low temperatures in liquid ethane which helps to preserve the native conformation of biological macromolecules. Two-dimensional EM images or micrographs collected at low electron doses are then used to create a 3D reconstruction of the virus using image processing packages like EMAN [62], RELION [63], cryoSPARC [64] etc. Cryo-EM has been used to determine the structures of a large number of pathogenic non-enveloped viruses like rotavirus, orthoreovirus, adenovirus etc. With the development of single-particle cryo-EM, it is now feasible to identify less stable or transient virus structures like disassembly/genome release intermediates. Single particle reconstruction of icosahedral viruses typically incorporates icosahedral geometric parameters in the protocol, which have the effect of averaging of density across all asymmetric units. Recently, several reconstructions of disassembly intermediates have been carried out without application of symmetry, in order to identify local alterations in the capsid. The Electron Microscopy Data Bank (EMDB) currently holds 2461 depositions in its repository for virus structures.

Electron cryo-tomography is a method that analyzes protein structures in situ, and can be applied to study virus structures in the context of host cells. This technique requires generation of ultrathin cell sections—either by cryo-sectioning or by Focused Ion Beam Milling (FIB milling)—of 50–100 nm thickness. For cryo-sectioning, the sample is flash frozen in a cryogenic medium, such as liquid nitrogen and transferred to a cryo-ultramicrotome where it is sectioned using a diamond knife pre-cooled to −180 °C. The resulting thin sections are collected on a supporting film, such as carbon or formvar, and can be stained with heavy metals, such as uranyl acetate and lead citrate, to increase contrast and improve imaging [65]. FIB milling involves the use of a beam of high-energy ions, typically gallium ions, to remove layers from vitrified samples that have been flash-frozen to preserve their native state. The sample is first mounted on a specialized holder, such as a cryo-FIB grid, which can maintain the sample at cryogenic temperatures during milling. The FIB is then used to selectively remove material from the sample in a precise and controlled manner, resulting in a thin section that can be imaged [66]. The sections are visualized in a cryo-electron microscope at different tilt angles, followed by alignment of the tilt series by utilization of fiducial markers and 3D reconstruction. Several computational methods have been developed for assignment of densities in the cellular reconstruction and identification of particles of interest. Electron cryo-tomography has been utilized for identification of structural alterations in rotavirus particles during the initial stages of cellular entry, however, the resolution remains significantly lower compared to SPR techniques. Cryotomography can be utilized in hybrid mode in combination with X-ray crystallography or SPR to get a better understanding of in situ structural alterations in viruses during disassembly.

Apart from experimental methods, in recent years, the development of high-performance computing and advancements in force field methods have led to increasingly sophisticated and accurate simulations of biological macromolecules, broadening the scope of problems that can be studied using Molecular Dynamics (MD) simulations [67]. MD simulation is a computational approach for studying the movements and interactions of atoms and molecules. The method involves solving the equations of motion for a system of interacting particles, typically using classical mechanics, to simulate the behaviour of the system over time. MD simulations can be used to study a wide range of materials, including liquids, solids, and biomolecules, as well as to investigate various properties such as thermodynamics, kinetics, and equilibrium behaviour. The ability to study the behaviour of complex systems, especially those that are difficult to study experimentally, has made MD an important tool in the fields of chemistry, physics, materials science, and biochemistry. In biology, this technique has been widely used to study the structural and dynamic properties of macromolecular biological complexes like viruses. MD simulation is being increasingly applied to understand virus structural dynamics, energetics and stability which is crucial for developing antiviral drugs and vaccines [22, 68]. It can also be used to study the interactions between viral capsid proteins, the viral genome, and host cell factors that are involved in viral disassembly. Whole virus simulations can also be used to study the effects of mutations and chemical modifications on capsid stability (Table 9.2). It can provide information on the specific interactions between atoms and molecules, as well as the energy changes that occur during the disassembly process. This can identify key residues and regions of the viral capsid that are key for structural stability. MD simulations can also be used to study the effects of different environmental conditions, such as pH, temperature, and the presence of ligands on the viral capsid. Interesting insights on several non-enveloped plant and animal viruses like the Southern Bean Mosaic Virus, Hepatitis B Virus, Triatoma Virus, Satellite Tobacco Mosaic Virus, Porcine Circovirus and others have been studied using MD simulation [69–73].

In addition to these structural techniques, a variety of biochemical and biophysical methods, the latter involving Small Angle X-ray Scattering (SAXS) [74, 75], Time resolved SAXS (TR-SAXS) [76], neutron scattering [77], fluorometric methods [78], atomic force microscopy [79] and mechanical fracture [79], and thermal shift [78] and calorimetric methods [80] have been utilized to study the physical and molecular features of virus disassembly.

Initiation of Disassembly and Generating Intermediates In Vitro

The mature, non-enveloped capsid generated post-assembly serves as a stable container to transport and protect the viral genome under harsh environmental conditions. In case of the vast majority of animal viruses, disassembly or uncoating is initiated when the metastable capsid is exposed to specific host cell factors such as receptors and co-receptors, endosomal compartments with low pH and reduced concentration of divalent cations, host cell proteases in cellular compartments like lysosomes etc. [1, 5]. The environment within host cells is considerably milder than the extracellular milieu, and the triggering of disassembly in a relatively mild setting is indicative of molecular switches programmed in during assembly/maturation that respond to specific conditions.

The exact steps involved in disassembly is unclear for most viruses. However, it is to be noted that the conformational changes involved in the virus assembly and disassembly processes are not simply reversible but unique to each process [81]. Among non-enveloped animal viruses with icosahedral geometry, the uncoating processes of nodaviruses, picornaviruses, polyomaviruses, adenoviruses and reoviruses have been studied in biochemical and structural detail [5, 82], as illustrated in Table 9.1. Generating disassembly intermediates of these viruses in vitro is usually achieved through association with recombinantly produced receptor (picornaviruses), exposure to factors that mimic intracellular environmental conditions such as low pH (picornavirus), varied ion concentrations (reovirus) and limited proteolysis (reovirus) [44, 83–88].

In case of picornaviruses, binding to the receptor appears to initiate disassembly, which in the early stages results in the exposure of membrane interacting peptides from the capsid interior, expansion and thinning of the capsid shell and an outward movement or gradual separation of the capsid proteins. Progression of conformational changes in the capsid eventually culminates in expansion of pores or openings at the symmetry axes or removal of one or multiple capsomeres from the capsid surface to allow genome escape [9, 83, 84, 89]. The disassembly pathway of members of the picornavirus family is one of the most investigated, probably because of the generation of stable and distinct disassembly intermediates by these viruses under in vitro conditions.

As described in the earlier section, picornavirus capsids consist of 60 copies each of the capsid proteins VP1, VP2, VP3 and VP4 arranged in a pseudo $T = 3$ icosahedral shell. While VP1-VP3 form the outer and inner capsid surfaces and contain the traditional β-barrel fold common to mammalian non-enveloped capsid proteins, VP4 is a small peptide which is located inside the capsid and is usually disordered in most of the crystal structures of picornaviruses. The first stable intermediate generated during disassembly, designated 135S (based on the sedimentation coefficient) or A particle, is a stain-permeable, slightly expanded form in which the N-terminus of VP1 is externalized and VP4 is partially or completely absent from the capsid [9, 84]. It has been shown in case of multiple picornaviruses that these components are hydrophobic or amphipathic, and are involved in cellular membrane penetration, although there are some exceptions [38, 84, 90–92]. The exposed hydrophobic sequences in the A particles are thought to facilitate delivery of the viral genomic RNA to the cytosol by forming pores in the host membrane [11, 93]. An additional function attributed to the exposed VP1 termini from poliovirus is to form membrane connectors, from the capsid to the host cell cytoplasm, to allow safe passage to the genome [85]. Altered picornavirus particles can be biochemically distinguished from the native particles by their increased hydrophobicity and ability to associate with synthetic or cellular membranes. A detailed review of membrane lytic components of non-enveloped viruses is available in ref. [4].

The exposure or loss of VP4 and VP1 N-terminus in the A particle likely indicates that externalization of these components for cellular membrane association constitutes the initial step in capsid conformational alterations during disassembly. As discussed previously, these components are also occasionally, transiently exposed from the resting capsid (the 160S particle)—during "breathing", which highlights their relative flexibility compared to the rest of the capsid [36]. Thus, the initial structural alterations during disassembly result in exposure of the more flexible parts of the capsid, which were proteolytically processed during maturation. The second intermediate, called the 80S or B particle, is an empty capsid with icosahedral features, that has already released the packaged genome [94, 95]. The majority of disassembly intermediates of picornaviruses, as well as other non-enveloped capsids, show exposure or loss of similar internal, flexible capsid components that are involved in plasma or endosomal membrane penetration. However, the somewhat confusing distinction in localization of these components on the surface of various intermediate structures reinforces their flexibility; and perhaps, the complex nature of disassembly intermediates, which could be a composite of rapidly altering conformational states [90].

Since in case of animal viruses, the disassembly pathway in host cells is likely triggered by receptor binding and low pH conditions in late endosomal compartments, a combination of these conditions have been utilized to generate stable intermediates in vitro. For example, the 135S and 80S particles of poliovirus have been successfully generated in vitro by incubating the native 160S virus particle with the purified receptor [85, 96, 97]. Application of low pH alone has been utilized to successfully generate intermediates of picornaviruses such as Echovirus 18 and Equine Rhinitis A Virus (ERAV) [19, 86]. Interestingly, application of heat to some non-enveloped animal viruses have been found to trigger disassembly-like structural alterations. This is likely a result of energy transfer to metastable particles programmed to go through specific conformational changes, which under in vivo or ex vivo conditions are typically an outcome of energy released due to receptor binding by the particle. Poliovirus disassembly intermediates can also be generated by heat *in lieu* of receptor binding, as are Triatoma Virus particles [98]. Human Rhinovirus (HRV2) disassembly intermediates can be generated by a combination of heating and low pH [99]. Disassembly intermediates of the picornavirus Israeli Acute Bee Paralysis Virus (IAPV), and the nodavirus, Flock House Virus (FHV) can be generated by supplying incremental heat to the particles [20, 100]. Incremental heating of FHV results in the generation of two, relatively stable, disassembly intermediates—the "eluted particle" with a negative stain permeable phenotype, and a "puffed particle" which had a cloud of genomic RNA exposed from the capsid. Interestingly, the eluted particle phenotype, which had externalized a fraction of the membrane active γ peptides, was also isolated from host cells one hour after initiation of viral infection [101]. The structural and biochemical similarity between the eluted particles isolated from ex vivo infection and in vitro incremental heating, establishes the feasibility of replicating viral disassembly intermediates outside the cellular milieu. The utilization of methods like incremental heating, low pH and other conditions also makes it possible to study disassembly-related conformational alterations in viruses whose host cell receptors are unknown.

Other methods that have been utilized for triggering disassembly in non-enveloped animal virus capsids in vitro include formaldehyde treatment (Enterovirus 71, Coxsackievirus A16), protease treatment (orthoreovirus, rotavirus) and incubation with monovalent cations (orthoreovirus) [102–104]. Disassembly intermediates of the mammalian Orthoreovirus can be generated by heating of the capsid as well as treatment with monovalent cations in vitro. The incomplete $T = 13$ icosahedral capsid of reovirus undergoes a series of disassembly-associated alterations in its outer protein layers, which consist of 200 heterohexamers of μ1—the receptor-binding and membrane-penetrating component, and σ3, the protector protein which conceals μ1 [88, 105, 106]. After receptor-mediated entry, σ3 is digested by cathepsin proteases in cellular compartments. This results in the exposure, rearrangement, and proteolytic maturation cleavage of the μ1 component into μ1N and Φ [40, 107]. The particle which has undergone these alterations is termed the Infectious Sub Viral Particle or ISVP. Further alterations result in the cleavage and release of the membrane penetrating peptide μ1N, which assists in the formation of membrane pores to allow release of genomic material [39, 108, 109]. These changes result in a discrete particle called ISVP* which incorporates the properties of hydrophobicity and membrane activity as seen in disassembly intermediates from other virus families [110, 111]. While the cellular triggers for these alterations in vivo or ex vivo are not identified yet, the progression of conformational alterations in the pathway to disassembly has been established primarily through the characterization of these intermediates generated in vitro, as well as through mutational studies [112]. A recent mutational study has established the role of μ1 residues 340–343, which form a loop in the β-barrel region of the protein, in maintaining capsid stability and regulating disassembly [113]. Interestingly, it has been shown that host cell membranes play an important role in the structural transition of reovirus particles during disassembly. It is thought that the transient exposure of μ1N during breathing results in the association of the peptides with host cell membranes, and the membrane associated μ1N can interact with particles and cause ISVP to ISVP* alterations in a positive feedback loop [114, 115]. In support of this hypothesis, synthetic membranes made from lipids like phosphotidylcholine and phosphotidylethanolamine can trigger ISVP to ISVP* conversion in vitro [114, 115].

Rotavirus capsids undergo major conformational changes, reminiscent of enveloped viruses, during their association with and disassembly inside host cells [45]. Rotavirus contains multiple layers of capsid proteins—the innermost layer being composed of 120 copies of VP2 which encloses the segmented dsRNA genome, which is surrounded by a T = 13 l icosahedral lattice formed by VP6. This particle containing VP2 and VP6 is replication competent and is termed the double layered particle (DLP) [116]. The VP2-VP6 layer is coated by capsid proteins VP4 and VP7, with VP4 forming prominent spikes on the outer surface and VP7 locking the spikes in place [48, 117]. The maturation cleavage of VP4 into VP5* and VP8* has been described in the previous section [45, 47]. The assembly, as well as the disassembly of the complete particle, called TLP (triple layered particle), is modulated by the availability of Ca^{2+}. Reduced Ca^{2+} concentration during entry of the particles into host cells triggers a series of molecular transformations, which results in the loss of VP7 and resultant exposure of VP4 spikes for establishing cellular interaction [118]. Eventually, the shedding of the VP4-VP7 layer during disassembly regenerates DLPs which do not disassemble further, and replication of viral genome is carried out within the compartment. The variable concentration of Ca^{2+} in the cytoplasm (~100 nM) and ER controls the processes of disassembly as well as assembly [118]. The early stages of rotavirus entry and structural alterations have been followed in BSC-1 cells using cryoelectron tomography [119]. CryoET and subtomogram averaging generated particle reconstructions at resolutions in the nm scale. However, some prominent structural features such as the presence of full-length VP4 spikes in some particles, and the decrease in spike length in another group indicating structural rearrangements in the VP5* region could be conjectured from the reconstructions [119].

Human papilloma virus 16 (HPV-16) uses receptor binding to initiate its disassembly process, which involves exposure of the viral internal protein L2 and furin-mediated cleavage of its N-terminus [120]. This altered HPV-16 virus then uses low pH in the late endosome to access the cytosol [121]. Another virus which undergoes stepwise disassembly is Adenovirus [122]. The disassembly of adenovirus is initiated upon receptor interaction, which results in the loss of the protruding fibres from the virus surface. This is followed by further loss of pentons from the capsid in the low pH conditions within the endosome, and eventual dismantling in the nucleus and externalization of genome. Also, imaging and tracking experiments have shown that host molecular motors help adenovirus to overcome diffusion barriers and employ a stepwise uncoating programme [123]. Immature adenovirus particles, which do not undergo maturation cleavage in protein VI, are unable to cause endosomal membrane disruption during disassembly and are incapable of causing infections [124]. Interestingly, mutated adenovirus particles which do not go through the maturation process are more stable, but not necessarily stiffer than the wild-type, infectious virions [125, 126]. A variety of conditions including incremental heating and mechanical stress have been utilized to initiate and study the stepwise dismantling of adenovirus capsids in vitro [125].

The non-enveloped polyomavirus SV40 (simian virus 40) binds to GM1 ganglioside receptors at the cell surface and enters into the host cells by a caveolae-mediated endocytic process that translocates SV40 to endoplasmic reticulum (ER) (12,941,687, 14,644,415, 2,556,405). Within ER, SV40 undergoes partial disassembly exposing VP2 and VP3, its internal capsid proteins (11,967,331). However, the packaged genomic DNA is released only after the partially uncoated virus reaches the cytoplasm (22,090,139). SV40 is also known to hijack the ER-associated degradation (ERAD) machinery for its penetration through the ER membrane to reach the cytosol (17,981,119, 19,002,207). The released DNA genome then enters the nucleus via VP_2 viroporin (22,929,056). Thus, SV40 follows a unique disassembly pathway with discrete uncoating steps that are separated temporally and topologically. Time-resolved experiments using cryo-EM and solution X-ray scattering have also detected the series of sequential changes occurring in the SV40 particles in vitro at pH 10 or above (32,104,873). These structural changes began by particle swelling, followed by the appearance of a hole in the capsid through which the genomic DNA escaped, then a slight shrinkage of the capsid, which finally proceeded to complete disintegration of the particle capsid (32,104,873).

The disassembly of the helical plant virus Tobacco Mosaic Virus (TMV) depends on the increase in pH and decrease in Ca^{2+} concentration [14]. TMV is highly resistant to temperature, and to pH conditions down to 1.0. The helical particles are generated from a single capsid protein (CP) of 17 kDa, comprising a four-helix bundle, which is tightly wound around the genomic RNA. High resolution structures of TMV have been resolved using cryo-EM and SPR in conditions with and without Ca^{2+}. From the structures, it is clear that the metastable switch in TMV capsid is provided by a group of Ca^{2+} sensitive Casper-carboxylate residues. This network is primarily formed by amino acids E95, E97, E106, E50 and D77, assisted by N98 and N101. E95 and E97, which are located within a radius of 20–60 Å from the radial axis of the capsid, are protonated at lower pH conditions, and stabilize the capsid [14]. Upon entry into the cell, and due to the rise in pH and consequent repulsion between the acidic residues, Ca^{2+} release is initiated from the lower radii regions and spreads to the middle radii regions [14]. However, the destabilization and disassembly of the particles is relatively limited, with the helical morphology being largely maintained. The RNA genome becomes accessible to the replisome machinery through its 5' end for downstream genome replication [127].

Brome Mosaic Virus (BMV) and Cowpea Chlorotic Mottle Virus (CCMV) are plant viruses that have a stability profile opposite to that observed in many non-enveloped animal viruses [12, 13]. Unlike animal viruses, but like other plant viruses such as TMV, the capsids of BMV and CCMV are stable at low pH conditions, but become destabilized at neutral and high pH conditions. This molecular switch has likely evolved due to the necessity of the particles to stay intact in the acidic pH of insect gut. Both BMV and CCMV disassemble under a combination of elevated pH and low Ca^{2+} concentration, and via a prominent "swelling" intermediate. During swelling, the particle diameter increases by ~10%, the capsid shell becomes more elastic, and pores of ~2 nm diameter open along the quasi threefold axes of symmetry of the icosahedral particles [128, 129]. Further increase in pH results in dissociation of the particles into the constituent protein and RNA components [12, 13].

Many in vitro methods for disassembly have also been tested in case of bacteriophages like the λ and T4 [15, 130]. The icosahedral heads of the bacteriophages are typically tightly packaged with genomic DNA and therefore highly pressurized [131–133]. Incremental heating has been utilized for studying structural transitions in the λ bacteriophage head, which shows that the genomic DNA escapes the capsid at a temperature of approx. 68 °C, that can be correlated with the rupture of the tail [15]. Also, lowering of salt concentration can lead to increase in genomic pressure, leading to rupturing of the head. It has been seen the in vitro addition of the λ phage receptor, LamB, can result in the genome escaping through the tail without structurally damaging the head of the phage [15, 134].

Progress of Disassembly-Related Structural Changes and Sites for Genome Release

Not all picornaviruses generate stable 135S and 80S intermediates like polio and rhinoviruses. The disassembly of the capsid appears to be relatively quicker in several picorna- or picorna-like viruses. For example, the 80S state is not entirely stable for Equine rhinitis A virus (ERAV), which quickly dissociates into smaller structural units like pentamers under low pH conditions [86]; while Triatoma virus disassembles directly into pentameric subunits or pentons, without formation of larger icosahedral intermediates [98].

For polio and rhinoviruses that generate stable disassembly intermediates, the presence of artificial membranes in conjunction with altered particles have been utilized to map the structural basis of membrane interaction and genome transfer [85, 135]. Structural studies of two intermediate particles, the empty 80S poliovirus particles, and Human Rhinovirus particles containing rod-shaped genome on the brink of release, have indicated the two-fold axis of symmetry to be the probable genome exit site [89, 136]. However, it is still not clear whether the twofold axis serves as a common genome release site for icosahedral capsids. Another pressing question that is yet to be answered is whether genome release occurs through one specific twofold axis, or through multiple axes simultaneously on the capsid surface.

A recent interesting and comprehensive, high-resolution structural study on Flock House Virus (FHV) disassembly intermediates detailed a series of local as well as global structural changes in the icosahedral capsid leading to genome release [20]. Two intermediate states, namely "eluted" and "puffed" particles were isolated by incrementally increasing temperature in vitro, and characterized (Fig. 9.3a). Structural characterization of these intermediates using cryoelectron microscopy suggested several intriguing similarities as well as differences with the previously reported disassembly-associated alterations in other non-enveloped viral capsids. The eluted particles of FHV showed ~4 nm reduction in particle diameter, which is contrary to the radial expansion (~4%) observed in the 135S or A particle of picornaviruses [96]. This is the first report of particle contraction during non-enveloped virus disassembly. Triatoma virus (TrV), which is a member of the insect virus family *Dicistroviridae*, also did not show any particle expansion during disassembly [98], however no reduction in diameter was reported. Thus, it seems likely that insect viruses may not follow the typical phenomenon of capsid expansion observed in picornaviruses during disassembly. However, structural studies on a larger population of disassembly intermediates from different virus families are required to reach a generalized conclusion regarding the disassembly process of insect viruses.

Some common disassembly related conformational changes were also observed between FHV and picornaviruses. A movement of helices lining the twofold axes resulted in the localized expansion of twofold axes in FHV during disassembly, supporting the growing evidence for the twofold axis to be the common fracture site or weak link on non-enveloped icosahedral virus capsids for genome release [89, 136]. Several picornaviruses have also displayed similar pore opening around the twofold axis due to the separation of α-helices lining that axis [137]. The presence of ordered RNA duplexes at the icosahedral two-fold contacts of the FHV particle in the crystal structure of FHV [138], also provide a reasonable explanation for the twofold axis to be the portal for genome release. Other noticeable

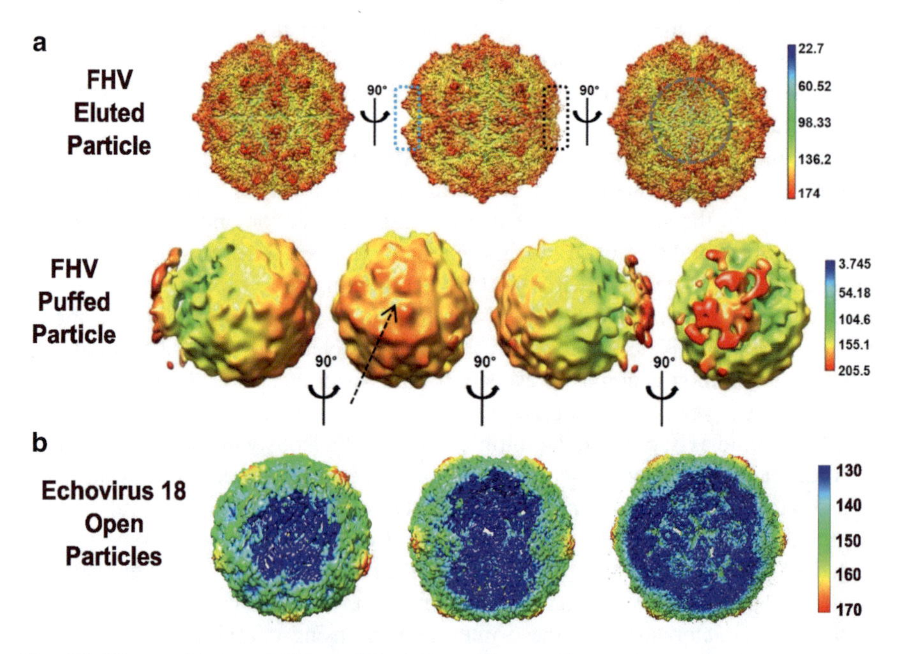

Fig. 9.3 Disassembly intermediates of Flock House Virus (FHV) and Echovirus 18. **a** Asymmetric cryo-EM 3D reconstructions of eluted (upper panel) and puffed (lower panel) particles of Flock House Virus (FHV), displayed as surface-rendered, radially coloured density maps in different orientations. Color keys (right) show the radial distance (in Å) from the particle centre. In the reconstruction of the eluted particle at the centre, the light blue and black dotted lines mark the icosahedral symmetric and asymmetric parts of the particle, respectively. The front view of the asymmetric part is also highlighted by a dotted grey circle on the eluted particle. A dotted arrow on the puffed particle density map indicates twofold symmetry axis. This figure has been reproduced with permission from Azad and Banerjee [20]. **b** Radially colored, asymmetric 3D reconstructions of echovirus 18 open particles lacking one (left panel; EMD-0184), two (middle panel; EMD-0186), or three (right panel; EMD-0187) adjacent pentamers. Color key at the right shows the radial distance (in Å) from the particle centre

common conformational alterations include global loss of the γ peptide, analogous to loss of VP4 during picornavirus disassembly [137] and localized loss of capsid protein components [19]. The individual capsid protein subunits in the FHV capsid showed differential movement, which has also been reported previously for the VP1, VP2 and VP3 capsid proteins of CAV7 during progression from native state to an empty capsid which has released genome [137]. One unique structural feature of FHV disassembly intermediates was partial disordering of the icosahedral cage following RNA release. The presence of genomic or non-genomic RNA is crucial for FHV capsid assembly [139]. This requirement of RNA for capsid stability may explain a drastic decrease in the stability of FHV capsid, leading to its disintegration following RNA release, unlike the observations in picornaviruses.

A fascinating step revealed during the disassembly of the HRV2 capsid is the compaction of the encapsidated genome [20] into a condensed, rod-shaped structure [136]. Conformational alterations in the genome may suggest that it contributes actively towards its own release from the capsid shell, rather than just being a passive entity during disassembly. Parallel studies on FHV captured a disassembly intermediate in the process of genome release, called the "puffed particle", in a relatively stable form. While a low resolution reconstruction (~26 Å) of the puffed particle clearly captures a particle releasing genomic RNA (Fig. 9.3a), it is impossible to clearly say whether the release occurs from a single or multiple open two-fold axes of symmetry.

Several recent studies suggest that highly symmetric, icosahedral particles contain local differences [140, 141]. These differences are usually lost during structure determination by single particle reconstruction as the result of the imposition of icosahedral symmetry. These local differences within a highly symmetric virus capsid may have biological relevance in specific steps in the virus life cycle including assembly, trafficking and genome release. Asymmetry in the particle may also be intensified due to the loss of capsid stability during particle disassembly. Asymmetric reconstruction implemented during the reconstruction of FHV intermediate particles [20], was crucial for deciphering both global as well as local conformational changes involved in disassembly and genome release. Thus, asymmetric 3D reconstruction is a more meaningful approach in studying symmetry loss and related structural changes that appears to occur during icosahedral virus capsid disassembly [142].

A recent example of asymmetric 3D reconstruction providing interesting insights into non-enveloped virus disassembly is the study by Buchta et al., which shows that the exit of RNA from human echovirus 18 particles results in the loss of 1–3 adjacent pentameric subunits from the capsid [19] (Fig. 9.3b). The subpopulations of the particles lacking capsid protein protomers, designated "open particles", have openings of ~120 Å diameter in the capsid. Such large pores enable the release of putative double-stranded RNA segments, without the need of unwinding of the genome. Similar capsid intermediates lacking pentamers during genome release were also observed for human echovirus 30 [19].

High-resolution reconstructions of the open particles of echovirus 18 revealed structural similarity to poliovirus 135S or rhinovirus A particles, except for the missing pentamers. The particles were characterized by an increased diameter compared to the native virus, absence of VP4 subunits, reduced inter-pentamer contacts and openings along the icosahedral twofold symmetry axes [19]. The N termini of VP1 were not resolved in the cryo-EM density map indicating their relative flexibility compared to the rest of the capsid. Interestingly, this study also proposed the re-structuring of the empty particles after genome release, by re-association of the discharged pentamers to the capsid openings after genome release. This model of capsid opening and genome release, followed by reassembly of the empty capsid, was also assessed through molecular dynamics simulations. Recently this model of capsid opening and genome release is also applicable to the Kashmir Bee Virus (KBV) from the *Dicistroviridae* family [143].

In sum, currently there are sparse structural studies conducted on disassembly intermediates from a limited number of non-enveloped virus families. The existing studies do not provide a sequential roadmap of disassembly-associated conformational changes. It is still debatable whether the viral RNA exits through the fivefold or twofold symmetry axes of the virion. Similarly, conformational changes in the packaged genome during release and the involvement of capsid proteins in this process still remain unresolved. A detailed and comprehensive map of disassembly-associated conformational changes, whether applicable broadly, or in context of a particular type of icosahedral capsid, is extremely essential since such knowledge can lead to the development of small molecule antivirals to block disassembly, a crucial step in the infection process.

Disassembly of Enveloped Virus Cores

Enveloped viruses contain a host-derived lipid envelope surrounding the nucleocapsid core, which consists of the genome encapsulated by the nucleocapsid protein. Receptor binding by these viruses is closely followed by the fusion of the virus envelope with cellular membrane which releases the nucleocapsid in host cell compartments [144, 145]. The nucleocapsid cores, depending on the virus type, may have regular, icosahedral or spherical symmetry, or may be helical and complex with associated enzymes and lateral bodies, in addition to the capsid protein [146]. Disassembly of the nucleocapsid cores and subsequent release of the genome from enveloped virions is also driven by cellular cues, and have been studied using the biophysical and structural techniques described in Chapter II (Table 9.3).

The HIV-1 capsid (L = ~120 nm, W = ~60 nm) encloses a single-stranded RNA genome which reverse transcribes to form dsDNA. Reverse transcription has been shown to be the initiator of capsid disassembly in vitro [147, 148]. The process of reverse transcription initiates prior to capsid disassembly and is completed within the capsid prior to its release in the host cell nucleus near the genome integration site [149, 150]. Analysis of reverse transcribing viral cores in vitro shows a steady increase of internal pressure with time. Eventually, this outward pressure leads to physical disruption of the capsid. Using high resolution mechanical mapping, local structural changes on the capsid surface were observed during this process [147]. At early time-points, the increasing pressure was found to be associated with simultaneous formation of distinct spiral patterns on the capsid surface. This was implicated as the existence of a coiled and stiff filamentous structure at the inner periphery of the capsid. However, at advanced time-points such surface patterns disappeared completely, along with a striking decrease in capsid stiffness. This was followed by the physical disruption of the nucleocapsid at its narrow end, resulting in the potential release of the reverse transcribed viral genome. Unique disassembly intermediates of reverse transcribing intermediates have been generated from purified HIV-1 cores in vitro [148]. Using a combination of cryo-electron tomography and lattice mapping discreet localized perturbations were observed in otherwise intact capsids containing external protrusions of nucleic acid loops. The length of such loops varied between 0.5 and 1.5 kb, which indicated that the majority of the viral genome remained packaged inside the core during this stage of disassembly. This highlighted the fact that the HIV-1 capsid can remain stable even when discrete sections are missing. However, no intermediates were observed where loops protruded from the narrow end of the lattice. It is thus likely that the narrow end serves as an exit portal for the bulk genome.

Unlike HIV-1 or herpesviruses, that do not undergo cytosolic disassembly, certain families of RNA viruses like Influenza and members of the Alphavirus family have evolved complex strategies to exploit cytosolic host factors to accomplish capsid disassembly. Dismantling of influenza A virus (IAV) capsid involves interaction with the aggresome formation and disassembly pathway (Fig. 9.4a). The acidic pH gradient in the early-late endosomal compartments triggers a sequential activation of the M2 proton channel followed by hemagglutinin (HA)-mediated fusion which expose the IAV capsid to the cytosol. Interestingly, IAV capsids display ubiquitin chains which mimic misfolded proteins. This results in association of histone deacetylase HDAC6, an adaptor molecule that shuttles misfolded proteins to the aggresome, with the IAV capsid via its C-terminal zinc finger ubiquitin binding domain (ZnF-UBP). Additionally, since HDAC6 also interacts with cytoskeletal motor proteins, myosin II, dynein and dynactin also aid in the disassembly process. This process generates opposing physical forces which ultimately result in capsid disassembly [151].

Another unique strategy for capsid disassembly has evolved in members of the alphavirus family, that includes important human pathogens like the Chikungunya Virus (CHIKV), Ross River Virus (RRV), Venezuelan Equine Encephalitis Virus (VEEV), Eastern Equine Encephalitis Virus (EEEV), O'nyong'nyong Virus (ONNV) and Mayaro virus (MAYV). Ribosome-mediated disassembly of the icosahedral nucleocapsid has been reported in the Alphavirus Semliki Forest Virus (SFV) and Sindbis Virus (SINV) [152, 153]. Studies have shown that at early time-points during infection, radioactively-labelled capsid proteins are transferred from viral core to the 60S large ribosomal subunit. Capsid-ribosome binding has also been observed in vitro upon exposure of the purified viral cores to the cell lysate. A 10–12 residue stretch (SINV: Gln94-Arg105; SFV: Lys101-Arg110) has been identified in the capsid that binds to 60S ribosomes [152]. This ribosome-binding site (RBS) lies in the linker

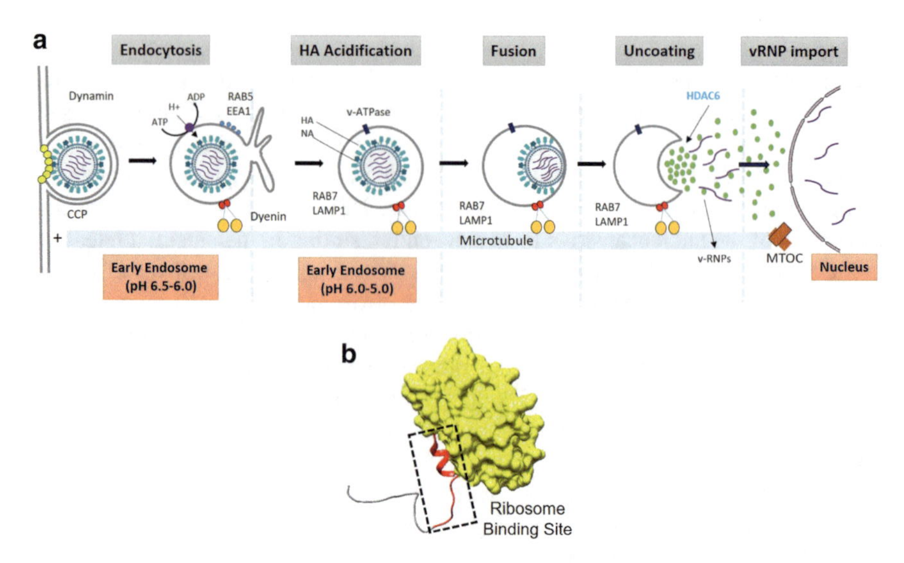

Fig. 9.4 Genome release mechanisms in enveloped viruses. **a** Schematic representation of the different stages of Influenza A Virus entry and HDAC6 dependent capsid disassembly. **b** Ribosome binding site of EEEV capsid (red) and its relative positioning relative to the capsid C-terminal region (yellow, surface)

region between the N and C-terminal of the capsid protein (Fig. 9.4b), and is characterized by the presence of a conserved tripeptide sequence "KPG". A high resolution cryo-EM structure of EEEV has provided valuable insight on the relative positioning of the RBS in the intact virion [154]. The structure shows that the RBS is composed of a loop and a short helix, and is located internally within the nucleocapsid core. This implies a critical conformational alteration post-fusion that partially opens up the core to allow externalization of the loop. Stabilizing interactions observed between the C-terminal outer nucleocapsid layer and the E2 endodomain are stripped off post-fusion. Hence, pH-driven virus fusion possibly transforms the nucleocapsid into its disassembly-competent conformation characterized by the externalized RBS loop.

Certain viral nucleocapsids undergo complete disassembly using a primary physiological trigger. For instance, the African Swine Fever Virus (ASFV), which causes a fatal disease in pigs, utilizes the lowering pH gradient of the endo-lysosomal system to disassemble and release its genome in the host cell cytosol [155]. The virus structure is uniquely organized with an outer membrane, an intermediate capsid and an internal membrane. The DNA genomic core is encapsulated by the inner membrane. ASFV enters immune cells by micropinocytosis and receptor-mediated endocytosis. The passage of the virus particles through the steadily acidifying endosomes shows a stepwise disassembly process. In early endosomes or macropinosomes, the virus particles appear intact with all three outer layers present. However, progression from early to late endosomes results in most particles loosing both the outer membrane and the intermediate capsid layer. Finally, the particles which are observed within lysosomal structures contain only the inner envelope. These particles are then observed to associate in close proximity to the inner lysosomal membrane and subsequently undergo a viral pE248R-dependent fusion event releasing the genomic core into the cytosol [155].

Virus core disassembly and genome release are events with possible spatio-temporal coupling. Similar to HIV-1 disassembly, nuclear delivery of the viral genome is a key step in the lifecycle for alphaherpesviruses like Herpes Simplex Virus type 1 and type 2 (HSV-1, HSV-2), Varicella Zoster Virus (VSV), Epstein-Barr Virus (EBV), Human Cytomegalovirus (HCMV) and Kaposi Sarcoma-associated Herpesvirus (KSHV). Unlike HIV-1, herpesviruses contain a DNA genome packaged inside the nucleocapsid which has two distinct outer layers, the tegument and the envelope. Post membrane fusion, the nucleocapsid enters the cytosol and migrates towards the nucleus along microtubules where the genome is injected through the nuclear pore complex. The 152 kbp viral DNA is known to generate a very high internal pressure in the nucleocapsid [156]. However, genome release is not associated with a global breakdown of the capsid structure and loss of symmetry. Instead, a single vertex, referred to as the packaging portal, serves as the exit passage of the genome [157]. This is in contrast to the disassembly intermediates observed in HIV-1 or non-enveloped viruses discussed before. In vitro, treatment of purified HSV-1 capsids with increased temperature, trypsinization, guanidium hydrochloride, cell lysates, an ATP-generating system or purified nuclei, can trigger genome release [156]. Furthermore, a positive correlation has been observed between the proteolytic cleavage of the portal protein UL6 and genome release, which implies that UL6 cleavage at the packaging portal is a prerequisite to make the capsid competent for the release of genome [157]. More recently, an interferon inducible protein, the cellular GTPase myxovirus resistance protein B (MxB) was shown to specifically interact with HSV-1, HSV-2 and VZV capsids [158]. GTPase activity and dimerization of MxB resulted in cytosolic disassembly of the herpesvirus capsid, and possibly acts as a host defence mechanism to restrict infection by early recognition of viral DNA. Electron microscopy of capsids exposed to MxB showed the presence of multiple morphologies that resembled intact icosahedral capsids, punched capsids with distinct holes at the vertices, or flats sheets wherein particles completely lost symmetry [158]. MxB-mediated herpesvirus capsid disassembly is an ideal example that shows the spatial–temporal importance of capsid disassembly for establishing a successful viral infection.

Presence of viral DNA in cytosol can be sensed by the STimulator of INterferon Genes (STING) which can activate the innate immune response. Thus cytosol is a particularly challenging environment for the disassembly of DNA viruses. However, poxviruses have evolved to undergo disassembly and replication in the cytosol. As such, the disassembly of the poxvirus capsid is a highly regulated process. Prior to disassembly, intact virus particles have an outer barrel-shaped conformation with a dumble-shaped core [159, 160]. Structures called lateral bodies flank the core and fill up the space between the outer membrane and the core, which encapsulates a 200 kbp dsDNA genome. Establishment of virus attachment with a target host cell triggers sequential structural alterations in the virus [160]. First, the contact points between the lateral bodies, and the outer membrane with the inner core, are disrupted.

Next, the core undergoes expansion and the overall structure changes from a dumble-shape to ovoid-shape. This is possibly due to reduction of disulphide bonds between the core structural proteins. The encapsulated condensed viral genome undergoes decondensation and fills up the entire internal space of the core. During this process, discrete tubular structures connected to the core are observed, along with disorder in the surface spike arrangement. An important observation was the presence of discrete channels of ~7 nm diameter, believed to be the exit route for early viral transcripts. These channels were maintained throughout these conformational changes [159]. Once internalized, the virus particle poised for disassembly is thought to release its bulk genome through a large opening at one broad side of the core. Although, smaller holes were observed in purified virus particles, these were found to be randomly positioned and considerably smaller.

Non-enveloped Capsid Disassembly and Intervention Methods

The requirement for understanding the structural basis of virus capsid disassembly is two-fold. A molecular roadmap for disassembly of human pathogens may be utilized for structure-based drug development specifically targeted to this step in virus-host interaction. The dissociation of the protective capsid and the release of genome within host cells constitutes a crucial early step in virus-host interaction. The existing knowledge of virus capsid structures at high resolution makes it possible to identify molecular switches that lead to metastability and dynamicity of capsids and to inactivate these switches by chemical intervention. It may be possible to alter the chemistry of these sites by employing small molecule inhibitors. It is also to be noted that a majority of non-enveloped viruses, as well as enveloped virus nucleocapsids, in spite of differences in structure, host specificity and downstream functionality, typically utilize common cues such as alteration in pH, cationic concentration, cellular proteases etc. for disassembly of the capsid and uncoating of genome [5, 161]. Thus, it may be possible to develop intervention strategies that are common to multiple groups of viruses.

Another important application of non-enveloped capsid disassembly is in the field of virus particle based drug delivery and nanomedicine [162–164]. A variety of plant and animal viruses, as well as bacteriophages, have been modified for usage as nano-containers [164, 165]. Engineering of these capsids have resulted in their utilization for packaging a variety of material such as drugs, imaging dyes and molecules, siRNA and other therapeutic material [164, 165]. Specific targeting moieties have been engineered on the outer surface of capsids in order to deposit the packaged material to different cell types [166]. However, disassembly of the capsid container in the correct cellular compartment and release of the therapeutic or imaging material is a primary requirement for the utilization of non-enveloped capsid based nanocontainers and nanodevices. A clear, molecular level understanding of disassembly is crucial in order to develop usable, smart, non-enveloped capsid based devices for biomedical delivery.

Table 9.2 Virus disassembly investigated by molecular dynamics (MD) simulation

Name	Triangulation number	Enveloped/ genome	Finding	Methodology	Reference
Cowpea Chlorotic Mottle Virus (CCMV)	3	Non-enveloped, RNA genome	Thermal dissociation results in capsid shrinkage and loss of symmetry, strong thermal fluctuations in surface monomers due to stronger hydrophobic interactions	CG simulations, MARTINI force field	[168]
Hepatitis B Virus (HBV)	4	Enveloped, DNA genome	Capsid destabilization is due to local bending and shifting of capsid proteins	AFM + SBCG simulations	[169]
			Dimeric interface of capsid hexamers exhibits structurally heterogeneous cracks	All-atom MD with applied isotropic pulling potential, Charmm36m force field	[71]
Porcine Circovirus (PCV)	1	Non-enveloped, DNA genome	Electrostatic interactions (Cl^- ions) stabilize capsid architecture	All Atom MD simulation, Amber03 force field	[170]
Satellite Tobacco Necrosis Virus (STNV)	1	Non-enveloped, RNA genome	Ca^{2+}-dependent process wherein removal of 2 critical Ca^{2+} ions near the threefold axis of symmetry initiates capsid swelling, water permeation and subsequent disassembly	All Atom MD simulation, Amber99sb-ILDN force field	[73]
Southern Bean Mosaic Virus (SBMV)	3	Non-enveloped, RNA genome	Quasi threefold and twofold axis of symmetry are the weakest points in the virus capsid	Atomistic force-probe molecular dynamics simulations	[70]
Triatoma virus (TrV)	Pseudo 3	Non-enveloped, RNA genome	pH-dependent proton channel formation at the fivefold axis of the virus capsid, proton permeation follows a unidirectional Grotthuss-like mechanism	Multiscale CG simulations, SIRAH force field	[72]
			pH-dependent deprotonation of RNA genome results in electrostatic repulsive force	Poisson-Boltzmann (PB) model based free energy analysis of multiscale simulations, SIRAH force field	[69]

Table 9.3 Disassembly of nucleocapsid cores from enveloped viruses

Virus	Genome	Disassembly site	Disassembly mechanism	Driving factor	References
African Swine Fever Virus	DNA	Late endosome	Not known	Low pH	[155]
Alphavirus	RNA	Cytosol	Ribosome (60S) binds to the ribosome binding site of capsid	Low pH-dependent membrane-E1 fusion	[152, 154]
Bacteriophage phi6 (φ6)	RNA	Not known	Not known	Ca^{2+} removal	[171]
Hepatitis B virus (HBV)	DNA	Nucleus	Capsid dissociate into assembly competent dimers	Not known	[172–174]
Human Immunodeficiency Virus-1 (HIV-1)	RNA	Nucleus	Physical Disruption	Initiation of Reverse Transcription	[147, 149, 150]
Influenza A	RNA	Cytosol	HDAC6 interaction with capsid associated unanchored ubiquitin chains	Acidic pH	[151, 175]

References

1. Tsai B (2007) Penetration of nonenveloped viruses into the cytoplasm. Annu Rev Cell Dev Biol 23:23–43. https://doi.org/10.1146/annurev.cellbio.23.090506.123454
2. Veesler D, Johnson JE (2012) Virus maturation. Annu Rev Biophys 41:473–496. https://doi.org/10.1146/annurev-biophys-042910-155407
3. Domitrovic T, Movahed N, Bothner B, Matsui T, Wang Q, Doerschuk PC et al (2013) Virus assembly and maturation: auto-regulation through allosteric molecular switches. J Mol Biol 425(9):1488–1496. https://doi.org/10.1016/j.jmb.2013.02.021
4. Banerjee M, Johnson JE (2008) Activation, exposure and penetration of virally encoded, membrane-active polypeptides during non-enveloped virus entry. Curr Protein Pept Sci 9(1):16–27. https://doi.org/10.2174/138920308783565732
5. Kumar CS, Dey D, Ghosh S, Banerjee M (2018) Breach: host membrane penetration and entry by nonenveloped viruses. Trends Microbiol 26(6):525–537. https://doi.org/10.1016/j.tim.2017.09.010
6. Spriggs CC, Harwood MC, Tsai B (2019) How non-enveloped viruses hijack host machineries to cause infection. Adv Virus Res 104:97–122. https://doi.org/10.1016/bs.aivir.2019.05.002
7. Dupzyk A, Tsai B (2016) How polyomaviruses exploit the ERAD machinery to cause infection. Viruses 8(9). https://doi.org/10.3390/v8090242

8. Das A, Barrientos R, Shiota T, Madigan V, Misumi I, McKnight KL et al (2020) Gangliosides are essential endosomal receptors for quasi-enveloped and naked hepatitis A virus. Nat Microbiol 5(9):1069–1078. https://doi.org/10.1038/s41564-020-0727-8
9. Tuthill TJ, Bubeck D, Rowlands DJ, Hogle JM (2006) Characterization of early steps in the poliovirus infection process: receptor-decorated liposomes induce conversion of the virus to membrane-anchored entry-intermediate particles. J Virol 80(1):172–180. https://doi.org/10.1128/JVI.80.1.172-180.2006
10. Bubeck D, Filman DJ, Hogle JM (2005) Cryo-electron microscopy reconstruction of a poliovirus-receptor-membrane complex. Nat Struct Mol Biol 12(7):615–618. https://doi.org/10.1038/nsmb955
11. Danthi P, Tosteson M, Li QH, Chow M (2003) Genome delivery and ion channel properties are altered in VP4 mutants of poliovirus. J Virol 77(9):5266–5274. https://doi.org/10.1128/jvi.77.9.5266-5274.2003
12. Wilts BD, Schaap IAT, Schmidt CF (2015) Swelling and softening of the cowpea chlorotic mottle virus in response to pH shifts. Biophys J 108(10):2541–2549. https://doi.org/10.1016/j.bpj.2015.04.019
13. Bond KM, Lyktey NA, Tsvetkova IB, Dragnea B, Jarrold MF (2020) Disassembly intermediates of the brome mosaic virus identified by charge detection mass spectrometry. J Phys Chem B 124(11):2124–2131. https://doi.org/10.1021/acs.jpcb.0c00008
14. Weis F, Beckers M, von der Hocht I, Sachse C (2019) Elucidation of the viral disassembly switch of tobacco mosaic virus. EMBO Rep 20(11):e48451. https://doi.org/10.15252/embr.201948451
15. Qiu X (2012) Heat induced capsid disassembly and DNA release of bacteriophage lambda. PLoS ONE 7(7):e39793. https://doi.org/10.1371/journal.pone.0039793
16. Badia-Martinez D, Oksanen HM, Stuart DI, Abrescia NG (2013) Combined approaches to study virus structures. Subcell Biochem 68:203–246. https://doi.org/10.1007/978-94-007-6552-8_7
17. Kaelber JT, Hryc CF, Chiu W (2017) Electron cryomicroscopy of viruses at near-atomic resolutions. Annu Rev Virol 4(1):287–308. https://doi.org/10.1146/annurev-virology-101416-041921
18. Doerschuk PC, Gong Y, Xu N, Domitrovic T, Johnson JE (2016) Virus particle dynamics derived from CryoEM studies. Curr Opin Virol 18:57–63. https://doi.org/10.1016/j.coviro.2016.02.011
19. Buchta D, Fuzik T, Hrebik D, Levdansky Y, Sukenik L, Mukhamedova L et al (2019) Enterovirus particles expel capsid pentamers to enable genome release. Nat Commun 10(1):1138. https://doi.org/10.1038/s41467-019-09132-x
20. Azad K, Banerjee M (2019) Structural dynamics of nonenveloped virus disassembly intermediates. J Virol 93(22). https://doi.org/10.1128/JVI.01115-19
21. Hong Y, Song Y, Zhang Z, Li S (2022) Cryo-electron tomography: the resolution revolution and a surge of in situ virological discoveries. Annu Rev Biophys. https://doi.org/10.1146/annurev-biophys-092022-100958
22. Borkotoky S, Dey D, Hazarika Z, Joshi A, Tripathi K (2022) Unravelling viral dynamics through molecular dynamics simulations—a brief overview. Biophys Chem 291:106908. https://doi.org/10.1016/j.bpc.2022.106908
23. Odegard A, Banerjee M, Johnson JE (2010) Flock house virus: a model system for understanding non-enveloped virus entry and membrane penetration. Curr Top Microbiol Immunol 343:1–22. https://doi.org/10.1007/82_2010_35
24. Schneemann A, Zhong W, Gallagher TM, Rueckert RR (1992) Maturation cleavage required for infectivity of a nodavirus. J Virol 66(11):6728–6734. https://doi.org/10.1128/JVI.66.11.6728-6734.1992
25. Maia LF, Soares MR, Valente AP, Almeida FC, Oliveira AC, Gomes AM et al (2006) Structure of a membrane-binding domain from a non-enveloped animal virus: insights into the mechanism of membrane permeability and cellular entry. J Biol Chem 281(39):29278–29286. https://doi.org/10.1074/jbc.M604689200

26. Zlotnick A, Reddy VS, Dasgupta R, Schneemann A, Ray WJ Jr, Rueckert RR et al (1994) Capsid assembly in a family of animal viruses primes an autoproteolytic maturation that depends on a single aspartic acid residue. J Biol Chem 269(18):13680–13684
27. Kearney BM, Johnson JE (2014) Assembly and maturation of a T = 4 quasi-equivalent virus is guided by electrostatic and mechanical forces. Viruses 6(8):3348–3362. https://doi.org/10.3390/v6083348
28. Canady MA, Tihova M, Hanzlik TN, Johnson JE, Yeager M (2000) Large conformational changes in the maturation of a simple RNA virus, nudaurelia capensis omega virus (NomegaV). J Mol Biol 299(3):573–584. https://doi.org/10.1006/jmbi.2000.3723
29. Matsui T, Tsuruta H, Johnson JE (2010) Balanced electrostatic and structural forces guide the large conformational change associated with maturation of T = 4 virus. Biophys J 98(7):1337–1343. https://doi.org/10.1016/j.bpj.2009.12.4283
30. Matsui T, Lander G, Johnson JE (2009) Characterization of large conformational changes and autoproteolysis in the maturation of a T=4 virus capsid. J Virol 83(2):1126–1134. https://doi.org/10.1128/JVI.01859-08
31. Matsui T, Lander GC, Khayat R, Johnson JE (2010) Subunits fold at position-dependent rates during maturation of a eukaryotic RNA virus. Proc Natl Acad Sci USA 107(32):14111–14115. https://doi.org/10.1073/pnas.1004221107
32. Domitrovic T, Matsui T, Johnson JE (2012) Dissecting quasi-equivalence in nonenveloped viruses: membrane disruption is promoted by lytic peptides released from subunit pentamers, not hexamers. J Virol 86(18):9976–9982. https://doi.org/10.1128/JVI.01089-12
33. Odegard AL, Kwan MH, Walukiewicz HE, Banerjee M, Schneemann A, Johnson JE (2009) Low endocytic pH and capsid protein autocleavage are critical components of Flock House virus cell entry. J Virol 83(17):8628–8637. https://doi.org/10.1128/JVI.00873-09
34. Banerjee M, Khayat R, Walukiewicz HE, Odegard AL, Schneemann A, Johnson JE (2009) Dissecting the functional domains of a nonenveloped virus membrane penetration peptide. J Virol 83(13):6929–6933. https://doi.org/10.1128/JVI.02299-08
35. Bothner B, Schneemann A, Marshall D, Reddy V, Johnson JE, Siuzdak G (1999) Crystallographically identical virus capsids display different properties in solution. Nat Struct Biol 6(2):114–116. https://doi.org/10.1038/5799
36. Lin J, Lee LY, Roivainen M, Filman DJ, Hogle JM, Belnap DM (2012) Structure of the Fab-labeled "breathing" state of native poliovirus. J Virol 86(10):5959–5962. https://doi.org/10.1128/JVI.05990-11
37. Basavappa R, Syed R, Flore O, Icenogle JP, Filman DJ, Hogle JM (1994) Role and mechanism of the maturation cleavage of VP0 in poliovirus assembly: structure of the empty capsid assembly intermediate at 2.9 A resolution. Protein Sci 3(10):1651–69. https://doi.org/10.1002/pro.5560031005
38. Shukla A, Padhi AK, Gomes J, Banerjee M (2014) The VP4 peptide of hepatitis A virus ruptures membranes through formation of discrete pores. J Virol 88(21):12409–12421. https://doi.org/10.1128/JVI.01896-14
39. Odegard AL, Chandran K, Zhang X, Parker JS, Baker TS, Nibert ML (2004) Putative autocleavage of outer capsid protein micro1, allowing release of myristoylated peptide micro1N during particle uncoating, is critical for cell entry by reovirus. J Virol 78(16):8732–8745. https://doi.org/10.1128/JVI.78.16.8732-8745.2004
40. Nibert ML, Odegard AL, Agosto MA, Chandran K, Schiff LA (2005) Putative autocleavage of reovirus mu1 protein in concert with outer-capsid disassembly and activation for membrane permeabilization. J Mol Biol 345(3):461–474. https://doi.org/10.1016/j.jmb.2004.10.026
41. Zhang X, Tang J, Walker SB, O'Hara D, Nibert ML, Duncan R et al (2005) Structure of avian orthoreovirus virion by electron cryomicroscopy and image reconstruction. Virology 343(1):25–35. https://doi.org/10.1016/j.virol.2005.08.002
42. Zhang X, Ji Y, Zhang L, Harrison SC, Marinescu DC, Nibert ML et al (2005) Features of reovirus outer capsid protein mu1 revealed by electron cryomicroscopy and image reconstruction of the virion at 7.0 Angstrom resolution. Structure 13(10):1545–57. https://doi.org/10.1016/j.str.2005.07.012

43. Sutton G, Sun D, Fu X, Kotecha A, Hecksel CW, Clare DK et al (2020) Assembly intermediates of orthoreovirus captured in the cell. Nat Commun 11(1):4445. https://doi.org/10.1038/s41467-020-18243-9
44. Wiethoff CM, Wodrich H, Gerace L, Nemerow GR (2005) Adenovirus protein VI mediates membrane disruption following capsid disassembly. J Virol 79(4):1992–2000. https://doi.org/10.1128/JVI.79.4.1992-2000.2005
45. Dormitzer PR, Nason EB, Prasad BV, Harrison SC (2004) Structural rearrangements in the membrane penetration protein of a non-enveloped virus. Nature 430(7003):1053–1058. https://doi.org/10.1038/nature02836
46. Trask SD, Kim IS, Harrison SC, Dormitzer PR (2010) A rotavirus spike protein conformational intermediate binds lipid bilayers. J Virol 84(4):1764–1770. https://doi.org/10.1128/JVI.01682-09
47. Yoder JD, Trask SD, Vo TP, Binka M, Feng N, Harrison SC et al (2009) VP5* rearranges when rotavirus uncoats. J Virol 83(21):11372–11377. https://doi.org/10.1128/JVI.01228-09
48. Chen JZ, Settembre EC, Aoki ST, Zhang X, Bellamy AR, Dormitzer PR et al (2009) Molecular interactions in rotavirus assembly and uncoating seen by high-resolution cryo-EM. Proc Natl Acad Sci USA 106(26):10644–10648. https://doi.org/10.1073/pnas.0904024106
49. Ludert JE, Michelangeli F, Gil F, Liprandi F, Esparza J (1987) Penetration and uncoating of rotaviruses in cultured cells. Intervirology 27(2):95–101. https://doi.org/10.1159/000149726
50. Sasaki M, Itakura Y, Kishimoto M, Tabata K, Uemura K, Ito N et al (2021) Host serine proteases TMPRSS2 and TMPRSS11D mediate proteolytic activation and trypsin-independent infection in group A rotaviruses. J Virol 95(11). https://doi.org/10.1128/JVI.00398-21
51. Silvestry M, Lindert S, Smith JG, Maier O, Wiethoff CM, Nemerow GR et al (2009) Cryo-electron microscopy structure of adenovirus type 2 temperature-sensitive mutant 1 reveals insight into the cell entry defect. J Virol 83(15):7375–7383. https://doi.org/10.1128/JVI.00331-09
52. Ansardi DC, Morrow CD (1995) Amino acid substitutions in the poliovirus maturation cleavage site affect assembly and result in accumulation of provirions. J Virol 69(3):1540–1547. https://doi.org/10.1128/JVI.69.3.1540-1547.1995
53. Hendrix RW, Johnson JE (2012) Bacteriophage HK97 capsid assembly and maturation. Adv Exp Med Biol 726:351–363. https://doi.org/10.1007/978-1-4614-0980-9_15
54. Gertsman I, Gan L, Guttman M, Lee K, Speir JA, Duda RL et al (2009) An unexpected twist in viral capsid maturation. Nature 458(7238):646–650. https://doi.org/10.1038/nature07686
55. Lander GC, Evilevitch A, Jeembaeva M, Potter CS, Carragher B, Johnson JE (2008) Bacteriophage lambda stabilization by auxiliary protein gpD: timing, location, and mechanism of attachment determined by cryo-EM. Structure 16(9):1399–1406. https://doi.org/10.1016/j.str.2008.05.016
56. Dai X, Zhou ZH (2018) Structure of the herpes simplex virus 1 capsid with associated tegument protein complexes. Science 360(6384). https://doi.org/10.1126/science.aao7298
57. Brown JC, Newcomb WW (2011) Herpesvirus capsid assembly: insights from structural analysis. Curr Opin Virol 1(2):142–149. https://doi.org/10.1016/j.coviro.2011.06.003
58. McPherson A, Gavira JA (2014) Introduction to protein crystallization. Acta Crystallogr F Struct Biol Commun 70(Pt 1):2–20. https://doi.org/10.1107/S2053230X13033141
59. Verdaguer N, Garriga D, Fita I (2013) X-ray crystallography of viruses. Subcell Biochem 68:117–144. https://doi.org/10.1007/978-94-007-6552-8_4
60. Hopper P, Harrison SC, Sauer RT (1984) Structure of tomato bushy stunt virus. V. Coat protein sequence determination and its structural implications. J Mol Biol 177(4):701–13. https://doi.org/10.1016/0022-2836(84)90045-7
61. Jiang W, Tang L (2017) Atomic cryo-EM structures of viruses. Curr Opin Struct Biol 46:122–129. https://doi.org/10.1016/j.sbi.2017.07.002
62. Tang G, Peng L, Baldwin PR, Mann DS, Jiang W, Rees I et al (2007) EMAN2: an extensible image processing suite for electron microscopy. J Struct Biol 157(1):38–46. https://doi.org/10.1016/j.jsb.2006.05.009

63. Scheres SH (2012) RELION: implementation of a Bayesian approach to cryo-EM structure determination. J Struct Biol 180(3):519–530. https://doi.org/10.1016/j.jsb.2012.09.006
64. Punjani A, Rubinstein JL, Fleet DJ, Brubaker MA (2017) cryoSPARC: algorithms for rapid unsupervised cryo-EM structure determination. Nat Methods 14(3):290–296. https://doi.org/10.1038/nmeth.4169
65. Bouchet-Marquis C, Hoenger A (2011) Cryo-electron tomography on vitrified sections: a critical analysis of benefits and limitations for structural cell biology. Micron 42(2):152–162. https://doi.org/10.1016/j.micron.2010.07.003
66. Wagner FR, Watanabe R, Schampers R, Singh D, Persoon H, Schaffer M et al (2020) Preparing samples from whole cells using focused-ion-beam milling for cryo-electron tomography. Nat Protoc 15(6):2041–2070. https://doi.org/10.1038/s41596-020-0320-x
67. Hollingsworth SA, Dror RO (2018) Molecular dynamics simulation for all. Neuron 99(6):1129–1143. https://doi.org/10.1016/j.neuron.2018.08.011
68. Ode H, Nakashima M, Kitamura S, Sugiura W, Sato H (2012) Molecular dynamics simulation in virus research. Front Microbiol 3:258. https://doi.org/10.3389/fmicb.2012.00258
69. Martinez M, Cooper CD, Poma AB, Guzman HV (2020) Free energies of the disassembly of viral capsids from a multiscale molecular simulation approach. J Chem Inf Model 60(2):974–981. https://doi.org/10.1021/acs.jcim.9b00883
70. Zink M, Grubmuller H (2009) Mechanical properties of the icosahedral shell of southern bean mosaic virus: a molecular dynamics study. Biophys J 96(4):1350–1363. https://doi.org/10.1016/j.bpj.2008.11.028
71. Ghaemi Z, Gruebele M, Tajkhorshid E (2021) Molecular mechanism of capsid disassembly in hepatitis B virus. Proc Natl Acad Sci USA 118(36). https://doi.org/10.1073/pnas.2102530118
72. Viso JF, Belelli P, Machado M, Gonzalez H, Pantano S, Amundarain MJ et al (2018) Multiscale modelization in a small virus: mechanism of proton channeling and its role in triggering capsid disassembly. PLoS Comput Biol 14(4):e1006082. https://doi.org/10.1371/journal.pcbi.1006082
73. Larsson DS, Liljas L, van der Spoel D (2012) Virus capsid dissolution studied by microsecond molecular dynamics simulations. PLoS Comput Biol 8(5):e1002502. https://doi.org/10.1371/journal.pcbi.1002502
74. Khaykelson D, Raviv U (2020) Studying viruses using solution X-ray scattering. Biophys Rev 12(1):41–48. https://doi.org/10.1007/s12551-020-00617-4
75. Chen J, Chevreuil M, Combet S, Lansac Y, Tresset G (2017) Investigating the thermal dissociation of viral capsid by lattice model. J Phys Condens Matter 29(47):474001. https://doi.org/10.1088/1361-648X/aa8d88
76. Chevreuil M, Lecoq L, Wang S, Gargowitsch L, Nhiri N, Jacquet E et al (2020) Nonsymmetrical dynamics of the HBV capsid assembly and disassembly evidenced by their transient species. J Phys Chem B 124(45):9987–9995. https://doi.org/10.1021/acs.jpcb.0c05024
77. Cuillel M, Berthet-Colominas C, Timmins PA, Zulauf M (1987) Reassembly of brome mosaic virus from dissociated virus. Eur Biophys J 15(3):169–176. https://doi.org/10.1007/BF00263681
78. Tresset G, Chen J, Chevreuil M, Nhiri N, Jacquet E, Lansac Y (2017) Two-dimensional phase transition of viral capsid gives insights into subunit interactions. Phys Rev Appl 7(1):014005. https://doi.org/10.1103/PhysRevApplied.7.014005
79. Castellanos M, Perez R, Carrillo PJ, de Pablo PJ, Mateu MG (2012) Mechanical disassembly of single virus particles reveals kinetic intermediates predicted by theory. Biophys J 102(11):2615–2624. https://doi.org/10.1016/j.bpj.2012.04.026
80. Ausar SF, Foubert TR, Hudson MH, Vedvick TS, Middaugh CR (2006) Conformational stability and disassembly of Norwalk virus-like particles. Effect of pH and temperature. J Biol Chem 281(28):19478–88. https://doi.org/10.1074/jbc.M603313200
81. Greber UF, Singh I, Helenius A (1994) Mechanisms of virus uncoating. Trends Microbiol 2(2):52–56. https://doi.org/10.1016/0966-842x(94)90126-0

82. Suomalainen M, Greber UF (2013) Uncoating of non-enveloped viruses. Curr Opin Virol 3(1):27–33. https://doi.org/10.1016/j.coviro.2012.12.004
83. Shah PNM, Filman DJ, Karunatilaka KS, Hesketh EL, Groppelli E, Strauss M et al (2020) Cryo-EM structures reveal two distinct conformational states in a picornavirus cell entry intermediate. PLoS Pathog 16(9):e1008920. https://doi.org/10.1371/journal.ppat.1008920
84. Butan C, Filman DJ, Hogle JM (2014) Cryo-electron microscopy reconstruction shows poliovirus 135S particles poised for membrane interaction and RNA release. J Virol 88(3):1758–1770. https://doi.org/10.1128/JVI.01949-13
85. Strauss M, Levy HC, Bostina M, Filman DJ, Hogle JM (2013) RNA transfer from poliovirus 135S particles across membranes is mediated by long umbilical connectors. J Virol 87(7):3903–3914. https://doi.org/10.1128/JVI.03209-12
86. Tuthill TJ, Harlos K, Walter TS, Knowles NJ, Groppelli E, Rowlands DJ et al (2009) Equine rhinitis A virus and its low pH empty particle: clues towards an aphthovirus entry mechanism? PLoS Pathog 5(10):e1000620. https://doi.org/10.1371/journal.ppat.1000620
87. Snyder AJ, Danthi P (2018) Infectious subviral particle to membrane penetration active particle (ISVP-to-ISVP*) conversion assay for mammalian orthoreovirus. Bio Protoc 8(2). https://doi.org/10.21769/BioProtoc.2700
88. Dryden KA, Wang G, Yeager M, Nibert ML, Coombs KM, Furlong DB et al (1993) Early steps in reovirus infection are associated with dramatic changes in supramolecular structure and protein conformation: analysis of virions and subviral particles by cryoelectron microscopy and image reconstruction. J Cell Biol 122(5):1023–1041. https://doi.org/10.1083/jcb.122.5.1023
89. Bostina M, Levy H, Filman DJ, Hogle JM (2011) Poliovirus RNA is released from the capsid near a twofold symmetry axis. J Virol 85(2):776–783. https://doi.org/10.1128/JVI.00531-10
90. Lin J, Cheng N, Chow M, Filman DJ, Steven AC, Hogle JM et al (2011) An externalized polypeptide partitions between two distinct sites on genome-released poliovirus particles. J Virol 85(19):9974–9983. https://doi.org/10.1128/JVI.05013-11
91. Panjwani A, Asfor AS, Tuthill TJ (2016) The conserved N-terminus of human rhinovirus capsid protein VP4 contains membrane pore-forming activity and is a target for neutralizing antibodies. J Gen Virol 97(12):3238–3242. https://doi.org/10.1099/jgv.0.000629
92. Davis MP, Bottley G, Beales LP, Killington RA, Rowlands DJ, Tuthill TJ (2008) Recombinant VP4 of human rhinovirus induces permeability in model membranes. J Virol 82(8):4169–4174. https://doi.org/10.1128/JVI.01070-07
93. Huang Y, Hogle JM, Chow M (2000) Is the 135S poliovirus particle an intermediate during cell entry? J Virol 74(18):8757–8761. https://doi.org/10.1128/jvi.74.18.8757-8761.2000
94. Hogle JM (2002) Poliovirus cell entry: common structural themes in viral cell entry pathways. Annu Rev Microbiol 56:677–702. https://doi.org/10.1146/annurev.micro.56.012302.160757
95. Lonberg-Holm K, Korant BD (1972) Early interaction of rhinoviruses with host cells. J Virol 9(1):29–40. https://doi.org/10.1128/JVI.9.1.29-40.1972
96. Strauss M, Schotte L, Karunatilaka KS, Filman DJ, Hogle JM (2017) Cryo-electron microscopy structures of expanded poliovirus with VHHs sample the conformational repertoire of the expanded state. J Virol 91(3). https://doi.org/10.1128/JVI.01443-16
97. Fricks CE, Hogle JM (1990) Cell-induced conformational change in poliovirus: externalization of the amino terminus of VP1 is responsible for liposome binding. J Virol 64(5):1934–1945. https://doi.org/10.1128/JVI.64.5.1934-1945.1990
98. Agirre J, Goret G, LeGoff M, Sanchez-Eugenia R, Marti GA, Navaza J et al (2013) Cryo-electron microscopy reconstructions of triatoma virus particles: a clue to unravel genome delivery and capsid disassembly. J Gen Virol 94(Pt 5):1058–1068. https://doi.org/10.1099/vir.0.048553-0
99. Garriga D, Pickl-Herk A, Luque D, Wruss J, Caston JR, Blaas D et al (2012) Insights into minor group rhinovirus uncoating: the X-ray structure of the HRV2 empty capsid. PLoS Pathog 8(1):e1002473. https://doi.org/10.1371/journal.ppat.1002473
100. Mullapudi E, Fuzik T, Pridal A, Plevka P (2017) Cryo-electron microscopy study of the genome release of the dicistrovirus israeli acute bee paralysis virus. J Virol 91(4). https://doi.org/10.1128/JVI.02060-16

101. Walukiewicz HE, Johnson JE, Schneemann A (2006) Morphological changes in the T = 3 capsid of Flock House virus during cell entry. J Virol 80(2):615–622. https://doi.org/10.1128/JVI.80.2.615-622.2006
102. Mellado MC, Mena JA, Lopes A, Ramirez OT, Carrondo MJ, Palomares LA et al (2009) Impact of physicochemical parameters on in vitro assembly and disassembly kinetics of recombinant triple-layered rotavirus-like particles. Biotechnol Bioeng 104(4):674–686. https://doi.org/10.1002/bit.22430
103. Ren J, Wang X, Hu Z, Gao Q, Sun Y, Li X et al (2013) Picornavirus uncoating intermediate captured in atomic detail. Nat Commun 4:1929. https://doi.org/10.1038/ncomms2889
104. Wang X, Peng W, Ren J, Hu Z, Xu J, Lou Z et al (2012) A sensor-adaptor mechanism for enterovirus uncoating from structures of EV71. Nat Struct Mol Biol 19(4):424–429. https://doi.org/10.1038/nsmb.2255
105. Coombs KM (1998) Stoichiometry of reovirus structural proteins in virus, ISVP, and core particles. Virology 243(1):218–228. https://doi.org/10.1006/viro.1998.9061
106. Liemann S, Chandran K, Baker TS, Nibert ML, Harrison SC (2002) Structure of the reovirus membrane-penetration protein, Mu1, in a complex with is protector protein, Sigma3. Cell 108(2):283–295. https://doi.org/10.1016/s0092-8674(02)00612-8
107. Zhang L, Chandran K, Nibert ML, Harrison SC (2006) Reovirus mu1 structural rearrangements that mediate membrane penetration. J Virol 80(24):12367–12376. https://doi.org/10.1128/JVI.01343-06
108. Chandran K, Farsetta DL, Nibert ML (2002) Strategy for nonenveloped virus entry: a hydrophobic conformer of the reovirus membrane penetration protein micro 1 mediates membrane disruption. J Virol 76(19):9920–9933. https://doi.org/10.1128/jvi.76.19.9920-9933.2002
109. Ivanovic T, Agosto MA, Zhang L, Chandran K, Harrison SC, Nibert ML (2008) Peptides released from reovirus outer capsid form membrane pores that recruit virus particles. EMBO J 27(8):1289–1298. https://doi.org/10.1038/emboj.2008.60
110. Chandran K, Nibert ML (2003) Animal cell invasion by a large nonenveloped virus: reovirus delivers the goods. Trends Microbiol 11(8):374–382. https://doi.org/10.1016/s0966-842x(03)00178-1
111. Middleton JK, Severson TF, Chandran K, Gillian AL, Yin J, Nibert ML (2002) Thermostability of reovirus disassembly intermediates (ISVPs) correlates with genetic, biochemical, and thermodynamic properties of major surface protein mu1. J Virol 76(3):1051–1061. https://doi.org/10.1128/jvi.76.3.1051-1061.2002
112. Gummersheimer SL, Snyder AJ, Danthi P (2021) Control of capsid transformations during reovirus entry. Viruses 13(2). https://doi.org/10.3390/v13020153
113. Snyder AJ, Danthi P (2017) The loop formed by residues 340 to 343 of reovirus mu1 controls entry-related conformational changes. J Virol 91(20). https://doi.org/10.1128/JVI.00898-17
114. Snyder AJ, Danthi P (2015) Lipid membranes facilitate conformational changes required for reovirus cell entry. J Virol 90(5):2628–2638. https://doi.org/10.1128/JVI.02997-15
115. Snyder AJ, Danthi P (2016) Lipids cooperate with the reovirus membrane penetration peptide to facilitate particle uncoating. J Biol Chem 291(52):26773–26785. https://doi.org/10.1074/jbc.M116.747477
116. Zhang X, Settembre E, Xu C, Dormitzer PR, Bellamy R, Harrison SC et al (2008) Near-atomic resolution using electron cryomicroscopy and single-particle reconstruction. Proc Natl Acad Sci USA 105(6):1867–1872. https://doi.org/10.1073/pnas.0711623105
117. Settembre EC, Chen JZ, Dormitzer PR, Grigorieff N, Harrison SC (2011) Atomic model of an infectious rotavirus particle. EMBO J 30(2):408–416. https://doi.org/10.1038/emboj.2010.322
118. Ruiz MC, Aristimuno OC, Diaz Y, Pena F, Chemello ME, Rojas H et al (2007) Intracellular disassembly of infectious rotavirus particles by depletion of Ca2+ sequestered in the endoplasmic reticulum at the end of virus cycle. Virus Res 130(1–2):140–150. https://doi.org/10.1016/j.virusres.2007.06.005

119. Abdelhakim AH, Salgado EN, Fu X, Pasham M, Nicastro D, Kirchhausen T et al (2014) Structural correlates of rotavirus cell entry. PLoS Pathog 10(9):e1004355. https://doi.org/10.1371/journal.ppat.1004355
120. Richards RM, Lowy DR, Schiller JT, Day PM (2006) Cleavage of the papillomavirus minor capsid protein, L2, at a furin consensus site is necessary for infection. Proc Natl Acad Sci USA 103(5):1522–1527. https://doi.org/10.1073/pnas.0508815103
121. Schelhaas M, Shah B, Holzer M, Blattmann P, Kuhling L, Day PM et al (2012) Entry of human papillomavirus type 16 by actin-dependent, clathrin- and lipid raft-independent endocytosis. PLoS Pathog 8(4):e1002657. https://doi.org/10.1371/journal.ppat.1002657
122. Kremer EJ, Nemerow GR (2015) Adenovirus tales: from the cell surface to the nuclear pore complex. PLoS Pathog 11(6):e1004821. https://doi.org/10.1371/journal.ppat.1004821
123. Burckhardt CJ, Suomalainen M, Schoenenberger P, Boucke K, Hemmi S, Greber UF (2011) Drifting motions of the adenovirus receptor CAR and immobile integrins initiate virus uncoating and membrane lytic protein exposure. Cell Host Microbe 10(2):105–117. https://doi.org/10.1016/j.chom.2011.07.006
124. Moyer CL, Besser ES, Nemerow GR (2016) A single maturation cleavage site in adenovirus impacts cell entry and capsid assembly. J Virol 90(1):521–532. https://doi.org/10.1128/JVI.02014-15
125. Ortega-Esteban A, Perez-Berna AJ, Menendez-Conejero R, Flint SJ, San Martin C, de Pablo PJ (2013) Monitoring dynamics of human adenovirus disassembly induced by mechanical fatigue. Sci Rep 3:1434. https://doi.org/10.1038/srep01434
126. van Rosmalen MGM, Nemerow GR, Wuite GJL, Roos WH (2018) A single point mutation in precursor protein VI doubles the mechanical strength of human adenovirus. J Biol Phys 44(2):119–132. https://doi.org/10.1007/s10867-017-9479-y
127. Liu N, Chen Y, Peng B, Lin Y, Wang Q, Su Z et al (2013) Single-molecule force spectroscopy study on the mechanism of RNA disassembly in tobacco mosaic virus. Biophys J 105(12):2790–2800. https://doi.org/10.1016/j.bpj.2013.10.005
128. Lavelle L, Michel JP, Gingery M (2007) The disassembly, reassembly and stability of CCMV protein capsids. J Virol Methods 146(1–2):311–316. https://doi.org/10.1016/j.jviromet.2007.07.020
129. Speir JA, Munshi S, Wang G, Baker TS, Johnson JE (1995) Structures of the native and swollen forms of cowpea chlorotic mottle virus determined by X-ray crystallography and cryo-electron microscopy. Structure 3(1):63–78. https://doi.org/10.1016/s0969-2126(01)00135-6
130. Arisaka F (2005) Assembly and infection process of bacteriophage T4. Chaos 15(4):047502. https://doi.org/10.1063/1.2142136
131. Ivanovska I, Wuite G, Jonsson B, Evilevitch A (2007) Internal DNA pressure modifies stability of WT phage. Proc Natl Acad Sci USA 104(23):9603–9608. https://doi.org/10.1073/pnas.0703166104
132. Purohit PK, Inamdar MM, Grayson PD, Squires TM, Kondev J, Phillips R (2005) Forces during bacteriophage DNA packaging and ejection. Biophys J 88(2):851–866. https://doi.org/10.1529/biophysj.104.047134
133. Sao-Jose C, de Frutos M, Raspaud E, Santos MA, Tavares P (2007) Pressure built by DNA packing inside virions: enough to drive DNA ejection in vitro, largely insufficient for delivery into the bacterial cytoplasm. J Mol Biol 374(2):346–355. https://doi.org/10.1016/j.jmb.2007.09.045
134. Evilevitch A, Lavelle L, Knobler CM, Raspaud E, Gelbart WM (2003) Osmotic pressure inhibition of DNA ejection from phage. Proc Natl Acad Sci USA 100(16):9292–9295. https://doi.org/10.1073/pnas.1233721100
135. Kumar M, Blaas D (2013) Human rhinovirus subviral a particle binds to lipid membranes over a twofold axis of icosahedral symmetry. J Virol 87(20):11309–11312. https://doi.org/10.1128/JVI.02055-13
136. Harutyunyan S, Kumar M, Sedivy A, Subirats X, Kowalski H, Kohler G et al (2013) Viral uncoating is directional: exit of the genomic RNA in a common cold virus starts with the poly-(A) tail at the 3'-end. PLoS Pathog 9(4):e1003270. https://doi.org/10.1371/journal.ppat.1003270

137. Seitsonen JJ, Shakeel S, Susi P, Pandurangan AP, Sinkovits RS, Hyvonen H et al (2012) Structural analysis of coxsackievirus A7 reveals conformational changes associated with uncoating. J Virol 86(13):7207–7215. https://doi.org/10.1128/JVI.06425-11
138. Fisher AJ, Johnson JE (1993) Ordered duplex RNA controls capsid architecture in an icosahedral animal virus. Nature 361(6408):176–179. https://doi.org/10.1038/361176a0
139. Venter PA, Schneemann A (2007) Assembly of two independent populations of flock house virus particles with distinct RNA packaging characteristics in the same cell. J Virol 81(2):613–619. https://doi.org/10.1128/JVI.01668-06
140. Wang JC, Mukhopadhyay S, Zlotnick A (2018) Geometric defects and icosahedral viruses. Viruses 10(1). https://doi.org/10.3390/v10010025
141. Therkelsen MD, Klose T, Vago F, Jiang W, Rossmann MG, Kuhn RJ (2018) Flaviviruses have imperfect icosahedral symmetry. Proc Natl Acad Sci USA 115(45):11608–11612. https://doi.org/10.1073/pnas.1809304115
142. Zhao Z, Zhang H, Shu D, Montemagno C, Ding B, Li J et al (2017) Construction of asymmetrical hexameric biomimetic motors with continuous single-directional motion by sequential coordination. Small 13(1). https://doi.org/10.1002/smll.201601600
143. Mukhamedova L, Fuzik T, Novacek J, Hrebik D, Pridal A, Marti GA et al (2021) Virion structure and in vitro genome release mechanism of dicistrovirus Kashmir bee virus. J Virol 95(11). https://doi.org/10.1128/JVI.01950-20
144. Cilliers T, Nhlapo J, Coetzer M, Orlovic D, Ketas T, Olson WC et al (2003) The CCR5 and CXCR4 coreceptors are both used by human immunodeficiency virus type 1 primary isolates from subtype C. J Virol 77(7):4449–4456. https://doi.org/10.1128/jvi.77.7.4449-4456.2003
145. Leung HS, Li OT, Chan RW, Chan MC, Nicholls JM, Poon LL (2012) Entry of influenza A Virus with a alpha2,6-linked sialic acid binding preference requires host fibronectin. J Virol 86(19):10704–10713. https://doi.org/10.1128/JVI.01166-12
146. Wulan WN, Heydet D, Walker EJ, Gahan ME, Ghildyal R (2015) Nucleocytoplasmic transport of nucleocapsid proteins of enveloped RNA viruses. Front Microbiol 6:553. https://doi.org/10.3389/fmicb.2015.00553
147. Rankovic S, Varadarajan J, Ramalho R, Aiken C, Rousso I (2017) Reverse transcription mechanically initiates HIV-1 capsid disassembly. J Virol 91(12). https://doi.org/10.1128/JVI.00289-17
148. Christensen DE, Ganser-Pornillos BK, Johnson JS, Pornillos O, Sundquist WI (2020) Reconstitution and visualization of HIV-1 capsid-dependent replication and integration in vitro. Science 370(6513). https://doi.org/10.1126/science.abc8420
149. Muller TG, Zila V, Peters K, Schifferdecker S, Stanic M, Lucic B et al (2021) HIV-1 uncoating by release of viral cDNA from capsid-like structures in the nucleus of infected cells. Elife 10. https://doi.org/10.7554/eLife.64776
150. Burdick RC, Li C, Munshi M, Rawson JMO, Nagashima K, Hu WS et al (2020) HIV-1 uncoats in the nucleus near sites of integration. Proc Natl Acad Sci USA 117(10):5486–5493. https://doi.org/10.1073/pnas.1920631117
151. Banerjee I, Miyake Y, Nobs SP, Schneider C, Horvath P, Kopf M et al (2014) Influenza A virus uses the aggresome processing machinery for host cell entry. Science 346(6208):473–477. https://doi.org/10.1126/science.1257037
152. Wengler G, Wurkner D, Wengler G (1992) Identification of a sequence element in the alphavirus core protein which mediates interaction of cores with ribosomes and the disassembly of cores. Virology 191(2):880–888. https://doi.org/10.1016/0042-6822(92)90263-o
153. Ulmanen I, Soderlund H, Kaariainen L (1976) Semliki Forest virus capsid protein associates with the 60S ribosomal subunit in infected cells. J Virol 20(1):203–210. https://doi.org/10.1128/JVI.20.1.203-210.1976
154. Hasan SS, Sun C, Kim AS, Watanabe Y, Chen CL, Klose T et al (2018) Cryo-EM structures of eastern equine encephalitis virus reveal mechanisms of virus disassembly and antibody neutralization. Cell Rep 25(11):3136–47e5. https://doi.org/10.1016/j.celrep.2018.11.067

155. Hernaez B, Guerra M, Salas ML, Andres G (2016) African Swine fever virus undergoes outer envelope disruption, capsid disassembly and inner envelope fusion before core release from multivesicular endosomes. PLoS Pathog 12(4):e1005595. https://doi.org/10.1371/journal.ppat.1005595
156. Bauer DW, Huffman JB, Homa FL, Evilevitch A (2013) Herpes virus genome, the pressure is on. J Am Chem Soc 135(30):11216–11221. https://doi.org/10.1021/ja404008r
157. Newcomb WW, Booy FP, Brown JC (2007) Uncoating the herpes simplex virus genome. J Mol Biol 370(4):633–642. https://doi.org/10.1016/j.jmb.2007.05.023
158. Serrero MC, Girault V, Weigang S, Greco TM, Ramos-Nascimento A, Anderson F et al (2022) The interferon-inducible GTPase MxB promotes capsid disassembly and genome release of herpesviruses. Elife 11. https://doi.org/10.7554/eLife.76804
159. Cyrklaff M, Risco C, Fernandez JJ, Jimenez MV, Esteban M, Baumeister W et al (2005) Cryo-electron tomography of vaccinia virus. Proc Natl Acad Sci USA 102(8):2772–2777. https://doi.org/10.1073/pnas.0409825102
160. Cyrklaff M, Linaroudis A, Boicu M, Chlanda P, Baumeister W, Griffiths G et al (2007) Whole cell cryo-electron tomography reveals distinct disassembly intermediates of vaccinia virus. PLoS ONE 2(5):e420. https://doi.org/10.1371/journal.pone.0000420
161. Pletan ML, Tsai B (2022) Non-enveloped virus membrane penetration: New advances leading to new insights. PLoS Pathog 18(12):e1010948. https://doi.org/10.1371/journal.ppat.1010948
162. Kim KR, Lee AS, Kim SM, Heo HR, Kim CS (2022) Virus-like nanoparticles as a theranostic platform for cancer. Front Bioeng Biotechnol 10:1106767. https://doi.org/10.3389/fbioe.2022.1106767
163. Obozina AS, Komedchikova EN, Kolesnikova OA, Iureva AM, Kovalenko VL, Zavalko FA et al (2023) Genetically encoded self-assembling protein nanoparticles for the targeted delivery in vitro and in vivo. Pharmaceutics 15(1). https://doi.org/10.3390/pharmaceutics15010231
164. Chung YH, Cai H, Steinmetz NF (2020) Viral nanoparticles for drug delivery, imaging, immunotherapy, and theranostic applications. Adv Drug Deliv Rev 156:214–235. https://doi.org/10.1016/j.addr.2020.06.024
165. Shukla S, Hu H, Cai H, Chan SK, Boone CE, Beiss V et al (2020) Plant viruses and bacteriophage-based reagents for diagnosis and therapy. Annu Rev Virol 7(1):559–587. https://doi.org/10.1146/annurev-virology-010720-052252
166. Bajaj S, Banerjee M (2015) Engineering virus capsids into biomedical delivery vehicles: structural engineering problems in nanoscale. J Biomed Nanotechnol 11(1):53–69. https://doi.org/10.1166/jbn.2015.1959
167. de Pablo PJ, San MC (2022) Seeing and touching adenovirus: complementary approaches for understanding assembly and disassembly of a complex virion. Curr Opin Virol 52:112–122. https://doi.org/10.1016/j.coviro.2021.11.006
168. Chen J, Lansac Y, Tresset G (2018) Interactions between the molecular components of the cowpea chlorotic mottle virus investigated by molecular dynamics simulations. J Phys Chem B 122(41):9490–9498. https://doi.org/10.1021/acs.jpcb.8b08026
169. Arkhipov A, Roos WH, Wuite GJ, Schulten K (2009) Elucidating the mechanism behind irreversible deformation of viral capsids. Biophys J 97(7):2061–2069. https://doi.org/10.1016/j.bpj.2009.07.039
170. Tarasova E, Farafonov V, Taiji M, Nerukh D (2018) Details of charge distribution in stable viral capsid. J Mol Liq 265:585–591. https://doi.org/10.1016/j.molliq.2018.06.019
171. Romantschuk M, Olkkonen VM, Bamford DH (1988) The nucleocapsid of bacteriophage phi 6 penetrates the host cytoplasmic membrane. EMBO J 7(6):1821–1829. https://doi.org/10.1002/j.1460-2075.1988.tb03014.x
172. Singh S, Zlotnick A (2003) Observed hysteresis of virus capsid disassembly is implicit in kinetic models of assembly. J Biol Chem 278(20):18249–18255. https://doi.org/10.1074/jbc.M211408200
173. Rabe B, Delaleau M, Bischof A, Foss M, Sominskaya I, Pumpens P et al (2009) Nuclear entry of hepatitis B virus capsids involves disintegration to protein dimers followed by nuclear reassociation to capsids. PLoS Pathog 5(8):e1000563. https://doi.org/10.1371/journal.ppat.1000563

174. Cui X, Ludgate L, Ning X, Hu J (2013) Maturation-associated destabilization of hepatitis B virus nucleocapsid. J Virol 87(21):11494–11503. https://doi.org/10.1128/JVI.01912-13
175. Zhirnov OP, Grigoriev VB (1994) Disassembly of influenza C viruses, distinct from that of influenza A and B viruses requires neutral-alkaline pH. Virology 200(1):284–291. https://doi.org/10.1006/viro.1994.1188

Chapter 10
Physical Virology with Atomic Force and Fluorescence Microscopies: Stability, Disassembly and Genome Release

María Jesús Rodríguez-Espinosa, Miguel Cantero, Klara Strobl, Pablo Ibáñez, Alejandro Díez-Martínez, Natalia Martín-González, Manuel Jiménez-Zaragoza, Alvaro Ortega-Esteban, and Pedro José de Pablo

Abstract The core of Atomic Force Microscopy (AFM) is a nanometric tip mounted at the extreme of a microcantilever that scans the surface where the virus particles are adsorbed. Beyond obtaining nanometric resolution of individual viruses in liquid environment, AFM allows the manipulation of single particles, the exploration of virus biomechanics and to monitor assembly/disassembly processes, including genome release in real time. This chapter starts providing some inputs about virus adsorption on surfaces and imaging, including an example of tip dilation artifacts. Later, we exemplify how to monitor the effects of changing the chemical environment of the liquid cell on TGEV coronavirus particles. We go on by describing approaches to study genome release, aging, and multilayered viruses with single indentation and mechanical fatigue assays. The chapter ends explaining an AFM/fluorescence combination to study the influence of crowding on GFP within P22 bacteriophage capsids.

M. J. Rodríguez-Espinosa · M. Cantero · K. Strobl · A. Díez-Martínez · M. Jiménez-Zaragoza · A. Ortega-Esteban · P. J. de Pablo (✉)
Departamento de Física de La Materia Condensada, Universidad Autónoma de Madrid, 28044 Madrid, Spain
e-mail: p.j.depablo@uam.es

M. J. Rodríguez-Espinosa
Department of Structure of Macromolecules, Centro Nacional de Biotecnología, (CNB–CSIC), Darwin 3, 28049 Madrid, Spain

P. Ibáñez
Department of Theoretical Physics of Condensed Matter, Universidad Autónoma de Madrid, Cantoblanco Campus, 28049 Madrid, Spain

N. Martín-González
AFMB UMR 7257, Aix Marseille Université, CNRS, Marseille, France

P. J. de Pablo
Solid Condensed Matter Institute IFIMAC, Universidad Autónoma de Madrid, 28049 Madrid, Spain

M. Comas-Garcia and S. Rosales-Mendoza (eds.), *Physical Virology*, Springer Series in Biophysics 24, https://doi.org/10.1007/978-3-031-36815-8_10

Keywords Atomic force microscopy · Force curve · Nanoindentation · Beam deflection · Tip · Cantilever · Stylus · Topography · Aqueous solution · Mechanical fatigue · DNA condensation · Assembly · Disassembly

Introduction

Viruses are outstanding examples of how the Nature evolution can achieve sophisticated systems designed to last with minimalistic resources [1]. They can be considered as nanomachines whose function and properties are not only interesting for understanding their biology, but also for applications in material science [2]. In this vein, the mechanical properties of biomolecular complexes are essential for their function, and viruses are not an exception [3]. Virus structures exhibit certain meta-stability whose modulation permits fulfilling each task of the viral cycle on time thanks to their physicochemical properties. The basic construction of non-enveloped viruses consists of the capsid, a shell made up of repeating protein subunits (capsomers) or lipids, packing within the viral genome [1]. Far from being static structures, both enveloped and non-enveloped viruses are highly dynamic complexes that shuttle and deliver their genome from host to host in a fully automatic process. These natural designed capabilities have impelled using viral capsids as protein containers of artificial cargoes (drugs, polymers, enzymes, minerals) [2] with applications in biomedical and materials sciences. Both natural and artificial protein cages have to protect their cargo against a variety of physicochemical aggressive environments, including molecular impacts in highly crowded media [4], thermal and chemical stresses [5], and osmotic shocks [6]. Thus, it is important to use methodologies that supply information about protein cages stability not only under different environments, but also its evolution upon structural changes. Structural biology techniques such as electron microscopy (EM) and X-ray are used to unveil the structure–function interplay, revealing high resolution structures of protein cages [7]. However, these methodologies require a heavy average of millions of particles present in the crystal (X-ray) or thousands of structures for the model reconstruction (cryo-EM). Thus, they provide limited information on possible structural differences between individual particles that differs from the average structure. In addition, these approaches require conditions (i.e., vacuum) far away of those where protein shells are functional (liquid), precluding the characterization of protein shells dynamics and properties in real time. Indeed, the advent of single molecule technologies has demonstrated that mechanical properties of biological molecular aggregates are essential to their function [8]. The exploration of these properties would complement the structural biology methodologies (EM and X-ray) to find the structure–function-property interplay of virus structures. Atomic Force Microscopy (AFM) may not only characterize the structure of individual protein-made particles in liquid milieu, but also to obtain physicochemical properties of each one [9]. In addition, the nano-dissection abilities of AFM allows the local manipulation of protein shells to learn about their assembly/disassembly [10]. In this chapter, we give a general overview of how to apply AFM methods to protein shells. Our tour starts with a basic review of the recipes for attaching protein cages to solid surfaces. Afterwards we describe some modes for imaging protein shells with AFM and comment on inherent artifacts, such as geometrical dilation. We describe the ability of AFM to explore virions under changing environments in the liquid cell. Subsequently we describe the nanoindentation methodology, which probes the stiffness, breaking force, brittleness, and cargo unpacking of individual protein shells. Afterwards we focus on the effects of cyclic loading on individual particles (disassembly), and the AFM/fluorescence combination for exploring protein cages packing GFP.

Insights on AFM Methodology

Cantilevers. In AFM a microcantilever with a nanometric tip located at the very extreme actuates as a force transducer which palpates the viruses adsorbed on a solid surface. Cantilevers with different spring constants are mounted on a chip. For example Olympus OMCL-RC include a variety of four spring constants: 0.39, 0.76, 0.05 and 0.10 N/m that can be used with different viruses, depending on their stability and the adsorption strength to the surface. Figure 10.1a shows and optical microscopy image of the 0.05 N/m cantilever where the pyramid tip is encircled. Olympus stopped the production of OMCL-RC cantilevers. PNP-DB and QP-BioAC tips (Nanoandmore™) with similar properties can be used. The chip holding the cantilever is attached to the quartz window of the liquid cell which is created between the surface and the cantilever holder (Fig. 10.1b). For the shake of comparison with viruses, Tetraspeck™ (Thermofisher) plastic spheres of 100 nm in diameter are spread on a glass surface that are seen as white blobs with a size of ~250 nm because of the diffraction limit.

Immobilization of virus particles on surfaces. Virus particles are adsorbed to the surface by using physical forces (physisorption) with the substrate, including polar, non-polar, an van der Waals interactions [11]. Physisorption traps virus particles without inducing chemical bonds that might alter their structure [12]. Each virus species has individualized features such as local charge densities or hydrophobic patches [13] that can be used for anchoring the particles, via hydrophobic and/or electrostatic interactions, on different materials, such as glass, mica and HOPG (Highly Oriented Pyrolytic Graphite). Mica and HOPG surfaces are layered materials whose preparation consist of removing the last layer with adhesive tape, exposing a fresh surface ready for experiments. Care should be taken on leaving a flat surface by avoiding dangling whiskers on the area of the meniscus because they might crash with the cantilever. HOPG presents a non-polar surface and protein cages adsorb via hydrophobic interactions [14]. In the case of human adenovirus (HAdv) virus particles can be adsorbed on HOPG, silanized glass and mica with different results. HOPG collapses most of HAdv particles, indicating a strong non-polar (hydrophobic) interaction. However, silanized glass [9] reduces the attachment interaction and

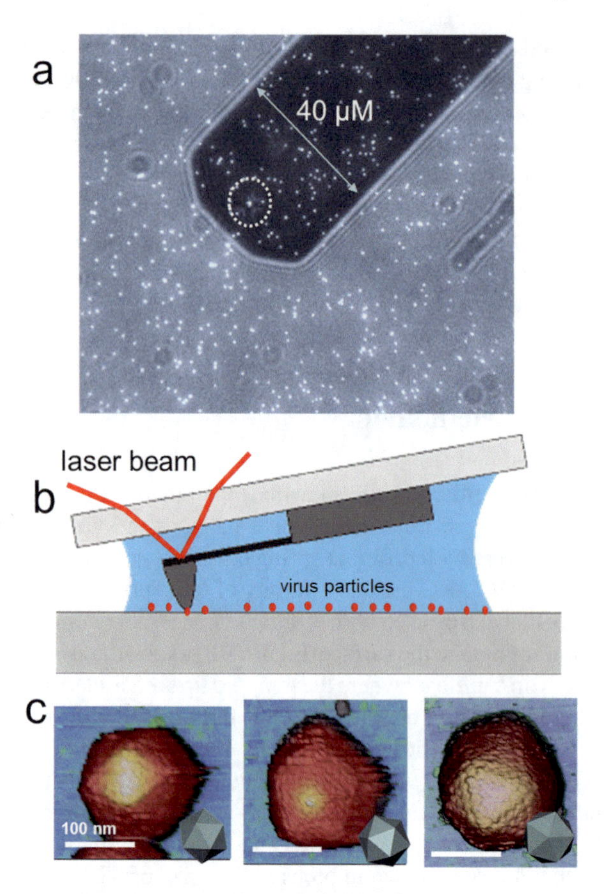

Fig. 10.1 Attaching protein shells on surfaces. **a** Optical microscopy image of an 0.05 N/m OMRC cantilever, with the encircled tip. Plastic bead of 100 nm in diameter are adsorbed on the surface for the sake of comparison. **b** Cartoon of the liquid cell. Protein cages and cantilever are not in scale. **c** HOPG, glass and mica bare substrates before attaching the viruses. **d** hAdV particles on HOPG, glass and mica. **e** Individual hAdV particles showing twofold, threefold and fivefold symmetry orientations

allows imaging intact icosahedral particles exhibiting fivefold, twofold and threefold symmetry orientations on the surface (Fig. 10.1c). Interestingly, using $NiCl_2$ 150 mM on mica [11] induce the adsorption of HAdv particles only at the threefold symmetry orientation, thus protein particles exhibit a triangular facet (Fig. 10.1c, right). Adsorption of virus particles on surfaces also may induce a reduction of the particle height [15, 16]. From a practical point of view predictions on proteins shells adsorption are difficult to make, and one uses the try-and-error methodology to find the best conditions.

Imaging. In AFM the tip scans the sample in x, y and z directions by using piezo actuators. While x and y scanners move in a pre-established way over a square region, the cantilever bends following the surface topography. Either the cantilever deflects perpendicularly to the surface applying a normal force (F_n) (Fig. 10.2a), or it bends laterally by torsion exerting a dragging force parallel to the surface (F_l) (Fig. 10.2b). Both F_n and F_l are monitored by focusing a laser beam at the end of the cantilever, whose reflection is registered in a four-quadrant photodiode. Thus, each pixel of the image located at a particular position of the planar coordinates (x, y), will be associated with certain bending values of the cantilever F_n and F_l. If the particle is not strongly enough attached or if it is too soft, it can be swept or modified under large bending forces. To avoid this effect as much as possible, a feedback loop is engaged to F_n to move the z piezo position in such way that F_n is kept constant. In this operational approach (contact mode, Fig. 10.2c), the AFM topography map will have x, y and z coordinates. The torsion F_l of the cantilever exerts about 40 times the perpendicular bending force F_n [17]. Individual virus particles are thus susceptible to undesired modifications by lateral forces. Their size of tens of nanometers offers a large topographical aspect ratio that is difficult to track by the feedback loop. A typical approach for surpassing this limitation is using fixation agents, such as glutaraldehyde. In such conditions AFM provide images whose resolution is comparable to that of some EM images [18]. Nevertheless, since glutaraldehyde structurally reinforces the specimens [19, 20], it precludes any characterization of dynamics or properties of intact native viruses, such as assembly/disassembly or physical properties [21]. Other approaches include developing imaging modes that avoid dragging forces as much as possible. In jumping mode (JM) [22], the lateral tip displacement occurs when the tip and sample are not in mechanical contact, thereby avoiding shear forces to a large extent (Fig. 10.2d). JM performs consecutive approach-release cycles at every pixel of the sample. In each cycle, known as force vs. distance z curve (FZ, Fig. 10.2e), the z-piezo approaches tip and sample from non-contact (label 1 at Fig. 10.2e) until establishing mechanical contact (label 2 at Fig. 10.2e) and reaching a certain feedback force (label 3 at Fig. 10.2e). After a few milliseconds, the z-piezo retracts about 100 nm until releasing the tip from the surface (label 1 at Fig. 10.2e). Subsequently the scanner moves laterally to the next pixel, and the process starts again. An AFM cantilever experiences a viscous drag while moving up and down in liquid, giving rise to a hysteresis loop (Fig. 10.2e) because forward and back curves do not coincide. Although AFM dynamic modes have also able of imaging protein shells in liquid conditions, it is difficult to control the applied force [23].

Tip dilation. The typical radius of the tip apex for usual cantilevers (OMCL-RC800PSA) is ~20 nm, and it is comparable to the curvature of the virus particles diameter. In this case, tip size plays an important role on the image resolution by inducing a lateral expansion, namely dilation, of the specimen [24]. Since dilation very often impairs high resolution in proteins, it is convenient to estimate how the tip-size is going to affect to AFM images. WSxM software [25] implements a geometrical dilation algorithm that allows simulating the dilation of protein shell's structure. By using Chimera software [26] it is possible to access to a particular protein shell structure, such as the electron microscopy model. The "Surface Color By Height" option generates a gray-scale image that captures the topography variation in a given orientation. The TIFF format of this image can be imported by WSxM software and calibrated (www.wsxm.eu). The dilation algorithm input is the tip radius, and the

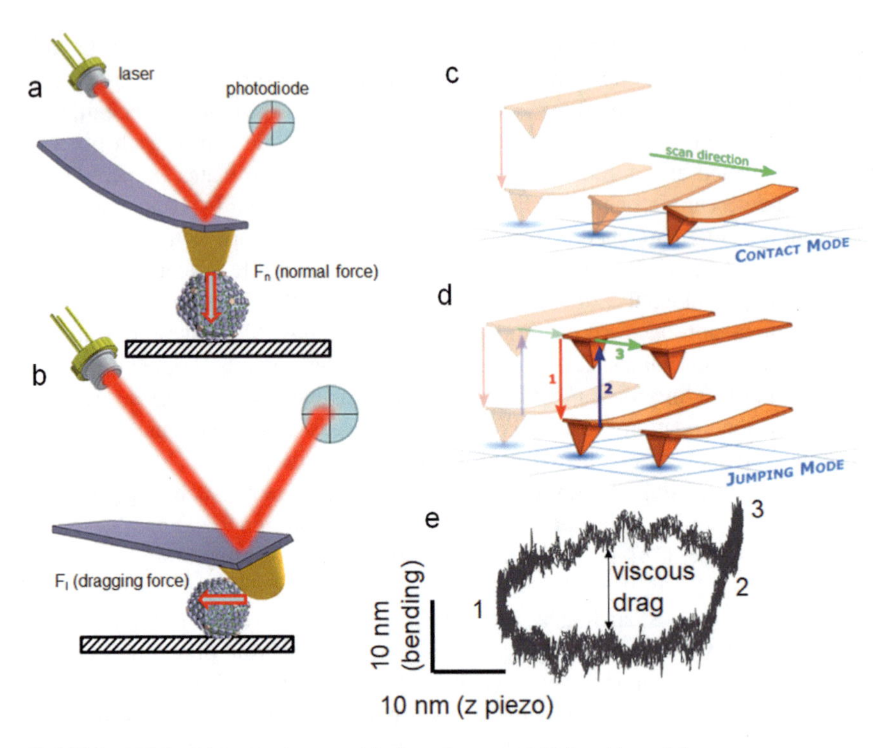

Fig. 10.2 AFM working modes. **a** and **b** show the normal (F_n) and lateral forces concepts, respectively. **c** and **d** indicate contact and jumping modes, respectively. **e** Force curve of jumping mode in liquids, indicating the viscous drag and the important stages (see text)

dilated structure is calculated. Figure 10.3 exemplifies the dilation of phi29 bacteriophage EM structure [27]. Figure 10.3a shows the 3D rendered topography data of a bacteriophage phi29 on HOPG. By using the dilation algorithm with a 10 nm in diameter tip on the PDB model of phi29 phage [27] it is possible to obtain the dilated topography (Fig. 10.3b). Dilation strongly depends on the tip size, as shown in Fig. 10.3c. The darker area of Fig. 10.3b represents the extra dilated geometry of phi29 around the electron microscopy model (light colored).

Applications

Imaging at different environment conditions. The conditions of the liquid environment in the liquid cell can be gradually changed for monitoring the stability of virus capsids over time. One pump connected to the inlet of the chamber inserts the buffer at the desired final concentration, and another pump connected to the outlet withdraws liquid at the same velocity of the inlet. In this particular case (Fig. 10.4) we observe the effect of a non-ionic detergent (IGEPAL® CA-630 (Sigma-Aldrich,

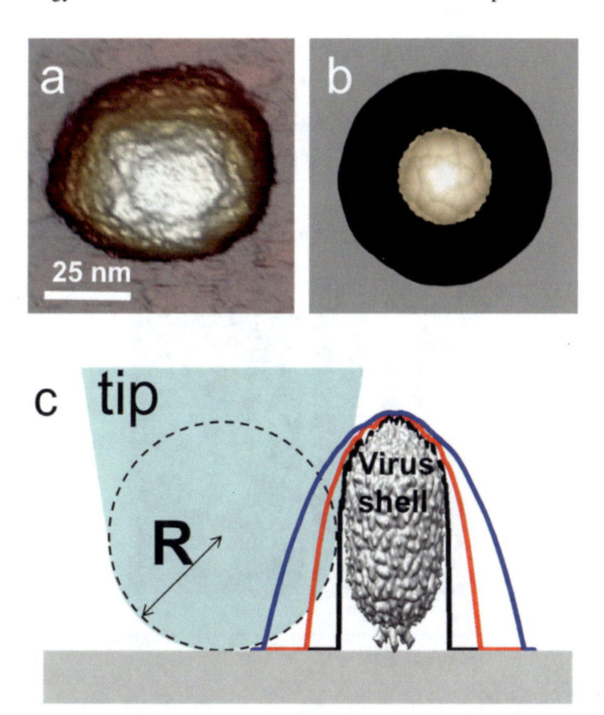

Fig. 10.3 Dilation effects in the phi29 bacteriophage shell. **a** represents the AFM image of a prohead. In **b** the bright color illustrates a EM structural model of phi29, and the dark area indicates the dilation corresponding to a tip of 10 nm in diameter. The cartoon of **c** indicates the dilation as a function of the tip size: dark, red and blue curves are the topographical profiles obtained with tips of 0.5 nm, 5 and 10 radius in diameter, respectively

Saint Louis, MI, USA, CAS: 9002-93-1)) on the structure of the coronavirus *transmissible gastroenteritis virus* (TGEV) [28], whose structure [29] and mechanical properties [30] are identical to SARS-CoV-2. Consecutive images of the same TGEV virus particle while increasing IGEPAL concentration in the liquid cell were taken (Fig. 10.4a). This virus was continuously imaged more than 40 times while IGEPAL was increased up to ~0.06% for 80 min. The virus particle remained stable at 0.020% IGEPAL, although at 0.023% some deterioration could be recognized. In frame #26, some material was ejected from the virus and a circular crack at the top appears. This damaged structure seems stable until frame #40. For statistical analysis, the height data evolution of eight particles subjected to similar process are plotted their individual (Fig. 10.4b, thin grey) and their average (Fig. 10.4b, thick dark) height variation over time. Consecutive images of virions in buffer without detergent were taken as control (Fig. 10.4b, blue), showing that their heights were not affected by AFM scanning. Figure 10.4c shows the effect of the pH variation on the stiffness of human adenovirus particles [31].

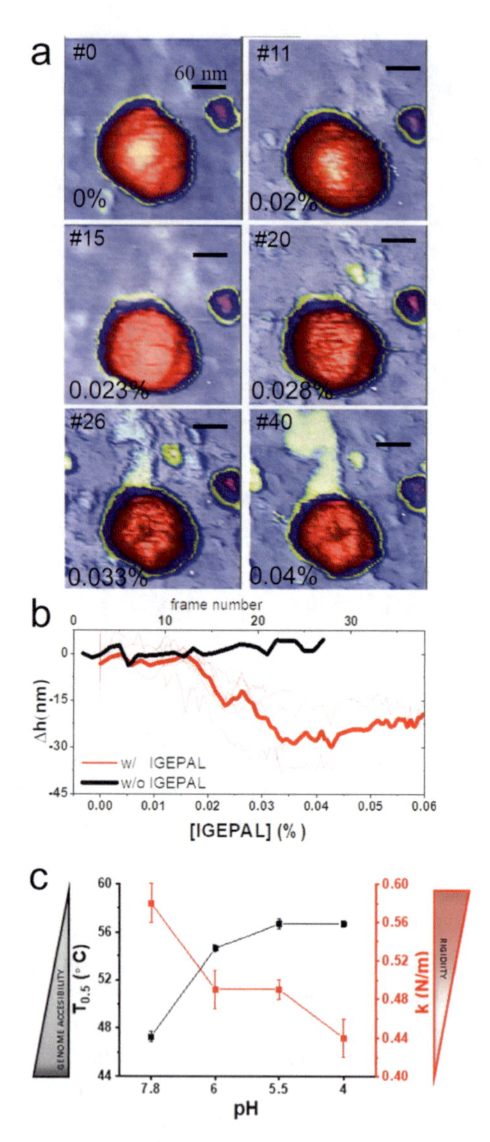

Fig. 10.4 Imaging while changing the environment in the liquid cell: effect of IGEPAL concentration on TGEV virions. **a** Topographical images were taken while the concentration of IGEPAL was increased over time. The frame number is shown in the upper left corner and the concentration of IGEPAL in the bottom-right. **b** Height loss during the time course assay. Experimental curves are shown in light red and the average curve of 8 observations is shown in red. As a control, the height of 1 virus particle without detergent is shown (black). Adapted from [32]. **c** Evolution of the spring constant with pH in human adenovirus. Adapted from [31]

Mechanical properties of viruses: nanoindentation. Single force curve (FZ) experiments consist on pushing on the top of a selected protein shell (Fig. 10.5a). The FZ is executed on the particle at a typical speed of 50 nm/s to allow the water leaving the virus when it is squeezed [33]. After the contact between tip and particle is stablished, FZ typically shows an approximate linear behavior, which corresponds to the elastic regime of the shell and ascribes to the mixed bending of the cantilever and sample deformation (Fig. 10.5a, b, label 2). When the z-piezo elongation surpasses the critical indentation, particle breaks (Fig. 10.5, 1.1 nN) inducing a drastic decrease of the force, that resemble the penetration of the tip apex trough the virus (Fig. 10.5a, b, label 3). Afterwards the FZ is linear again and represents the cantilever bending on the solid substrate. By performing a FZ on the substrate (Fig. 10.5b, dotted line) and assuming that it is much more rigid that the cantilever, we can obtain the cantilever deformation. In this case, the human picobirnavirus particle [34] yields and breaks after the nanoindentation experiment (Fig. 10.5c). The subtraction of sample from substrate curves allows isolating the deformation of the virus cage (Fig. 10.5d). From these data we can obtain a few mechanical parameters: Fitting of the elastic part from 0 to 5 nm results in the stiffness or spring constant of virus shell (k = 0.25 N/m). The breaking or yield force is the force value when the elastic regime finish at 5 nm of indentation (F_b = 1.1 nN). The critical indentation δ_c, is the deformation of the virus when it breaks (5 nm). Thin shell theory relates the protein shell stiffness with the Young's modulus as $k \approx E\frac{h^2}{R}$, where h is the thickness of the shell and R its radius [35]. The area enclosed between forward and backward curves from indentation 0 up to 5 nm is the energy used to break the cage. In this case, it is about 8.8 nm × nN, i. e. 8.8 × 10^{-18} J or 2140 k_BT, which approaches the order of magnitude of the total energy used for assembling the virus [36]. In addition, the critical strain $\varepsilon_c = \delta_c/h$, where h is the initial height of the protein cage as measured with AFM, informs about the brittleness or the mechanical stability of protein cages [15]. In this case, ε_c = 5/36 = 0.14, saying that the particle breaks when it is deformed 14%.

Genome externalization and mechanical properties of cargo. Nanoindentations at low force can induce cracks, allowing the access to the inner cargo of viruses. For instance, the consecutive application of nanoindentation cycles in HAdv cracks-open the shell in a controlled fashion to probe the mechanical properties of the core [37, 38]. These mechanical properties relate with the condensation state of dsDNA. In particular, the HAdv core is formed by 35 kbp dsDNA molecule bound to positively charged viral proteins that constitute ~50% of the core molecular weight. Approximately about 500–800 copies of protein VII (pVII) contributes with 22 kDa to the core, becoming the majority protein. Consecutive nanoindentation experiments (Fig. 10.6) performed on wild type (HAdv-wt) and mutant lacking pVII (HAdv-PVII-) human adenovirus particles provide information on the pVII function. Figure 10.6a presents the typical evolution of HAdv-wt and HAdv-VII– particles in a multiple indentation experiment. The first indentation (FIC#0) on the intact shell (Fig. 10.6a, #0) opens a crack at the icosahedral facet with a similar size to the apex of the AFM tip (~20 nm) (Fig. 10.6a, #1), allowing a direct access of the tip to the

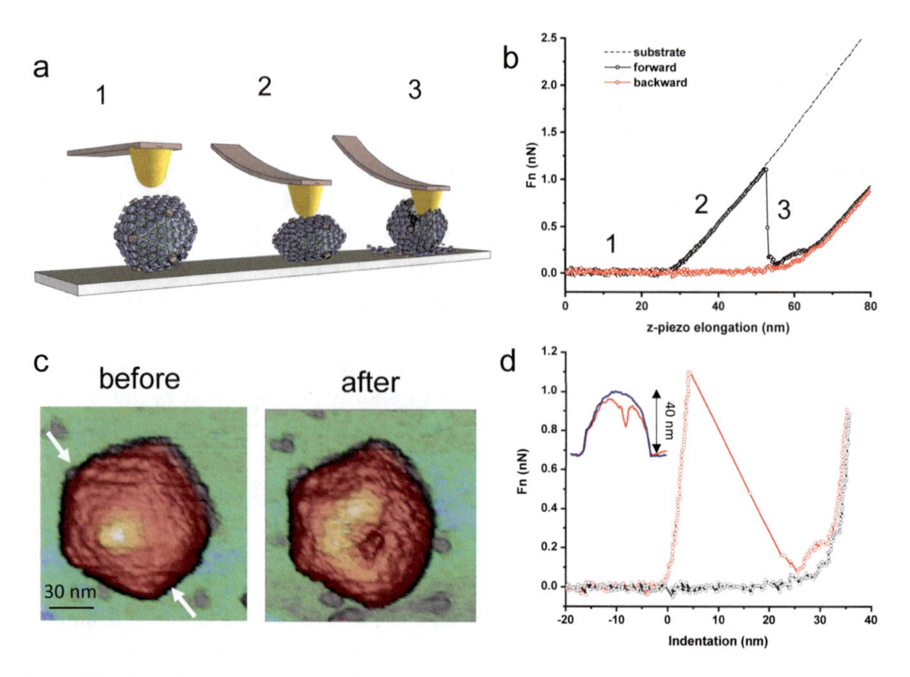

Fig. 10.5 Single indentation assay. **a** Cartoon showing the three main stages during nanoindentation experiment on a protein cage: before contact (1), during deformation (2) and after breaking (3). **b** Evolution of F_n along the z-piezo elongation. Forward curve exhibits the three stages commented in **a**. **c** AFM topographies before and after nanoindentation showing a crack. **d** Nanoindentation data extracted from **b**, showing the shell deformation. Inset compares topographical profiles before (blue) and after (red) de experiment obtained at the line marked by the arrows in **c**. Inset shows the profile of the particle

core. Subsequent indentations present a non-linear Hertzian behavior in contrast with the linear deformation found in intact particles (Fig. 10.5). During the first indentation after capsid cracking (Fig. 10.6a, #1), HAdv-wt cores are not deformed beyond 40 nm at 3 nN, while HAdv-VII– cores undergo larger indentations, up to ~60 nm, at forces as low as 2 nN (Fig. 10.6b). This difference indicates that it is easier for the AFM tip to penetrate/deform HAdv-VII– than HAdv-wt cores. That is, in the absence of protein VII the HAdv core is softer and less condensed. Elongated structures with a height compatible with dsDNA strands (Fig. 10.6a, yellow) appeared on the substrate surrounding HAdv-VII– particles after the first fracture of the shell (Fig. 10.6a, #1, top), but not in HAdv-wt (Fig. 10.6a, #1, bottom). Material consistent with unpacked DNA seemed to be more abundant in HAdv-VII– particles that in HAdv-wt for the equivalent indentation number (Fig. 10.6a). This qualitative observation suggests that it is easier to exit the disrupted shell for the VII-free genome than for the VII-bound one. An alternative possibility would be that the VII-free genome has a larger tendency to adsorb to the mica substrate than the VII-bound DNA. Measurements of the topographical profile across the crater produced by the successive indentations (Fig. 10.6c) support the first possibility. Because

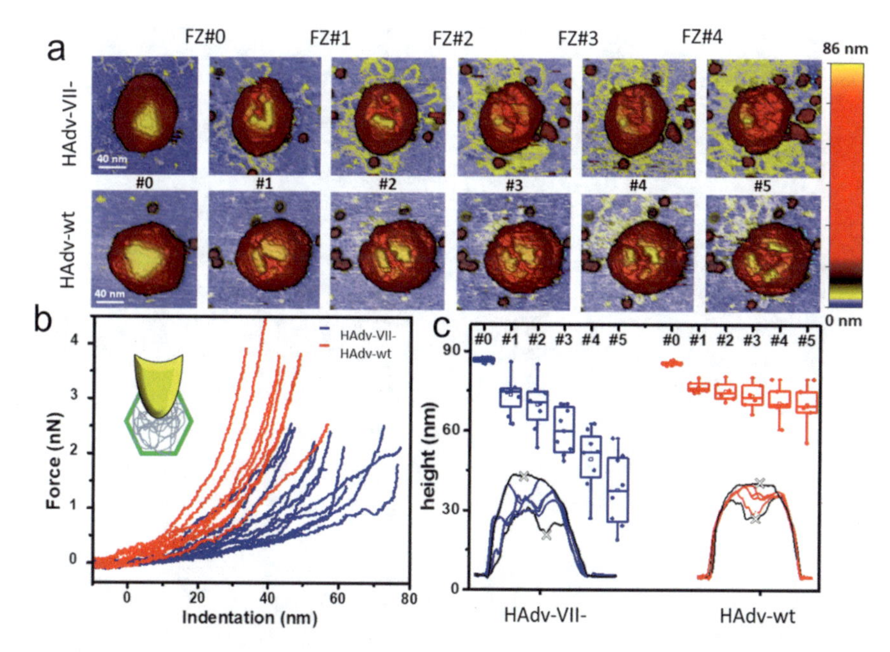

Fig. 10.6 Mechanical properties of human adenovirus core: the role of protein VII. **a** Images showing the evolution HAdv-wt and HAdv-VII– particles during multiple indentation assays. Images are colored by height, as indicated by the color bar at the right hand side. **b** FZ curves corresponding to the first indentation performed on the core of 8 HAdv-wt (red) and 12 HAdv-VII– (blue) particles. **c** Evolution of the core height (as indicated by the minimum height in the crater, grey crosses) for eight HAdv-VII– and six HAdv-wt particles. Insets: Examples of topographical profiles obtained through the crater opened by the indentation Adapted from [38]

topographical profiles include both the crater and rims, we used the lowest height inside the crack as an indicator of the remaining core contents (Fig. 10.6c, insets). Plotting the evolution of this parameter along six consecutive indentations for eight HAdv-VII– and six HAdv-wt particles (Fig. 10.6c) showed that, indeed, the core components are leaving faster the HAdv-VII– than the HAdv-wt cracked particles [38]. The application of consecutive indentations to coronavirus structures (TGEV), also inform about stability and genome release [32]. These experiments unveiled two different behaviors of TGEV particles: In the first case (Fig. 10.7a), a virus particle was probed seven times with maximum forces ranging from 0.6 to 3 nN, remaining unaltered through the alternating AFM images. In fact, ~65% of the explored virus particles (N = 69) kept their height constant (Fig. 10.7c, black), indicating that their structure was elastic and not affected by the nanoindentations. However, in the second case (Fig. 10.7b) and using the same FZ number and parameters, the TGEV particle showed evident structural changes consisting of appearing a circular crater after the first FZ that enlarged after consecutive deformations until the virus appears to have been ripped open, releasing its content (Fig. 10.7, #7). This behavior was observed in ~35% of the explored particles and was accompanied by a gradual decrease in height

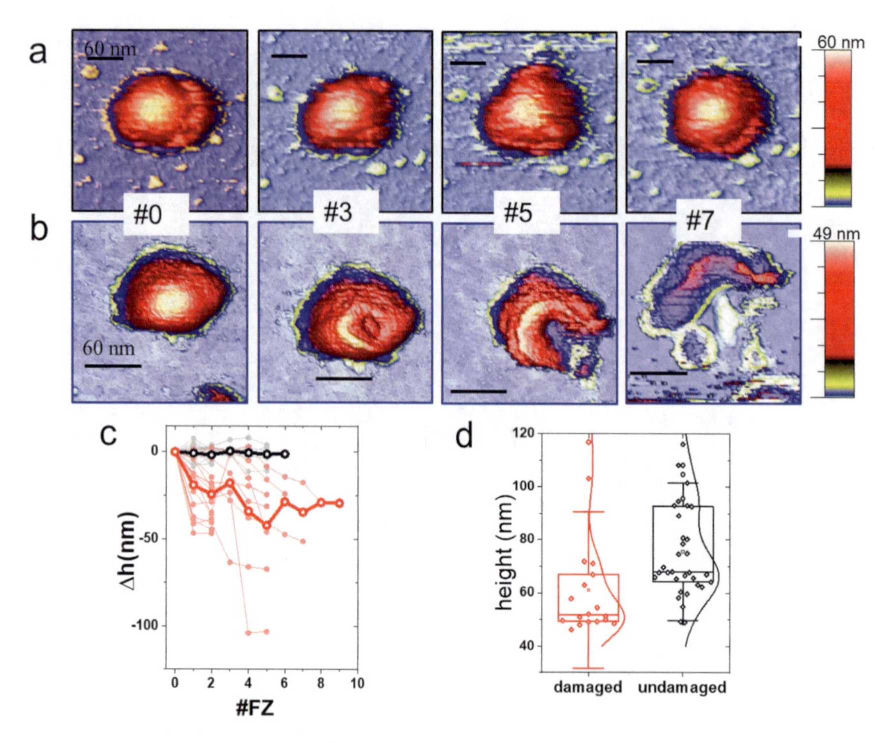

Fig. 10.7 Mechanical properties of TGEV coronavirus: the role of packing fraction. **a** and **b** present topographical images of TGEV virions after consecutive indentations. Particles can be classified in two groups: indentation resistive (**a**) and indentation sensitive (**b**) as shown in the topo images. Frame number indicates how many indentations were performed. **c** Chart evolution of virus height as a function of the indentation number (FZ#). The values of height are taken from the indented region. Black and red charts represent indentation resistive an indentation sensitive, respectively. **d** Height box-plot distribution of intact virus particles. Adapted from [32]

after each indentation (Fig. 10.7c, red). It is important to remark that the initial height of the particles that remained undamaged (indentation resistive) was larger than the damaged ones (indentation sensitive) (Fig. 10.7d), a difference that was significantly different at a 95% level of confidence. By assuming that all viruses package the same amount of genome, it is likely that everyone contains a similar number of RNPs, no matter their size. Therefore, the larger height of these particles (Fig. 10.7d) suggests a loosened core state represented by the 'eggs-in-a-nest' assembly [39]. In this case, when the tip retracts, RNPs recover their original positions and the flexible membrane returns to its previous height. In contrast, indentation-sensitive particles (Fig. 10.7b) exhibited lower heights than the resistive ones (Fig. 10.7d). The particle crack-opened with a permanent fracture (Fig. 10.7b) of ~25 nm in depth (Fig. 10.7c, blue). The lower height indicates a higher packing fraction ('pyramid' packing) [39] without internal free room, preventing the RNPs from rearranging under the mechanical stress induced by the tip. In fact, structural data of SARS-CoV-2 indicate this 'pyramid' packing occupies ~36% of the cavity, very close to viruses with high packing fraction values, such as lambda and phi29 bacteriophages [40]. Together with the AFM tip, solidly packed RNPs act as a "hammer and anvil" on the membrane, concentrating the mechanical stress on the virus envelope and causing permanent fractures (Fig. 10.7b). Consecutive imaging of the virus particles after each nanoindentation (Fig. 10.7b) reported increasing damage, with virus height loss from 25 to 100 nm (average decrease ~40 nm; Fig. 10.7c, red).

Mechanical fatigue, disassembly and aging. The breaking force describes the largest force that the virus can survive. Single indentation experiments collapse the particle by inducing large and uncontrollable changes in its structure. It is thus difficult to derive consequences about disassembly, since in the cycle of many virus shells, for instance, disassembly takes place by losing individual capsomers in an gradual way [41]. A protein cage must also resist a constant barrage of sub-lethal collisions in crowded environments [42]. Equipartition theorem provides an estimation of the energy transferred in a molecular collision to be $\sim\frac{3}{2}k_BT$, which is far below that the energy supplied by single indentation assay experiments. Imaging of individual protein shells with AFM in jumping mode requires thousands of load cycles (FZs) at low force (~100 pN per pixel, Fig. 10.2c) [22]. A rough estimation indicates that $\sim 10\ k_BT$ is transferred to the particle at every cycle [43], very close to the molecular collisions value. The continuous imaging of a particle enables the evaluation of any structural alteration while subjected to cycle load at low forces. Mechanical fatigue experiments have demonstrated to be a disassembly agent able of recapitulating the natural pathway of adenovirus uncoating [22]. Therefore fatigue provides additional information by providing information on shell stability under multiple deformation cycles at low force (~100 pN) [44], well below the breaking force (Fig. 10.5d). Let us exemplify the mechanical fatigue methodology in the case study of HAdv [45]. This virus travels to the cytoplasm until the nuclear pore to deliver its genome. Inside the host cell, the virus finely tunes the sequential loss of pentons (proteins located at the icosahedron corners) to render a semi disrupted capsid at the nuclear pore. Viruses with too many pentons would not be able to release their DNA through the cell's nuclear pore, whereas those with too few pentons would liberate their genome before reaching the nucleus. Mechanical fatigue induces penton failure [44] (Fig. 10.8a), which mimics the stresses the virus sustains during its journey to the nucleus. This approach allows to study the transition kinetics of penton release as a two-state process, from where it is possible derive the spontaneous escape rate and the free energy barrier of a penton [45]. Moreover, fitting the survival probability P of penton N (Fig. 10.8b) under different fatigue forces to Weibull analysis: $P(N) = 1 - \exp\left[-\left(\frac{N+1}{\lambda}\right)^{\beta}\right]$, where λ is the half-life and β determines the nature of the penton's dynamics release. In particular, $\beta < 1$ would imply that failure rate decreases over time (*Lindy effect*). This effect can be found in ancient monuments, since the older something is, the more likely it will survive. The case of $\beta = 1$ would imply a Poisson process with a constant probability of penton failure over time. However, $\beta > 1$ (Fig. 10.8b) involves the phenomenon termed in materials science as *aging*, which means that the failure rate increases over time. This indicates that the probability for the virus of losing a penton increases if other

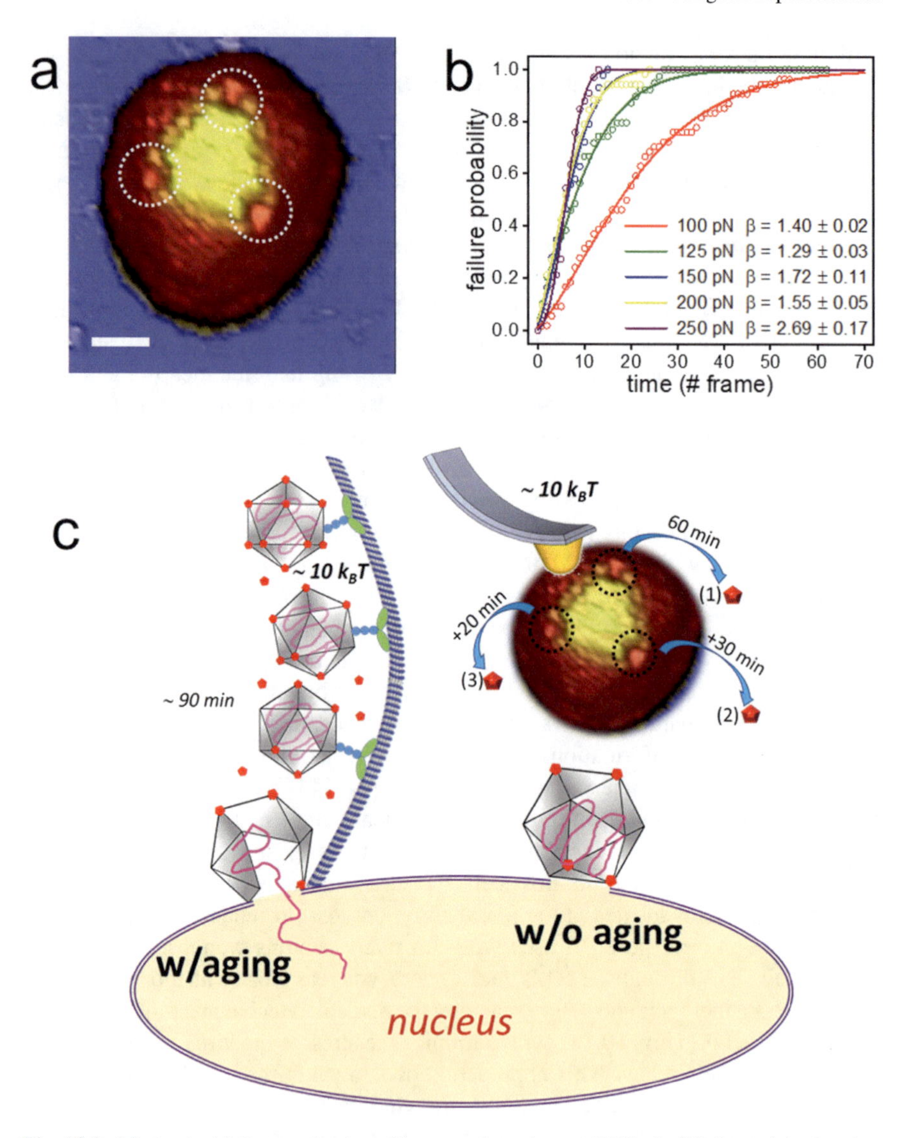

Fig. 10.8 Mechanical fatigue and aging of human adenovirus. **a** AFM of a HAdv particle showing three vacancies of pentons on the triangular facet (encircled). **b** Failure probability of pentons overt time for different forces. Inset show the fitting parameters of Weibull statistics. **c** Diagram showing the biological implications of aging on the virus infectivity (see text). Adapted from [45]

penton has been lost before. This aging process ($\beta > 1$) accelerates the overall penton escape rate by about 50% (Fig. 10.8c) with respect to a sequence of independent escape events ($\beta = 1$). Specifically, while the first penton resists about 60 min, the second and third would last 30 and 20 min more, respectively (Fig. 10.8c). In this way, *aging* guarantees to have the virus particle just disrupted enough to release the genome trough the nuclear pore. Without aging, virus particle would be still quite complete and the genome could not escape (Fig. 10.8c), with negative consequences for infection.

Layer-by-layer disassembly of a multilayered virus with mechanical fatigue. Several viruses wrap their genome in several layers of protective casing, resulting in an onion-like structure. For example, the rotavirus (RV), that can induce severe diarrhea in young children, have three layers built of different proteins with various functionalities. Rotavirus subparticles may exist with only one or two of these coats, which allows to study each layer in detail [46]. Mechanical fatigue allows to explore the strength and binding of each layer (Fig. 10.9a). The RV infectious particle is a 100 nm non-enveloped triple-layered particle (TLP) composed of three concentric protein shells enclosing the dsRNA genome and the viral RNA polymerase and capping enzyme. The inner layer surrounds the eleven dsRNA genomic segments associated with the RNA-dependent RNA-polymerase VP1 (125 kDa) and the RNA-capping enzyme VP3 (88 kDa) at the pentameric positions. This thin single-layered particle (SLP), an intermediate structure that is involved in the packing and replication of the viral genome, is surrounded by a thick layer formed by 260 VP6 pear-shaped trimers (45 kDa) in the double-layered particle (DLP). This particle, which does not disassemble during the infection, constitutes the transcriptional machinery that initiates the core steps of the viral replication cycle once delivered in the host cell cytoplasm. Cyclic imaging of the TLP (Fig. 10.9b, left) at forces between 100 to 200 pN per pixel shows that, while the VP4 spikes are removed from the particle surface in a few frames (Fig. 10.9b, middle), the VP7 layer remains mostly intact (Fig. 10.9b, right) during 80 frames (light red in Fig. 10.9f). These results illustrate that the spikes are easily removed by the AFM tip and are not strongly anchored. However, the VP7 layer displays a strong resistance against fatigue. A strong binding energy between capsomers would not only result in a high resistance of individual proteins against fatigue, but also will contribute to a high breaking force when all capsomers are probed in a single indentation assay experiment [46]. The current model proposes a calcium concentration drop in endosomal compartments during RV entry as the factor that triggers VP7 disassembly and membrane penetration. In fact, calcium depletion by chelating agents (as EDTA) is used to uncoat TLP to DLP by inducing VP7 trimer dissociation. To explore the structural consequences of this process in real time, fatigue assays are carried out on TLP while EDTA simultaneously flowed in the AFM liquid chamber, to induce the gradual depletion of Ca ions of the particles (Fig. 10.9c). In these conditions, fatigue induces the neat VP7 detachment from the VP6 subjacent layer (indicated by a circle in Fig. 10.9c, #19) even before the spikes are removed. Indeed, the evolution of the topographic profiles (dark red in Fig. 10.9f) show abrupt downwards steps very close to the VP7 thickness (red arrow of Fig. 10.9f) indicating that TLP particle loses VP7 completely while keeping VP6. These results not only suggest that Ca ions mediate the interaction between VP7 and VP6 layers, but also that the absence of ions weakens the interaction between VP7 subunits. If fatigue continues, VP6 subunits are neatly removed from VP2 layer (circle in Fig. 10.9c, #29). Therefore, VP6 layer appears forming a weak shell whose interaction with the beneath VP2 layer is not very strong, since it peels off rapidly to reveal the SLP. Similar results are found on DLP. Again, fatigue induced a clean VP6 disassembly after less than 10 frames (circle in Fig. 10.9d, #8). In this case the evolution of the topographic profiles (blue in Fig. 10.9f) undergoes sharp reductions very close to the VP6 thickness, inducing the gradual uncovering of the innermost VP2 (blue arrow in Fig. 10.9f). These experiments not only illustrate a weak interaction between VP6 and VP2 layers, but also a very feeble VP6-VP6 binding energy. Finally, the thin SLP VP2 is highly unstable under fatigue (Fig. 10.9e) experiments collapsing well before reaching 10 frames (green in Fig. 10.9f).

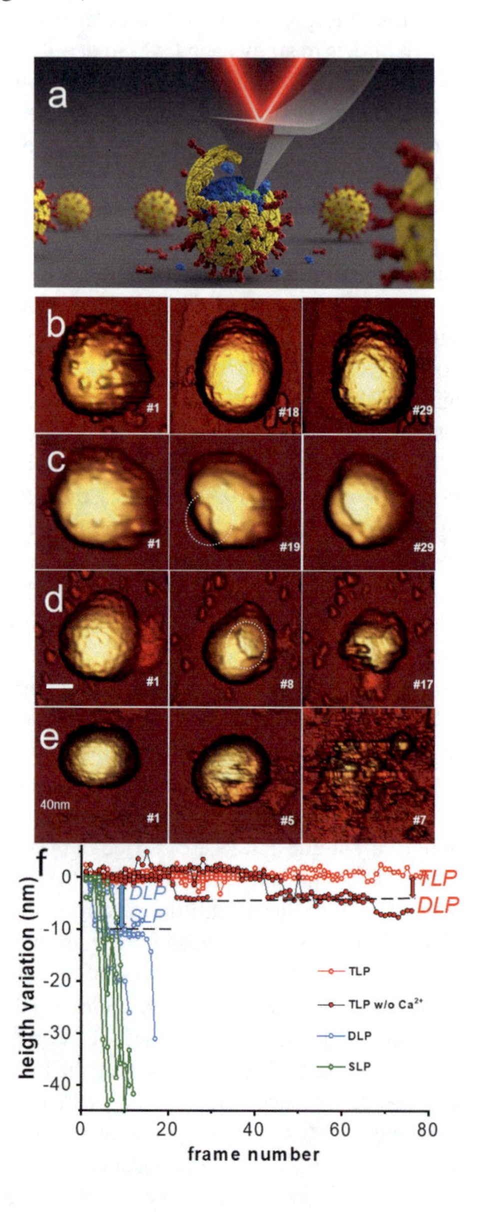

Fig. 10.9 Peeling a virus layer by layer with mechanical fatigue. **a** cartoon of the peeling experiment. Topographic evolution of TLP (**b**), TLP + EDTA (**c**), DLP (**d**) and SLP (**e**) during continuous imaging at low force (~60–120 pN) indicating the corresponding frame. **f** Topographic evolutions obtained at center of each virus particle (**b**–**e**) in TLP (red), DLP (blue) and SLP (green) particles. Dark red color indicates fatigue of TLP + EDTA. Red and blue arrows indicate the loss of height from TLP to DLP and for DLP to SLP, respectively. Adapted from [46]

AFM/fluorescence combination. Here we discuss the methodology for studying the activity of individual nanoreactors based on a protein container. About 220 GFPs are packed inside P22 bacteriophage capsids [47] and its functionality is explored as a function of the mechanical stress applied with the AFM tip to the virus particle while monitoring fluorescence with Total Internal Reflection Fluorescence Microscopy (TIRFM) [48]. The integration of a single molecule fluorescence microscope with AFM requires to monitor the fluorescence signal at the surface to avoid not only the background signal of the AFM probe itself, but also the light coming from the bulk solution (Fig. 10.10a). In this approach [49] P22 bacteriophages are immobilized on a glass surface, in such way that a region with particles visualized with AFM (Fig. 10.10b, left) is correlated with their TIRFM pattern (Fig. 10.10b, right). The diffraction-limited resolution of the optical system restricted the size of the smallest spot to $\lambda/2$ ($\approx$250 nm), which is larger than the P22 VLPs. Consequently, the particles that were too close together (arrow, Fig. 10.10b, left) appeared as a single fluorescence spot (bottom right, Fig. 10.10b, right). The inset of Fig. 10.10b (right) shows the topographic (white) and light (green) profiles of two VLPs (Fig. 10.10b, white and green dotted lines) where the particle width in the fluorescence signal is $\approx$500 nm, approximately five times larger than in the AFM topography. After localizing an intact virus showing fluorescence signal, it is possible to execute a nanoindentation experiment (Fig. 10.10c, black) by using a Si_3N_4 tip to avoid electronic quenching of GFP and isolate pure mechanical effects [49]. The fluorescence remained constant (Fig. 10.10c) during the tip approach (Fig. 10.5a, label 1). Furthermore, the light signal stayed stable even when the tip established contact with the VLP at 1 s and linearly deformed the VLP structure (Fig. 10.10c). Fluorescence quenching started at 1.2 s, coinciding with the force steps that indicated the yielding and/or breaking of the protein shell, reaching a minimum value once the particle stopped collapsing (2 s). At 2.6 s the AFM tip moved backwards, away from the surface, and the fluorescence signal remained stable until the tip was released from the VLP (3 s), coming back to a similar value to the initial magnitude (4 s). This fluoresce reduction is most likely due to changes in the integrity of the packed GFPs. In particular, GFPs can oligomerize inside the P22 cavity (internal radius of 22.2 nm) due to crowding phenomena. In the P22 prohead system there are $\approx$209 GFPs attached internally to the capsid with an, resulting in a nearest neighbor distance of $\approx$3 nm between the protein centers. Since the GFP Stokes radius is 2.8 nm, we derived an average distance of 2 Å between fluorophores. It is possible to simulate the compression of a single P22 particle between two planes (Fig. 10.10d, left). The compressed particle shows that there are around 10 unfolded GFPs inside (Fig. 10.10d, right –red-) which account for the amount of quenched light (Fig. 10.10c, green).

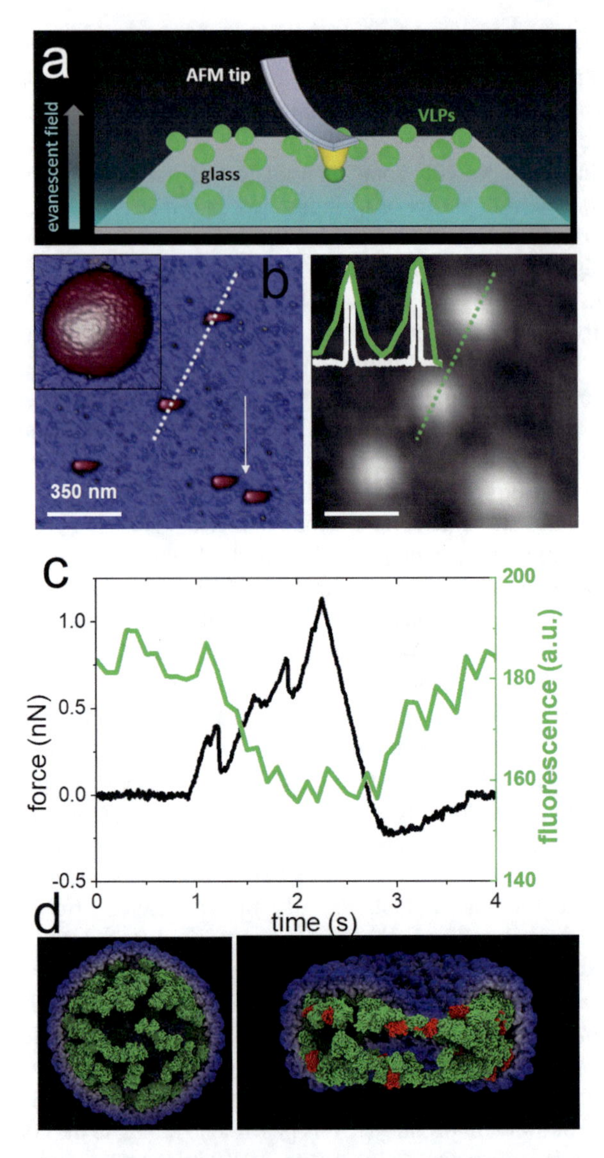

Fig. 10.10 AFM/fluorescence combination: quenching of packed GFP inside P22 protein cages. **a** Presents a diagram of the AFM/TIRF system, showing the P22 VLPs (light green) and the AFM probed particle in dark green. **b** Simultaneous AFM and fluorescence images (right and left, respectively) of individual VLPs on the glass surface. The white arrow (left) indicates a couple of particles appearing as one spot in fluorescence (right). The inset in b-left shows a high-resolution AFM image of P22 VLP resolving the proteinaceous structure of the virus. Inset of b-right compares the lateral resolution of both microscopies by plotting the signal profile of the AFM topography (white) and the fluorescence (green) obtained at the lines indicated by the dashed white and green lines, respectively. **c** Chart representing the simultaneous signals of force (black) and light (green). **d** A simulation of the GFP aggregates (green) inside a P22 procapsid (blue) before deformation (left). In this 3D image, one half of the capsid has been removed to show the internal configuration of the GFPs located below the virus lumen. The right panel shows a deformed virus where the partially unfolded GFPs are highlighted in red. Adapted from [49]

Conclusion and Perspective

Bacteriophage phi29 was the first virus structure to be studied with a scanning probe technique [50], one year before the invention of AFM [51]. A few years later AFM was a generalized technique for studying biological samples [52] at single molecule level. AFM was initially applied just for imaging viruses [53, 54], but it was also used to study the biophysical properties of viruses [9]. During the last 20 years AFM is gradually being considered as a tool for virus research, although is not yet at the level of the classical structure techniques such as electron or X-ray microscopies. Using AFM for measuring the binding forces of virus to cells [55] or for the assembly of 2D virus-like assemblies in real time [56] are also promising applications of AFM that, hopefully, will convince virologists that AFM is an equally valid approach for studying viruses. We hope that the applications of AFM exposed in this chapter will help in this task.

Acknowledgements The authors acknowledge to support by grants received from the Spanish Ministry of Science and Innovation projects (FIS2017-89549-R, FIS2017-90701-REDT and PID2021-126608OB-I00) and the Human Frontiers Science Program (HFSPO RGP0012/2018). IFIMAC is a Center of Excellence "María de Maeztu". J.R.C. acknowledges the Spanish Ministry of Science and Innovation (PID2020-113287RB-I00) and the Comunidad Autónoma de Madrid (P2018/NMT-4389).

References

1. Flint SJ, Enquist LW, Racaniello VR, Skalka AM (2004) Principles of virology. ASM Press, Washington D.C.
2. Douglas T, Young M (1998) Host–guest encapsulation of materials by assembled virus protein cages. Nature 393:152–155. https://doi.org/10.1038/30211
3. Mateu MG (2013) Assembly, stability and dynamics of virus capsids. Arch Biochem Biophys 531:65–79. https://doi.org/10.1016/j.abb.2012.10.015
4. Minton AP (2006) How can biochemical reactions within cells differ from those in test tubes? J Cell Sci 119:2863–2869. https://doi.org/10.1242/jcs.03063
5. Agirre J, Aloria K, Arizmendi JM et al (2011) Capsid protein identification and analysis of mature Triatoma virus (TrV) virions and naturally occurring empty particles. Virology 409:91–101. https://doi.org/10.1016/j.virol.2010.09.034
6. Cordova A, Deserno M, Gelbart WM, Ben-Shaul A (2003) Osmotic shock and the strength of viral capsids. Biophys J 85:70–74. https://doi.org/10.1016/S0006-3495(03)74455-5
7. Baker TS, Olson NH, Fuller SD (1999) Adding the third dimension to virus life cycles: three-dimensional reconstruction of icosahedral viruses from cryo-electron micrographs. Microbiol Mol Biol Rev. https://doi.org/10.1128/MMBR.63.4.862-922.1999
8. Egan P, Sinko R, LeDuc PR, Keten S (2015) The role of mechanics in biological and bio-inspired systems. Nat Commun 6:7418. https://doi.org/10.1038/ncomms8418
9. Ivanovska IL, de Pablo PJ, Ibarra B et al (2004) Bacteriophage capsids: tough nanoshells with complex elastic properties. Proc Natl Acad Sci 101:7600–7605. https://doi.org/10.1073/pnas.0308198101
10. de Pablo PJ (2018) Atomic force microscopy of virus shells. In: Seminars in cell & developmental biology. Academic Press, pp 199–208

11. Müller DJ, Amrein M, Engel A (1997) Adsorption of biological molecules to a solid support for scanning probe microscopy. J Struct Biol 119:172–188. https://doi.org/10.1006/jsbi.1997.3875
12. Zeng C, Hernando-Pérez M, Dragnea B et al (2017) Contact mechanics of a small icosahedral virus. Phys Rev Lett 119:038102. https://doi.org/10.1103/PhysRevLett.119.038102
13. Armanious A, Aeppli M, Jacak R et al (2016) Viruses at solid-water interfaces: a systematic assessment of interactions driving adsorption. Environ Sci Technol 50:732–743. https://doi.org/10.1021/acs.est.5b04644
14. Llauró A, Guerra P, Irigoyen N et al (2014) Mechanical stability and reversible fracture of vault particles. Biophys J 106:687–695. https://doi.org/10.1016/j.bpj.2013.12.035
15. Llauró A, Luque D, Edwards E et al (2016) Cargo–shell and cargo–cargo couplings govern the mechanics of artificially loaded virus-derived cages. Nanoscale 8:9328–9336
16. Zeng C, Hernando-Perez M, Dragnea B et al (2017) Contact Mechanics of a Small Icosahedral Virus. Phys Rev Lett 119:038102. https://doi.org/10.1103/PhysRevLett.119.038102
17. Carpick RW, Ogletree DF, Salmeron M (1997) Lateral stiffness: A new nanomechanical measurement for the determination of shear strengths with friction force microscopy. Appl Phys Lett 70:1548–1550. https://doi.org/10.1063/1.118639
18. Kuznetsov Y, Gershon PD, McPherson A (2008) Atomic force microscopy investigation of vaccinia virus structure. J Virol 82:7551–7566. https://doi.org/10.1128/jvi.00016-08
19. Vinckier A, Heyvaert I, D'Hoore A et al (1995) Immobilizing and imaging microtubules by atomic force microscopy. Ultramicroscopy 57:337–343. https://doi.org/10.1016/0304-3991(94)00194-r
20. Carrasco C, Luque A, Hernando-Pérez M et al (2011) Built-in mechanical stress in viral shells. Biophys J 100:1100–1108. https://doi.org/10.1016/j.bpj.2011.01.008
21. Roos WH (2018) AFM nanoindentation of protein shells, expanding the approach beyond viruses. Semin Cell Dev Biol 73:145–152. https://doi.org/10.1016/j.semcdb.2017.07.044
22. Ortega-Esteban A, Horcas I, Hernando-Perez M et al (2012) Minimizing tip-sample forces in jumping mode atomic force microscopy in liquid. Ultramicroscopy 114:56–61
23. Legleiter J, Park M, Cusick B, Kowalewski T (2006) Scanning probe acceleration microscopy (SPAM) in fluids: mapping mechanical properties of surfaces at the nanoscale. Proc Natl Acad Sci 103:4813–4818. https://doi.org/10.1073/pnas.0505628103
24. Villarrubia JS (1997) Algorithms for scanned probe microscope image simulation, surface reconstruction, and tip estimation. J Res Natl Inst Stand Technol 102:425–454. https://doi.org/10.6028/jres.102.030
25. Horcas I, Fernández R, Gomez-Rodriguez JM, Colchero JW, Gómez-Herrero JW, Baro AM (2007) WSXM: a software for scanning probe microscopy and a tool for nanotechnology. Rev Sci Instrum 78(1):013705. https://doi.org/10.1063/1.2432410
26. Pettersen EF, Goddard TD, Huang CC et al (2004) UCSF Chimera—a visualization system for exploratory research and analysis. J Comput Chem 25:1605–1612. https://doi.org/10.1002/jcc.20084
27. Tang J, Olson N, Jardine PJ et al (2008) DNA poised for release in bacteriophage ø29. Structure 16:935–943. https://doi.org/10.1016/j.str.2008.02.024
28. Risco C, Antón IM, Enjuanes L, Carrascosa JL (1996) The transmissible gastroenteritis coronavirus contains a spherical core shell consisting of M and N proteins. J Virol 70:4773–4777
29. Casanova L, Rutala WA, Weber DJ, Sobsey MD (2009) Survival of surrogate coronaviruses in water. Water Res 43:1893–1898. https://doi.org/10.1016/j.watres.2009.02.002
30. Kiss B, Kis Z, Pályi B, Kellermayer MSZ (2021) Topography, spike dynamics, and nanomechanics of individual native SARS-CoV-2 virions. Nano Lett 21:2675–2680. https://doi.org/10.1021/acs.nanolett.0c04465
31. Perez-Illana M, Martin-Gonzalez N, Hernando-Perez M et al (2021) Acidification induces condensation of the adenovirus core. Acta Biomater 135:534–542. https://doi.org/10.1016/j.actbio.2021.08.019
32. Cantero M, Carlero D, Chichón FJ et al (2022) Monitoring SARS-CoV-2 surrogate TGEV individual virions structure survival under harsh physicochemical environments. Cells 11:1759. https://doi.org/10.3390/cells11111759

33. Zink M, Grubmüller H (2009) Mechanical properties of the icosahedral shell of southern bean mosaic virus: a molecular dynamics study. Biophys J 96:1350–1363. https://doi.org/10.1016/j.bpj.2008.11.028
34. Ortega-Esteban Á, Mata CP, Rodríguez-Espinosa MJ et al (2020) Cryo-electron microscopy structure, assembly, and mechanics show morphogenesis and evolution of human picobirnavirus. J Virol 94:e01542-e1620. https://doi.org/10.1128/JVI.01542-20
35. Landau Theory of Elasticity - 3rd Edition. https://www.elsevier.com/books/theory-of-elasticity/landau/978-0-08-057069-3?country=ES&format=print&utm_source=google_ads&utm_medium=paid_search&utm_campaign=spainshopping&gclid=CjwKCAiAv9ucBhBXEiwA6N8nYFtCfz0HTtRL1L1syWMEHTtv-UorKaSJ__wX7OwicfFg6RH2LwSclRoCE24QAvD_BwE&gclsrc=aw.ds. Accessed 12 Dec 2022
36. Katen S, Zlotnick A (2009) The Thermodynamics of virus capsid assembly. Methods Enzymol 455:395–417. https://doi.org/10.1016/S0076-6879(08)04214-6
37. Ortega-Esteban A, Condezo GN, Pérez-Berná AJ et al (2015) Mechanics of viral chromatin reveals the pressurization of human adenovirus. ACS Nano 9:10826–10833
38. Martín-González N, Hernando-Pérez M, Condezo GN et al (2019) Adenovirus major core protein condenses DNA in clusters and bundles, modulating genome release and capsid internal pressure. Nucleic Acids Res 47:9231–9242
39. Yao H, Song Y, Chen Y et al (2020) Molecular architecture of the SARS-CoV-2 virus. Cell 183:730-738.e13. https://doi.org/10.1016/j.cell.2020.09.018
40. Petrov AS, Boz MB, Harvey SC (2007) The conformation of double-stranded DNA inside bacteriophages depends on capsid size and shape. J Struct Biol 160:241–248. https://doi.org/10.1016/j.jsb.2007.08.012
41. Greber UF, Willetts M, Webster P, Helenius A (1993) Stepwise dismantling of adenovirus 2 during entry into cells. Cell 75:477–486
42. Zhou H-X, Rivas G, Minton AP (2008) Macromolecular crowding and confinement: biochemical, biophysical, and potential physiological consequences. Annu Rev Biophys 37:375–397. https://doi.org/10.1146/annurev.biophys.37.032807.125817
43. Hernando-Pérez M, Lambert S, Nakatani-Webster E et al (2014) Cementing proteins provide extra mechanical stabilization to viral cages. Nat Commun 5:1–8
44. Ortega-Esteban A, Pérez-Berná AJ, Menéndez-Conejero R et al (2013) Monitoring dynamics of human adenovirus disassembly induced by mechanical fatigue. Sci Rep 3:1434. https://doi.org/10.1038/srep01434
45. Martín-González N, Delgado-Buscalioni R, de Pablo PJ (2021) Long-range cooperative disassembly and aging during adenovirus uncoating. Phys Rev X 11:021025
46. Jiménez-Zaragoza M, Yubero MP, Martín-Forero E et al (2018) Biophysical properties of single rotavirus particles account for the functions of protein shells in a multilayered virus. eLife 7:e37295. https://doi.org/10.7554/eLife.37295
47. O'Neil A, Prevelige PE, Basu G, Douglas T (2012) Coconfinement of fluorescent proteins: spatially enforced communication of GFP and mCherry encapsulated within the P22 capsid. Biomacromol 13:3902–3907. https://doi.org/10.1021/bm301347x
48. Ortega-Esteban A, Bodensiek K, San Martín C et al (2015) Fluorescence tracking of genome release during mechanical unpacking of single viruses. ACS Nano 9:10571–10579
49. Strobl K, Selivanovitch E, Ibáñez-Freire P et al (2022) Electromechanical photophysics of GFP packed inside viral protein cages probed by force-fluorescence hybrid single-molecule microscopy. Small 18:2200059. https://doi.org/10.1002/smll.202200059
50. Baró AM, Miranda R, Alamán J et al (1985) Determination of surface topography of biological specimens at high resolution by scanning tunnelling microscopy. Nature 315:253–254. https://doi.org/10.1038/315253a0
51. Binnig G, Quate CF, Gerber Ch (1986) Atomic force microscope. Phys Rev Lett 56:930–933. https://doi.org/10.1103/PhysRevLett.56.930
52. Bustamante C, Vesenka J, Tang CL et al (1992) Circular DNA molecules imaged in air by scanning force microscopy. Biochemistry 31:22–26. https://doi.org/10.1021/bi00116a005

53. Day J, Kuznetsov YG, Larson SB et al (2001) Biophysical studies on the RNA cores of satellite tobacco mosaic virus. Biophys J 80:2364–2371. https://doi.org/10.1016/S0006-3495(01)76206-6
54. Drygin YF, Bordunova OA, Gallyamov MO, Yaminsky IV (1998) Atomic force microscopy examination of tobacco mosaic virus and virion RNA. FEBS Lett 425:217–221. https://doi.org/10.1016/S0014-5793(98)00232-4
55. Alsteens D, Newton R, Schubert R et al (2017) Nanomechanical mapping of first binding steps of a virus to animal cells. Nat Nanotechnol 12:177–183. https://doi.org/10.1038/nnano.2016.228
56. Valbuena A, Maity S, Mateu MG, Roos WH (2020) Visualization of single molecules building a viral capsid protein lattice through stochastic pathways. ACS Nano 14:8724–8734

Chapter 11
Virus Mechanics: A Structure-Based Biological Perspective

Mauricio G. Mateu

Abstract A virus particle possesses a certain degree of stiffness, brittleness, strength against disruption by mechanical force, and resistance to material fatigue. Atomic force microscopy (AFM) is being increasingly used to experimentally study the mechanical properties of viruses, their physical foundations, and their biological implications. This chapter attempts to provide a molecular structure-based, biology-oriented interpretation of the results of AFM studies on virus mechanics. More specifically, it reviews the current evidence that relates the mechanical response of viruses under load to their atomic structure, intra- and inter-molecular interactions, conformational stability and dynamics, and biological adaptation.

Keywords Virus structure · Conformational stability and dynamics · Mechanical properties · Atomic force microscopy

Introduction

Viruses have interested physicists since the times of the *Phage Group*, an informal network of scientists gathered around Max Delbrück in the 1940s to study bacteriophages from an interdisciplinary perspective [130]. Their seminal studies on viruses trascended virology and contributed decisively to the foundation of modern molecular biology. Much more recently, the advent of nanoscience and the development of single-molecule experimental techniques and theoretical approaches have elicited a renewed interest in the study of viruses from a physicist's perspective. The term *physical virology* has been coined to encompass such studies [13, 34, 66, 70, 112, 128].

One emergent area within physical virology concerns the mechanical properties of viruses. As other solid-state objects, a virus particle possesses a certain degree of stiffness/elasticity, brittleness, strength against disruption by point forces, and resistance to material fatigue. In the last ~15 years, several research groups have

M. G. Mateu (✉)
Centro de Biología Molecular "Severo Ochoa" (CSIC-UAM) and Department of Molecular Biology, Universidad Autónoma de Madrid, Campus de Cantoblanco, 28049 Madrid, Spain
e-mail: mgarcia@cbm.csic.es

M. Comas-Garcia and S. Rosales-Mendoza (eds.), *Physical Virology*, Springer Series in Biophysics 24, https://doi.org/10.1007/978-3-031-36815-8_11

started fundamental studies at the interface between physics and biology on the mechanical properties of virions and their protein–based capsids.

Very few experimental techniques have been employed so far to investigate the mechanical response of virus particles under load. Atomic force microscopy (AFM) has been used in the vast majority of cases [26, 85].

The first AFM-based study on the mechanical behavior of a virus is possibly the one by Falvo et al., who used the cantilever tip of an atomic force microscope to physically manipulate single virions of the rod-like tobacco mosaic virus (TMV) in air [39]. The current AFM-based experimental approach to study the mechanical properties of virus particles in a liquid medium was established in a trend-setting study by Ivanovska, de Pablo, Schmidt, Wuite and their collaborators [57]. They adapted procedures that were being used to determine the mechanical properties of other nano-objects (such as carbon nanotubes or protein microtubules) to quantify the mechanical properties of the bacteriophage Φ29 capsid, which they chose as a model virus particle. Their approach was based on the indentation, under controlled conditions, of individual viral particles using an AFM tip (see Chap. 10). The sample and AFM tip were submerged in a buffered aqueous solution. Thus, different mechanical properties of the virus particle could be determined in close to physiological conditions, including temperature, pH and ionic strength, either in the absence or presence of added viral or cellular components or other factors.

The same AFM-based approach has been applied since then by a growing number of groups to study the mechanical properties of many viruses, including some biomedically and/or economically important ones. Together, their results are providing quantitative descriptions of virus mechanics; unveiling its structural foundations; providing new insights into the molecular mechanisms that govern different stages in the viral infectious cycle; revealing the adaptive nature of the mechanical behavior of some viruses; and contributing to the development of biomedical or technological applications of virus-based particles (reviewed by [13, 25–28, 85, 88, 90, 109, 110, 112, 116]).

Virus mechanics is being intensively investigated not only by experiment, but also theoretically using models and simulations based on idealized virus-shaped objects. Theoretical studies on virus mechanics started at about the same time than AFM-based studies [68, 77, 94, 129, 140, 147]. Since then, a steadily growing number of theoretical studies, in parallel or in combination with experimental analyses, have contributed fundamental physics-based explanations for observed mechanical features of virus particles. They, have also predicted mechanical responses of viruses that have not been experimentally studied so far. Theoretical studies are of the utmost importance to capture the physical essence of virus structure, mechanics and dynamics [83].

The present chapter is not focused on the physics-based foundations of virus mechanics. This chapter is written from the perspective of a biochemist/molecular virologist, and attempts to provide an atomic structure-based, biology-oriented interpretation of the results from AFM studies on virus mechanics.

Mechanical Properties of Virus Particles

AFM-based indentation assays have been used to quantify the response to mechanical force of virions or capsids from over two-dozen virus species so far. Many of the viruses analyzed are quite different from others regarding size, shape, biochemical composition, molecular structure, conformational dynamics, and biological mechanisms of action during the infectious cycle (Fig. 11.1). They include bacterial viruses (bacteriophages), plant viruses and animal viruses of very different lifestyles; non-enveloped or enveloped viruses (Fig. 11.1a); rod-like or quasi-spherical viruses (Fig. 11.1b); symmetric or pleomorphic viruses (Fig. 11.1c); and the protein capsids of those viruses, including helical, icosahedral, modified icosahedral or hexagonal lattice-based capsids (compare Fig. 11.1). Quasi-spherical virus particles used for mechanical studies are also widely different in size (from 20 to 120 nm or more in diameter, a difference of over 200-fold in volume; Fig. 11.1). Thus, the viral particles whose mechanical properties are being probed may constitute an adequate sample of the remarkable variety that can be found in the virosphere regarding structure, properties, or function. Mechanical properties quantified in indentation experiments using AFM include stiffness, intrinsic elasticity, mechanical strength, brittleness, and/or resistance to material fatigue (Fig. 11.2).

Stiffness and Intrinsic Elasticity

A certain degree of elasticity or stiffness constitutes a salient mechanical feature of virus particles. Stiffness can be quantified by the value of the elastic constant (spring constant) of the particle, k_p. The viral particle is indented using an AFM cantilever with a known elastic constant k_c, ended in a tip of adequate radius. The indentations are kept shallow enough to achieve a reversible deformation of the particle. Force *versus* distance (Fz) curves are obtained under the elastic regime, and Hooke's law is applied to two linear springs (cantilever plus particle) in series to obtain the k_p value (Fig. 11.2; [57]).

The k_p value depends not only on the material the particle is made of, but also on the size and geometry of the particle. The value of the modulus of elasticity (Young's modulus), E_p, may provide a biologically more meaningful comparison of elasticity between virus particles, as E_p does not depend on the size or geometry of the object.

When considering the calculated E_p for a viral particle, or for a specific region or component, one should be aware that it represents an averaged value for the elastic behavior of an inhomogeneous material, or even of a composite material. Averaging may blur biologically important differences in the local mechanical behavior. Even so, calculated E_p values can be useful to identify differences or changes in the average elasticity of the viral material, especially between structurally similar virus particles, such as different conformational states or point mutants of a same virus particle.

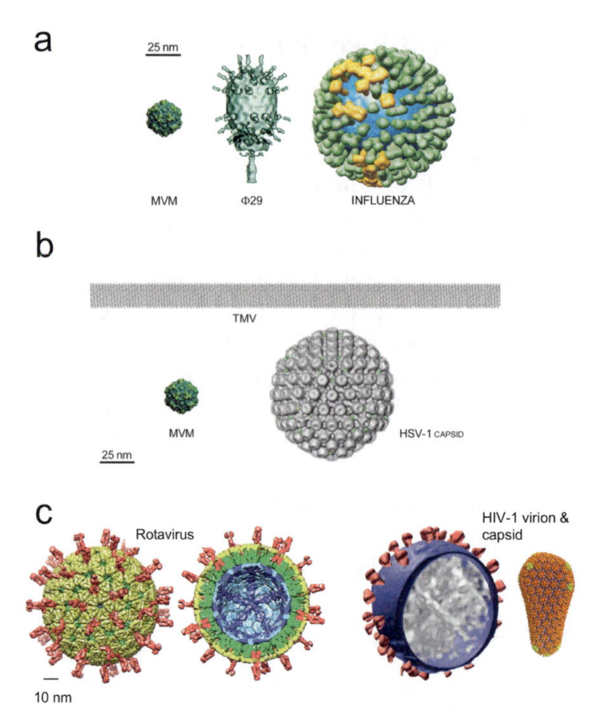

Fig. 11.1 Diversity of virions and capsids used in AFM-based studies on virus mechanics. **a** non-enveloped or enveloped virions. Left to right: the minute virus of mice (MVM), a structurally simple nonenveloped virus; bacteriophage Φ29, a complex nonenveloped virus; influenza virus, an enveloped virus. **b** helical or icosahedral virions. Top, tobacco mosaic virus (TMV), a virus with a helical capsid; bottom left, MVM, a virus with a simple icosahedral capsid; bottom right, herpex simplex 1 virus (HSV-1) (devoid of its lipid envelope), a virus with a complex icosahedral capsid. **c** symmetric versus pleiotropic viruses. Left to right: rotavirus, a symmetric nonenveloped virus (a cross-section of the viral particle shows the three capsid concentric layers colored blue, green or yellow); the mature human immunodeficiency virus type 1 (HIV-1), a pleiomorhic enveloped virus (the mature HIV-1 capsid is shown at right). All viruses are represented at approximately the same scale. A scale bar is included in each panel. Figures reproduced from [90]

When different virions were compared, very large (over 100-fold) differences were found in k_p values. For the enveloped severe acute respiratory syndrome coronavirus 2 (SARS-CoV-2) virion, the k_p value was 0.013 N/m [67], whereas for the non-enveloped minute virus of mice (MVM) virion the k_p value reached 1.4 N/m [15]). Large differences were found also for E_p values. For example, a 20-fold difference in E_p was found between enveloped virions: ~0.05 GPa for the influenza virion under certain conditions [35] versus 1.0 GPa for the murine leukemia virus (MLV) immature virion [71].

Likewise, large differences (over 100-fold) in both stiffness and intrinsic elasticity were observed for protein-only capsids of similar size and shape, such as the icosahedral capsids of different virus species. For example, the norovirus (NV) capsid yielded $k_p = 0.01$–0.06 N/m (depending on pH) and $E_p = 0.03$ GPa (at acidic pH)

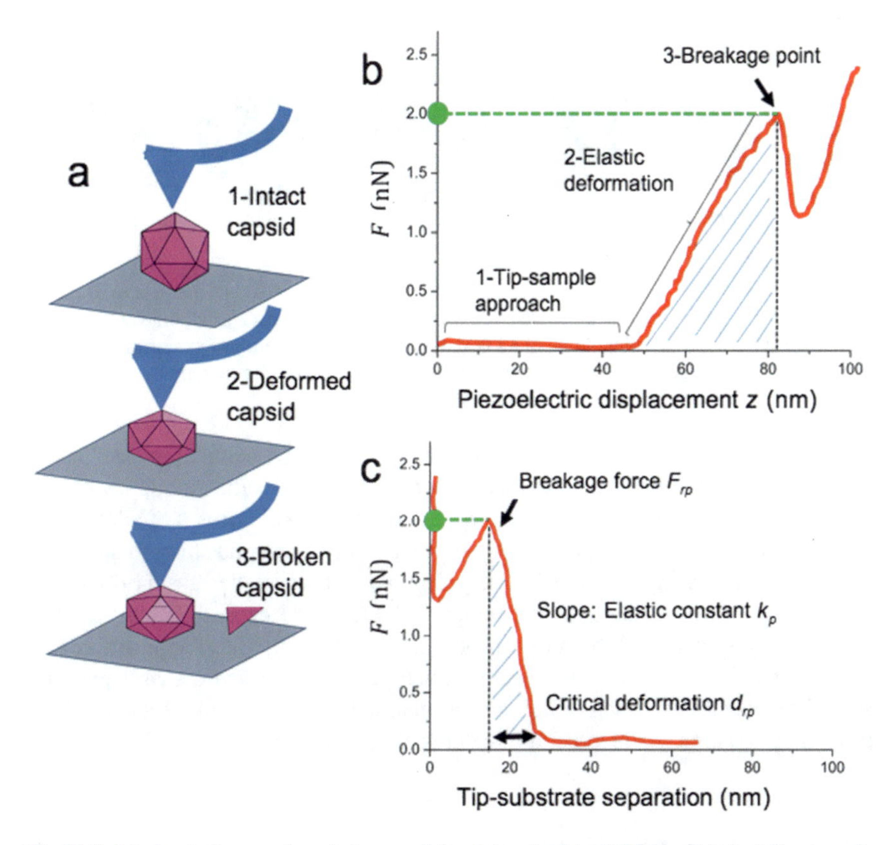

Fig. 11.2 Mechanical properties of virus particles determined in AFM-based indentation experiments. **a** three stages during indentation of a virus particle with the AFM tip: 1, intact capsid before indentation; 2, deformed capsid during indentation under the elastic regime; 3, broken capsid after its elastic limit was surpassed. **b** force versus z-displacement during the three stages indicated in panel (**a**). **c** force versus tip-substrate separation derived from data shown in panel (**b**). The elastic constant k_p can be determined from the slope of the linear part of the curve; the breakage (yield) force F_{rp} is the force at which a nonlinear event associated to capsid disruption occurred; the critical deformation d_{rp} is the deformation at which the particle was broken

[24], in contrast, for some mutant MVM capsids the corresponding values were up to 1.3 N/m and 2.8 GPa [20].

It could be argued that viral capsids are all made of the same material (protein) and that the large differences in E_p values could have arisen mostly from differences in the model assumptions needed for the calculation. However large, genuine differences in E_p were found when this parameter was calculated using the same procedure for virus particles of similar size and geometry. An example was provided by comparison of the intrinsic elasticity of capsid intermediate states (prohead I, prohead-II, EI, head-II) during maturation of phage HK97 [113] (Fig. 11.3).

Remarkably, large differences in stiffness and intrinsic elasticity could be found even between mutants of a same virion or capsid that differed in just one amino

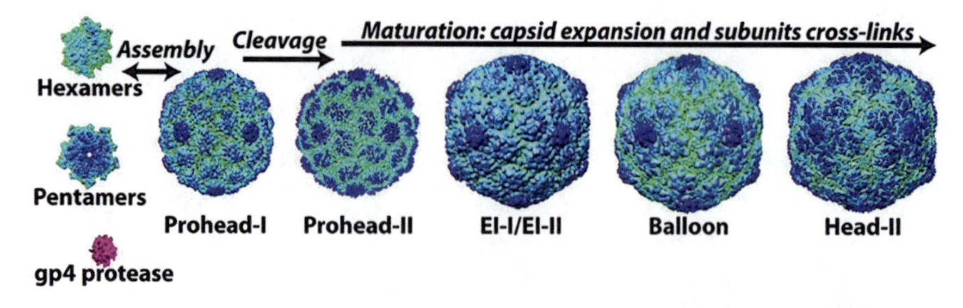

Fig. 11.3 Maturation of the bacteriophage HK97 capsid. The immature capsid (prohead I) of HK97 is assembled from the capsid building blocks. Maturation involves a series of biochemical changes and structural rearrangements to produce the mature capsid (Head-II). Figure reproduced from [56]

acid residue per capsid subunit; these mutants were nearly identical in size, shape, architecture, molecular structure, and conformation (even at the atomic level). For example, a natural mutation in the cowpea chlorotic mosaic virus (CCMV) capsid had minor effects on its structure, but it increased its stiffness and strength due to the establishment of many additional noncovalent interactions in the capsid [93], [127]. Structurally nearly identical single mutants of the MVM capsid [49, 84] showed up to nearly threefold differences in both k_p and E_p [20]. The large differences in mechanical properties between some structurally almost identical virus particles should be considered when searching for quantitative relationships between those properties and large-scale structural features of virus particles.

Strength and Brittleness

When viral particles were subjected to deep enough indentations, characteristic nonlinear events in their mechanical response were detected in the Fz curves (Fig. 11.2; see Chap. 10). Comparison of AFM images taken immediately before or after the nonlinear event revealed that the event was, in some cases, associated to an irreversible deformation of the viral particle that involved buckling, bending and/or shifting of subunits (e.g., [4, 57, 111]). Fractures that separated capsid subunits were sometimes observed, and the damaged viral particles showed a plastic response under load, instead of an elastic response [57, 59]. In many other cases, nonlinear events involved the loss of particle subunits, or catastrophic breakage of the particle into smaller substructures or components (e.g., [21, 57, 58]). The resistance to disruption under load, termed here mechanical strength, can be quantified by the value of the breaking force (yield force) F_{rp}, defined as the force required to disrupt the physical integrity of the viral particle (Fig. 11.2).

A strength-related mechanical property of virus particles is brittleness. Two viral particles indented with an AFM tip may show the same mechanical strength (i.e., they may break under the same yield force F_{rp}); but one of them (more brittle) may break when moderately deformed, while the other (less brittle) may break only when it is greatly deformed. The relative brittleness of quasi-spherical virus particles can be compared using the $(d_r/D)_p$ ratio, d_{rp} being the critical deformation (the maximum deformation accepted by the particle without any disruption), and D_p being the average external diameter of the particle (or, alternatively, the particle height determined by AFM). A higher $(d_r/D)_p$ value indicates a lower brittleness (Fig. 11.2).

Virions or capsids from different species can differ widely also in mechanical strength and/or brittleness. For example, icosahedral capsids were disrupted by yield forces that differed by over tenfold. Yield forces reached up to 6 nN for nucleic acid-filled herpes simplex virus type 1 (HSV-1) particles devoid of their lipoprotein envelope [76], or 7 nN for phage T7 virions [141]. Also, some icosahedral capsids (e.g., MVM, adeno-associated virus type 2 (AAV-2), NV at alkaline pH, Φ29, the mature phage λ capsid, HK97 prohead-II) are relatively brittle, as they were disrupted at moderate deformations ($(d_r/D)_p$ ~0.1–0.3); in contrast, other icosahedral capsids (e.g., CCMV, NV under non-alkaline conditions, HK97 prohead-I, hepatitis B virus (HBV)) withstood very large deformations. Some capsids (those of CCMV at pH = 6 [93] or NV at acidic or neutral pH [24], and some enveloped virions (SARS-CoV-2, [67], recovered their native shape even after wall-to-wall deformation under load.

Resistance to Material Fatigue and Self-healing

Virus particles are susceptible to material fatigue when cyclically indented using forces well below the yield force. Depending on both the maximum load and indentation frequency, no visible damage was detected after several indentation cycles. However, after a high enough number of cycles, disruption of the viral particle and loss of components were observed (Fig. 11.4).

Material fatigue by cyclic indentation with an AFM tip has been detected for structurally very different particles from non-enveloped or enveloped viruses. Examples include the capsids of phages Φ29 [57], λ [53, 58] and HK97 [113]; the phage T7 capsid [54] and virion [141] the mature human immunodeficiency virus type 1 (HIV-1) capsid protein lattice [132] rotavirus particles [62] adenovirus (AdV) capsids or virions [31, 96, 97], (Fig. 11.4); and the SARS-CoV-2 virion [14]. Thus, susceptibility to material fatigue is probably a general feature of virus particles.

Fatigue of structurally simple virus particles (e.g., the mature HIV-1 capsid protein lattice) led to the removal of capsid subunits or clusters of subunits; fatigue of more complex viral particles (e.g., AdV or T7) frequently led to the gradual, differentiated removal of some specific capsid subunits or subassemblies, auxiliary proteins, the viral nucleic acid and/or other components (Fig. 11.4). Irrespective of particle complexity, catastrophic failure could be observed after the gradual removal of some components, or even without any intermediate state being detected.

Differences in the experimental setup (applied load and frequency), and gradual or abrupt dissociation through different pathways upon cyclic indentation, may preclude a meaningful comparison of the relative propensity of different viruses to fatigue.

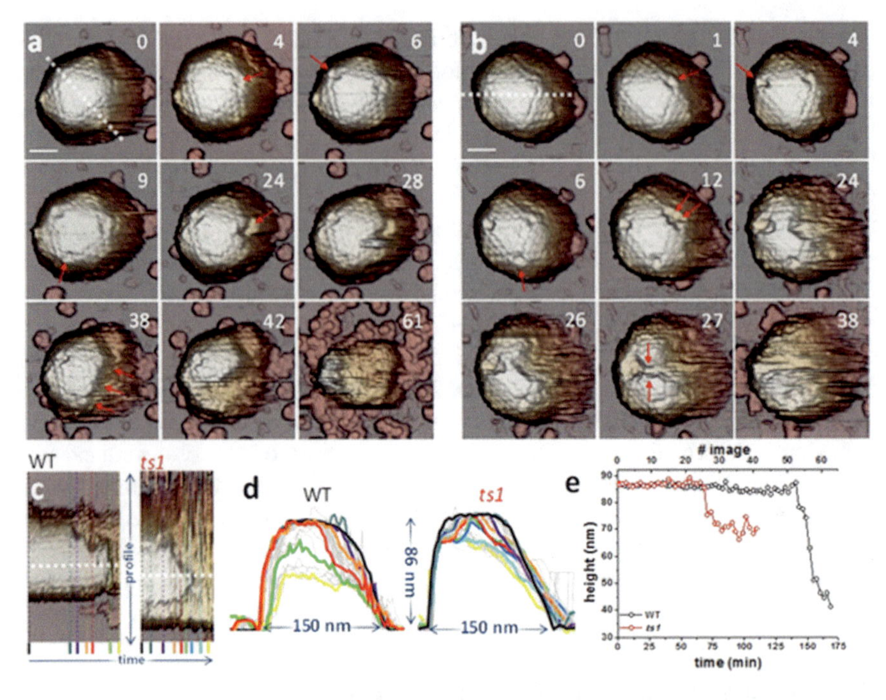

Fig. 11.4 Gradual disassembly of mature or immature AdV by material fatigue. **a**, **b** selected AFM images taken during the mechanically-induced gradual disassembly of mature AdV (**a**) or of an immature AdV mutant (**b**). Higher numbers correspond to images taken at later times during the disassembly process. Positions in the viral particles where pentons were released are indicated by red arrows. Scale bars in (**a**) or (**b**) correspond to 46 nm or 40 nm, respectively. **c** kymographs showing the variation in particle height over time for mature AdV (left) or immature AdV (right) along the white dotted lines traced in images numbered as "0" in panels (**a**) or (**b**), respectively. **d** transversal height profiles obtained form the data in panel (**c**). Colored curves correspond to the points in time indicated by vertical lines in panel (**c**) using the same color key. **e** time-dependent evolution of the maximum height of mature AdV (black) or immature AdV (red), determined along the white dotted lines traced in the kymographs in panel (**c**). Figure reproduced from [96]

However, the results obtained so far do suggest that virus particles from different species may substantially differ in their resistance to fatigue.

Some virus particles showed a remarkable propensity for self-healing after being damaged by the application of mechanical force. After unloading, the damage (e.g., fractures or dislocation of some subunits) was spontaneously reduced, or even disappeared altogether. Examples include the NV capsid at acidic or neutral pH [24], the HIV-1 capsid protein lattice [32, 131], the T7 capsid and virion [29, 141], and the SARS-CoV-2 virion [67]. Self-healing has also been detected in non-viral protein complexes, such as vaults [79]. Comparative results are not available yet to assess the differences among virus particles in their self-healing response to mechanically induced damage.

Mechanical Anisotropy

The orientation of elongated virus particles adsorbed on a substrate could be readily determined by AFM (e.g., the phage Φ29 capsid; [57], Fig. 11.1a, center). In addition, the orientation of some icosahedral virus particles could be ascertained through the observation of conspicuous topographic features at or around a symmetry axis type (e.g., for the parvoviruses MVM [15], Fig. 11.5) and AAV-2 [148]). For larger particles, orientation was revealed by height differences and/or visualization of icosahedral facets (e.g., for the capsids of phages T7 [54] or P22 [81], AdV [100], (Fig. 11.4), or HSV-1 particles [126].

Comparison of k_p values obtained by indentation of different, identified regions of a virus particle revealed that some virions or capsids are nearly isotropic in stiffness (e.g., the wild-type (wt) MVM capsid [15]). However, others are largely anisotropic (e.g., the MVM virion [15], or the capsids of phages Φ29 [57] or T7 [54]. The quantification of differences in stiffness when different regions in a virus particle are indented has revealed connections between local changes in structure and stiffness that modulate virus infectivity. In addition to anisotropic stiffness, virus particles, even some that differ in just one amino acid residue, can show different degrees of anisotropy regarding strength and/or brittleness [91].

Anisotropic stiffness of a virus particle may be an inescapable physical consequence of a particular molecular structure but, at the same time, it may also have arisen

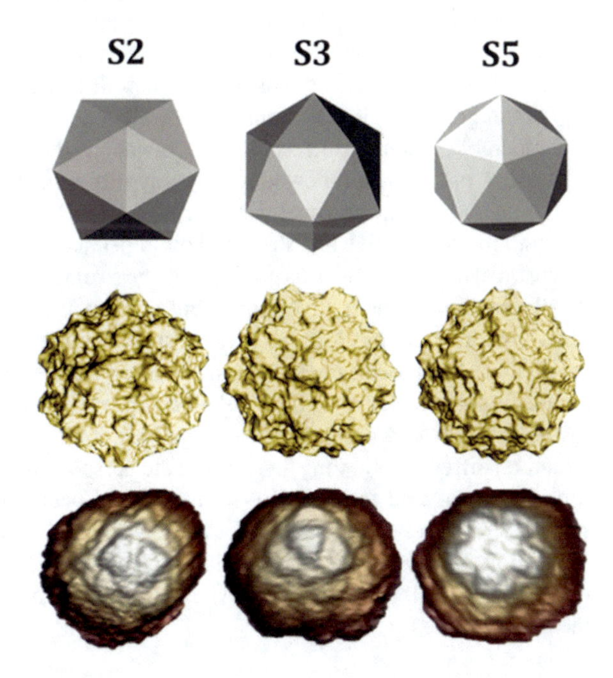

Fig. 11.5 Determination of the orientation of adsorbed individual icosahedral virus particles by AFM imaging. Left to right: icosahedral objects observed along a S2, S3 or S5 symmetry axis. Top row: an icosahedron. Middle row: a structural model of MVM. Bottom row: individual MVM particles imaged by AFM. Figure reproduced from [18]

as a result of biological adaptation [15, 16, 18, 20, 22] (see Sect. "Anisotropic Stiffness/Conformational Dynamics as an Adaptive Trait of MVM to Impair a Heat-Induced Deleterious Transition Without Impairing an Infectivity-Determining Transition"). Likewise, anisotropic strength and brittleness may be biologically relevant [104, 105, 146].

Marker topographic features and orientation were not clearly identifiable in AFM images of many virus particles. Thus, only averaged k_p, F_{rp} and/or $(d_r/D)_p$ values could be obtained. In those cases, mechanical anisotropy (if present), and its possible consequences, remained undetermined.

Comparing the Mechanical Properties of Viruses: A Summary

Virions or capsids from a highly diverse sample of virus species showed equally diverse local and/or global mechanical responses under load. Stiffness, intrinsic elasticity, mechanical strength, brittleness and, perhaps, resistance to material fatigue of either virions or their protein-only capsids differed by over one or two orders of magnitude. Some trends were experimentally observed. For example, some enveloped virions are relatively soft (e.g., influenza virus, mature HIV-1, SARS-CoV-2), whereas the much smaller, non-enveloped virions of the parvoviruses MVM and AAV-2 are much stiffer. These large mechanical differences are largely the result of very different size, composition, architecture and/or molecular structure. However, many exceptions to those trends were found, not only when comparing stiffness, but also regarding intrinsic elasticity (which does not depend on size and geometry), strength or brittleness. Some structurally very different virus particles showed a comparable mechanical response under load; whereas some structurally nearly identical virus particles showed remarkably different stiffness, strength, brittleness and/or anisotropic mechanical behavior.

From the perspective of a structural biologist, an important observation is that the mechanical properties of virus particles can be strongly dependent on minor structural details, such as single atomic groups in their capsids. Relationships between large-scale structural features (e.g., size and/or shape) and mechanical properties of viruses are predicted by models or simulations that may capture the physical essence of virus architecture [83]. However, those relationships may be greatly blurred in real viruses by the strong effects on mechanical properties of small-scale (down to atomic level) structural differences among them. From the perspective of an evolutionary biologist, it could be guessed that the diverse mechanical behavior of evolutionarily closely related viruses, even mutants of the same virus is, in part, the result of biological adaptation. It may have partly arisen through the gradual introduction, by mutation, of relatively small structural changes in response to specific selective pressures (see the next sections).

Modulation of the Mechanical Properties of Virus Particles

Since the times of the Phage Group 80 years ago, countless qualitative or quantitative biological, biochemical, or biophysical studies on viruses have revealed that virus particles are structurally highly dynamic (see [89], and references therein). During the different stages of the infectious cycle (Fig. 11.6), viruses and their components respond to specific physical or (bio)chemical cues through specific structural changes. Virus particles are first assembled in an infected cell from their molecular components in a complex morphogenetic pathway. They later undergo several controlled changes in molecular composition, covalent structure, conformation and/or motions of structural elements. Ultimately, those changes mediate virion maturation, exit from the cell; survival in the extracellular medium; entry into a new host cell; intracellular trafficking; particle disassembly and/or genome ejection; expression and replication of the viral genome; and assembly of progeny viruses, closing the infectious cycle (Fig. 11.6).

A widely diverse ensemble of physical or chemical effectors has been shown to promote or interfere with virus infectivity by inducing specific changes in the structure or conformational dynamics of the viral particle during some stage of the

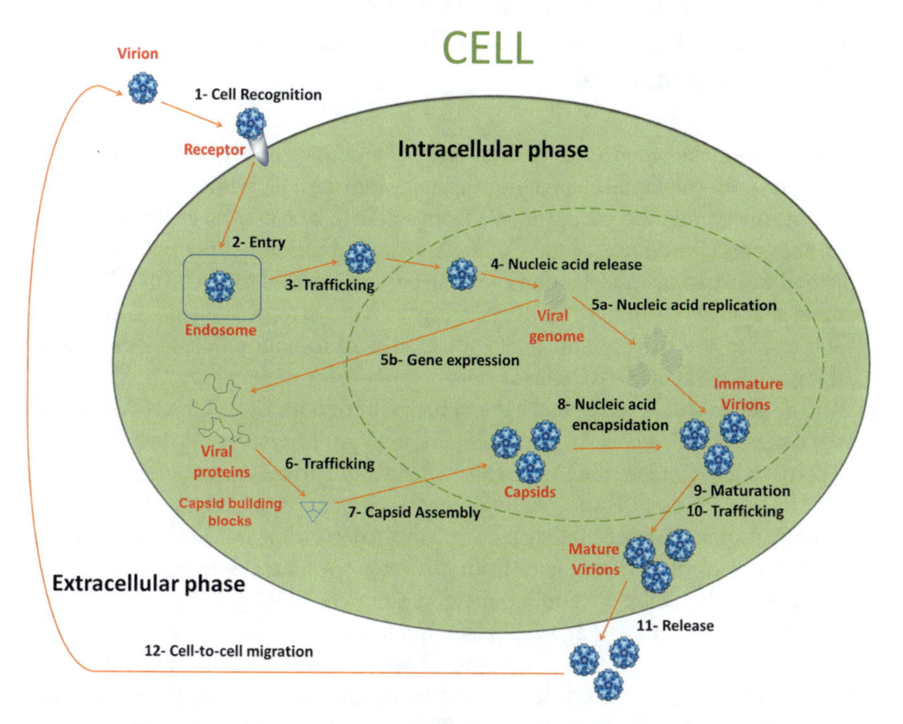

Fig. 11.6 A typical infectious cycle of an animal virus. Main stages in the infectious cycle of an animal virus (the parvovirus MVM) are indicated. Major differences in the infectious cycles of different viruses are not indicated in this scheme. Figure reproduced from [90]

infectious cycle. Depending on virus species, those effectors include: (i) physico-chemical conditions (pH, ionic strength, temperature, macromolecular crowding, osmotic pressure); (ii) virus particle-binding ligands (ions, organic molecules, host cell proteins, other non-viral biomolecules); (iii) the viral nucleic acid or structural viral proteins recruited by a virus particle (e.g., during morphogenesis or maturation), or removed from the particle (e.g., during maturation, genome uncoating or particle disassembly); (iv) chemical modifications of viral components (proteolytic processing, introduction of disulfide bridges or other covalent bonds, other post-translational modifications); or v) during virus multiplication in the cell, substitutions, additions, or deletions of nucleotides in the viral nucleic acid or amino acids in viral proteins, that may contribute to virus adaptation for survival and evolution.

In the last decade and a half, AFM studies have clearly revealed that many effector-dependent, virus infectivity-determining changes in the structure or dynamics of virus particles are reflected in a variation of their mechanical behavior under load, as determined in indentation experiments. Examples are provided in the following three subsections.

Changes in Mechanical Properties of Virus Particles in Response to Physico-Chemical Conditions, Ions or Non-viral Compounds

pH, ionic strength, temperature. Variations in pH modulate infection by many viruses by eliciting conformational changes during entry into the cell, intracellular trafficking or genome uncoating. Increasing the pH from acidic to near neutral greatly decreased the stiffness and brittleness of the CCMV capsid and virion. At pH = 6, the capsid withstood wall-to-wall deformations without being disrupted [68, 93, 144]. The stiffness and mechanical strength of the NV capsid were also pH-dependent: at acidic to neutral pH, it was very soft and withstood wall-to-wall deformations without breaking,whereas at pH = 10 it broke at very small deformations ($(d_r/D)_p$ ~0.2) [24]. The pH-dependent changes in mechanical behavior of both CCMV and NV particles did not correlate with any large morphological change. They were attributed to weakened or rearranged interactions in the capsid that would prime the particle for swelling (in CCMV), or for other conformational rearrangements required for infection.

Acidification decreased the stiffness of the AdV virion but not the capsid, an effect that was attributed to neutralization of negative charges by protonation and compaction of the DNA-containing nucleoprotein core [101]. Changes in pH that affected the infectivity of phage C22 also modulated its stiffness, and the same was observed upon changes in ionic strength or temperature [118, 119]. Heating also modified the strength and brittleness of the T7 capsid [142]. In these and other cases, the observed changes in particle stiffness or strength signaled the existence of effector-induced structural changes that could modulate virus infectivity.

Macromolecular crowding. The effects that the macromolecularly crowded environment in the cell or inside a virus particle may have on different stages of the viral infectious cycle are poorly known. Macromolecular crowding dramatically promoted mature HIV-1 capsid assembly under physiological ionic strength conditions in vitro, whereas in the absence of crowding no assembly could be detected at all under the same conditions [30]. Macromolecular crowding can also modulate the mechanical properties of virus particles. For example, the presence of a crowding agent modulated the stiffness and strength of brome mosaic virus (BMV) in a complex fashion that was dependent on the concentration of the crowding agent and the absence or presence of the single-stranded (ss) RNA in the viral particle. The results were consistent with the RNA molecule exerting a small negative internal pressure that prestresses the virion [149]. Macromolecular crowding also enhanced dsDNA ejection from phage λ in vitro, due to a pulling force resulting from DNA condensation in the crowded medium induced by osmotic stress [61]. These and other results indicate that macromolecular crowding constitutes an important determinant for viral infection.

Osmotic pressure. An increase in internal osmotic pressure led to further condensation of the viral dsDNA in both phage λ and HSV-1. This effect removed the internal pressure exerted on the capsid wall, relieved the mechanical stress, and increased the strength of the particle against an applied force. Relieving the internal pressure inhibited genome ejection and impaired infectivity [7, 8, 10, 11, 36, 37, 58]. Thus, for some viruses osmotic pressure can determine infection (see Sect. "Balanced Mechanical Strength as an Adaptive Trait of Tailed Bacteriophages and HSV-1 for Pressure-Driven Genome Injection in the Host Cell").

Bound ions and polyelectrolytes. Specific binding of metal ions to some virus particles is known to modulate virus infectivity. For both tomato bushy stunt virus (TBSV) [80] and simian virus 40 (SV40) [137], Ca^{2+} binding increased capsid stiffness and strength against irreversible deformation or disruption. In both cases, the authors suggested that Ca^{2+} bridges were responsible for the observed mechanical effects.

(Poly)cations such as spermine, spermidine or Mg^{2+} decreased the stiffness of dsDNA-containing virions, including phages λ [36, 37, 58] and Φ29 (52), AdV [98], as well as the dsDNA-filled HSV-1 capsid [11]. The effect of cations on the stiffness of those virions was generally traced to increased condensation of the DNA molecule or compaction of the nucleoprotein core that followed the neutralization of negative charges. This effect reduced the internal pressure on the capsid wall and controlled virus stability (see Sect. "Balanced Mechanical Strength as an Adaptive Trait of Tailed Bacteriophages and HSV-1 for Pressure-Driven Genome Injection in the Host Cell").

Mg^{2+} also decreased the stiffness of the ssRNA-containing triatoma virus (TrV). Both RNA compaction due to charge neutralization and screened RNA-capsid interactions were contemplated as likely causes [123].

Bound host cell proteins. Binding of integrins to AdV promoted virus infectivity, whereas binding of defensins to AdV impaired infection. AFM studies revealed that their binding also resulted in opposite anisotropic changes in virion stiffness:

integrin softened the vertex capsid regions, whereas defensin stiffened them. These effects were respectively associated with a facilitated or impaired release of individual pentons from the capsid vertices during the chemically-induced disassembly of the virus particle [124]. As another example, binding of proviral cellular protein CypA to the mature HIV-1 capsid modulated its stiffness [78]. Changes in the mechanical response of viral particles upon binding of cellular proteins are providing further insights into structural changes controlled by interactions between host molecules and viruses.

Bound organic molecules. Binding to virus particles of natural or synthetic organic compounds that inhibit or promote virus infectivity may lead to substantial changes in the mechanical properties of the particle. For example, antiviral compounds that fill hydrophobic pockets in the human rhinovirus (HRV) capsid (pleconaril, pirodavir) increased virion stiffness [133]. Binding of the antiviral compound tannin to the helical TMV virion increased its strength against disruption when the viral ssRNA was pulled out of the capsid [143].

Binding of specific antiviral or proviral compounds to different sites on the mature HIV-1 capsid exerted different mechanical effects. Binding of PF74 stiffened the truncated cone-shaped capsid [105], but it did not stiffen the hexameric protein lattice the capsid is made of [33]. These results indicate that the stiffening effect of PF74 depends on capsid geometry, including the presence of pentameric "defects" in the hexameric lattice that determine capsid curvature and closing. PF74 also decreased the mechanical strength and fatigue resistance of the HIV-1 capsid protein lattice [33]. In contrast, antiviral compounds CAP-1 or 55 bound to specific sites on the mature HIV-1 capsid protein lattice had no effect on its strength or fatigue resistance, but they decreased its stiffness and intrinsic elasticity [33]. Binding of the proviral cellular compound IP6 to the HIV-1 core either stiffened or softened the viral particle, depending on the activation of reverse transcription of the RNA inside the core [106, 146] (see Sect. "Ligand-Mediated Modulation of Mechanical Strength as an Adaptive Trait of HIV-1 for Efficient Reverse Transcription of the Viral Genome and Controlled Capsid Disassembly").

The above results revealed an association between small-molecule binding to virus particles, complex changes in the mechanical response of the latter due to changes in structure, molecular interactions or conformational dynamics, and modulation of virus infectivity [2, 33, 133]. For HRV, a clear quantitative correlation was found between increased stiffness (measured by the elastic constant k_p) and reduced infectivity (determined by viral titration) [133]. These findings opened up the possibility to develop novel antiviral drugs based on the modulation of stiffness or other infectivity-determining mechanical properties of virus particles.

Changes in Mechanical Properties of Virus Particles in Response to Removal, Addition or Modification of Viral Components

The viral nucleic acid. Morphogenesis of many ssRNA viruses involves the co-assembly of capsid and viral nucleic acid in a single condensation process. In contrast, during morphogenesis of some dsDNA viruses (e.g., tailed phages), the capsid is assembled first, with or without the help of scaffolding proteins (e.g., [56], Fig. 11.3), and the viral dsDNA is packaged in the preformed capsid during virus maturation. Parvoviruses (MVM, AAV-2) also package their ssDNA into a preformed capsid.

Irrespective of nucleic acid type and assembly/maturation/packaging mechanisms, in most (but not all) tested cases the presence of the viral nucleic acid affected the stiffness, strength and/or brittleness of the virus particle, at least under certain conditions. The viral ssDNA stiffened MVM [15, 16], but not AAV-1 [148]. The viral dsDNA stiffened phages λ [58], Φ29 [52], T7 [141] and PRD-1 [5], as well as AdV virions [98] and nonenveloped HSV-1 particles [76].

The mechanical properties of some RNA viruses were also dependent on the presence of the nucleic acid. The ssRNA genome stiffened CCMV [93]. Identical brome mosaic virus (BMV) capsids containing one of three different viral ssRNA molecules showed different stiffness and strength [138]. The stiffness of avian infectious bursal disease virus (IBDV) particles increased as a function of the number of ribonucleoprotein (RNP) complexes they contained, from 0 to 4 [92]. The RNPs also stiffened the enveloped influenza virion [75].

The mechanism by which the viral nucleic acid stiffens a virion, however, does depend on the virus and nucleic acid type. The ssDNA-filled MVM virion has nearly the same atomic structure than the empty capsid (except for the presence of the DNA molecule), but the regions around S2 and S3 symmetry axes are substantially stiffer in the virion. Wedge-shaped ssDNA segments are non-covalently bound to pockets at the capsid inner wall close to the S2 axes [1] (Fig. 11.7). These ssDNA wedges act like inner molecular buttresses that anisotropically stiffen the S2 and S3 regions [15, 16]. Specific removal through point mutation of some capsid-ssDNA interactions in the virion decreased the stiffness of the S2 and S3 regions, and it made the virion as soft as the empty capsid, even though the full ssDNA molecule was still inside [16]. The ssDNA-mediated anisotropic stiffening of the virion appears to be a biological adaptation for improving extracellular virion survival without impairing its infectivity (see Sect. "Anisotropic Stiffness/Conformational Dynamics as an Adaptive Trait of MVM to Impair a Heat-Induced Deleterious Transition Without Impairing an Infectivity-Determining Transition").

The softening effect of cations and/or positively charged auxiliary viral proteins on dsDNA-containing tailed phages, HSV-1 or AdV is consistent with a direct effect of the nucleic acid or nucleoprotein core on viral particle stiffness. In all those viruses, the stiff dsDNA molecule is tightly packed in the capsid. Cations and/or "histone-like" condensing proteins partially neutralize the negative charge of the DNA molecule. Still, non-neutralized charges and the long persistence length of the

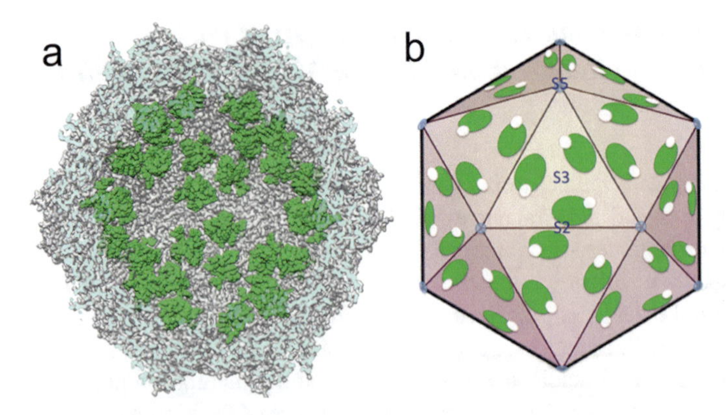

Fig. 11.7 Capsid-bound ssDNA segments in the MVM virion. **a** Cross-section of the atomic structure of the MVM virion [1]. ssDNA segments bound to equivalent sites at the capsid inner wall are colored green. **b** Scheme of the icosahedral MVM virion. The approximate location of capsid-bound ssDNA segments or nearby cavities are respectively indicated by green or white ovals. The positions of one S2, one S3 and one S5 axis in the capsid are indicated. Figure reproduced from [18]

dsDNA molecule generate a substantial internal pressure that stresses the capsid wall. For some of those viruses, a high internal pressure may stiffen the DNA-filled capsid like the pressurized air stiffens a football [52]. For phage λ, a combination of experiment and theory indicated that the stiffening effect of a high internal pressure could be slight at best, resembling the minute effect of the water vapor pressure on the stiffness of a pressure cooker. The dsDNA-mediated stiffening of the λ capsid was traced to osmotic effects caused by DNA hydration [37, 58]. Virion pressurization and the high mechanical strength of dsDNA viruses provide a clear example of the adaptive value of the mechanical properties of viruses (see Sect. "Biological Adaptation of the Mechanical Properties of Viruses in Response to Forces Acting in Vivo"). The mechanisms by which the free ssRNA, or ssRNA-containing RNP complexes stiffen RNA virus particles are not well understood yet.

To sum up, the observed stiffening and/or mechanical strengthening of virus particles by the packed viral nucleic acid have provided new insights into different stages of the infectious cycle, including virus morphogenesis and maturation, disassembly and/or genome uncoating. Nucleic acid-mediated mechanical strengthening or stiffening of virions can also be directly relevant for the survival of some viruses.

Specific capsid protein subunits or capsomers. Structurally complex virus capsids include a number of different components. For example, pentons or hexons form different, position specific subassemblies that constitute the icosahedral capsid of tailed phages, HSV-1 (Fig. 11.1b, bottom right) or AdV. Removal of pentons at the vertices of the AdV virion by either heating or induced material fatigue (Fig. 11.4) reduced particle stiffness and strength and promoted capsid disassembly and genome release [87, 96, 100]. A similar change in mechanical behavior was observed upon the release of pentons from the capsids of HSV-1 [69, 111] or phage P22 [64, 81].

AFM studies in which individual virus particles are imaged and, in some cases, indented, together with structural, biochemical and/or genetic analyses have revealed that pentons of AdV, HSV-1, P22 and others are weakly bound to the rest of the capsid. In vitro, chemical or physical effectors, including mechanical force, modulated their gradual release from the capsid, a situation that may recapitulate AdV disassembly and genome release in vivo. These and other observations have provided new insights into the mechanisms by which these processes are controlled during infection (see Sect. "Balanced Mechanical Strength as an Adaptative Trait of AdV for Pressure-Mediated, Gradual Disassembly and Uncoating").

Auxiliary viral capsid-binding proteins. The capsids of some complex viruses recruit auxilliary viral proteins during their assembly and maturation. Viral cementing proteins bound to λ [53] or P22 [81] capsids generally increased both stiffness and strength. Auxiliary proteins also stiffened the HSV-1 capsid at the vertices where pentons are located [38, 44, 117, 126]. By contrast, recruitment of positively charged viral proteins that condense the viral dsDNA decreased the stiffness of the AdV virion [86].

Introduction of covalent bonds in viral particles. Introduction of covalent bonds between capsid subunits during phage HK97 maturation increased the mechanical strength of the capsid, but it had no effect on its intrinsic elasticity [113] (Fig. 11.3). Intersubunit disulfide bonds increased the mechanical strength of the SV40 capsid without affecting its elastic behavior during the first indentation [137]. Introduction of intersubunit covalent bonds constitutes an alternative strategy to cementing proteins to strengthen the viral capsid during maturation. Likewise, engineering in the laboratory of non-natural disulfide bonds between and within hexamers in the mature HIV-1 capsid protein lattice greatly increased its mechanical strength and fatigue resistance, but it had no effect on stiffness or intrinsic elasticity [32].

Proteolytic processing of viral proteins. Proteolysis of capsid proteins or auxiliary structural proteins has a major role during during maturation of many viruses, including dsDNA viruses like phage HK97 [56], (Fig. 11.3), HSV-1, and AdV. Proteolysis can trigger large conformational rearrangements in the maturing viral particle, and it can also mediate substantial changes in its mechanical behavior. For example, during maturation of the HSV-1 capsid, proteolytic removal of the scaffolding protein had no effect on stiffness, but it increased the mechanical strength of the viral particle [111]. Also, during the complex maturation process of the phage HK97 capsid, proteolysis of the N-terminal domain of the capsid protein (that serves as a scaffolding domain during procapsid assembly) led to large increases in both stiffness and brittleness. Together these effects, and those of other structural changes during maturation of HK97, contributed to yield a mechanically more resilient mature virus particle [113] (Fig. 11.3).

Increased robustness achieved by proteolysis-induced conformational rearrangements, cementing proteins, and/or intrasubunit covalent bonds provides another clear example for the adaptive value of the mechanical properties of viruses. It helps withstand the internal pressure exerted by the viral nucleic acid on the capsid of dsDNA viruses (see Sect. "Biological Adaptation of the Mechanical Properties of Viruses in Response to Forces Acting in Vivo"). It may also contribute to virus survival when confronted with mechanical aggressions in the extracellular environment.

Multiple capsid layers. Some nonenveloped viruses contain several concentric protein layers (e.g., rotaviruses; Fig. 11.1c, left), or include an internal lipoprotein layer (e.g., phage PRD-1). Rotavirus particles containing only the innermost protein layer, the two innermost layers, or all three protein layers were increasingly stiffer, stronger, and less brittle [62]. Phage PRD-1 particles that contained only the innermost lipoprotein layer surrounding the viral dsDNA were very soft, whereas PRD-1 capsids that contained an additional protein layer were much stiffer [5].

Likewise, in the NV capsid the protruding (P) domain of the capsid protein forms a non-continuous outermost protein layer on top of the continuous, basal capsid protein layer. Removal of the P domain did not preclude capsid assembly but decreased capsid stiffness and mechanical strength. The observed stiffening of the NV capsid mediated by the P domain was attributed to the introduction of isotropic prestress that may help stabilize the capsid [6].

Together, those studies indicate that, in multi-layered virus particles, the outer protein layers that surround a relatively soft inner shell tend to stiffen and/or strengthen the particle. Like maturation, the increased robustness conferred by additional protein layers may help resist internal or external stresses during the infectious cycle.

The outer lipoprotein layer. In enveloped viruses, both the outer lipid envelope and the envelope-embedded proteins can be major determinants of the mechanical properties of the virus particle. The protein-made matrix layer located immediately below the lipid envelope can also influence the mechanical behavior of the particle, partly through its interaction with envelope proteins.

For example, removal of proteins embedded in the lipid envelope reduced the stiffness of the ssRNA-containing influenza virion (Fig. 11.1a, right), which became as soft as pure liposomes. Removal of the matrix layer also softened the viral particle. The observed softening events helped to propose a sequence of structural events during influenza virus fusion to the cell membrane and RNP release [46, 74, 75, 120].

As another example, maturation softened the enveloped, ssRNA-containing particles of the retroviruses MLV [71] and HIV-1 [72, 99]. The C-terminal (Ct) domain of the envelope-embedded Env protein of HIV-1 (Fig. 11.1c, right) was responsible for the higher stiffness of the immature HIV-1 virion. The maturation-induced softening of the HIV-1 virion facilitated entry into the host cell and is biologically relevant [99, 116] (see Sect. “Mechanical Softening as an Adaptive Trait of HIV-1 to Enable Virus Entry into the Host Cell”).

Changes in Mechanical Properties of Virus Particles in Response to Point Mutations

The mechanical properties of virus particles are highly sensitive to even the smallest, genetically introduced chemical changes in the capsid protein subunits. For example, a naturally occurring single amino acid substitution (per capsid subunit) that increased the stability of the CCMV capsid and virion in a high-salt medium also increased their stiffness and mechanical strength [93].

The individual effects on MVM mechanics of over 40 engineered amino acid substitutions in different capsid regions have been analyzed so far (Fig. 11.8). Most of those mutations led to substantial, anisotropic changes in stiffness, mechanical strength and/or brittleness of the capsid and/or the virion. In many cases, changes in the mechanical behavior of mutant MVM particles could be linked to different mechanisms by which those mutations impair virus infectivity [16, 18, 20, 22, 91] (see Sect. "Changes in Mechanical Strength or Stiffness of Virus Particles, and Linked Changes in Structural Stability or Conformational Dynamics").

Mutant HRV virions carrying cavity-filling amino acid substitutions in capsid pockets revealed an inverse correlation between infectivity and stiffness, dependent on the size of the substituted side chain [133]. Four single amino acid substitutions that impaired disassembly of the mature HIV-1 capsid increased its stiffness [102].

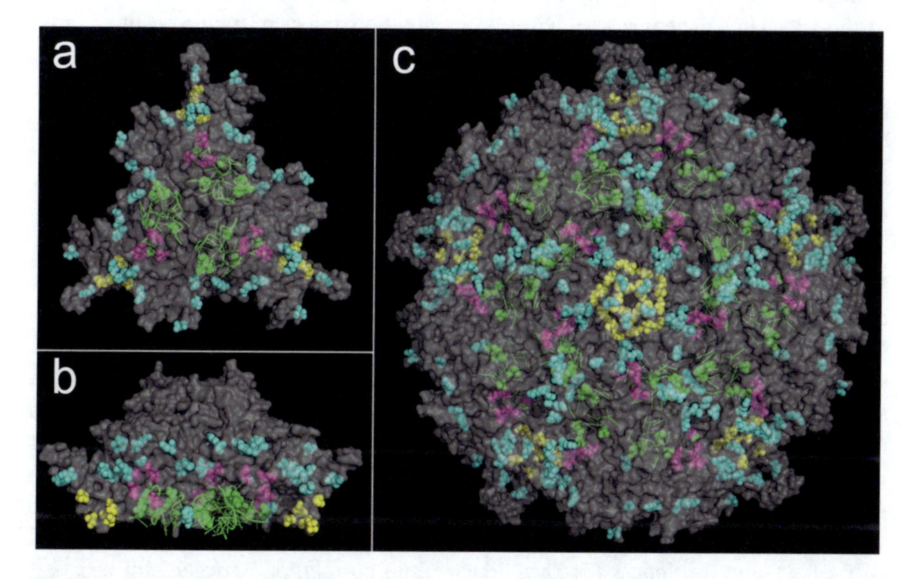

Fig. 11.8 Localization in the MVM virion of some amino acid residues whose individual role on the mechanical properties of MVM particles has been analyzed. **a**, **b**, front view (**a**) or side view (**b**) of a trimeric capsid building block in the atomic structure of the MVM virion. **c** front view of an ensemble of five trimers around a S5 axies in the atomic structure of the MVM virion [1]. A semi-transparent depiction is used to reveal buried amino acid residues. Thirty two residues per capsid subunit that were individually mutated to investigate mechanical properties and related changes in infectivity, conformational stability, or dynamics, are represented as spacefilling models and color coded as follows: cyan, residues at the interfaces between trimeric capsid building blocks; yellow, residues surrounding the base of the capsid pores around each S5 symmetry axis; magenta, residues delimiting capsid small cavities close to ssDNA-binding sites; green, residues involved in interactions with genomic ssDNA segments in the virion. ssDNA segments that interact with those capsid residues are represented as green ribbons. Figure reproduced from [18]

A single amino acid substitution in an auxiliary capsid protein stiffened the AdV capsid [136].

The detailed relationships between local chemical changes introduced by single mutations in virus particles, changes in atomic structure and/or conformational dynamics, changes in mechanical properties, and modulation of virus infectivity have begun to be investigated. Single amino acid substitutions may result in very small changes in the equilibrium (minimum free energy) structure of the viral particle. However, small structural changes may be translated into substantial changes in the number and/or energy of intersubunit interactions, leading to remarkable changes in the mechanical strength of the particle [91, 93]. Moreover, small changes in many atomic positions and interatomic interaction energies may be translated into very large local or global changes in the conformational dynamics of the viral particle, leading to large changes in stiffness and intrinsic elasticity [18, 20, 22, 49, 84, 133] (see Sect. "Changes in Mechanical Strength or Stiffness of Virus Particles, and Linked Changes in Structural Stability or Conformational Dynamics"). The high sensitivity of the mechanical behavior of viruses to mutation may contribute to their rapid adaptation and evolution.

Is There a Correlation Between Different Mechanical Properties of Virus Particles?

Some effectors either increased or decreased both the strength and the stiffness of certain virus particles. However, several studies have shown that the stiffness (hard to bend) and the mechanical strength (hard to break) of a virus particle are not necessarily linked. This fact was clearly supported by a detailed comparison of the effects of many single amino acid substitutions in the MVM capsid on its mechanical properties. These substitutions were all located at the capsid intersubunit interfaces and did not change capsid size or shape. For each capsid mutant, the value determined for one mechanical parameter was plotted against that determined for another parameter. A very good quantitative correlation was found between capsid stiffness k_p and brittleness $(d_r/D)_p$; however, no correlation between stiffness (k_p) and mechanical strength (F_{rp}) was observed. Moreover, the amino acid residues that constituted major determinants of capsid stiffness were different from those residues that were major contributors to mechanical strength [91]. Stiffness and mechanical strength have different structural, thermodynamic, and kinetic foundations (see Sect. "Pulling, Pushing, Mobility of Atomic Groups, and Mechanical Properties of Virus Capsids").

Modulation of the Mechanical Properties of Virus Particles: A Summary

Effector-induced changes in molecular composition, atomic structure, intraparticle interactions and/or conformational dynamics of virions or capsids control different stages of the viral infectious cycle. AFM-based indentation studies have revealed that stiffness, intrinsic elasticity, strength, brittleness and/or fatigue resistance of virus particles can be highly sensitive to effector-induced structural modifications, even minor ones. Large mechanical differences have frequently arisen upon removal of a single molecular component, replacement of a single amino acid residue, or as a consequence of other small chemical modifications or conformational changes of a virus particle.

The observed changes in the mechanical properties of virions and capsids upon the action of infectivity-determining effectors have provided new insights into structural modifications and mechanisms that control different stages of the viral infectious cycle. They have also contributed to the discovery of previously undetected structural changes or mechanisms required for infection. Moreover, AFM indentation and/or imaging of individual virus particles have uniquely allowed the detection of sequences of events in position-dependent and component-specific structural changes mediated by different cues.

To reiterate and sum up, biologically relevant, effector-mediated changes in the structure or dynamics of virions or capsids can be sensitively probed by AFM through the detection of changes in their mechanical properties. The results obtained, combined with those of other experimental or theoretical approaches, are improving our understanding of the physicochemical mechanisms that govern different stages of the infectious cycle: virion assembly, maturation, exit from the host cell, survival in the extracellular environment, entry into a new cell, intracellular trafficking, virion disassembly and genome release.

Changes in Mechanical Strength or Stiffness of Virus Particles, and Linked Changes in Structural Stability or Conformational Dynamics

What determines the observed associations between effector-mediated changes in structure and/or dynamics of virus particles, and changes in their mechanical properties? This Section describes a unifying, simple model based on structural, thermodynamic, or kinetic considerations that is consistent with theoretical and experimental results. Structurally simple protein capsids (either empty or filled with the viral nucleic acid) are considered here, to avoid dealing with complications introduced by other components present in more complex virus particles.

Pulling, Pushing, Mobility of Atomic Groups, and Mechanical Properties of Virus Capsids

Structural changes have been mechanically induced in single protein molecules or supramolecular assemblies (including viruses) by "pulling" on them [3, 12, 40, 41, 43, 60, 95, 103]. For example, in pulling experiments to study protein unfolding, a protein molecule is stretched using mechanical force. If the unfolding reaction was two-state, and the stretching did not reach a certain limit, an elastic behavior was generally detected. Non-linear events (the unfolding transition) were not observed and, after unloading, the molecule returned to its equilibrium position. All-atom MD simulations indicated that many noncovalent bonds were gradually stressed and eventually broken, but they fully recovered when the force was removed. From the standpoint of thermodynamics and reaction kinetics, the system moved upwards in the free energy well of the native state, along the mechanical reaction coordinate, but it did not reach the free energy peak that corresponds to the transition state of the reaction.

However, if enough force was applied to stretch the molecule past a certain length, a non-linear event was observed, and the protein was (reversibly) unfolded. The system moved along the mechanical reaction coordinate enough to reach the free energy peak (the transition state) that separates the free energy well of the native state from that of the denatured state. In pulling experiments, mechanical force (helped by thermal energy) is used to overcome the free energy barrier along the mechanical reaction coordinate.

When "pushing" on a quasi-spherical protein shell (a virus capsid) to determine its stiffness, brittleness, strength or fatigue, some similarities to what happens in "pulling" experiments regarding structure, thermodynamics, and kinetics are apparent, as described next (compare Figs. 11.2 and 11.9; [22, 33 88]).

Stiffness. As indentation increasingly deforms the capsid "crust", atomic groups and larger structural elements at or under the contact area are gradually displaced from their equilibrium positions. The high density and complex network of intra- or inter-subunit interactions will propagate the effect to other capsid regions. Each weak noncovalent bond between atomic groups in the capsid can be imagined as a virtual spring. These springs are relaxed at equilibrium, but they will become increasingly stretched as the indentation progresses. As a result, the weaker intersubunit interactions will eventually be broken.

Coulombic interactions will accept some stretching before being considerably weakened. Directional hydrogen bonds will be drastically weakened when "bent" or stretched. Non-directional van der Waals interactions would resist some "sliding" between the contacting atoms, but they will be broken upon minor stretching. Changes in the position of so many atoms will lead also to steric clashes that will have to be relieved by continuous changes in atomic displacements. These changes, in turn, will stress, and eventually break, other weak interactions. Increased separation between apolar groups will also weaken the hydrophobic effect that contributes to stabilize and hold together the capsid protein subunits. As a net result, the free

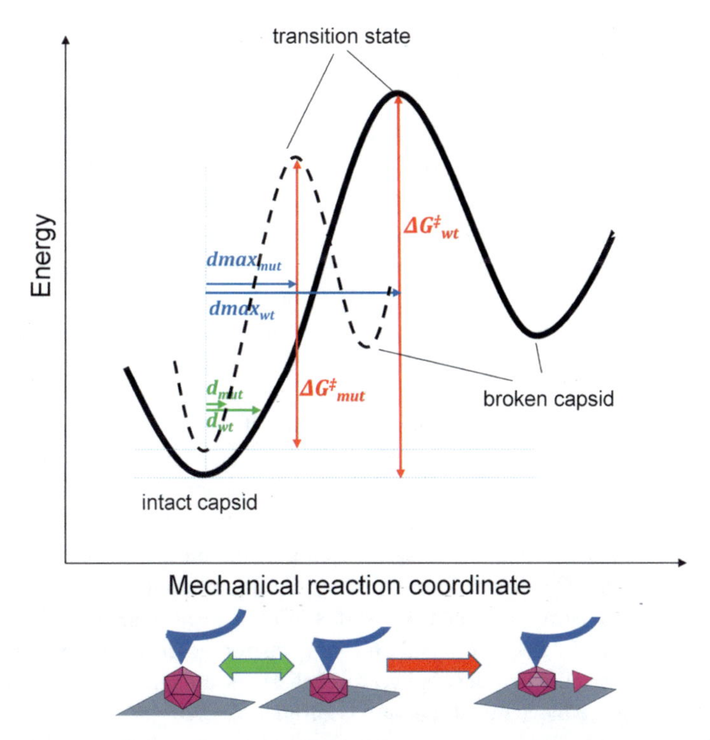

Fig. 11.9 Simplified energy diagram during indentation of a virus capsid in the absence or presence of an effector that modifies the capsid mechanical behavior. The energy of the system during elastic deformation of the capsid, followed by capsid disruption in a two-state reaction, is represented along the mechanical reaction coordinate. The thick line corresponds to a virus capsid in the absence of any mechanics-determining effector (*wt* subscript). The thin, dashed line corresponds to the same virus capsid in the presence of a mechanics-determining effector (e.g., an amino acid substitution introduced in the capsid; or a ligand that had been either bound to or removed from the capsid, see Sect. Modulation of the Mechanical Properties of Virus Particles) (*mut* subscript). In this example, the effector stiffens the capsid and makes it more brittle and less strong, as respectively indicated by the following symbols and color code. Green horizontal arrows: distance d_{mut} is shorter than d_{wt}. In mechanical terms, under the elastic regime a same applied force achieves a lower deformation for the *mut* capsid than for the *wt* capsid; the elastic constant k_p will be higher for *mut* than for *wt*. From a thermodynamics/kinetics perspective, the slope of the free energy well of the initial state (intact capsid) along the mechanical reaction coordinate is steeper for *mut* than for *wt*. -Blue horizontal arrows: distance $dmax_{mut}$ is shorter than $dmax_{wt}$. In mechanical terms, the *mut* capsid is disrupted at a lower deformation than the *wt* capsid; the critical deformation d_{rp} will be lower for *mut* than for *wt*. From a thermodynamics/kinetics perspective, the position along the mechanical coordinate of the transition state ‡ for capsid disruption is closer to the position of the initial state for *mut* than for *wt*. -Red vertical arrows: the $\Delta G^{\ddagger}{}_{mut}$ value is lower than the $\Delta G^{\ddagger}{}_{wt}$ value. In mechanical terms, the *mut* capsid requires less applied force than the *wt* capsid to become disrupted; the yield force F_{rp} will be lower for *mut* than for *wt*. From a thermodynamics/kinetics perspective, the free energy barrier along the mechanical coordinate for capsid disruption is lower for *mut* than for *wt*

energy will increase, and the system will gradually move along the mechanical reaction coordinate, away from its equilibrium position at the bottom of the free energy well of the native state (Fig. 11.9).

If deformation is kept within a certain limit, and the AFM tip is then retracted, the broken or stretched noncovalent bonds will usually recover in a microscopically reversible process, and the virus capsid will return to its minimum free energy conformation at equilibrium. From a mechanical standpoint, the fully reversible stretching, bending and disruption of multiple noncovalent interactions will be detected, in a typical indentation experiment, as an elastic deformation of the capsid. The gradual, reversible breaking of many weak interactions could lead to multiple nonlinearities along the slope of the *Fz* curve, but they would be too small to be resolved.

The stiffness, i.e., the mechanical force required to achieve a certain (elastic) deformation of the capsid (the k_p value; Fig. 11.2) will be related to the energy needed to displace the atoms a certain distance along the mechanical reaction coordinate from their equilibrium positions. i.e., k_p will be related to the steepness of the free energy well of the native state along the reaction coordinate (Fig. 11.9).

Mechanical strength and brittleness. When pushing on a viral capsid with high enough force, nonlinear event(s) are typically observed. Each of those events may correspond to a transition between two states of the capsid (e.g., a conformational rearrangement, buckling, loss of a subunit, or catastrophic disassembly. The elastic limit will be surpassed when the applied force disrupts enough interactions to move the system along the mechanical reaction coordinate, out of the free energy well of the native state, and up into the free energy peak of the transition state. From there, the system will fall into the free energy well that corresponds to the final state of the reaction (e.g., a conformationally rearranged capsid, a deformed capsid, a capsid with a missing subunit, or a broken capsid) (Fig. 11.9).

If a transition involves loss of one subunit only, or catastrophic disruption of the capsid in a two-state reaction, the force required to overcome the single free energy barrier of the reaction will correspond to the yield force F_{rp} that defines the capsid mechanical strength. The distance between the native state and the transition state along the mechanical reaction coordinate will correspond to the critical deformation d_{rp}, and will determine the $(d_r/D)_p$ value that defines capsid brittleness (compare Figs. 11.2 and 11.9).

In most cases, capsid disruption/disassembly could hardly be described as a two-state reaction. In fact, the reaction may proceed gradually through a series of intermediates in a complex free energy landscape. However, each non-linear step along the *Fz* curve detected by indentation may correspond to one populated disassembly intermediate, separated from the previous intermediate along the reaction coordinate by a transition state. The simplified considerations mentioned above could be applied to each of the individual transitions between populated intermediate states along a gradual disruption/disassembly process.

Material fatigue. During cyclic indentation of a viral particle using a force well below the breaking force F_{rp}, small "cracks" between capsid subunits may be initiated by disruption of a small number of non-covalent interactions at some weaker spots of the intersubunit interfaces. The exquisite chemical and sterical complementarity between the interacting capsid subunits will favor a fully reversible process. As mentioned above, once the load is removed, and given enough time, each lost noncovalent interaction may faithfully recover, and an elastic deformation of the viral particle will be observed. However, if the indentation frequency is high enough, not every broken interaction in every crack will have time to recover between load cycles. Subsequent indentations will lead to the gradual enlargement and merging of several cracks, until complete interfaces between subunits are disrupted, leading to loss of viral components and, eventually, to particle disassembly.

The mechanism leading to material fatigue and the gradual loss of subunits from a virus capsid could be seen as a kinetic ratchet. Cyclic indentation using small forces will gradually move the system, through many small steps, upwards along the mechanical reaction coordinate. It will eventually reach, one after the other, the free energy peaks (the transition states) between disassembly intermediates. The final outcome will depend on many variables, including applied force; indentation frequency; and the energy, density and distribution of the non-covalent interactions that hold the capsid subunits together.

Force-induced conformational rearrangements. Irreversible buckling, loss of subunits or breakage, or plastic deformation of a broken viral particle are not the only possible mechanical outcomes if the elastic limit of the particle is surpassed. Mechanical force can also provide in vitro the energy needed to trigger reversible transitions between different conformational states. For example, indentation of the T7 virion or capsid under the elastic regime revealed stepwise, nonlinear events that were associated to relatively large, but fully reversible changes in particle structure and dynamics [141]. Transitions induced by mechanical force in vitro may also occur in vivo if enough energy is provided by some physical or (bio)chemical cue, including mechanical forces acting during viral infection.

In the following three subsections the simplified structural/thermodynamic/kinetic interpretation of the mechanical behavior of a virus capsid under load described above (Fig. 11.9) is invoked to justify the observed associations between effector-mediated changes in structural stability or dynamics of virus particles, and changes in their mechanical properties.

Linked Changes in Capsid Stiffness and Equilibrium Dynamics

Virus particles are conformationally highly dynamic even at equilibrium [9, 121]. For a virus particle, the free energy well of the native state (Fig. 11.9) tends to be relatively "flat". Very low thermal energy is frequently enough to allow large displacements of many atomic groups and larger structural elements from their equilibrium positions in the minimum free energy conformation of the capsid (the bottom of the energy well). The fluctuating positions of atomic groups and larger structural elements may lead to conspicuous "breathing" of the virus particle at equilibrium [9]. Effector-mediated changes in the local or global equilibrium dynamics of a virus particle have been shown to modulate virus infectivity through different mechanisms [9].

Equilibrium dynamics of virus particles in solution has been probed by a number of biophysical techniques [9]. Hydrogen–deuterium exchange determined by mass spectrometry (HDX-MS) is a gold standard technique to reveal the equilibrium dynamics in solution of different regions in a supramolecular complex [145]. Normalized B factors in structural models obtained by X-ray crystallography, or local resolution in electron density maps obtained by cryo-electron microscopy, can also provide high-resolution information on the equilibrium dynamics of virus particles [84]. The amplitude of the relative movements ("breathing") of protein hexamers in the HIV-1 capsid protein lattice at equilibrium has been determined by AFM imaging [131].

The model described in the previous subsection predicts that an effector-mediated increase in stiffness of a virus particle is inextricably linked to a reduction in its equilibrium dynamics, and vice versa. Strong experimental support for this prediction is provided by the effects of individual amino acid substitutions on the equilibrium dynamics and stiffness of the MVM capsid.

Comparison of the atomic structures of wt and three mutant MVM capsids carrying single amino acid substitutions at different positions (N170A, F55A, D263A) revealed exceedingly small structural differences, even at the atomic level. However, such minor structural differences were translated into a substantial reduction of the movements of many atomic groups around their equilibrium positions, both around the mutated residue and in distant capsid regions. The reduced dynamics of many regions in the mutant capsids was revealed by reductions in normalized B-factors obtained by X-ray crystallography, or increases in local resolution in electron density maps obtained by cryo-electron microscopy; the validity of these parameters as a signature of equilibrium dynamics was supported by their correlation with hydrogen–deuterium exchange rates determined for the reference (wt) capsid by HDX-MS [49, 84, 135].

The reduced mobility of atomic groups in different regions of the mutant MVM capsids indicates that more thermal energy was required to displace them a certain distance from their equilibrium positions (at the bottom of the free energy well) (Fig. 11.9). It could be expected that a higher mechanical force should also be applied through indentation to displace those atomic groups, along the mechanical reaction coordinate, a certain distance from their equilibrium positions; in mechanical terms, to deform the capsid by a certain value. Determination of k_p values verified that the three mutant capsids were locally and globally stiffer than the wt (e.g., from 50 to 80% for N170A, depending on the indented region) [49, 84].

A similar linkage between changes in equilibrium dynamics and mechanical stiffness was observed for the HRV virion. Binding of antiviral drugs to hydrophobic pockets in the capsid, or individual amino acid substitutions that partially filled the cavity, led both to a reduction in the equilibrium dynamics of the virion and to an increase in its stiffness [133].

Likewise, betaine or some antiviral compounds (CAP1, compound 55) bound to specific sites in the mature HIV-1 capsid protein lattice led to both increased "breathing" (amplitude of the relative movements of capsid protein hexamers) and reduced stiffness [33, 131]. The above results, obtained with very different virus models, provided strong support for an inextricable relationship between changes in the equilibrium dynamics of a virus capsid and changes in its mechanical stiffness.

It must be emphasized that an inextricable linkage between an effector-mediated change in equilibrium dynamics and stiffness may be expected only if no significant alterations in virus particle size or geometry occur, because the latter would influence the k_p value. In fact, what is actually predicted is a linkage between changes in equilibrium dynamics and intrinsic elasticity (E_p value), which depends on the structure of the capsid material alone, and not on geometrical considerations.

It should be noted also that the linkage between capsid equilibrium dynamics and stiffness (or, rather, intrinsic elasticity) could be blurred if the change in breathing and stiffness were strongly directional. HDX-MS and most other biophysical methods will provide a direction-averaged value related to the amplitude of atomic displacements in a capsid region. In contrast, indentation experiments will probe the resistance to atomic displacements along the direction of the applied force (the mechanical coordinate). Thus, the obtained k_p value could sometimes fail to provide a good estimation of the deformability along other directions. A significant anisotropic deformability can be expected for small enough regions of proteins; especially, for elements such as "mechanical clamps" in shock-absorbing proteins [19]. However, indentation with an AFM tip will necessarily affect a very large region of a virus capsid, which contains a huge number of atoms and atomic interactions. Thus, a significant direction-dependent deformability of any indented region may not be generally expected.

As already observed for MVM, HRV and HIV-1, infectivity-related, effector-mediated changes in the stiffness of other virus particles may be linked to changes in their equilibrium dynamics. For example, the stiffening of some virus capsids by the internal pressure generated by the packed dsDNA could be expected to restrain the equilibrium dynamics of many capsid regions. Effector-mediated changes in breathing (determined by HDX-MS or other biophysical techniques) or stiffness (determined by AFM) are different manifestations of the same physical phenomenon: a structure-based variation in the energy required to move atomic groups and larger structural elements in a virus particle a certain distance from their equilibrium positions.

Linked Changes in Capsid Stiffness and Propensity for Conformational Rearrangements

Several stages of the viral infectious cycle are mediated by reversible or irreversible, effector-mediated transitions between different conformations of the virus particle separated by a free energy barrier [9, 63, 89, 121, 139]. For example, for many viruses, capsid assembly is followed by irreversible structural modification(s) (Fig. 11.3).

The result of this maturation process is frequently a metastable virion: a particle that can resist physical or chemical aggressions, but that can also be weakened through controlled conformational changes in response to specific cues in the host cell, to release the viral genome [139].

Several studies have detected a connection for a virus particle between an effector-mediated change in its propensity to undergo a conformational change, and a change in its stiffness (see Sect. "Modulation of the Mechanical Properties of Virus Particles"). Examples provided by extensive mutational studies of two independent conformational transitions in MVM are summarized next.

A conformational transition of the MVM capsid associated to translocation of infectivity-determining signals. Exit of MVM infectivity-determining peptide signals through pores at the S5 axes of the capsid [134] was facilitated by a reversible conformational rearrangement of the capsid. This transition was promoted in vitro by moderate heating, and was detected by a subtle change in the exposure of some capsid tryptophans to solvent [20, 107] and by local changes in capsid equilibrium dynamics [135].

Individual substitutions of 6 infectivity-determining amino acid residues that surrounded the base of the capsid pores abolished the pore-associated conformational rearrangement ([107], yellow residues in Fig. 11.8). Two of these substitutions (N170A and D263A) were analyzed for a possible effect on the capsid equilibrium dynamics. Both substitutions impaired the dynamics of the S5 regions around the pores and other regions as well [49, 84]. Each one of those 6 mutations stiffened the S5 regions. A crosslinking agent abolished the same conformational transition and stiffened the S5 regions. Pseudo-reversions of two of those mutations, L172A (A172I) or D263A (A263N) restored the pore-associated conformational transition and softened the S5 regions. The correlation between impaired conformational change and stiffening was specific for the amino acid residues around the pores: individual substitution of many amino acid residues in other capsid regions neither impaired the pore-related transition, nor stiffened the S5 regions (cyan, magenta, or green residues in Fig. 11.8). The effects of the tested mutations on the stiffness of other capsid regions (S2 or S3) showed no correlation with their effects on the pore-related transition. Together, these results revealed an inextricable linkage in MVM between the impairment of a biologically relevant conformational transition associated to signal translocation through capsid pores, and the stiffening of the pore regions [20].

A conformational transition related to heat-induced inactivation of the MVM virion. The ssDNA genome of MVM is uncoated through a capsid pore without capsid disassembly. *In vitro,* this transition can be promoted by moderate heating [114]. Twelve individual amino acid substitutions that either removed noncovalent interactions between the viral ssDNA and the capsid, or that distorted capsid cavities close to the DNA-binding sites (Fig. 11.7, green or magenta residues in Fig. 11.8), increased to different extents the rate of a heat-induced, virus-inactivating transition [22, 108], most likely related to unproductive DNA release. For the tested mutants, an excellent quantitative correlation was found between the virus inactivation rate, and the stiffening of some capsid regions. The rate constant of the virus-inactivating transition decreased exponentially with the increase in k_p value (stiffness) of the S2 and S3 regions, close to the DNA-binding sites [22].

The intimate correlation found for two conformational transitions in MVM between stiffening of capsid regions involved in the transition, and impairment of the transition itself, was justified in the light of the model described in Fig. 11.9 and transition state theory [22]. It was assumed that a direct linear relationship exists between the degree of mechanical stiffness (k_p value) of a region in a virus particle, and the free energy barrier of a structural transition in which that region is involved ($\Delta G^{\ddagger} = Ck_p$, where C is an empirical proportionality factor). Substituting Ck_p for $\Delta G^{\ddagger}$ in the Eyring equation results in an equation that describes an exponential relationship between the reaction rate constant and stiffness (the k_p value), as experimentally observed for the heat-induced, MVM-inactivating transition [22] (Fig. 11.10).

Why a direct relationship between $\Delta G^{\ddagger}$ and k_p? Consider a region in a virus capsid that is involved in a transition between two conformations separated by a free energy barrier. What could happen to that transition if a mutation introduces additional interatomic interactions that increase the resistance of atomic groups to move from their equilibrium positions? In terms of reaction kinetics, if the distance between the native state and the transition state along the reaction coordinate remains the same, more energy will be required to disrupt enough interactions and allow the atomic groups to reach their positions in the transition state. The free energy barrier $\Delta G^{\ddagger}$ will increase, and the reaction rate will decrease (Fig. 11.10). In mechanical terms, a higher force will be required to deform the capsid along the mechanical reaction coordinate enough to reach the position of the transition state. When the

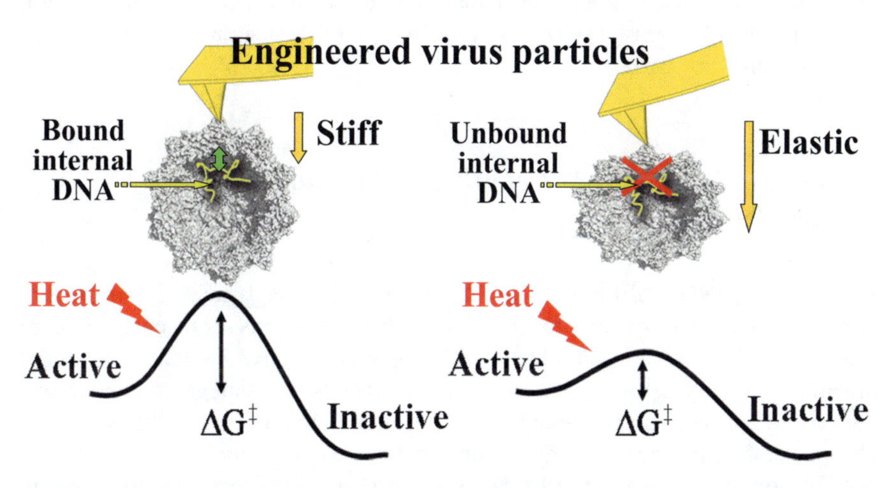

Fig. 11.10 A relationship between stiffening of the MVM virion mediated by specific ssDNA-capsid interactions, and decreased propensity for a heat-induced, virion-inactivating transition. Left, the capsid-bound ssDNA anisotropically stiffens the MVM virion and increases the free energy barrier of a heat-induced conformational transition that inactivates the virus. Right, specific removal by mutation of ssDNA-capsid interactions softens the virion and reduces the free energy barrier of the virus-inactivating transition. Figure reproduced from [22]

region involved in the transition is indented, the stiffness (k_p value) determined for the mutant capsid will be higher than that of the nonmutated capsid, as experimentally observed.

It should be emphasized that the linkage between changes in virus particle stiffness and changes in its propensity for a conformational transition will apply only when a same virus species, viral particle and particle region are considered. For different particle regions or virus species, the proportionality factor C between $\Delta G^{\ddagger}$ and k_p may be different. It cannot be predicted that, if the capsid of virus X is stiffer than that of virus Y, the former will have a lower propensity to undergo conformational rearrangements. What it could, in principle, be predicted is that if a mutation or chemical effector stiffens a particular region of the virus X capsid, its propensity to undergo a conformational rearrangement that involves that region will be reduced.

The biological outcome will of course depend on the effect of the transition on virus infectivity. If the conformational rearrangement favors viral infection (e.g., the pore-associated transition in MVM), stiffening will be associated to impaired virus survival. If the conformational rearrangement is detrimental for infection (e.g., the heat-induced, virus inactivating transition in MVM), stiffening will be associated to improved virus survival.

It must be also noted that the linkage between changes in stiffness and propensity for a conformational rearrangement may be blurred if the change in stiffness was strongly directional. However, as already discussed, such situation may not be frequently expected.

To sum up, strong evidence indicates that effector-mediated structural changes in a virus particle that either impair or promote a conformational transition will also respectively increase or decrease the stiffness of the region(s) involved in the transition, and vice versa.

Changes in Mechanical Strength and Structural Stability of Virus Particles

Effector-mediated stabilization or destabilization of a virus particle against dissociation have been analyzed in vitro using heat, chemical agents, or mechanical force (see Sect. "Modulation of the Mechanical Properties of Virus Particles"). Changes in particle stability against irreversible loss of subunits or disintegration under load (i.e., changes in mechanical strength) frequently correlated, at least qualitatively, with changes in thermal and/or chemical stability. Such a correlation could be expected based on the simple model described in Fig. 11.9. If, for example, a cementing protein introduces additional capsid intersubunit interactions, the free energy barrier for particle disruption will be higher. Additional energy should be provided to break all those extra interactions, but, in principle, the means to provide that energy (thermal, chemical, or mechanical) could be irrelevant.

However, in some cases no correspondence was found between changes in mechanical strength and changes in thermal or chemical stability against irreversible particle disruption. For example, certain mutations increased the thermal stability of the mature HIV-1 capsid protein lattice against disassembly, but they had no effect on its mechanical strength or fatigue resistance [32]. Several thermodynamic or kinetic considerations could explain the different outcomes. For example, it should be noted again that, in indentation experiments, energy is applied on a viral particle through the action of a vectorial force, which is not the case when heat or a chemical compound are used. In certain cases, the energy required to disrupt the intersubunit interactions by pushing in a single direction (i.e., the free energy barrier along the mechanical reaction coordinate) may substantially differ from the free energy barrier for thermal or chemical dissociation. Interestingly, in some of those cases, changes in mechanical strength could be more relevant than changes in thermal or chemical stability to understand capsid disassembly in vivo (i.e. when disassembly is promoted by internal pressure pushing radially on the capsid wall).

Changes in Mechanical Strength or Stiffness of Virus Particles, and Linked Changes in Structural Stability or Conformational Dynamics: A Summary

Someone could argue that, in nature, viruses may not be subjected to strong, vectorial mechanical forces, and thus their mechanical response under high load in AFM experiments may be biologically irrelevant. In fact, in most studies on virus mechanics, applied force was not used to reproduce in vitro any mechanical force that could be acting on virus particles in vivo. Instead, application of mechanical force to virus particles, and quantification of effector-mediated changes in their mechanical properties, have frequently been used to detect and/or understand changes in structure, stability or conformational dynamics related to virus infectivity. Effector-mediated changes in stiffness of a virus particle may provide a signature for changes in its equilibrium dynamics or propensity for conformational rearrangements; changes in mechanical strength and/or fatigue are related to changes in structural stability or propensity for disassembly and genome uncoating. Studies on virus mechanics are relevant to understand the biology of viruses, irrespective of whether virus particles are actually subjected to mechanical forces in vivo.

Similar considerations apply to many other studies that have investigated virus structure, properties, or function by subjecting viruses to high temperature, chemical denaturants, pH extremes, high ionic strength, or other non-physiological agents. Those agents, like mechanical force, are generally used in vitro to influence the thermodynamics or kinetics of virus particles and usually provide adequate substitutes for in vivo effectors, which are frequently unknown or poorly characterized [89].

Biological Adaptation of the Mechanical Properties of Viruses in Response to Forces Acting in Vivo

The biological relevance of many studies on virus mechanics, even if viruses were not subjected to any mechanical force in vivo, has been reviewed and justified in the previous sections. But, in addition, it must be realized that viruses in nature can be subjected to mechanical stress, in or out of infected cells. Viruses are biological entities confronted with many selective pressures for survival, and they can readily adapt to those pressures through mutation and natural selection. Thus, the specific degree of stiffness, intrinsic elasticity, mechanical strength, brittleness, or fatigue resistance exhibited by a virus particle may have been tailored by evolution to withstand, or even benefit from the forces it may encounter during the infectious cycle.

Virus particles in nature may be subjected to mechanical stress under many circumstances that, depending on virus species, include the following: (i) when exposed to desiccation or osmotic shock; (ii) by the action of shear forces, for example during circulation in the bloodstream; (iii) by deformation of the viral particle during entry into the host cell, for example during attachment of nonenveloped virions to multiple cell receptors and internalization, or during the fusion of enveloped virions and cellular membranes; (iv) as a result of squeezing during interaction with nuclear pores; (v) by the internal forces exerted on the virus capsid wall by the nucleic acid molecule packed inside; (vi) by forces generated during genome replication or transcription within the viral capsid; etc.

In addition, most virus particles may be subjected in vivo to smaller, but repeated mechanical forces that could result in material fatigue. Fatigue could be induced, for example, by frequent collisions with other (macro)molecules, especially in crowded extracellular or intracellular environments [53]; or by repeated force strokes during their intracellular transport by molecular motors [76, 87].

However, caution should be exercised before accepting that any mechanical property of a virus particle constitutes a biological adaptation in response to a mechanical force that may be exerted on that particle in vivo. In most cases, the forces a virus particle may experience in different stages of the infectious cycle have not been quantified. Some of those forces may be too weak to exert a selective pressure on a virus particle; the particle could be strong, stiff, or elastic enough based on physicochemical considerations alone. For example, cellular protein cages, or even a protein cage designed and engineered in the laboratory, showed mechanical properties that resembled those of some virus capsids. Examples include vaults, bacterial encapsulin (whose protein subunits have the same fold than phage HK97 subunits), lumazine synthase, or a non-natural protein cage. Moreover, non-natural effectors, including mutations designed in the laboratory or synthetic cargos, did modulate the mechanical properties of non-viral protein cages or virus capsids into which they were introduced [50, 51, 79, 82, 125].

Several experimental studies have carefully addressed the question of whether a mechanical property of a virus particle (e.g., a certain degree of strength or stiffness) could be the result of biological adaptation for survival in response to forces acting on that particle in vivo. The results have provided strong support for an affirmative answer to that question. Some relevant examples are reviewed in the next subsections.

Balanced Mechanical Strength as an Adaptive Trait of Tailed Bacteriophages and HSV-1 for Pressure-Driven Genome Injection in the Host Cell

Maturation strengthened and/or increased the fatigue resistance of viral particles from dsDNA phages, including HK97 (Fig. 11.3; [113]), λ [53, 117], T7 [54, 141] and P22 [64, 64, 81]. Maturation also affected the strength and/or fatigue resistance of dsDNA eukaryotic viruses, including HSV-1 particles devoid of their lipoprotein envelope (Fig. 11.1b, bottom right) [38, 111, 126].

In the viruses cited above, the presence of the stiff dsDNA molecule tightly packed inside the viral particle leads to a very high internal pressure that can reach tens of atmospheres [7, 36, 45, 52, 58, 122]. Such a high internal pressure imposes a strong selective pressure for evolving a mechanically strong capsid.

The anisotropic stiffness of the phage Φ29 capsid (Fig. 11.1a, center) indicated the presence of prestress [57] that could help resist the internal pressure exerted by the packaged dsDNA [17, 65]. Binding of cementing proteins during maturation of the phage λ capsid greatly increased its mechanical strength and fatigue resistance [53]. Likewise, binding of cementing proteins during maturation of the phage P22 capsid strengthened the capsid at the vertices, impairing the release of pentons [64, 81]. During maturation of the phage HK97 capsid ([56], Fig. 11.3), introduction of covalent bonds between capsid subunits greatly strengthened the capsid [113]. Likewise, maturation of the T7 capsid increased its resistance to fatigue [54], and the T7 virion was stronger than the mature empty capsid [141].

One might wonder why all those dsDNA phages evolved a mechanically very strong capsid, instead of reducing the internal pressure by making the capsid slightly larger, or the DNA molecule somewhat shorter. In fact, a reduced pressure in the phage λ virion impaired genome ejection and, thus, virus infectivity [8, 36, 61, 73]. The very tight packaging of the dsDNA in those viruses, and the resulting high internal pressure, appear to constitute a critical evolutionary trait. It provides a mechanism to store in the virion enough mechanical energy to drive injection of their genome during infection of the host bacterium. This feature, in turn, requires the evolution of a mechanically strong mature capsid.

During HSV-1 maturation, the virus particle becomes mechanically stronger through the binding of auxiliary proteins that strengthen the capsid, mainly at the vertices (Fig. 11.1b, bottom right), which constitute the most stressed parts of the capsid [38, 69, 117, 126]. Envelope-free HSV-1 particles enter the host cell and are transported to nuclear pores, where they translocate their genome into the cell nucleus [42]. Pressurization of the HSV-1 virion facilitates injection of its dsDNA into the nucleus trough the nuclear pore, which is required for successful infection [7, 10].

Another remarkable finding is that the portal vertex, through which the DNA molecule is ejected during infection, constitutes the mechanically weakest point in both dsDNA phages (λ, P22) and HSV-1 [8].

To sum up, strong evidence indicates that dsDNA-containing tailed phages and HSV-1 have evolved a pressurized virion to drive genome injection during infection of the host cell; a mechanically strong mature capsid able to withstand the high internal pressure required for infection; and a portal element mechanically strong enough to keep the DNA confined during virus circulation, but weak enough to allow its release during infection.

Balanced Mechanical Strength as an Adaptative Trait of AdV for Pressure-Mediated, Gradual Disassembly and Uncoating

In recent years, mechanical analysis using AFM have complemented structural, biochemical, and genetic studies [47, 48] to provide a detailed mechanochemical model of AdV maturation, disassembly and uncoating ([28], see also Chap. 10).

Maturation of AdV, contrary to that of tailed phages, is not accompanied by large structural rearrangements of the viral particle. Auxiliary viral proteins reinforce the capsid vertices where weakly bound pentons are located. The mechanism by which the dsDNA is packaged is still unclear, but recent evidence supports a co-assembly process, in which capsid subassemblies bind and help encapsidate a condensed core made of dsDNA and packaging proteins [23, 28].

DNA condensation is mediated by positively charged viral proteins that neutralize a substantial part of the negative charges in the nucleic acid molecule. During maturation, a viral protease cleaves several viral proteins on the capsid inner surface, which leads to some decondensation of the nucleoprotein core, and the generation of considerable internal pressure. AdV maturation diminishes the mechanical strength of the capsid and primes the virion for DNA release mediated by the internal pressure exerted on the capsid wall. Under these circumstances, the weakly bound pentons at the capsid vertices are gradually released (Fig. 11.4). Penton release is also modulated by host cell proteins, such as integrins and defensins [124]. Loss of pentons leads to capsid cracking, a gradual exposure of the nucleoprotein core through a cascade of dismantling events and, finally, delivery of the viral genome into the cell nucleus [28, 31, 55, 86, 96–98, 100].

The pressure exerted on the capsid wall by the decondensed nucleic acid core may not be the only biologically relevant mechanical force acting on the AdV virion in the infected cell. During cytoplasmic transport of the virion on microtubules by a cellular motor, weak but frequent force strokes could contribute to the controlled release of pentons, facilitating the gradual disassembly required for genome uncoating [87].

To sum up, strong evidence indicates that AdV has evolved a mechanically finely balanced capsid. The AdV capsid is mechanically strong enough to maintain virion integrity during the extracellular stage, but weak enough to allow its pressure-induced gradual disassembly to release the viral genome into the cell nucleus.

Mechanical Softening as an Adaptive Trait of HIV-1 to Enable Virus Entry into the Host Cell

The immature HIV-1 virion is unable to enter the host cell and is non-infectious. During maturation, the virion undergoes a dramatic structural rearrangement and becomes infectious (Fig. 11.1c, right). Infection by HIV-1 involves the internalization of the virus core (the mature capsid containing the viral RNP complex), which occurs through fusion of the virion lipid envelope and the host cell membrane. Membrane fusion is an energetically costly process that involves a mechanical pulling force that results from a dramatic conformational rearrangement of the Env protein, embedded in the viral envelope. As a result, the viral and cellular membranes are gradually deformed and brought closer together, eventually completing their fusion.

Indentation experiments revealed that the stiffness of the mature HIV-1 virion is dramatically reduced (by 14-fold) compared to the immature virion. Immature HIV-1 virions in which the Ct domain of Env had been deleted were nearly as soft as mature virions and were also competent for membrane fusion and cell entry [72]. Immature HIV-1 virions containing the Env Ct domain, but lacking the rest of the Env protein, were as stiff as the intact immature virion, and incompetent for cell entry [99]. In carefully designed experiments, immature HIV-1 virions that incorporated a fixed amount of Env protein missing the Ct domain (to enable cell entry) and increasing amounts of the Env Ct domain (lacking the rest of the Env protein) were tested for stiffness and efficiency for entry into cells. The results revealed a neat correlation between Env Ct-mediated stiffening of the HIV-1 virion, and loss of competence for cell entry [99]. The authors suggested, as a likely mechanism, that the decreased stiffness of the HIV-1 virion during maturation would lower the free energy barrier to deformation of the viral envelope during its fusion with the cell membrane [99, 116].

To sum up, strong evidence indicates that the reduced stiffness (high deformability) of the mature HIV-1 virion may be a biological adaptation that facilitates the mechanically driven fusion of the viral and cellular membranes during infection.

Ligand-Mediated Modulation of Mechanical Strength as an Adaptive Trait of HIV-1 for Efficient Reverse Transcription of the Viral Genome and Controlled Capsid Disassembly

Recent evidence indicates that, during infection by HIV-1, the mature capsid (Fig. 11.1c, far right) is used as a container for endogenous reverse transcription (ERT) of the viral RNA genome. Binding of the cellular factor IP6 stabilizes the capsid and increases ERT efficiency. In turn, as ERT progresses the capsid is destabilized, leading to genome uncoating [2].

Indentations on the HIV-1 core were performed as ERT progressed, either in the absence or presence of bound IP6. The results revealed a transient stiffening of the HIV-1 core. Peaks of stiffness of the HIV-1 core bound to IP6 during ERT were temporally associated to specific stages of viral DNA synthesis in the core. ERT-mediated capsid stiffness peaks were strongly suggestive of the action of transient point forces that would push on the capsid wall. Further incubation led to softening of the core, eventual disassembly and uncoating [2, 104, 106].

The ensemble of these and other results led the authors to propose a model in which ERT-generated, transient mechanical forces induce, in several discrete steps, localized cracks in the mature HIV-1 capsid. As ERT is completed, the accumulated cracks lead to partial disruption of the capsid [2, 106]. This model implies that IP6 binding to the mature HIV-1 capsid may constitute an adaptive trait that (among other effects) modulates capsid strength to optimize virus infectivity.

To sum up, the mechanical strength of the IP6-bound HIV-1 capsid may have been adjusted through evolution to preserve its integrity for efficient ERT, but also to allow its mechanically induced disruption for uncoating as the reverse transcription process is completed [2, 116].

Anisotropic Stiffness/Conformational Dynamics as an Adaptive Trait of MVM to Impair a Heat-Induced Deleterious Transition Without Impairing an Infectivity-Determining Transition

The MVM capsid is quite resistant to thermal, chemical, or mechanical disruption, but it is conformationally highly dynamic [115, 134]. During MVM morphogenesis in vivo, the ssDNA genome is packaged in a previously assembled capsid by a mechanism that may involve a viral motor bound to a capsid S5 region. As the ssDNA molecule is internalized through a capsid pore, the N-terminal (Nt) segments of many VP2 capsid subunits are externalized through other pores. The externalized VP2 Nt segments provide a signal for nuclear export of the mature virion and are required for infectivity. In vitro, the VP2 Nts can be externalized in the absence of the viral DNA by moderate heating or addition of a chemical denaturant [134]. VP2 Nt externalization is associated to a subtle change in capsid conformation [107, 135]. The results suggest that, as the viral ssDNA molecule is encapsidated in vivo and fills the capsid, it exerts enough mechanical force to push the VP2 Nts out of the capsid through S5 pores.

Uncoating of the ssDNA genome in the host cell occurs also through a capsid pore and involves a different transition that, in vitro, can be promoted by moderate heating [114] or mechanical force (Strobl and de Pablo, personal communication). During propagation between hosts, the virion may encounter relatively high temperatures that will promote that transition. In addition, frequent molecular collisions, especially in macromolecularly crowded environments [53], could lower the temperature required to induce the transition. The resulting premature release of the viral genome will lead to virus inactivation.

The MVM virion is, thus, under antagonic selective pressures: On one hand, it must be conformationally stable enough to impair the transition that leads to the release of its genome out of the host cell and virus inactivation; on the other hand, it must be conformationally labile enough to facilitate the transition associated to externalization of VP2 Nts (and other signals) required for infectivity.

A model on how MVM may have biologically adapted to those conflicting demands is strongly supported by the results reviewed in the subsection "Linked Changes in Capsid Stiffness and Propensity for Conformational Rearrangements". MVM appears to have evolved, by positive selection, ssDNA-binding sites at the capsid inner wall, close to S2 and S3 regions but not to S5 regions (Fig. 11.7). The capsid-bound ssDNA segments act like molecular buttresses that mechanically stiffen/impair the conformational dynamics of the capsid S2/S3 regions. The heat/force-induced transition that would lead to untimely DNA release is, thus, impaired, increasing the chances of virion survival during propagation between hosts. At the same time, negative selection may have kept the capsid S5 regions around the pores free of bound DNA, thus maintaining their low stiffness/high conformational dynamism. The transition that is required for externalization of VP2 Nt signals through the capsid pores (mechanically induced by the encapsidating ssDNA) is not impaired, and virus infectivity is preserved.

To sum up, strong evidence indicates that the anisotropic distribution of stiffness/conformational dynamism in the MVM virion may be the result of biological adaptation. Some capsid regions are stiff enough to impair a heat-induced virus inactivation associated to untimely viral nucleic acid release; other capsid regions are soft enough to allow, during genome encapsidation, a mechanically-induced conformational rearrangement required for virus infectivity [22, 88].

Conclusions and Perspectives

An important goal of mechanical virology is the discovery of physics-based principles that determine common features in the mechanical behavior of viruses. However, understanding the structural bases and biological relevance of virus-specific differences in mechanical behavior constitutes an equally important goal.

Some viruses may have totally different evolutionary origins. Moreover, viruses have uniquely adapted through different evolutionary pathways to almost any environment in the Earth´s Biosphere. Many biological, biochemical, or biophysical studies over many decades have discovered striking structural and mechanistic differences between virus species, even between mutants of a same species. Likewise, recent AFM-based studies have revealed that the mechanical properties of virus particles can be strongly dependent on small-scale (down to atomic-level) structural details. Virus-specific responses to mechanical force under load are not anecdotal. They likely reflect profound adaptive solutions for survival in the face of different selective pressures in nature.

In most AFM-based studies on virus mechanics so far, the mechanical forces applied on virus particles were not intended to mimic in vitro the mechanical forces that could be acting on those particles in vivo. Changes in the mechanical behavior of virus particles in response to virus infectivity-determining effectors were probed to explore structural modifications, conformational transitions and physicochemical mechanisms that control different stages of the viral cycle: virion morphogenesis, maturation, exit from the host cell, survival in the extracellular environment, entry into a new cell, intracellular trafficking, disassembly and genome uncoating. The results already obtained using the above-mentioned approach have proven to be of the utmost relevance to understand virus biology, irrespective of whether virus particles are subjected to mechanical forces in nature.

A unifying model based on simple structural, thermodynamic, and kinetic considerations may justify the observed associations between effector-mediated changes in stability or conformational dynamics of virus particles, and changes in their mechanical behavior. Effector-mediated changes in mechanical strength and/or fatigue are related to changes in structural stability or propensity for disassembly and genome uncoating. Changes in stiffness provide signatures for changes in equilibrium dynamics and/or propensity for conformational rearrangements; effector-mediated changes in equilibrium dynamics or stiffness are manifestations of the same physical phenomenon.

Some AFM-based studies have addressed the important question of whether the degree of mechanical strength or stiffness exhibited by a virus particle constitutes a biologically adaptive trait, selected in response to mechanical forces acting on that particle in vivo. Several studies so far have provided strong support for the biological tuning of the mechanical strength or stiffness of specific viruses confronted with internal or external mechanical forces during some stage of their life cycle. The high mechanical strength exhibited by mature dsDNA-containing virions may be the result of biological adaptation for pressure-driven genome injection in the host cell (tailed phages, HSV-1), or gradual disassembly and uncoating (AdV). The mechanical strength of the HIV-1 core may be modulated by a bound ligand for efficient endogenous reverse transcription of its genome and controlled nucleic acid release. The low stiffness (high deformability) exhibited by the mature HIV-1 virion may be the result of biological adaptation to facilitate entry into the host cell. The anisotropic stiffness/conformational dynamics exhibited by the MVM virion may be the result of adaptation to impair a heat-induced deleterious transition, without impairing an infectivity-determining transition.

Research on the mechanical responses of different viruses may also constitute a key for developing novel strategies to combat specific viral diseases, or for choosing a specific virus particle to develop a particular bio/nanotechnological application [90]. Some small organic compounds inhibit virus infectivity by modifying the stiffness or mechanical strength of the virus particles [33, 105, 133, 143]. This discovery may lead to the development of new antiviral drugs that modify mechanical properties of virus particles related to infection-determining mechanisms. Also, the discovery that, during maturation, some viruses are mechanically strengthened by intersubunit covalent bonds inspired the engineering of a mechanically improved protein material. A complete network of intersubunit disulfide bonds were genetically introduced in the HIV-1 capsid protein lattice. The result was a chain mail-like nanosheet with genetically improved thermostability, and resistance to mechanical stress and fatigue, with no unwanted stiffening [32].

Acknowledgements I gratefully acknowledge Drs. P.J. de Pablo, J. Gómez-Herrero, C. Carrasco, M. Hernando-Pérez, W.H. Roos, N. Verdaguer, J.R. Castón and A. Valbuena for advice, discussion and/or collaboration related to virus mechanics. I also gratefully acknowledge former or current researchers in my group including Drs. A. Valbuena, M. Castellanos, R. Pérez, P.J.P. Carrillo, M. Medrano and S. Domínguez-Zotes, for excellent work on virus mechanics using AFM, current predoctoral students J. Escrig and L. Valiente for AFM and other biophysical work on viruses, and M.A. Fuertes and A. Rodríguez-Huete for great technical assistance. The author is an associate member of the Institute for Biocomputation and Physics of Complex Systems, Zaragoza, Spain. This study was funded by grants from MICINN/FEDER EU (Spain, RTI2018-096635-B-I00 and PID2021-126973OB-I00) to M.G.M., and institutional support from the Ramon Areces Foundation.

References

1. Agbandje-McKenna M, Llamas-Saiz AL, Wang F, Tattersall P, Rossmann MG (1998) Functional implications of the structure of the murine parvovirus, minute virus of mice. Structure 6:1369–1381
2. Aiken C, Rousso I (2021) The HIV-1 capsid and reverse transcription. Retrovirology 18:20
3. Alsteens D, Pesavento E, Cheuvart G, Dupres V, Trabelsi H, Soumillion P, Dufrêne YF (2009) Controlled manipulation of bacteriophages using single-virus force spectroscopy. ACS Nano 3:3063–3068
4. Arkhipov A, Roos WH, Wuite GJL, Schulten K (2009) Elucidating the mechanism behind irreversible deformation of viral capsids. Biophys J 97:2061–2069

5. Azinas S, Bano F, Torca I, Bamford DH, Schwartz GA, Esnaola J, Oksanen HM, Richter RP, Abrescia NG (2018) Membrane-containing virus particles exhibit the mechanics of a composite material for genome protection. Nanoscale 10:7769–7779
6. Baclayon M, Shoemaker GK, Uetrecht C, Crawford SE, Estes MK, Prasad BVV, Heck AJR, Wuite GJL (2011) Roos WH 2011: Prestress strengthens the shell of Norwalk virus nanoparticles. Nano Lett 11:4865–4869
7. Bauer DW, Huffman JB, Homa FL, Evilevitch A (2013) Herpes virus genome, the pressure is on. J Am Chem Soc 135:11216–11221
8. Bauer DW, Li D, Huffman J, Homa FL, Wilson K, Leavitt JC, Casjens SR, Baines J, Evilevitch A (2015) Exploring the balance between DNA pressure and capsid stability in herpes viruses and phages. J Virol 89:9288–9298
9. Bothner B, Hilmer JK (2011) Probing viral capsids in solution. In: Agbandje-McKenna M, McKenna R (eds) Structural virology. RSC Publishing, Cambridge, pp 41–61
10. Brandariz-Nuñez A, Liu T, Du T, Evilevitch A (2019) Pressure-driven release of a viral genome into a host nucleus is a mechanism leading to herpes infection. eLife 8:e47212
11. Brandariz-Nuñez A, Robinson SJ, Evilevitch A (2020) Pressurized state inside herpes capsids-a novel antiviral target. PLoS Pathog 16:e1008604
12. Bustamante C, Chemla YR, Forde NR, Izhaky D (2004) Mechanical processes in biochemistry. Annu Rev Biochem 73:705–748
13. Buzón P, Maity S, Roos WH (2020) Physical virology: from virus self-assembly to particle mechanics. WIREs Nanomed Nanobiotechnol 12:e1613
14. Cardoso-Lima R, Noronha Souza PF, Florindo Guedes MI, Santos-Oliveira R, Rebelo Alencar LM (2021) SARS-CoV-2 unrevealed: ultrastructural and nanomechanical analysis. Langmuir 37:10762–10769
15. Carrasco C, Carreira A, Schaap I, Serena P, Gómez-Herrero J, Mateu MG, de Pablo PJ (2006) DNA-mediated anisotropic mechanical reinforcement of a virus. Proc Natl Acad Sci USA 103:13706–13711
16. Carrasco C, Castellanos M, de Pablo PJ, Mateu MG (2008) Manipulation of the mechanical properties of a virus by protein engineering. Proc Natl Acad Sci USA 105:4150–4155
17. Carrasco C, Luque A, Hernando-Pérez M, Miranda R, Carrascosa JL, Serena PA, de Ridder M, Raman A, Gómez-Herrero J, Schaap IAT, Reguera D, de Pablo PJ (2011) Built-in mechanical stress in viral shells. Biophys J 100:1100–1108
18. Carrillo PJP, Medrano M, Valbuena A, Rodríguez-Huete A, Castellanos M, Pérez R, Mateu MG (2017) Amino acid side chains buried along intersubunit interfaces in a viral capsid preserve low mechanical stiffness associated with virus infectivity. ACS Nano 11:2194–2208
19. Carrión-Vázquez M, Li H, Lu H, Marszalek PE, Oberhauser AF, Fernandez JM (2003) The mechanical stability of ubiquitin is linkage-dependent. Nat Struct Biol 10:738–743
20. Castellanos M, Pérez R, Carrasco C, Hernando-Pérez M, Gómez-Herrero J, de Pablo PJ, Mateu MG (2012) Mechanical elasticity as a physical signature of conformational dynamics in a virus particle. Proc Natl Acad Sci USA 109:12028–12033
21. Castellanos M, Pérez R, Carrillo PJP, de Pablo PJ, Mateu MG (2012) Mechanical diassembly of single virus particles reveals kinetic intermediates predicted by theorym. Biophys J 102:2615–2624
22. Castellanos M, Carrillo PJP, Mateu MG (2015) Quantitatively probing propensity for structural transitions in engineered virus nanoparticles by single-molecule mechanical analysis. Nanoscale 7:5654–5664
23. Condezo GN, San Martín C (2017) Localization of adenovirus morphogenesis players, together with visualization of assembly intermediates and failed products, favor a model where assembly and packaging occur concurrently at the periphery of the replication center. PLoS Pathog 13:e1006320
24. Cuellar JL, Meinhoevel F, Hoehne M, Donath E (2010) Size and mechanical stability of norovirus capsids depend on pH: a nanoindentation study. J Gen Virol 91:2449–2456
25. de Pablo PJ (2018) Atomic force microscopy of virus shells. Sem Cell Dev Biol 73:199–208

26. de Pablo PJ (2019) The application of atomic force microscopy for viruses and protein shells: imaging and spectroscopy. Adv Virus Res 105:161–187
27. de Pablo PJ, Mateu MG (2013) Mechanical properties of viruses. In: Mateu MG (ed) Structure and physics of viruses. Springer, Dordrecht, The Netherlands, pp 519–551
28. de Pablo PJ, San Martín C (2022) Seeing and touching adenovirus: complementary approaches for understanding assembly and disassembly of a complex virion. Curr Opin Virol 52:112–122
29. de Pablo PJ, Hernando-Pérez M, Carrasco C, Carrascosa JL (2018) Direct visualization of single virus restoration after damage in real time. J Biol Phys 44:225–235
30. del Álamo M, Rivas G, Mateu MG (2005) Effect of macromolecular crowding agents on human immunodeficiency virus type 1 capsid protein assembly in vitro. J Virol 79:14271–14281
31. Denning D, Bennet S, Mullen T, Moyer C, Vorselen D, Wuite GJL, Nemerow G, Roos WH (2019) Maturation of adenovirus primes the protein nanoshell for successful endosomal escape. Nanoscale 11:4015–4024
32. Domínguez-Zotes S, Fuertes MA, Rodríguez-Huete A, Valbuena A, Mateu MG (2022a) A genetically engineered, chain mail-like nanostructured protein material with increased fatigue resistance and enhanced self-healing, Small 2105456
33. Domínguez-Zotes S, Valbuena A, Mateu MG (2022b) Antiviral compounds modulate elasticity, strength and material fatigue of a virus capsid framework. Biophys J 121:919–931
34. Dragnea B (2022) Viruses: a physical chemistry perspective. J Phys Chem B 126:4411–4414
35. Eghiaian F, Schaap IA, des Georges A, Skehel JJ, Veigel C (2009) The influenza virus mechanical properties are dominated by its lipid envelope. Biophys J 96:15a
36. Evilevitch A, Lavelle L, Knobler CM, Raspaud E, Gelbart WM (2003) Osmotic pressure inhibition of DNA ejection from phage. Proc Natl Acad Sci USA 100:9292–9295
37. Evilevitch A, Roos WH, Ivanovska IL, Jeembaeva M, Jönsson B, Wuite GJL (2011) Effects of salt on internal DNA pressure and mechanical properties of phage capsids. J Mol Biol 405:18–23
38. Evilevitch A, Sae-Ueng U (2022) Mechanical capsid maturation facilitates the resolution of conflicting requirements for herpesvirus assembly. J Virol 96:e01831-e1921
39. Falvo MR, Washburn S, Superfine R, Finch M, Brooks FP Jr, Chi V, Taylor RM II (1997) Manipulation of individual viruses: friction and mechanical properties. Biophys J 72:1396–1403
40. Ferreon AC, Deniz AA (2011) Protein folding at single-molecule resolution, Biochim Biophys Acta 1814:1021–1029
41. Fisher TE, Marszalek PE, Fernandez JM (2000) Stretching single molecules into novel conformations using the atomic force microscope. Nat Struct Biol 7:719–724
42. Flatt JW, Greber UF (2017) viral mechanisms for docking and delivering at nuclear pore complexes. Sem Cell Dev Biol 68:59–71
43. Forman JR, Clarke J (2007) Mechanical unfolding of proteins: insights into biology, structure and folding. Curr Opin Struct Biol 17:58–66
44. Freeman KG, Huffman JB, Homa FL, Evilevitch A (2021) UL25 capsid binding facilitates mechanical maturation of the herpesvirus capsid and allows retention of pressurized DNA. J Virol 95:e00755-e821
45. Gelbart WM, Knobler CM (2009) Pressurized viruses. Science 323:1682–1683
46. Greber UF (2014) How cells tune viral mechanics-insights from biophysical measurements of influenza virus. Biophys J 106:2317–2321
47. Greber UF (2016) Virus and host mechanics support membrane penetration and cell entry. J Virol 90:3802–3805
48. Greber UF, Suomalainen M (2022) Adenovirus entry: stability, uncoating and nuclear import. Mol Microbiol 00:1–12
49. Guerra P, Valbuena A, Querol-Audí J, Silva C, Castellanos M, Rodríguez-Huete A, Garriga D, Mateu MG, Verdaguer N (2017) Structural basis for biologically relevant mechanical stiffening of a virus capsid by cavity-creating or spacefilling mutations. Sci Rep 7:4101

50. Guerra P, González-Alamos M, Llauró A, Casañas A, Querol-Audí J, de Pablo PJ, Verdaguer N (2022) Symmetry disruption commits vault particles to disassembly. Sci Adv 8:eabj7795
51. Heinze K, Sasaki E, King NP, Baker D, Hilvert D, Wuite GJL, Roos WH (2016) Protein nanocontainers from nonviral origin: testing the mechanics of artificial and natural protein cages by AFM. J Phys Chem B 120:5945–5952
52. Hernando-Pérez M, Miranda R, Aznar M, Carrascosa JL, Schaap IAT, Reguera D, de Pablo PJ (2012) Direct measurement of phage phi29 stiffness provides evidence of internal pressure. Small 8:2366–2370
53. Hernando-Pérez M, Lambert S, Nakatani-Webster E, Catalano CE, de Pablo PJ (2014) Cementing proteins provide extra mechanical stabilization to viral cages. Nat Comm 5:4520
54. Hernando-Pérez M, Pascual E, Aznar M, Ionel A, Castón JR, Luque A, Carrascosa JL, Reguera D, de Pablo PJ (2014) The interplay between mechanics and stability of viral cages. Nanoscale 6:2702–2709
55. Hernando-Pérez M, Martín-González N, Pérez-Illana M, Suomalainen M, Condezo G, Ostapchuk P, Gallardo J, Menéndez M, Greber UF, Hearing P, de Pablo PJ, San Martín C (2020) Dynamic competition for hexon binding between vore protein VII and lytic protein VI promotes adenovirus maturation and entry. Proc Natl Acad Sci USA 117:13699–13707
56. Huang RK, Khayat R, Lee KK, Gertsman I, Duda RL, Hendrix RW, Johnson JE (2011) The prohead-I structure of bacteriophage HK97: implications for scaffold-mediated control of particle assembly and maturation. J Mol Biol 408:541–544
57. Ivanovska IL, de Pablo PJ, Ibarra B, Sgalari G, MacKintosh FC, Carrascosa JL, Schmidt CF, Wuite GJL (2004) Bacteriophage capsids: tough nanoshells with complex elastic properties. Proc Natl Acad Sci USA 101:7600–7605
58. Ivanovska IL, Wuite G, Jönsson B, Evilevitch A (2007) Internal DNA pressure modifies stability of wt phage. Proc Natl Acad Sci USA 104:9603–9608
59. Ivanovska IL, Miranda R, Carrascosa JL, Wuite GJL, Schmidt CF (2011) Discrete fracture patterns of virus shells reveal mechanical building blocks. Proc Natl Acad Sci USA 108:12611–12616
60. Javadi Y, Fernandez JM, Perez Jimenez R (2013) Protein folding under mechanical forces: a physiological view. Physiology (Bethesda) 28:1–17
61. Jeembaeva M, Castelnovo M, Larsson F, Evilevitch A (2008) Osmotic pressure: resisting or promoting DNA ejection from phage? J Mol Biol 381:310–323
62. Jiménez-Zaragoza M, Yubero MPL, Martín-Forero E, Castón JR, Reguera D, Luque D, de Pablo PJ, Rodríguez JM (2018) Biophysical properties of single rotavirus particles account for the functions of protein shells in a multilayered virus. eLife 7:e37295
63. Johnson JE (2003) Virus particle dynamics. Adv Protein Chem 64:197–218
64. Kant R, Llauró A, Rayaprolu V, Qazi S, de Pablo PJ, Douglas T, Bothner B (2018) Changes in the stability and biomechanics of P2 bacteriophage capsid during maturation. Biochem Biophys Acta 1862:1492–1504
65. Keller N, Berndsen ZT, Jardine PJ, Smith DE (2017) Experimental comparison of forces resisting viral DNA packaging and driving DNA ejection. Phys Rev E 95:052408
66. Kiss B, Mudra D, Török G, Mártonfalvi Z, Csík G, Herényi L, Kellermayer M (2020) Single-particle virology. Biophys Rev 12:1141–1154
67. Kiss B, Kis Z, Pályi B, Kellermayer MSZ (2021) Topography, spike dynamics, and nanomechanics of individual native SARS-CoV-2 virions. Nano Lett 21:2675–2680
68. Klug WS, Bruinsma RF, Michel J-P. Knobler CM, Ivanovska IL, Schmidt CF, Wuite GJL (2006) Failure of viral shells. Phys Rev Lett 97:228101
69. Klug WS, Roos WH, Wuite GJL (2012) unlocking internal prestress from protein nanoshells. Phys Rev Lett 109:168104
70. Knobler CM, Gelbart WM (2009) Physical chemistry of DNA viruses. Annu Rev Phys Chem 60:367–383
71. Kol N, Gladnikoff M, Barlam D, Shneck RZ, Rein A, Rousso I (2006) Mechanical properties of murine leukemia virus particles: effect of maturation. Biophys J 91:767–774

72. Kol N, Shi Y, Tsvitov M, Barlam D, Shneck RZ, Kay MS, Rousso I (2007) A stiffness switch in human immunodeficiency virus. Biophys J 92:1777–1783
73. Köster S, Evilevitch A, Jeembaeva M, Weitz DA (2009) Influence of internal capsid pressure on viral infection by phage λ. Biophys J 97:1525–1529
74. Li S, Eghiaian F, Sieben C, Herrmann A, Schaap IA (2011) Bending and puncturing the influenza virus envelope. Biophys J 100:637–645
75. Li S, Sieben C, Ludwig K, Höfer CT, Chiantia S, Hermann A, Eghiaian F, Schaap IAT (2014) pH-controlled two-step uncoating of influenza virus. Biophys J 106:1447–1456
76. Liashkovich I, Hafezi W, Kühn JE, Oberleithner H, Kramer A, Shahin V (2008) Exceptional mechanical and structural stability of HSV-1 unveiled with fluid atomic force microscopy. J Cell Sci 121:2287–2292
77. Lidmar J, Mirny L, Nelson DR (2003) Virus shapes and buckling transitions in spherical shells. Phys Rev E 68:051910
78. Liu C, Perilla JR, Ning J, Lu M, Hou G, Ramalho R, Himes BA, Zhao G, Bedwell GJ, Byeon I-J, Ahn J, Gronenborn AM, Prevelige PE, Rousso I, Aiken C, Polenova T, Schulten K, Zhang P (2016) Cyclophilin A stabilizes the HIV-1 capsid through a novel non-canonical binding site. Nat Comm 7:10714
79. Llauró A, Guerra P, Irigoyen N, Rodríguez JF, Verdaguer N, de Pablo PJ (2014) Mechanical stability and reversible fracture of vault particles. Biophys J 106:687–695
80. Llauró A, Coppari E, Imperatori F, Bizzarri AR, Castón JR, Santi L, Cannistraro S, de Pablo PJ (2015) Calcium ions modulate the mechanics of tomato bushy stunt virus. Biophys J 109:390–397
81. Llauró A, Schwarz B, Koliyatt R, de Pablo PJ, Douglas T (2016) Tuning capsid nanoparticle stability with symmetrical morphogenesis. ACS Nano 10:8465–8473
82. Llauró A, Guerra P, Kant R, Bothner B, Verdaguer N, de Pablo PJ (2016) Decrease in pH destabilizes individual vault nanocages by weakening the inter-protein lateral interaction. Sci Rep 6:34143
83. Luque A, Reguera D (2013) Theoretical studies on assembly, physical stability and dynamics of viruses. In: Mateu MG (ed) Structure and physics of viruses. Springer, Dordrecht, The Netherlands, pp 553–5595
84. Luque D, Ortega-Esteban A, Valbuena A, Vilas JL, Rodríguez-Huete A, Mateu MG, Castón JR (2022) Equlibrium dynamics of a biomolecular complex analyzed at single amino acid resolution by cryo-electron microscopy. J Mol Biol 435:168024
85. Marchetti M, Wuite GJL, Roos WH (2016) Atomic force microscopy observation and characterization of single virions and virus-like particles by nano-indentation. Curr Opin Virol 18:82–88
86. Martín-González N, Hernando-Pérez M, Condezo GN, Pérez-Illana M, Siber A, Reguera D, Ostapchuk P, Hearing P, San Martín C, de Pablo PJ (2019) Adenovirus major core protein condenses DNA in clusters and bundles, modulating genome release and capsid internal pressure. Nucleic Acids Res 47:9231–9242
87. Martín-González N, Ibáñez-Freire P, Ortega-Esteban A, Laguna-Castro M, San Martín C, Valbuena A, Delgado-Buscalioni R, de Pablo PJ (2021) Long-range cooperative disassembly and aging during adenovirus uncoating. Phys Rev X 11:021025
88. Mateu MG (2012) Mechanical properties of viruses analyzed by atomic force microscopy: a virological perspective. Virus Res 168:1–22
89. Mateu MG (ed) (2013) Structure and Physics of Viruses. Springer, Dordrecht, The Netherlands
90. Mateu MG (2017) Assembly, engineering and applications of virus-based protein nanoparticles. In: Cortajarena AL Grove TZ (eds) Protein-based Engineered Nanostructures, Adv Exp Med Biol Vol. 940, Switzerland 2017, Springer, pp 83–120
91. Medrano M, Valbuena A, Rodríguez-Huete A, Mateu MG (2019) Structural determinants of mechanical resistance against breakage of a virus-based protein nanoparticle at a resolution of single amino acids. Nanoscale 11:9369–9383
92. Mertens J, Casado S, Mata CP, Hernando-Pérez M, de Pablo PJ, Carrascosa JL, Castón JR (2015) A protein with simultaneous capsid scaffolding and dsRNA binding activities enhances the birnavirus capsid mechanical stability. Sci Rep 5:13486

93. Michel JP, Ivanovska IL, Gibbons MM, Klug WS, Knobler CM, Wuite GJL, Schmidt CF (2006) Nanoindentation studies of full and empty viral capsids and the effects of capsid protein mutations on elasticity and strength. Proc Natl Acad Sci USA 103:6184–6189
94. Nguyen TT, Bruinsma RJ, Gelbart WM (2005) Elasticity theory and shape transitions of viral shells. Phys Rev E 72:051923
95. Oberhauser AF, Carrión-Vázquez M (2008) Mechanical biochemistry of proteins one molecule at a time. J Biol Chem 283:6617–6621
96. Ortega-Esteban A, Pérez-Berná AJ, Menéndez-Conejero R, Flint SJ, San Martín C, de Pablo PJ (2013) Monitoring dynamics of human adenovirus disassembly induced by mechanical fatigue. Sci Rep 3:1434
97. Ortega-Esteban A, Bodensiek K, San Martín C, Suomalainen M, Greber UF, de Pablo PJ, Schaap IAT (2015) Fluorescence tracking of genome release during mechanical unpacking of single viruses. ACS Nano 9:10571–10579
98. Ortega-Esteban A, Condezo GN, Pérez-Berná AJ, Chillón M, Flint SJ, Reguera D, San Martín C, de Pablo PJ (2015) Mechanics of viral chromatin reveals the pressurization of human adenovirus. ACS Nano 11:10826–10833
99. Pang H-B, Hevroni L, Kol N, Eckert DM, Tsvitov M, Kay MS, Rousso I (2013) Virion stiffness regulates mature HIV-1 entry. Retrovirology 10:4
100. Pérez-Berná AJ, Ortega-Esteban A, Menéndez-Conejero R, Winkler DC, Menéndez M, Steven AC, Flint SJ, de Pablo PJ, San Martín C (2012) The role of capsid maturation on adenovirus priming for sequential uncoating. J Biol Chem 287:31582–31595
101. Pérez-Illana M, Martín-González N, Hernando-Pérez M, Condezo GN, Gallardo J, Menéndez M, San Martín C, de Pablo PJ (2021) Acidification induces condensation of the adenovirus core. Acta Biomater 135:534–542
102. Ramalho R, Rankovic S, Zhou J, Aiken C, Rousso I (2016) Analysis of the mechanical properties of wild type and hyperstable mutants of the HIV-1 capsid. Retrovirology 13:17
103. Rankl C, Kienberger F, Wildling L, Wruss J, Gruber HJ, Blaas D, Hinterdorfer P (2008) Multiple receptors involved in human rhinovirus attachment to live cells. Proc Natl Acad Sci USA 105:17778–17783
104. Rankovic S, Varadarajan J, Ramalho R, Aiken C, Rousso I (2017) Reverse transcription mechanically initiates HIV-1 capsid disassembly. J Virol 91:e00289-e317
105. Rankovic S, Ramalho R, Aiken C, Rousso I (2018) PF74 reinforces the HIV-1 capsid to impair reverse transcription-induced uncoating. J Virol 92:e00845-e918
106. Rankovic S, Deshpande A, Harel S, Aiken C, Rousso I (2021) HIV-1 uncoating occurs via a series of rapid biomechanical changes in the core related to individual stages of reverse transcription. J Virol 95:e00166-e221
107. Reguera J, Carreira A, Riolobos L, Almendral JM, Mateu MG (2004) Role of interfacial amino acid residues in assembly, stability and conformation of a spherical virus capsid. Proc Natl Acad Sci USA 101:2724–2729
108. Reguera J, Grueso E, Carreira A, Sánchez-Martínez C, Almendral JM, Mateu MG (2005) Functional relevance of amino acid residues involved in interactions with ordered nucleic acid in a spherical virus. J Biol Chem 280:17969–17977
109. Roos WH, Wuite GJL (2009) Nanoindentation studies reveal material properties of viruses. Adv Mater 21:1187–1192
110. Roos WH, Ivanovska IL, Evilevitch A, Wuite GJL (2007) Viral capsids: mechanical characteristics, genome packaging and delivery mechanisms. Cell Mol Life Sci 64:1484–1497
111. Roos WH, Radtke K, Kniesmeijer E, Geertsema H, Sodeik B, Wuite GJL (2009) Scaffold expulsion and genome packaging trigger stabilization of herpes simplex virus capsids. Proc Natl Acad Sci USA 106:9673–9678
112. Roos WH, Bruinsma R, Wuite GJL (2010) Physical Virology. Nat Phys 6:733–743
113. Roos WH, Gertsman I, May ER, Brooks CL III, Johnson JE, Wuite GJL (2012) Mechanics of bacteriophage maturation. Proc Natl Acad Sci USA 109:2342–2347
114. Ros C, Baltzer C, Mani B, Kempf C (2006) Parvovirus uncoating in vitro reveals a mechanism of DNA release without capsid disassembly and striking differences in encapsidated DNA stability. Virology 345:137–147

115. Ros C, Bayat N, Wolfisberg R, Almendral JM (2017) Protoparvovirus cell entry. Viruses 9:313
116. Rousso I, Deshpande A (2022) Applications of atomic force microscopy in HIV-1 research. Viruses 14:648
117. Sae-Ueng U, Liu T, Catalano CE, Huffman JB, Homa FL, Evilevitch A (2014) Major capsid reinforcement by a minor protein in herpesviruses and phage. Nucleic Acids Res 42:9096–9107
118. Sae-Ueng U, Bhunchoth A, Phironrit N, Treetong A, Sapcharoenkun C, Chatchawankanphanich O, Leartsakulpanich U, Chitnumsub P (2020) C22 podovirus infectivity is associated with intermediate stiffness. Sci Rep 10:12604
119. Sae-Ueng U, Bhunchoth A, Phironrit N, Treetong A, Sapcharoenkun C, Chatchawankanphanich O, Leartsakulpanich U, Chitnumsub P (2022) Thermoresponsive C22 phage stiffness modulates the phage infectivity. Sci Rep 12:13001
120. Schaap IAT, Eghiaian F (2012) des georges A, Veigel C: Effect of envelope proteins on the mechanical properties of influenza virus. J Biol Chem 287:41078–41088
121. Sherman MB, Smith HQ, Smith TJ (2020) The dynamic life of virus capsids. Viruses 12:618
122. Smith DE, Tans SJ, Smith SB, Grimes S, Anderson DL, Bustamante C (2001) The bateriophage Φ29 portal motor can package DNA against a large internal force. Nature 413:748–752
123. Snijder J, Uetrecht C, Rose RJ, Sanchez-Eugenia R, Marti GA, Agirre J, Guérin DMA, Wuite GJL, Heck AJR, Roos WH (2013) Probing the biophysical interplay between a viral genome and its capsid. Nature Chem 5:502–509
124. Snijder J, Reddy VS, May ER, Roos WH, Nemerow GR, Wuite GJL (2013) Integrin and defensin modulate the mechanical properties of adenovirus. J Virol 87:2756–2766
125. Snijder J, Kononova O, Barbu IM, Uetrecht C, Ryrup WF, Burnley RJ, Koay MST, Cornelissen JJLM, Roos WH, Barsegov V, Wuite GJL, Heck AJR (2016) Assembly and mechanical properties of the cargo-free and cargo-loaded bacterial nanocompartment encapsulin. Biomacromol 17:2522–2529
126. Snijder J, Radtke K, Anderson F, Scholtes L, Corradini E, Baines J, Heck AJR, Wuite GJL, Sodeik B, Roos WH (2017) Vertex-specific proteins pUL17 and pUL25 mechanically reinforce herpes simplex virus capsids. J Virol 91:e00123-e217
127. Speir JA, Bothner B, Qu C, Willits DA, Young MJ, Johnson JE: Enhanced local symmetry interactions globally stabilize a mutant virus capsid that maintains infectivity and capsid dynamics, J Virol 80:3582–3591
128. Stockley PG, Twarock R (eds) (2010) Emerging Topics in Physical Virology, London 2010, Imperial College Press.
129. Tama F, Brooks CL III (2005) Diversity and identity of mechanical properties of icosahedral viral capsids studied with elastic network normal mode analysis. J Mol Biol 345:299–314
130. Ullmann A (2011) Escherichia coli and the emergence of Molecular Biology, Ecosal Plus 4(2)
131. Valbuena A, Mateu MG (2015) Quantification and modification of the equilibrium dynamics and mechanics of a viral capsid lattice, self-assembled as a protein nanocoating. Nanoscale 7:14953–14964
132. Valbuena A, Mateu MG (2017) Kinetics of surface-driven self-assembly and fatigue-induced disassembly of a virus-based nanocoating. Biophys J 112:663–673
133. Valbuena A, Rodríguez-Huete A, Mateu MG (2018) Mechanical stiffening of human rhinovirus by cavity-filling antiviral drugs. Nanoscale 10:1440–1452
134. Valle N, Riolobos L, Almendral JM (2006) Synthesis, post-translational modification and trafficking of the parvovirus structural polypeptides. In: Kerr JR, Cotmore SF, Bloom ME, Linden RM, Parrish CR (eds) Parvoviruses. Edward Arnold, London, pp 291–304
135. van de Waterbeemd M, Llauró A, Snijder J, Valbuena A, Rodríguez-Huete A, Fuertes MA, de Pablo PJ, Mateu MG, Heck AJR (2017) Structural analysis of a temperature-induced transition in a viral capsid probed by HDX-MS. Biophys J 112:1157–1165
136. van Rosmalen MGM, Nemerow GR, Wuite GJL, Roos WH (2017) A single point mutaton in precursor protein VI doubles the mechanical strength of human adenovirus. J Biol Phys. https://doi.org/10.1007/s10867-017-9479-y

137. van Rosmalen MGM, Li C, Zlotnick A, Wuite GJL, Roos WH (2018) Effect of dsDNA on the assembly pathway and mechanical strength of SV40 VP1 virus-like particles. Biophys J 115:1656–1665
138. Vaughan R, Tragesser B, Ni P, Ma X, Dragnea B, Kao CC (2014) The tripartite virions of the brome mosaic virus have distinct physical properties that affect the timing of the infection process. J Virol 88:6483–6491
139. Veesler D, Johnson JE (2012) Virus maturation. Annu Rev Biophys 41:473–496
140. Vliegenthart GA, Gompper G (2006) Mechanical deformation of spherical viruses with icosahedral symmetry. Biophys J 91:834–841
141. Vörös Z, Csík G, Herényi L, Kellermayer MSZ (2017) Stepwise reversible nanomechanical buckling in a viral capsid. Nanoscale 9:1136–1143
142. Vörös Z, Csík G, Herényi L, Kellermayer M (2018) Temperature-dependent nanomechanics and topography of bacteriophage T7. J Virol 92:e01236-e1318
143. Wang H, Chen Y, Zhang W (2019) A single-molecule atomic force microscopy study reveals the antiviral mechanism of tannin and its derivatives. Nanoscale 11:16368–16376
144. Wilts BD, Schaap IAT, Young M, Douglas T, Knobler CM, Schmidt CF (2010) Swelling and softening of the CCMV plant virus capsid in response to pH shifts. Biophys J 98:656a
145. Worner TP, Shamorkina TM, Snijder J, Heck AJR (2021) Mass spectrometry-based structural virology. Anal Chem 93:620–640
146. Xu C, Fischer DK, Rankovic S, Li W, Dick RA, Runge B, Zadorozhnyi R, Ahn J, Aiken C, Polenova T, Engelman AN, Ambrose Z, Rousso I, Perilla JR (2020) Permeability of the HIV-1 capsid to metabolites modulates viral DNA synthesis. PLoS Biol 18:e3001015
147. Zandi R, Reguera D (2005) Mechanical properties of viral capsids, Phys Rev E 72:021917
148. Zeng C, Moller-tank S, Asokan A, Dragnea B (2017) Probing the link among genomic cargo, contact mechanics, and nanoindentation in recombinant adeno-associated virus 2. J Phys Chem 121:1843–1853
149. Zeng C, Scott L, Malyutin A, Zandi R, Van der Schoot P, Dragnea B (2021) Virus mechanics under molecular crowding. J Phys Chem B 125:1790–1798

Chapter 12
Cryo-Electron Microscopy and Cryo-Electron Tomography of Viruses

Daniel Luque and José R. Castón

Abstract When viruses are viewed as dynamic containers of an infectious genome, their structural, physical, and biochemical analyses become necessary to understand the molecular mechanisms that control their successful life cycle. Information on virus structures at the highest possible resolution is essential for identifying the principles of their structure–function relationship, and could lead to development of antivirals, vaccines, and the advancement of new platforms for virus-based nanotechnology. Cryogenic electron microscopy (cryo-EM), which has revolutionized structural biology, is central to determining high-resolution structures of many viral assemblies, within a feasible time frame and in near-native conditions. In addition, cryo-EM allows dynamic studies of functional complexes that are often flexible or transient. State-of-the-art approaches in structural virology now extend beyond purified symmetric capsids and focus on the asymmetric components such as the packaged genome and minor structural proteins that were previously missed. A variation of cryo-EM, cryo-electron tomography (cryo-ET), can handle pleomorphic and complex viruses as well as viruses in the cellular context at unprecedented resolution. These and other emerging methods will support studies to address viral entry, assembly, replication and egress within the cellular host. This review describes the use of cryo-EM and cryo-ET in structural virology, and provides a few recent examples of how these techniques have been applied successfully in basic research to decipher fundamental aspects of virus biology and to investigate threatening viruses, including SARS-CoV-2, responsible for the COVID-19 pandemic.

Keywords Capsid · Cryo-electron microscopy · Cryo-electron tomography · Three-dimensional reconstruction · Viral assemblies · Virus life cycle

---

D. Luque
Spanish National Microbiology Centre, Institute of Health Carlos III, Madrid, Spain

J. R. Castón (✉)
Department of Structure of Macromolecules, Centro Nacional de Biotecnología (CNB-CSIC), Campus Cantoblanco, Madrid, Spain
e-mail: jrcaston@cnb.csic.es

Nanobiotechnology Associated Unit CNB-CSIC-IMDEA, Campus de Cantoblanco, Madrid, Spain

M. Comas-Garcia and S. Rosales-Mendoza (eds.), *Physical Virology*, Springer Series in Biophysics 24, https://doi.org/10.1007/978-3-031-36815-8_12

Introduction

To understand the principles of the structure, function, and evolutionary relationships of virus and viral macromolecular assemblies, structural information at the highest possible resolution is essential. This analysis can also lead to the development and design of antiviral drugs and vaccines [1], and to advancement of new platforms for virus-based nanotechnology [2]. The three major techniques in current use for structural determination are X-ray crystallography, nuclear magnetic resonance (NMR) spectroscopy, and cryogenic electron microscopy (cryo-EM). Three-dimensional (3D) cryo-EM provides an effective means of determining the structure of many macromolecular assemblies at atomic or near-atomic resolution [3, 4], as highlighted by the 2017 Nobel Prize in Chemistry, awarded in recognition of the extraordinary impact of cryo-EM on many scientific disciplines [5].

For 3D cryo-EM, rapid freezing immobilizes a radiation-sensitive sample in vitreous ice (water molecules in an amorphous state) in a near-physiological environment, and protects it from radiation damage during imaging. A unique application of vitrification is its ability to capture complexes in their multiple native states and trap intermediates at fixed time points (also termed time-resolved cryo-EM). Whereas X-ray crystallography is limited by the difficulty of crystallizing large, flexible complexes, as well as by the large quantities of material needed for crystallization trials, NMR is limited by the size of the macromolecules under study. In structural biology, 3D cryo-EM is currently a go-to technique for three main reasons, (i) the broad molecular weight range available for study (~50–100,000 kDa), (ii) the need for a (relatively) small amount of sample, and (iii) the ability to deal with heterogeneous samples to discern multiple conformational (or compositional) states [6–8].

3D cryo-EM can be used to study samples from single proteins [9, 10] to large macromolecular complexes such as virus particles [7, 8]. Cell organelles, entire eukaryotic cells, bacteria, and tissue sections can be analyzed by an analogous method known as cryogenic electron tomography [11, 12] (cryo-ET). In virology, it also enables rapid structural analysis of emerging viruses [13]; for example, cryo-EM was pivotal in combating several viruses implicated in recent deadly epidemics, such as Ebola (2014–2016) [14], Zika (2015–2016) [13], dengue (2019–2020) [15], MERS-CoV (2012–2015) [16, 17], and recently, SARS-CoV-2.

Owing to their large mass and high symmetry, viruses and viral assemblies were (and are) a major driving force in cryo-EM development. Cryo-EM includes several imaging modes through a transmission electron microscope (TEM); single-particle analysis (SPA) and electron cryo-tomography (cryo-ET) are the methods most extensively used, as they can address a wide range of biological problems. SPA is ideal for structure determination at near-atomic resolution of infectious virions of a broad size range, viral macromolecular assemblies that are large and dynamic, with compositional heterogeneity (states that present intermediate functional conformations during assembly), soluble proteins such as viral polymerases or surface spikes, and the genome that follows an asymmetric organization. The structure of apoferritin, a 24-subunit protein nanocage (similar to a closed icosahedral capsid) was recently solved at 1.2 Å resolution using SPA cryo-EM, and individual hydrogen atoms in the protein were clearly located [18, 19]. Cryo-ET enables analysis at unprecedented resolution of non-symmetric, pleomorphic, and/or complex viruses, and of viruses not only in their native physiological state, but also in their natural cell environment (in situ). Sub-volume or sub-tomogram averaging [STA, averaging of repeating structures from tomographic reconstructions, or tomograms [20] can be further exploited to resolve higher resolution detail of any repeating structures such as the viral surface glycoproteins of SARS-CoV-2 [21] and the internal HIV nucleocapsid [22].

Here we describe the basis of the cryo-EM and cryo-ET to the study of viruses, discuss some recent examples and approaches that highlight the strength of cryo-EM, and describe how SPA and cryo-ET are revolutionizing our understanding of virology across the resolution scale, from cells to atoms. The development of direct electron detectors (DED) for recording images was a key factor driving the "resolution revolution" [23, 24], in addition to improvements in preparation methods, the development of dedicated cryo-electron microscopes, and computational methods to deal with structural/compositional heterogeneity [25]. As a result, many challenging topics in virology previously unsuitable for cryo-EM analysis, such as packaged viral genome structure, jumbo viruses, heterogeneous viruses, or high-dynamic viral surface assemblies, can now be determined feasibly, many at near-atomic resolution (2–4 Å). Finally, we briefly describe recent advances in visualizing the virus life cycle in situ using cryo-ET.

Sample Preparation and Grids

Specimens are initially examined by negative staining, in which samples are dehydrated and embedded in a heavy metal salt cast that replicates specimen shape [26]. This quick, simple method is used to assess sample homogeneity and quality, as it provides high-contrast images.

For cryo-EM, the sample is immobilized in a thin film of vitreous (amorphous) ice that prevents structural alterations in the specimen caused by ice crystal formation [27]. To prepare these vitrified samples, an aliquot (~3 μL) of the specimen in its buffer is applied to an electron microscopy grid coated with a holey carbon support film (for example, Cu/Rh Quantifoil grids). It is then blotted with filter paper to yield a very thin film of the particle suspension in the holes. The grid is then plunged into liquid ethane cooled by liquid nitrogen [28, 29]. Vitreous ice will ideally be uniform across the entire grid, only slightly thicker than the assembly/molecule of interest, and will contain well-distributed particles at sufficient concentration. To improve reproducibility, this is done using robotic plunge-freezer devices such as a Vitrobot (Thermo Fisher Scientific, TFS), Cryo-plunge (Gatan), or EM GP2 (Leica

Microsystems). At −180 °C, vitrified ice does not sublimate significantly in the high-vacuum conditions of the electron microscope column, and protects the sample from radiation damage [30–32].

In recent years, it was shown that that macromolecules under study can interact preferentially with the air–water interface (AWI) and/or can (fully or partially) denature on exposure to it. Vitrification by formation of a thin solvent film using blotting paper followed by plunging into cryogens usually takes some seconds in which the sample could contact AWI [33, 34]. Different approaches are being developed to improve cryo-EM sample quality and reproducibility, including (i) direct sample deposit onto self-wicking grids using an inkjet piezo dispenser (Spotion/Chameleon) [35], (ii) using microcapillarity for sample deposit (cryoWriter) [36], (iii) spraying sample onto the grid with an ultrasonic humidifier [37], (iv) pin printing followed by vitrification with a cryogen stream (Vitrojet) [38], and (v) time-resolved by microfluidic spraying (TED) [39].

The buffered sample concentration is critical for obtaining a thin film with uniform specimen distribution after blotting (confirmed by negative staining analysis). When sample concentration is limited, a continuous carbon support foil is used rather than the holey film. UltrAuFoil grids (holey gold support foil on gold grids) provide another alternative. The presence of sucrose or cesium chloride, used in sample purification by cushion and density gradient ultracentrifugation, reduce vitrification efficiency, as does the glycerol used to preserve frozen samples. These compounds bubble when exposed to the electron beam and are therefore incompatible with high-resolution imaging [25]. When carried out correctly, this method provides a near-physiological, water-like environment for the specimen, leaving the native structure preserved.

Sample stability and homogeneity should be checked by conventional biochemical and biophysical analyses prior to cryo-EM analysis [40]. Size exclusion chromatography separates specimen subpopulations based on the hydrodynamic radius of their particles, and is an appropriate method for preparing homogeneous samples [41, 42]. In addition to contaminant complexes, specimen heterogeneity can result from a sample in distinct compositional or conformational states. The GraFix technique can also be used to help 'purify' the sample [43]; here, a chemical crosslinker (e.g., glutaraldehyde) is used to stabilize complexes from dissociated elements and aggregations.

Image Acquisition

The major factor that limits resolution of the cryo-EM structure is radiation damage caused by energy deposition from electron beams to the sample. Use low electron doses to prevent radiation damage nonetheless leads to noisy low-contrast images with a very poor signal-to-noise ratio (SNR) for subsequent image processing, for example hindering the accuracy of particle alignment and atomic-resolution reconstruction. Hence many different views of the same macromolecular complex are averaged to enhance the SNR and calculate a 3D reconstruction.

DED, such as the Falcon camera (from FEI Thermo Fisher), the K camera (Gatan), and the DE camera (Direct Electron Inc.) became pivotal in achieving improved achievable resolution. DED are complementary metal oxide semiconductor (CMOS)-based sensors that convert electrons directly into an electrical current with high detective quantum efficiency (DQE), a measure of detection efficiency as a function of spatial frequency [44, 45] that increases the SNR. DED can operate in integration or counting mode. In integration mode, the signal is either integrated over the entire exposure time or dose-fractionated into multiple movie frames. In counting mode, single electron scattering events are detected. Although integrative mode was faster and counting optimized the acquisition quality, more recent detectors (like Gatan K3 and TFS Falcon 4) are able to record high quality counting acquisitions at high speed (up to 900 exposures/hour).

DED have high frame-rate acquisition and images are collected as movies or a collection of frames; this allows dose fractionation into multiple frames per second [46, 47] (each frame has an extremely low electron dose, 0.5-1e^-/pixel). More importantly, the beam-induced movement of the specimen and support foil (and/or the mechanical drift of the stage) that results in motion-induced image blurring can be tracked and corrected by computational alignment of a series of low-dose frames taken in one area before averaging [48] (Fig. 12.1). Subframes with optimized doses can also be selected for subsequent image processing [49] (later frames would be of lower quality due to radiation damage). The combination of dose fractionation and motion correction greatly improves data acquisition efficiency—nearly all images are of a quality suitable for recovering high resolution information [50]. Beam-induced motion is greatly reduced by using holey gold support foil [51] (for example, UltrAuFoil grids). HexAuFoil grids (hexagonal-pattern gold grids of gold foil with $< 0.3\ \mu m$ holes) [52] almost eliminate beam-induced motion during imaging.

Because of the large depth of field in the electron microscope, each micrograph is a projection through the specimen, yielding 2D information. These 2D projection images are modulated by the contrast transfer function (CTF) of the microscope, and image defocus correction is required for accurate interpretation of specimen structure. The CTF is a characteristic function of each microscope (the spherical aberration coefficient of the objective lens, Cs) and the imaging settings (defocus level, beam coherence, and accelerating voltage). By choosing different defocus settings, specific image frequencies can be accentuated at the expense of others. In appropriate conditions, a contrast-enhancing effect can thus be obtained. CTF-modulated information implies the existence of frequency ranges in which there are no data (zeros in the CTF), as well as frequency regions in which the information has reversed contrast (between the first and the second zero, between the third and fourth zero, and so on). It also considers the attenuation of useful information at higher frequencies, the so-called envelope function, which is dependent mainly on the beam coherence.

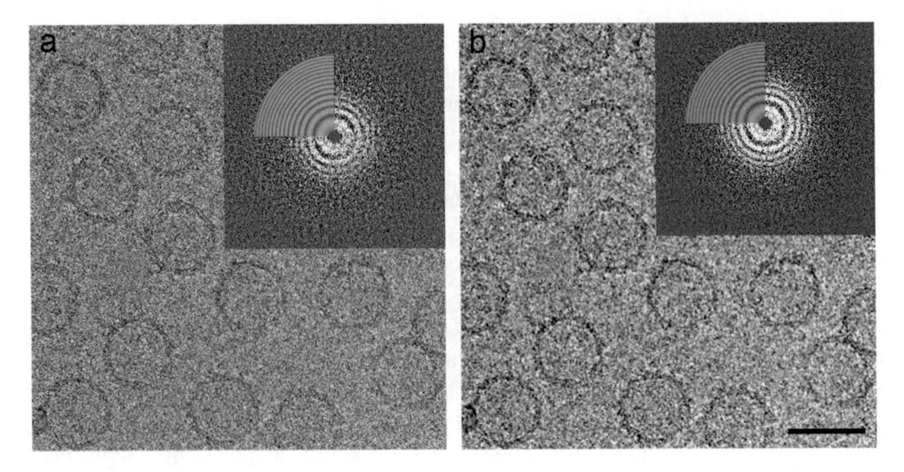

Fig. 12.1 Correction of beam-induced motion and recovery of high-resolution information. **a** Uncorrected two-dimensional movie average. **b** Two-dimensional movie average after translational alignment. Cryo-EM image corresponds to vitrified empty capsids of Rosellinia necatrix quadrivirus 1, a ~450-Å-diameter, fungal virus. Barr = 500 Å. Power spectra calculated from the averaged movie before (**a**) and after (**b**) motion correction are shown. Motion correction restores Thon rings from ~6–7 (**a**) till 3–4 Å (**b**) resolution. The top left corner of each power spectrum shows a CTF model fit to the experimental Thon rings

During data acquisition, images are recorded at a range of defocus settings to enhance different features of the specimen and to fill in missing information when they are combined into a 3D reconstruction. Averaged movie quality can be analyzed by inspection of the CTF zeros or Thon rings in the images' computed diffraction patterns. Large symmetrical assemblies (such as viruses) are relatively easy for a computer to motion-correct and can be imaged at small defocus levels with good image contrast and preserved high-resolution signals. For small (100–300 kDa) and asymmetric assemblies, higher defocus settings are needed, which limit the resolution that can be achieved.

A number of automated data collection software packages such as free SerialEM [53] and Leginon [54], as well as the commercial EPU (TFS), JADAS (JEOL) [55], and Latitude S (Gatan) systems are available for collecting large data sets for reliable statistical analysis.

Structure Determination by SPA

The general SPA workflow involves quality assessment of motion-corrected average images, particle picking (arduous with low molecular weight specimens), image defocus estimation and CTF correction, 2D particle classification and alignment, estimation of particle orientation and refinement, image reconstruction and 3D refinement, resolution assessment, and cryo-EM map validation [56] (Fig. 12.2).

Projection subtraction and local reconstruction are used routinely to deal with non-symmetric or flexible components of the complex analyzed (below). A number of image processing packages can perform these tasks, provide improved classification of multiple conformational or compositional states, and sort images into different 2D and 3D image classes. All this has become computationally feasible with improved central processing units (CPU) and, more recently, graphic processing units (GPU). These software packages include RELION [57], EMAN2 [58], cryoSPARC [59] and FREALIGN [60]. SCIPION [61] provides a framework for integrating software packages via a workflow-based procedure designed for less experienced users, and offers user-friendly graphic interfaces. Most macromolecular complexes are intrinsically dynamic and normally contain many different conformational (or compositional) states. This structural heterogeneity becomes the major limiting factor for attainable resolution in cryo-EM. Additional methods such as 3D-classification of particle conformations [62–64], non-homogeneous refinement, multibody refinement, refinement of subparticle orientations, per-particle focus and motion refinement, and correction of Ewald sphere effects for large viruses have recently been applied to overcome limitations in the conventional data processing pipeline [65, 66].

SPA provides the 3D structure of examined specimens by computationally merging images of many (easily exceeding tens or hundreds of thousands) individual macromolecules of a homogeneous subset that have been aligned with one another at high precision. Each particle image is a randomly orientated 2D projection image that contains all the structural detail of the 3D specimen. Angular orientation parameters are determined by comparing the 2D projections with spatially defined reprojections of an initial 3D model filtered at low resolution. A new 3D map of the macromolecular complex is then calculated from these 2D projections by "back-projection", that is, the combination of all views into a single 3D map [67]. This projection-matching process is performed iteratively to obtain new 3D reconstructions with improved resolution until no further improvement is possible.

Resolution Estimation, Model Building and Validation

When generating a cryo-EM map, a standard approach is to split the data randomly into two independent sets at the beginning of the data processing procedure and then compute, by iterative refinements and reconstructions, two independent reconstructions from each half-set. These two reconstructions are then correlated as a function of the spatial frequency to determine the extent to which structural features have been reliably reproduced. This is termed the "gold standard" Fourier shell correlation (FSC) method [68, 69]. As the resolution is often non-isotropic in different map regions, a visual inspection of map quality in key regions is needed to establish local conformations and their functional implications [70]. Local variations in resolution

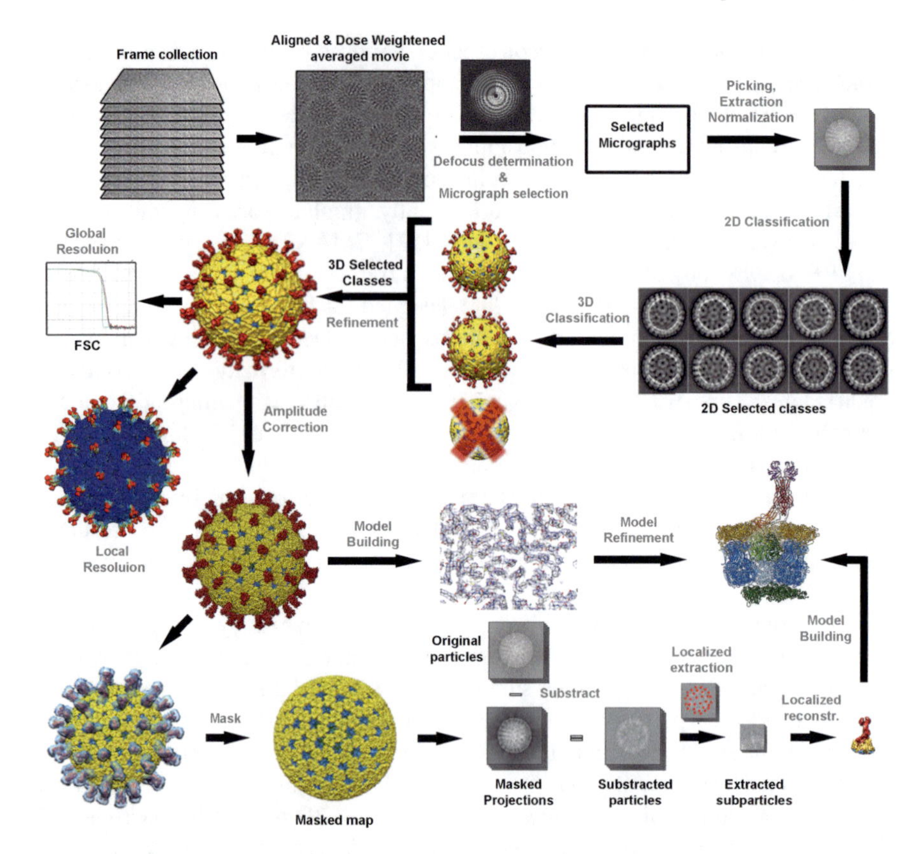

Fig. 12.2 Overview of the single-particle cryo-EM workflow, from data collection to 3D model. The vitrified sample is imaged by collecting movie frames that are aligned, dose weightened and averaged. Using motion-corrected images, image defocus is calculated, good quality micrographs are selected, and individual particles (such as protein cages) are picked. Next, particles are extracted, normalized, and then are subjected to 2D classification and averaging, to get class averages according to their similarity and clean up the data set of bad particles. Selected particles can be used to obtain an ab initio 3D model. This low-resolution model is then used as a reference for 3D classification, a necessary step for identifying distinct conformations, macromolecular heterogeneities or particle subsets with different structural integrity. Orientation refinement is iteratively done until the structure converges, as indicated in the resolution analysis by the FSC method. The final map is amplitude corrected to reveal high-resolution details. Finally, the protein sequence is fitted into the 3D map to build a de novo 3D model of the protein which is refined to obtain a final model. If necessary, particles with the density corresponding to certain regions of the 3D structure subtracted can be obtained by subtracting the projections of the final map after mask the volumes of interest. In order to analyze non symmetric components of the structures, once final orientations has been assigned to each particle, it is also possible to extract subparticles the corresponds to a 3D coordinate (and their symmetry related mates) in the reconstructed map. 2D images and 3D maps correspond to the cryo-EM structure of the rotavirus triple layered particle

can be assessed using Resmap [71] or MonoRes [72] programs. Differences in cryo-EM local resolution values can be taken as a signature of conformational fluctuations at equilibrium in solution [73].

Once a 2–4 Å cryo-EM map has been obtained, the next stage is to build and refine a model. The known protein sequence is fitted into the 3D map by positioning the bulky amino acid side chains first, allowing de novo structure determination. The 3D atomic model built is then subjected to further refinement. Homology models of atomic structures obtained by X-ray crystallography and NMR might also be fitted into a low resolution cryo-EM map (termed the hybrid approach). There are several model-building and validation tools, most of them based on X-ray crystallography analysis, including CCP-EM [74], Phenix [75], Rosetta [76], UCSF Chimera [77], ISOLDE [78], and Coot [79]. Prediction of protein structures from primary sequence information with AlphaFold [80] or RoseTTAFold [81] provide preliminary models that can be built into cryo-EM data.

Cryo-Electron Tomography (Cryo-ET)

Cryo-ET [11, 82] is based on the acquisition of series of projection images collected by tilting the specimen around an axis perpendicular to the electron beam inside the microscope, and combining the resulting "tilt-series" into a 3D volume. The 3D reconstruction of the field of interest, or tomogram, is calculated as a back-projection in real space; assembly of correctly aligned projections is reverse-projected into 3D space. Cryo-ET has allowed the study of samples ranging in size from large (pleomorphic) complexes to organelles, prokaryotic cells, and eukaryotic cell lamellae [thin cell sections obtained by focused ion beam (FIB) milling [83], that is, samples not amenable to SPA averaging method. Similarly to SPA cryo-EM, multiple copies of repeating structures in a tomogram can be extracted as a volume, aligned, and averaged. This STA approach delivers structures of large macromolecular complexes at subnanometer resolution in their native cellular environment [84] and in some cases, by averaging 3D structures instead of 2D images as in cryo-EM, at high resolution [22]. The introduction of correlative light and electron microscopy (CLEM) and focused ion beam (FIB) milling have further empowered cryo-ET to reveal native ultrastructures of large eukaryotic cells or even tissues, with unprecedented spatiotemporal resolution [20, 85].

The structural information within a reconstructed tomogram is limited by three notable factors: low signal in each tilt image to limit accumulated electron dose, missing information due to mechanical limitations in the microscope stage (termed the "missing wedge"), and accuracy of alignments between each successive tilt image prior to reconstructing a 3D volume [11].

Some Recent Major Contributions of Cryo-EM and Cryo-ET in Structural Virology

Due to the ordered nature, structural integrity, and homogeneity of many viral assemblies, cryo-EM is used extensively to solve near-atomic resolution (2–4 Å) structures of infectious virions of a broad size range, with helical or icosahedral symmetry, tailed phages, virus-like particles (VLP), hybrid VLP (with heterologous cargos), and viral complexes that represent intermediate functional states during assembly. Even isolated proteins such as viral polymerases (purified from viruses) or non-structural proteins are analyzed by cryo-EM (cryo-EM ex vivo or ex virio), although these structural targets require much more expertise for cryo-EM sample preparation and optimization. The symmetry-based averaging imposed nevertheless results in the loss of unique (but ubiquitous) asymmetric features. Recent new approaches extend these analyses to the asymmetric organization of the packaged genome and minor structural proteins such as viral polymerases that were previously missed (cryo-EM in situ) [86]. Asymmetric structures have important functions in numerous steps of the virus replication cycle, and many of these might be key targets for the development of new antiviral drugs.

Here we will consider some recent studies showing asymmetric structural features in viruses and the reconstruction methods used to identify these novel or unexpected components associated with essential functions. We will briefly discuss other challenging viral systems such as giant (or jumbo) viruses, and the cryo-EM and cryo-ET contributions in the battle against the COVID-19 pandemic. Finally, we will show the latest structural insights disclosed by cryo-ET into complex and pleomorphic viruses, as well as the mechanisms of its entry, replication, assembly, and budding—all the dynamic steps during the life cycle in the cellular context.

Cryo-EM in situ: Structural Analysis of Asymmetric Components of Icosahedral Viruses

Probing asymmetry in icosahedral and helical viruses can be difficult if the asymmetric feature is small or has a weak signal. An exception is the tailed bacteriophage, with a large assembly that allows correct alignment of the asymmetric assembly. A variety of image processing-based strategies is used to study non-symmetric capsid features; these approaches are (i) standard asymmetric refinement, (ii) symmetry relaxation, (iii) symmetry expansion, and (iv) subparticle classification/refinement [65, 66, 87]. For asymmetric refinement, no symmetry is applied during data processing (C1 symmetry), and orientation sampling is performed on the full orientation space. In the relaxed symmetry approach, the icosahedral orientations determined are relaxed and iterative sampling is tested for the 60 symmetry-related orientations. For expanded symmetry, each of the 60 symmetry-related orientations is assigned to each particle before focused (or masked) classification/

refinement. In subparticle classification and/or refinement, subregions or subvolumes are reextracted and allowed to deviate from icosahedral symmetry.

Analysis of the viral genome is central to understanding dynamic processes of the life cycle such as assembly, infection, and genome replication. Bacteriophage MS2 and Qβ capsids incorporate a single copy of the maturation protein, responsible for binding both viral ssRNA and host receptors and the genome, which is relatively rigid. Asymmetric reconstruction/refinement shows that ssRNA has multiple stem-loops termed packaging signals (PS) that bind cooperatively to a capsid protein recognition motif. The PS act as allosteric switches, control assembly efficiency, and are responsible for in vivo packaging specificity [88, 89]. High resolution asymmetric cryo-EM reconstructions (standard asymmetric refinements) of MS2, which traces 80% of the genome with 16 stem-loops, confirm this assembly model [90–92] (Fig. 12.3a).

Whereas dsDNA viruses in the extracellular life cycle are in a quiescent state, dsRNA virus capsids in the host cytoplasm are dynamic nanocages, and carry out mRNA synthesis using their genome segments as templates for the transcriptional enzyme complexes (TEC; they include an RdRp and other enzymes). In cypovirus, the 10 genomic dsRNA segments have a non-spooled organization and there are 10 TEC anchored to the inner surface of the capsid shell around the five-fold axis, leaving two of the 12 possible positions vacant [93, 96] (Fig. 12.3b). RdRp, resolved to 3.3 Å resolution, are in different conformations in the transcribing and non-transcribing states. These studies relied on subtracting capsid density from the original images, followed either by standard asymmetric refinement [93] or refinement based on relaxing symmetry [96]. Similar in situ structural analyses with rotavirus, reovirus, aquareovirus and bluetongue virus show the mechanism of RNA transcription and replication [97–100]. In bacteriophage ϕ6, with a trisegmented genome, the dsRNA adopts a dsDNA-like single-spooled genome organization and RdRp are detached from the capsid inner surface. These variations might be a consequence of the differences in transcription mechanisms between the two virus families (semi-conservative vs. conservative transcription) [101].

Herpesvirus consists of an inner icosahedral capsid that contains the linear dsDNA genome, a middle proteinaceous tegument layer, and an outer envelope containing transmembrane glycoproteins. The viral genome is packaged/ejected by an ATP-driven terminase through the asymmetric portal complex located at a predetermined fivefold symmetry axis of the capsid. Processing of cryo-EM images of HSV-1 virions using a sequential localized classification and symmetry relaxation allowed decoupling and reconstruction of heterosymmetric and asymmetric capsid elements. The HSV1 map shows the in situ structures of the unique portal vertex, genomic termini, and ordered dsDNA coils [94] (Fig. 12.3c). A similar approach resolved the human cytomegalovirus structures of the portal and the capsid vertex-specific components [102].

Feline calicivirus (FCV) has a positive-sense ssRNA genome encapsidated within a $T = 3$ icosahedral capsid. After binding its receptor, focused classification leads to a portal-like structure formed of 12 copies of VP2 (a minor structural protein) located at a unique threefold symmetry axis [95]. The conformational changes observed in

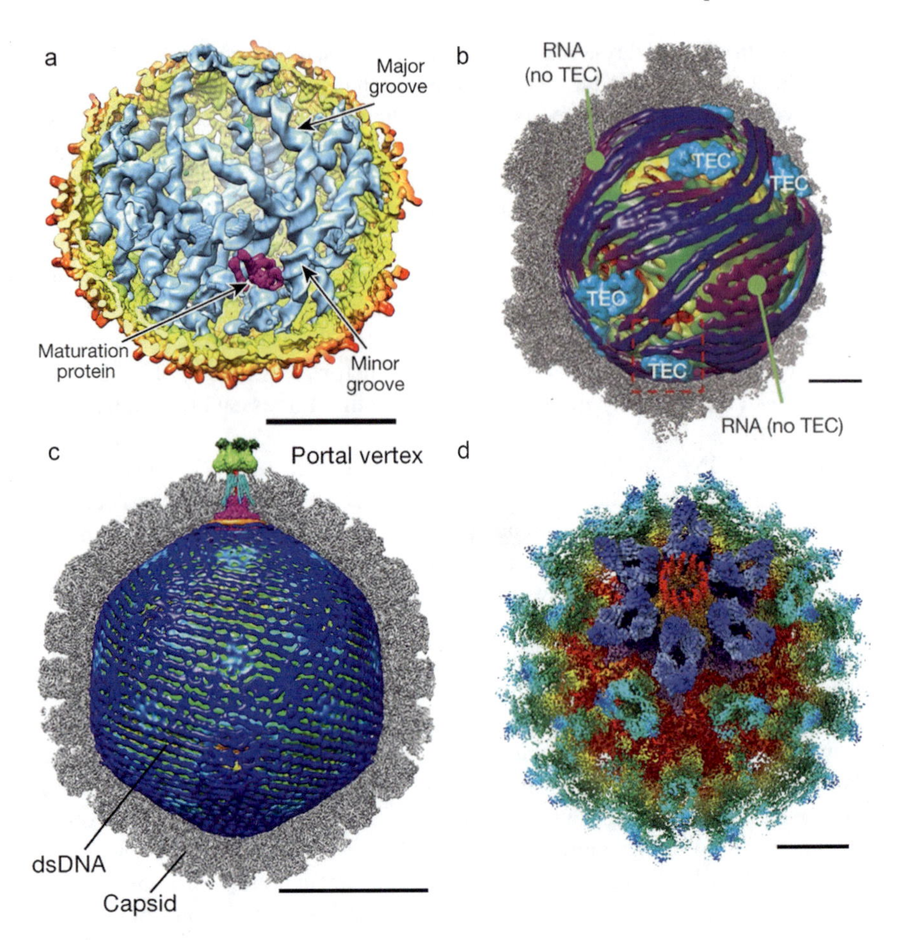

Fig. 12.3 Cryo-EM and asymmetric reconstructions of icosahedral viruses for visualization of the viral genomes and other asymmetric components in situ. **a** A cut-open view of bacteriophage MS2 to show a single copy of the maturation protein (magenta) and the ssRNA viral genome (blue) mostly organized as dsRNA (major and minor grooves of double-stranded regions are indicated). Reproduced with permission from Ref. [90]. **b** Cryo-EM asymmetric reconstruction of cypovirus, with the front half of the icosahedral capsid removed to show the asymmetric components, the dsRNA genome (dark blue) and the transcriptional enzyme complexes (TEC, cyan). Reproduced with permission from Ref. [93]. **c** Herpes simplex virus type 1 (HSV-1) with the front half of the protein shell removed to visualize the portal vertex elements (portal, magenta; portal cap, green) and ordered dsDNA coils (blue) spooled around a disordered dsDNA core. Reproduced with permission from Ref. [94]. **d** Radially colored full feline calicivirus (FCV) virion to highlight the receptor proteins (blue) and the VP2 portal-like assembly (red) at a unique threefold symmetry axis. Reproduced with permission from Ref. [95]. Scale bars 100 Å (a, b and d) and 450 Å (**c**)

the capsid protein VP1 allow for extrusion of the internalized VP2 proteins as well as the formation of a small pore in the capsid shell. The VP2 portal-like assembly might function in a mechanism for endosomal escape and genome delivery, a process not well understood in viruses without membranes (Fig. 12.3d).

Atomic Structures of Jumbo Viruses (>100 nm)

Giant viruses belong to the nucleocytoplasmic large DNA viruses (NCLDV) that include nine families, *Ascoviridae, Asfarviridae, Iridoviridae, Marseilleviridae, Mimiviridae, Pandoraviridae, Phycodnaviridae, Pithoviridae*, and *Poxviridae*, as well as some independent species. These viruses, with the dsDNA genome encapsulated in a lipid bilayer [103], range in size from ~150 nm to 2 μm, and include high-symmetry icosahedral viruses such as Paramecium bursaria chlorella virus 1 (PBCV-1) and amphora-shaped (asymmetric) viruses like Pithovirus.

Two main strategies are applied to calculate high resolution cryo-EM structures of large viruses [104]: (i) use of electron microscopes with high acceleration voltage, e.g., one megavolt cryo high-voltage electron microscopy (1MV cryo-HVEM), which improves sample penetration and overcomes the limitations imposed by electro-optical physics at lower voltages [105], and (ii) the 'block-based reconstruction' method [106].

For thick samples, the influence of depth of field causes an internal focus shift (a focus gradient within a single virus particle), imposing a hard limit on the attainable resolution [107]. Standard CTF correction methods, which assume a single constant defocus for the entire particle, are insufficient for large, high-resolution structures. Increasing the accelerating voltage increases depth of field, however, and improves the (electron) optical conditions in thick samples. For 300 kV electron microscopes, recent advances in image processing, such as Ewald sphere correction to tackle the focus gradient within a single virus particle, can offer great improvements in attainable resolutions. 'Block-based' reconstruction [106] focuses on sub-sections ("blocks") of the virus to allow localized defocus refinement across each particle, and reduces the size of the box required.

For high-resolution 3D SPA of giant viruses, there are currently seven structures that exceed 1 nm resolution: two for PBCV-1 [108, 109], two for African swine fever virus (ASFV) [110, 111], the marseillevirus Singapore grouper iridovirus [112] and Tokyovirus [105]. All were analyzed using 'block-based' reconstruction (with 300 kV microscopes) except Tokyovirus (with a maximum diameter of 250 nm), which was analyzed as a complete viral particle using 1 MV cryo-HVEM. PBCV-1 at 3.5 Å resolution [108] and ASFV at 4.1 Å resolution [111] were solved by cryo-EM SPA. The PBCV-1 and ASFV capsids are composed of one "major" capsid protein (MCP) and a combination of 14 (PBCV-1) or four kinds (ASFV) of "minor" capsid proteins (mCP). The MCP of PBCV-1 and ASFV are trimeric and share the double "jelly roll" motif, each of which consists of eight β strands [113].

SARS-CoV-2 and Viral Spikes

The coronavirus disease 2019 (COVID-19) pandemic, caused by the severe acute respiratory syndrome coronavirus 2 (SARS-CoV-2), is associated with over ~760 million cases and ~6.9 million deaths as of March 2023. Cryo-EM (and cryo-ET) studies played a crucial role, in record time, in developing therapeutics targeted to the spike protein (S protein) that decorates the SARS-CoV-2 viral surface. Not only was the atomic structure of many S protein specimens established by SPA cryo-EM analysis, but also its multiple conformational states, essential for understanding its mechanism of action, as well as the spike complexes with the human angiotensin-converting enzyme-2 (ACE2) receptor and neutralizing antibodies [114]. Basic research on previous limited outbreaks of the closely similar coronaviruses SARS-CoV-1 (SARS-CoV) in 2002 and Middle Eastern respiratory syndrome coronavirus (MERS-CoV) in 2012, as well as accumulated experience with other emergent viral pandemics (HIV and influenza viruses), were also fundamental to the development of mRNA-based vaccines and effective antibody therapies [16, 17]. This rapid structural analysis of SARS-CoV-2 indicates that our global society is prepared to apply similar strategies to combat future emerging viruses.

The immature S protein is proteolyzed into two subunits, S1 and S2, which form a heterodimer, which in turn assembles into a trimer that results in the spike on the virion that coats the viral surface [115] (Fig. 12.4). The S trimeric protein is a metastable fusion nanomachine; in situ analysis of S trimers by cryo-ET/STA of intact virions showed that most spikes are in the prefusion conformation, with only 3% in the postfusion conformation [21, 116]. The receptor-binding domain (RBD), located in the S1 subunit, can be in a "down" or "up" conformation that corresponds to ACE2-inaccessible or -accessible conformations, respectively. During infection, RBD of the prefusion form in the up conformation binds to ACE2 before fusion of viral and host membrane, mediated by the fusion peptide in S2. The S protein, which elicits a strong immune response, is the most important target antigen for the development of therapeutics and current vaccines. The Moderna and Pfizer mRNA vaccines are based on S protein stabilized in the prefusion conformation by two-proline substitutions [117]. Neutralizing antibodies to specific S protein regions and recombinant ACE2-derived inhibitors are also promising therapeutic agents for treating emergent variants. Another ~25 proteins have been described; many of these, some still structurally uncharacterized, are under study as candidate targets for antiviral drug development [118].

Cryo-ET in the Study of Enveloped Viruses and the Viral Life Cycle in the Host

Cryo-ET has allowed 3D analysis of complex and/or pleomorphic enveloped viral particles in vitro, many of them responsible for human diseases such as

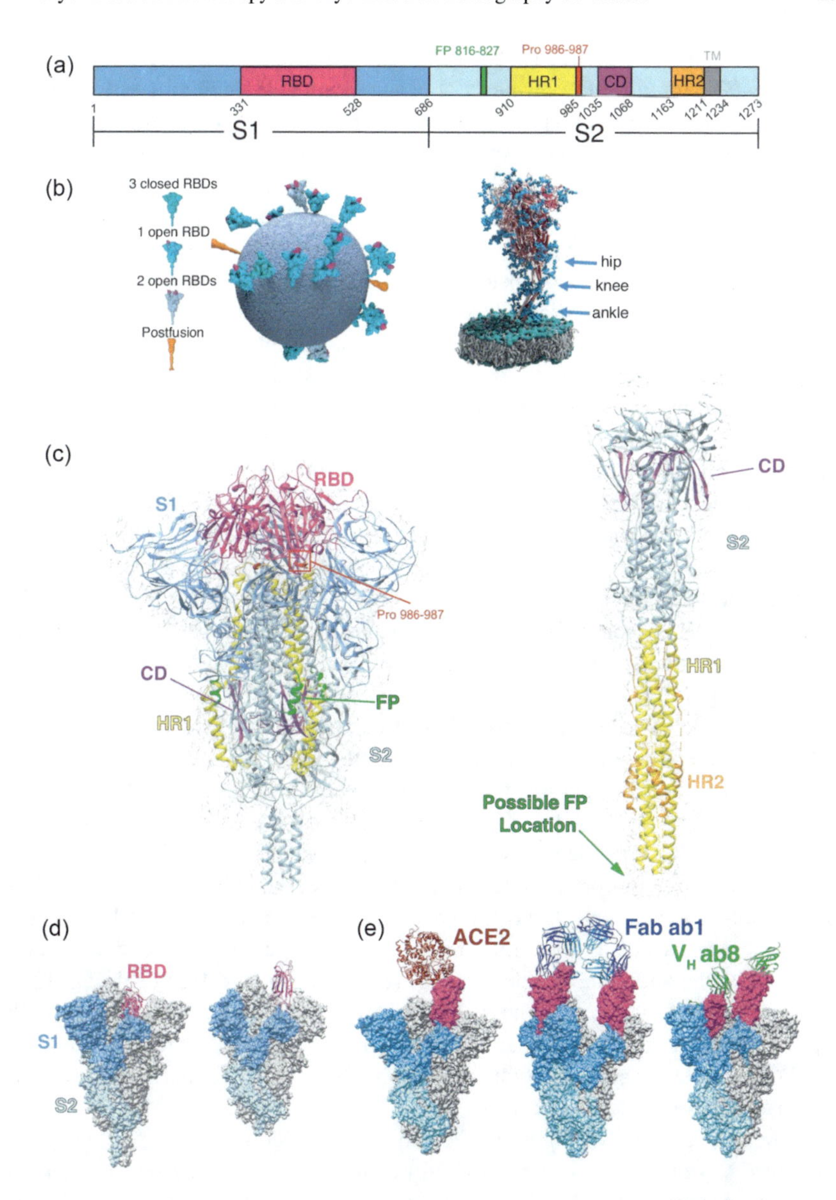

Fig. 12.4 Structures of SARS-CoV-2 spikes. **a** Scheme of the SARS-CoV-2 S domains showing the S1 subunit, S2 subunit, Receptor binding domain (RBD), fusion peptide (FP), heptad repeat 1 (HR1), connector domain (CD), heptad repeat 2 (HR2) and the transmembrane region (TM). **b** 3D model of a SARS-CoV-2 virion showing the conformations and flexibility of S on the virion surface. Three flexible hinges are marked by arrows (right). **c** SPA cryo-EM structure of the S trimer in its pre-fusion (left) and post-fusion (right) conformations [structural elements colored as in (**a**)]. **d** Comparison of S with RBD in the down (left) and up (right) conformations. **e** Cryo-EM structures of S bound to ACE2 with RBD in the up conformation (left), SN501Y mutant bound by Fab ab1 with RBD in the up conformation (middle), and SN501Y bound by VH ab8 in both the RBD up and down conformations (right). Reproduced with permission from Ref. [118]

HIV-1 [22, 119], herpesviruses [65, 120], and influenza virus [66, 121]. In addition, cryo-ET permits study of the viral replication cycle, including entry, replication, assembly, maturation, and egress within the native environment –the cell– at unprecedented resolution. STA typically improves resolution to ~2–3 nm, and in some cases is able to achieve ~4 Å resolution, that is, the cryo-ET resolution revolution [122]. We highlight a few examples for which cryo-ET has yielded important (and in some cases unanticipated) discoveries of complex viruses and in situ viral infection. The helical nucleocapsid (NC) of Ebola virus (with a non-segmented, negative-sense ssRNA) was solved by a subtomogram averaging at 6.6-Å resolution [123]; the NP carboxy-terminal extended α-helix is important for RNP assembly and recruitment of accessory proteins (Fig. 12.5a). In the study of HIV-1, a 7 Å resolution cryo-ET structure of the capsid from intact virions confirmed the hollow cone shape of the capsid, and allowed specific placement of each individual capsid hexamer and pentamer within the lattice structure [124]. The structure of the capsid domain and spacer peptide 1 (CA-SP1) of Gag (the precursor that includes MA, CA, and NC) in immature capsids was solved at 3.9-Å resolution [22], a value previously feasible only by single-particle cryo-EM. In influenza A virus (IAV), the matrix protein (M1) forms an endoskeleton beneath the virus membrane within intact virions. Cryo-ET/STA at ~8 Å resolution of M1 within purified virus particles, as well as the structure of M1 oligomers reconstituted in vitro, offered insights into IAV assembly and disassembly [125].

Although the icosahedral viruses that infect Archaea belong to the HK97 and PRD1 lineages, other archaeal viruses, such as the bottle-, lemon- or spindle-shaped viruses, have unique morphologies and are optimal assemblies for study using cryo-ET [127]. Construction of tailed spindle viruses involves a metastable multistart helical assembly of variable width that extends through the lemon-shaped capsid and into the tail. These conformational capsid dynamics could be used to drive genome ejection into the host cell. Cryo-ET analysis of the nodavirus flock house virus, a positive ssRNA eukaryotic virus, has identified a genome-replication complex associated with membrane vesicles at 8.5 Å resolution by cryo-ET [126] (Fig. 12.5b) similar to that of many other (+) ssRNA viruses (including SARS-CoV-2) [128]. In the remodeled host membranes, these complexes resemble a crown (determined by STA), a dodecameric ring with RNA polymerase activity that gates release of progeny ssRNA molecules [129].

Cryogenic correlative light and electron microscopy (cryo-CLEM), which combines spatiotemporal information from fluorescence microscopy (proteins of interest are tagged with fluorescent labels) with structural data from cryo-ET, is very appropriate for studying viral assembly and/or infrequent and short-lived events in selected regions of virus-infected cells [130, 131]. *Pseudomonas* phages assemble a compartment for DNA replication. Before cell lysis to release phages, assembled empty capsids on the host cell membrane migrate along the PhuZ spindles to the compartment surface for DNA packaging before release of viral progeny [132, 133]. A phase plate cryo-ET study of cyanophage Syn5, which infects the cyanobacterium *Synechococcus*, identifies distinct assembly intermediates in situ such as procapsids with SP, expanded empty and full capsids, and complete virions with the tail [134].

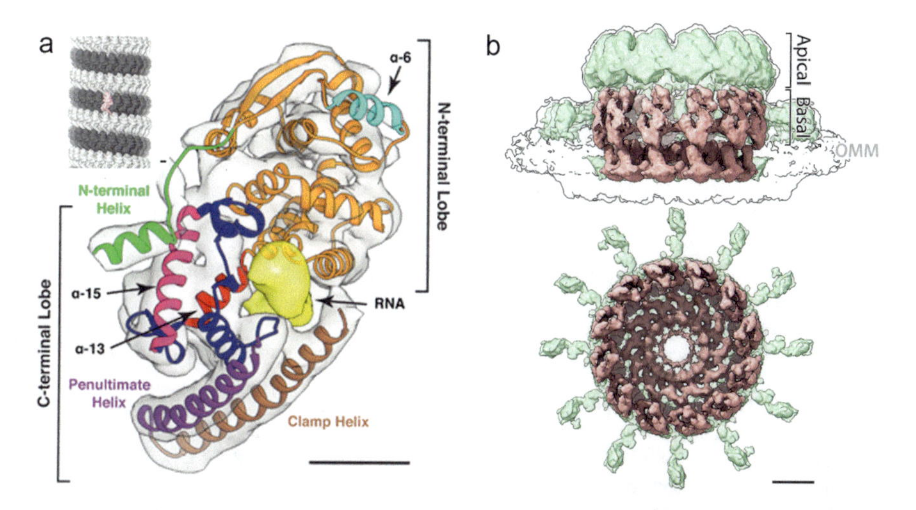

Fig. 12.5 Cryo-ET of pleomorphic viruses and intracellular viral assembly. **a** Structure of the helical nucleocapsid of Ebola virus solved by STA, indicating N- and C-terminal regions of nucleocapsid and the putative RNA density (yellow). Model of the helical RNP calculated by docking of NC in a tomogram (inset, a single NC is highlighted in pink). Scale bar, 20 Å. Adapted with permission from Ref. [123]. **b** The mature nodavirus crown resolved by cryo-ET STA comprises two stacked 12-mer rings (pink and green) of the viral RNA replication protein A. The lower ring, termed "proto-crown" (pink), is a precursor in RNA replication complex assembly, and protein A has alternate conformation in each ring. Reproduced with permission from Ref. [126]

In summary, advances in cryo-ET and associated techniques such as cryo-FIB milling have provided key insights that have expanded our knowledge of intracellular activities of viruses at unprecedented resolution.

Liquid Electron Microscopy (Liquid-EM)

Liquid electron microscopy (liquid-EM) is a relatively new technique that allows to image biological samples in their native liquid state. Unlike traditional transmission EM, which requires samples to be fixed (EM) or vitrified (cryo-EM) in a vacuum environment, liquid-EM delivers real-time data of dynamic processes in solution. Although liquid-EM results may not reach the resolution achieved with cryo-EM studies, it provides complementary information to cryo-EM/ET analyses [135]. Liquid-EM have been used to image multiple dynamic states of Adeno-associated virus in solution [136] and real-time host–pathogen interactions of phages with bacteria in solution [137].

Acknowledgements We apologize to our colleagues with outstanding contributions who were not mentioned due to space limitations. The authors thank C. Mark for editorial assistance. This work was supported by a grant of the Instituto de Salud Carlos III (PI20CIII-00014) to DL, and by grants from the Spanish Ministry of Science and Innovation (PID2020-113287RB-I00) and the Comunidad Autónoma de Madrid (P2018/NMT- 4389) to JRC.

References

1. Schlicksup CJ et al (2018) Hepatitis B virus core protein allosteric modulators can distort and disrupt intact capsids. Elife 7. https://doi.org/10.7554/eLife.31473
2. Schwarz B, Uchida M, Douglas T (2017) Biomedical and catalytic opportunities of virus-like particles in nanotechnology. Adv Virus Res 97:1–60. https://doi.org/10.1016/bs.aivir.2016.09.002
3. Chua EYD et al (2022) Better, faster, cheaper: recent advances in cryo-electron microscopy. Annu Rev Biochem 91:1–32. https://doi.org/10.1146/annurev-biochem-032620-110705
4. Chmielewski D, Chiu W (2021) Encyclopedia of virology. In: Bamford DH, Zuckerman M 4th edn, vol. 1. Academic Press, pp 233–241
5. Bertozzi C (2017) Atoms out of blobs: CryoEM takes the nobel prize in chemistry. ACS Cent Sci 3:1056. https://doi.org/10.1021/acscentsci.7b00494
6. Cianfrocco MA, Kellogg EH (2020) What could go wrong? a practical guide to single-particle cryo-EM: from biochemistry to atomic models. J Chem Inf Model 60:2458–2469. https://doi.org/10.1021/acs.jcim.9b01178
7. Jiang W, Tang L (2017) Atomic cryo-EM structures of viruses. Curr Opin Struct Biol 46:122–129. https://doi.org/10.1016/j.sbi.2017.07.002
8. Kaelber JT, Hryc CF, Chiu W (2017) Electron cryomicroscopy of viruses at near-atomic resolutions. Annu Rev Virol 4:287–308. https://doi.org/10.1146/annurev-virology-101416-041921
9. Khoshouei M, Radjainia M, Baumeister W, Danev R (2017) Cryo-EM structure of haemoglobin at 3.2 A determined with the Volta phase plate. Nat Commun 8:16099. https://doi.org/10.1038/ncomms16099
10. Merk A et al (2016) Breaking cryo-EM resolution barriers to facilitate drug discovery. Cell 165:1698–1707. https://doi.org/10.1016/j.cell.2016.05.040
11. Lucic V, Rigort A, Baumeister W (2013) Cryo-electron tomography: the challenge of doing structural biology in situ. J Cell Biol 202:407–419. https://doi.org/10.1083/jcb.201304193
12. Wan W, Briggs JA (2016) Cryo-electron tomography and subtomogram averaging. Methods Enzymol 579:329–367. https://doi.org/10.1016/bs.mie.2016.04.014
13. Sirohi D et al (2016) The 3.8 A resolution cryo-EM structure of Zika virus. Science 352:467–470. https://doi.org/10.1126/science.aaf5316
14. Sugita Y, Matsunami H, Kawaoka Y, Noda T, Wolf M (2018) Cryo-EM structure of the Ebola virus nucleoprotein-RNA complex at 3.6 A resolution. Nature 563:137–140. https://doi.org/10.1038/s41586-018-0630-0
15. Zhang X et al (2013) Cryo-EM structure of the mature dengue virus at 3.5-A resolution. Nat Struct Mol Biol 20:105–110. https://doi.org/10.1038/nsmb.2463
16. Yuan Y et al (2017) Cryo-EM structures of MERS-CoV and SARS-CoV spike glycoproteins reveal the dynamic receptor binding domains. Nat Commun 8:15092. https://doi.org/10.1038/ncomms15092
17. Pallesen J et al (2017) Immunogenicity and structures of a rationally designed prefusion MERS-CoV spike antigen. Proc Natl Acad Sci U S A 114:E7348–E7357. https://doi.org/10.1073/pnas.1707304114
18. Nakane T et al (2020) Single-particle cryo-EM at atomic resolution. Nature 587:152–156. https://doi.org/10.1038/s41586-020-2829-0
19. Yip KM, Fischer N, Paknia E, Chari A, Stark H (2020) Atomic-resolution protein structure determination by cryo-EM. Nature 587:157–161. https://doi.org/10.1038/s41586-020-2833-4

20. Quemin ERJ et al (2020) Cellular electron cryo-tomography to study virus-host interactions. Annu Rev Virol 7:239–262. https://doi.org/10.1146/annurev-virology-021920-115935
21. Yao H et al (2020) Molecular architecture of the SARS-CoV-2 Virus. Cell. https://doi.org/10.1016/j.cell.2020.09.018
22. Schur FK et al (2016) An atomic model of HIV-1 capsid-SP1 reveals structures regulating assembly and maturation. Science 353:506–508. https://doi.org/10.1126/science.aaf9620
23. Kuhlbrandt W (2014) Biochemistry. The resolution revolution. Science 343:1443–1444. https://doi.org/10.1126/science.1251652
24. Nogales E, Scheres SH (2015) Cryo-EM: a unique tool for the visualization of macromolecular complexity. Mol Cell 58:677–689. https://doi.org/10.1016/j.molcel.2015.02.019
25. Elmlund D, Le SN, Elmlund H (2017) High-resolution cryo-EM: the nuts and bolts. Curr Opin Struct Biol 46:1–6. https://doi.org/10.1016/j.sbi.2017.03.003
26. Harris JR (1997) Negative staining and cryoelectron microscopy: the thin film techniques. BIOS Scientific Publishers Ltd.
27. Dubochet J et al (1988) Cryo-electron microscopy of vitrified specimens. Q Rev Biophys 21:129–228
28. Dobro MJ, Melanson LA, Jensen GJ, McDowall AW (2010) Plunge freezing for electron cryomicroscopy. Methods Enzymol 481:63–82. https://doi.org/10.1016/S0076-6879(10)81003-1
29. Grassucci RA, Taylor DJ, Frank J (2007) Preparation of macromolecular complexes for cryo-electron microscopy. Nat Protoc 2:3239–3246. https://doi.org/10.1038/nprot.2007.452
30. Henderson R (1992) Image contrast in high-resolution electron microscopy of biological macromolecules: TMV in ice. Ultramicroscopy 46:1–18
31. Conway JF et al (1993) The effects of radiation damage on the structure of frozen hydrated HSV-1 capsids. J Struct Biol 111:222–233. S1047-8477(83)71052-X [pii] https://doi.org/10.1006/jsbi.1993.1052
32. Bammes BE, Jakana J, Schmid MF, Chiu W (2010) Radiation damage effects at four specimen temperatures from 4 to 100 K. J Struct Biol 169:331–341. https://doi.org/10.1016/j.jsb.2009.11.001
33. D'Imprima E et al (2019) Protein denaturation at the air-water interface and how to prevent it. eLife 8. https://doi.org/10.7554/eLife.42747
34. Klebl DP et al (2020) Need for speed: examining protein behavior during CryoEM grid preparation at different timescales. Structure 28:1238–1248 e1234. https://doi.org/10.1016/j.str.2020.07.018
35. Wei H et al (2018) Optimizing "self-wicking" nanowire grids. J Struct Biol 202:170–174. https://doi.org/10.1016/j.jsb.2018.01.001
36. Schmidli C et al (2019) Microfluidic protein isolation and sample preparation for high-resolution cryo-EM. Proc Natl Acad Sci U S A 116:15007–15012. https://doi.org/10.1073/pnas.1907214116
37. Rubinstein JL et al (2019) Shake-it-off: a simple ultrasonic cryo-EM specimen-preparation device. Acta Crystallogr D Struct Biol 75:1063–1070. https://doi.org/10.1107/S2059798319014372
38. Ravelli RBG et al (2020) Cryo-EM structures from sub-nl volumes using pin-printing and jet vitrification. Nat Commun 11:2563. https://doi.org/10.1038/s41467-020-16392-5
39. Kontziampasis D et al (2019) A cryo-EM grid preparation device for time-resolved structural studies. IUCrJ 6:1024–1031. https://doi.org/10.1107/S2052252519011345
40. Passmore LA, Russo CJ (2016) Specimen preparation for high-resolution Cryo-EM. Methods Enzymol 579:51–86. https://doi.org/10.1016/bs.mie.2016.04.011
41. Duong-Ly KC, Gabelli SB (2014) Gel filtration chromatography (size exclusion chromatography) of proteins. Methods Enzymol 541:105–114. https://doi.org/10.1016/B978-0-12-420119-4.00009-4
42. Skiniotis G, Southworth DR (2016) Single-particle cryo-electron microscopy of macromolecular complexes. Microscopy (Oxf) 65:9–22. https://doi.org/10.1093/jmicro/dfv366

43. Kastner B et al (2008) GraFix: sample preparation for single-particle electron cryomicroscopy. Nat Methods 5:53–55. https://doi.org/10.1038/nmeth1139
44. McMullan G, Faruqi AR, Clare D, Henderson R (2014) Comparison of optimal performance at 300keV of three direct electron detectors for use in low dose electron microscopy. Ultramicroscopy 147:156–163. https://doi.org/10.1016/j.ultramic.2014.08.002
45. McMullan G, Faruqi AR, Henderson R (2016) Direct electron detectors. Methods Enzymol 579:1–17. https://doi.org/10.1016/bs.mie.2016.05.056
46. Brilot AF et al (2012) Beam-induced motion of vitrified specimen on holey carbon film. J Struct Biol 177:630–637. https://doi.org/10.1016/j.jsb.2012.02.003
47. Scheres SH (2014) Beam-induced motion correction for sub-megadalton cryo-EM particles. Elife 3:e03665. https://doi.org/10.7554/eLife.03665
48. Li X et al (2013) Electron counting and beam-induced motion correction enable near-atomic-resolution single-particle cryo-EM. Nat Methods 10:584–590. https://doi.org/10.1038/nmeth.2472
49. Grant T, Grigorieff N (2015) Measuring the optimal exposure for single particle cryo-EM using a 2.6 A reconstruction of rotavirus VP6. Elife 4:e06980. https://doi.org/10.7554/eLife.06980
50. Ripstein ZA, Rubinstein JL (2016) Processing of cryo-EM movie data. Methods Enzymol 579:103–124. https://doi.org/10.1016/bs.mie.2016.04.009
51. Russo CJ, Passmore LA (2014) Electron microscopy: ultrastable gold substrates for electron cryomicroscopy. Science 346:1377–1380. https://doi.org/10.1126/science.1259530
52. Naydenova K, Russo CJ (2022) Integrated wafer-scale manufacturing of electron cryomicroscopy specimen supports. Ultramicroscopy 232:113396. https://doi.org/10.1016/j.ultramic.2021.113396
53. Mastronarde DN (2005) Automated electron microscope tomography using robust prediction of specimen movements. J Struct Biol 152:36–51. https://doi.org/10.1016/j.jsb.2005.07.007
54. Cheng A et al (2021) Leginon: new features and applications. Protein Sci 30:136–150. https://doi.org/10.1002/pro.3967
55. Zhang J et al (2009) JADAS: a customizable automated data acquisition system and its application to ice-embedded single particles. J Struct Biol 165:1–9. https://doi.org/10.1016/j.jsb.2008.09.006
56. Cheng Y, Grigorieff N, Penczek PA, Walz T (2015) A primer to single-particle cryo-electron microscopy. Cell 161:438–449. https://doi.org/10.1016/j.cell.2015.03.050
57. Scheres SH (2016) Processing of structurally heterogeneous cryo-EM Data in Relion. Methods Enzymol 579:125–157. https://doi.org/10.1016/bs.mie.2016.04.012
58. Ludtke SJ (2016) Single-particle refinement and variability analysis in EMAN2.1. Methods Enzymol 579:159–189. https://doi.org/10.1016/bs.mie.2016.05.001
59. Punjani A, Rubinstein JL, Fleet DJ, Brubaker MA (2017) cryoSPARC: algorithms for rapid unsupervised cryo-EM structure determination. Nat Methods 14:290–296. https://doi.org/10.1038/nmeth.4169
60. Grigorieff N (2016) Frealign: an exploratory tool for single-particle Cryo-EM. Methods Enzymol 579:191–226. https://doi.org/10.1016/bs.mie.2016.04.013
61. de la Rosa-Trevin JM et al (2016) Scipion: a software framework toward integration, reproducibility and validation in 3D electron microscopy. J Struct Biol 195:93–99. https://doi.org/10.1016/j.jsb.2016.04.010
62. Scheres SH (2012) RELION: implementation of a Bayesian approach to cryo-EM structure determination. J Struct Biol 180:519–530. https://doi.org/10.1016/j.jsb.2012.09.006
63. Scheres SH et al (2007) Disentangling conformational states of macromolecules in 3D-EM through likelihood optimization. Nat Methods 4:27–29. https://doi.org/10.1038/nmeth992
64. Lyumkis D, Brilot AF, Theobald DL, Grigorieff N (2013) Likelihood-based classification of cryo-EM images using FREALIGN. J Struct Biol 183:377–388. https://doi.org/10.1016/j.jsb.2013.07.005
65. Goetschius DJ, Lee H, Hafenstein S (2019) CryoEM reconstruction approaches to resolve asymmetric features. Adv Virus Res 105:73–91. https://doi.org/10.1016/bs.aivir.2019.07.007

66. Huiskonen JT (2018) Image processing for cryogenic transmission electron microscopy of symmetry-mismatched complexes. Biosci Rep 38. https://doi.org/10.1042/BSR20170203
67. Penczek PA (2010) Fundamentals of three-dimensional reconstruction from projections. Methods Enzymol 482:1–33. https://doi.org/10.1016/S0076-6879(10)82001-4
68. Henderson R et al (2012) Outcome of the first electron microscopy validation task force meeting. Structure 20:205–214. https://doi.org/10.1016/j.str.2011.12.014
69. Scheres SH, Chen S (2012) Prevention of overfitting in cryo-EM structure determination. Nat Methods 9:853–854. https://doi.org/10.1038/nmeth.2115
70. Subramaniam S, Earl LA, Falconieri V, Milne JL, Egelman EH (2016) Resolution advances in cryo-EM enable application to drug discovery. Curr Opin Struct Biol 41:194–202. https://doi.org/10.1016/j.sbi.2016.07.009
71. Kucukelbir A, Sigworth FJ, Tagare HD (2014) Quantifying the local resolution of cryo-EM density maps. Nat Methods 11:63–65. https://doi.org/10.1038/nmeth.2727
72. Vilas JL et al (2018) MonoRes: automatic and accurate estimation of local resolution for electron microscopy maps. Structure 26:337–344 e334 (2018). https://doi.org/10.1016/j.str.2017.12.018
73. Luque D et al (2023) Equilibrium dynamics of a biomolecular complex analyzed at single-amino acid resolution by cryo-electron microscopy. J Mol Biol 435:168024. https://doi.org/10.1016/j.jmb.2023.168024
74. Burnley T, Palmer CM, Winn M (2017) Recent developments in the CCP-EM software suite. Acta Crystallogr D Struct Biol 73:469–477. https://doi.org/10.1107/S2059798317007859
75. Afonine PV et al (2018) Real-space refinement in PHENIX for cryo-EM and crystallography. Acta Crystallogr D Struct Biol 74:531–544. https://doi.org/10.1107/S2059798318006551
76. Wang RY et al (2016) Automated structure refinement of macromolecular assemblies from cryo-EM maps using Rosetta. eLife 5. https://doi.org/10.7554/eLife.17219
77. Pettersen EF et al (2021) UCSF ChimeraX: structure visualization for researchers, educators, and developers. Protein Sci 30:70–82. https://doi.org/10.1002/pro.3943
78. Croll TI (2018) ISOLDE: a physically realistic environment for model building into low-resolution electron-density maps. Acta Crystallogr D Struct Biol 74:519–530. https://doi.org/10.1107/S2059798318002425
79. Casanal A, Lohkamp B, Emsley P (2020) Current developments in Coot for macromolecular model building of Electron Cryo-microscopy and crystallographic data. Protein Sci 29:1069–1078. https://doi.org/10.1002/pro.3791
80. Jumper J et al (2021) Highly accurate protein structure prediction with AlphaFold. Nature 596:583–589. https://doi.org/10.1038/s41586-021-03819-2
81. Baek M et al (2021) Accurate prediction of protein structures and interactions using a three-track neural network. Science 373:871–876. https://doi.org/10.1126/science.abj8754
82. Zhang P, Mendonça L (2021) Encyclopedia of virology. In: Bamford DH, Zuckerman M (eds) 4th edn. vol. 1, Academic Press, pp 242–247
83. Schaffer M et al (2017) Optimized cryo-focused ion beam sample preparation aimed at in situ structural studies of membrane proteins. J Struct Biol 197:73–82. https://doi.org/10.1016/j.jsb.2016.07.010
84. Oikonomou CM, Jensen GJ (2017) Cellular electron cryotomography: toward structural biology in situ. Annu Rev Biochem 86:873–896. https://doi.org/10.1146/annurev-biochem-061516-044741
85. Turk M, Baumeister W (2020) The promise and the challenges of cryo-electron tomography. FEBS Lett 594:3243–3261. https://doi.org/10.1002/1873-3468.13948
86. Stass R, Ilca SL, Huiskonen JT (2018) Beyond structures of highly symmetric purified viral capsids by cryo-EM. Curr Opin Struct Biol 52:25–31. https://doi.org/10.1016/j.sbi.2018.07.011
87. Jose J, Hafenstein SL (2022) Asymmetry in icosahedral viruses. Curr Opin Virol 54:101230. https://doi.org/10.1016/j.coviro.2022.101230
88. Twarock R, Bingham RJ, Dykeman EC, Stockley PG (2018) A modelling paradigm for RNA virus assembly. Curr Opin Virol 31:74–81. https://doi.org/10.1016/j.coviro.2018.07.003

89. Twarock R, Stockley PG (2019) RNA-mediated virus assembly: mechanisms and consequences for viral evolution and therapy. Annu Rev Biophys 48:495–514. https://doi.org/10.1146/annurev-biophys-052118-115611
90. Dai X et al (2017) In situ structures of the genome and genome-delivery apparatus in a single-stranded RNA virus. Nature 541:112–116. https://doi.org/10.1038/nature20589
91. Koning RI et al (2016) Asymmetric cryo-EM reconstruction of phage MS2 reveals genome structure in situ. Nat Commun 7:12524. https://doi.org/10.1038/ncomms12524
92. Gorzelnik KV et al (2016) Asymmetric cryo-EM structure of the canonical Allolevivirus Qbeta reveals a single maturation protein and the genomic ssRNA in situ. Proc Natl Acad Sci U S A 113:11519–11524. https://doi.org/10.1073/pnas.1609482113
93. Zhang X et al (2015) In situ structures of the segmented genome and RNA polymerase complex inside a dsRNA virus. Nature 527:531–534. https://doi.org/10.1038/nature15767
94. Liu YT, Jih J, Dai X, Bi GQ, Zhou ZH (2019) Cryo-EM structures of herpes simplex virus type 1 portal vertex and packaged genome. Nature 570:257–261. https://doi.org/10.1038/s41586-019-1248-6
95. Conley MJ et al (2019) Calicivirus VP2 forms a portal-like assembly following receptor engagement. Nature 565:377–381. https://doi.org/10.1038/s41586-018-0852-1
96. Liu H, Cheng L (2015) Cryo-EM shows the polymerase structures and a nonspooled genome within a dsRNA virus. Science 349:1347–1350.
97. Pan M, Alvarez-Cabrera AL, Kang JS, Wang L, Fan C, Zhou ZH (2021) Asymmetric reconstruction of mammalian reovirus reveals interactions among RNA, transcriptional factor micro2 and capsid proteins. Nat Commun 12:4176.
98. Ding K, Celma CC, Zhang X, Chang T, Shen W, Atanasov I, Roy P, Zhou ZH (2019) In situ structures of rotavirus polymerase in action and mechanism of mRNA transcription and release. Nat Commun 10: 2216.
99. Cui Y, Zhang Y, Zhou K, Sun J, Zhou ZH (2019) Conservative transcription in three steps visualized in a double-stranded RNA virus. Nat Struct Mol Biol 26:1023–1034.
100. He Y, Shivakoti S, Ding K, Cui Y, Roy P, Zhou ZH (2019) In situ structures of RNA-dependent RNA polymerase inside bluetongue virus before and after uncoating. Proc Natl Acad Sci 116:16535–16540.
101. Ilca SL et al (2019) Multiple liquid crystalline geometries of highly compacted nucleic acid in a dsRNA virus. Nature 570:252–256. https://doi.org/10.1038/s41586-019-1229-9
102. Li Z, Pang J, Dong L, Yu X (2021) Structural basis for genome packaging, retention, and ejection in human cytomegalovirus. Nat Commun 12:4538. https://doi.org/10.1038/s41467-021-24820-3
103. Abergel C, Legendre M, Claverie JM (2015) The rapidly expanding universe of giant viruses: Mimivirus, Pandoravirus Pithovirus and Mollivirus. FEMS Microbiol Rev 39:779–796. https://doi.org/10.1093/femsre/fuv037
104. Burton-Smith RN, Murata K (2021) Cryo-electron microscopy of the giant viruses. Microscopy (Oxf) 70:477–486. https://doi.org/10.1093/jmicro/dfab036
105. Chihara A et al (2022) A novel capsid protein network allows the characteristic internal membrane structure of Marseilleviridae giant viruses. Sci Rep 12:21428. https://doi.org/10.1038/s41598-022-24651-2
106. Zhu D et al (2018) Pushing the resolution limit by correcting the Ewald sphere effect in single-particle Cryo-EM reconstructions. Nat Commun 9:1552. https://doi.org/10.1038/s41467-018-04051-9
107. Downing KH, Glaeser RM (2018) Estimating the effect of finite depth of field in single-particle cryo-EM. Ultramicroscopy 184:94–99. https://doi.org/10.1016/j.ultramic.2017.08.007
108. Fang Q et al (2019) Near-atomic structure of a giant virus. Nat Commun 10:388. https://doi.org/10.1038/s41467-019-08319-6
109. Zhang X et al (2011) Three-dimensional structure and function of the Paramecium bursaria chlorella virus capsid. Proc Natl Acad Sci U S A 108:14837–14842. https://doi.org/10.1073/pnas.1107847108

110. Liu S et al (2019) Cryo-EM structure of the african swine fever virus. Cell Host Microbe 26:836–843 e833. https://doi.org/10.1016/j.chom.2019.11.004
111. Wang N et al (2019) Architecture of African swine fever virus and implications for viral assembly. Science 366:640–644. https://doi.org/10.1126/science.aaz1439
112. Pintilie G et al (2019) Segmentation and comparative modeling in an 8.6-A Cryo-EM map of the Singapore grouper iridovirus. Structure 27:1561–1569 e1564 (2019). https://doi.org/10.1016/j.str.2019.08.002
113. Benson SD, Bamford JK, Bamford DH, Burnett RM (2004) Does common architecture reveal a viral lineage spanning all three domains of life? Mol Cell 16:673–685. S1097276504007099 [pii] https://doi.org/10.1016/j.molcel.2004.11.016
114. Rapp M, Shapiro L, Frank J (2022) Contributions of single-particle cryoelectron microscopy toward fighting COVID-19. Trends Biochem Sci 47:117–123. https://doi.org/10.1016/j.tibs.2021.10.005
115. Cai Y et al (2020) Distinct conformational states of SARS-CoV-2 spike protein. Science 369:1586–1592. https://doi.org/10.1126/science.abd4251
116. Ke Z et al (2020) Structures and distributions of SARS-CoV-2 spike proteins on intact virions. Nature 588:498–502. https://doi.org/10.1038/s41586-020-2665-2
117. Wrapp D et al (2020) Cryo-EM Structure of the 2019-nCoV spike in the prefusion conformation. bioRxiv. https://doi.org/10.1101/2020.02.11.944462
118. Hardenbrook NJ, Zhang P (2022) A structural view of the SARS-CoV-2 virus and its assembly. Curr Opin Virol 52:123–134. https://doi.org/10.1016/j.coviro.2021.11.011
119. Vankadari N, Shepherd DC, Carter SD, Ghosal D (2022) Three-dimensional insights into human enveloped viruses in vitro and in situ. Biochem Soc Trans 50:95–105. https://doi.org/10.1042/BST20210433
120. Zeev-Ben-Mordehai T et al (2016) Two distinct trimeric conformations of natively membrane-anchored full-length herpes simplex virus 1 glycoprotein B. Proc Natl Acad Sci U S A 113:4176–4181. https://doi.org/10.1073/pnas.1523234113
121. Arranz R et al (2012) The structure of native influenza virion ribonucleoproteins. Science 338:1634–1637. https://doi.org/10.1126/science.1228172
122. Hong Y, Song Y, Zhang Z, Li S (2023) Cryo-electron tomography: the resolution revolution and a surge of in situ virological discoveries. Annu Rev Biophys. https://doi.org/10.1146/annurev-biophys-092022-100958
123. Wan W et al (2017) Structure and assembly of the Ebola virus nucleocapsid. Nature 551:394–397. https://doi.org/10.1038/nature24490
124. Mattei S, Glass B, Hagen WJ, Krausslich HG, Briggs JA (2016) The structure and flexibility of conical HIV-1 capsids determined within intact virions. Science 354:1434–1437. https://doi.org/10.1126/science.aah4972
125. Peukes J et al (2020) The native structure of the assembled matrix protein 1 of influenza a virus. Nature 587:495–498. https://doi.org/10.1038/s41586-020-2696-8
126. Zhan H et al (2023) Nodavirus RNA replication crown architecture reveals proto-crown precursor and viral protein a conformational switching. Proc Natl Acad Sci U S A 120:e2217412120. https://doi.org/10.1073/pnas.2217412120
127. Hochstein R et al (2018) Structural studies of Acidianus tailed spindle virus reveal a structural paradigm used in the assembly of spindle-shaped viruses. Proc Natl Acad Sci U S A 115:2120–2125. https://doi.org/10.1073/pnas.1719180115
128. Wolff G et al (2020) A molecular pore spans the double membrane of the coronavirus replication organelle. Science 369:1395–1398. https://doi.org/10.1126/science.abd3629
129. Unchwaniwala N, Zhan H, den Boon JA, Ahlquist P (2021) Cryo-electron microscopy of nodavirus RNA replication organelles illuminates positive-strand RNA virus genome replication. Curr Opin Virol 51:74–79. https://doi.org/10.1016/j.coviro.2021.09.008
130. Hampton CM et al (2017) Correlated fluorescence microscopy and cryo-electron tomography of virus-infected or transfected mammalian cells. Nat Protoc 12:150–167. https://doi.org/10.1038/nprot.2016.168
131. Chaikeeratisak V et al (2017) Assembly of a nucleus-like structure during viral replication in bacteria. Science 355:194–197. https://doi.org/10.1126/science.aal2130

132. Chaikeeratisak V et al (2019) Viral capsid trafficking along treadmilling tubulin filaments in Bacteria. Cell 177:1771–1780 e1712 (2019). https://doi.org/10.1016/j.cell.2019.05.032
133. Chaikeeratisak V et al (2017) The phage nucleus and tubulin spindle are conserved among large pseudomonas phages. Cell Rep 20:1563–1571. https://doi.org/10.1016/j.celrep.2017.07.064
134. Dai W et al (2013) Visualizing virus assembly intermediates inside marine cyanobacteria. Nature 502:707–710. https://doi.org/10.1038/nature12604
135. Kelly DF et al (2022) Liquid-EM goes viral - visualizing structure and dynamics. Curr Opin Struct Biol 75:102426. https://doi.org/10.1016/j.sbi.2022.102426
136. Jonaid GM et al (2021) High-resolution imaging of human viruses in liquid droplets. Adv Mater 33:e2103221. https://doi.org/10.1002/adma.202103221
137. Kennedy E, Nelson EM, Tanaka T, Damiano J, Timp G (2016) Live Bacterial physiology visualized with 5 nm resolution using scanning transmission electron microscopy. ACS Nano 10:2669–2677. https://doi.org/10.1021/acsnano.5b07697

Chapter 13
Bacteriophage Lambda as a Nano Theranostic Platform

Carlos Enrique Catalano

Abstract A variety of nanoparticles have been developed for use in therapeutic and diagnostic (theranostic) applications. In this review we describe a "designer" nanoparticle platform based on the phage lambda capsid. The shell can be decorated with a variety of synthetic, organic, and biological ligands, alone and in combination, and at rigorously defined surface densities. The particles retain structural and physical integrity, they possess physiochemical properties compatible with pharmaceutical standards and are amenable to formulation as thermostable, single shot preparations. We describe their potential applications including targeted delivery of therapeutic antibodies to cancer cells, intracellular delivery of biologic agents and as a nimble platform for vaccine development. The ability to package DNA and the potential for selective incorporation of protein cargos into the shell interior, simultaneously decorated with multiple ligands in a defined manner provide a nimble platform that can be rapidly adapted to a variety of user-defined theranostic applications.

Keywords Bacteriophage lambda · Phage-like particles · Designer nanoparticles · Theranostic nanoparticles · Intracellular delivery of biologics · Vaccine platform

Introduction

Nanoparticles show great promise in a variety of applications including sensors, diagnostics, and therapeutic agents [1, 84, 32, 41]. A variety of materials have been developed as nanoparticles, from metals [31, 80, 16] to synthetic polymers [36, 39, 42, 71] to biological agents, including viral [9, 35, 44, 49, 66, 69, 74, 73, 89] and even bacterial [64] platforms. In this chapter, we focus on bacteriophage lambda as a semi-synthetic platform for "theranostic" nanoparticle development. We first describe the developmental pathway for the virus in *Escherichia coli* and biochemical/biophysical characterization of capsid assembly mechanisms in vitro. We next

C. E. Catalano (✉)
Department of Pharmaceutical Sciences, Skaggs School of Pharmacy and Pharmaceutical Sciences, University of Colorado, Anschutz Medical Campus, Aurora, CO 80045, USA
e-mail: carlos.catalano@cuanschutz.edu

M. Comas-Garcia and S. Rosales-Mendoza (eds.), *Physical Virology*, Springer Series in Biophysics 24, https://doi.org/10.1007/978-3-031-36815-8_13

discuss the conceptual cooptation of the process towards a "designer" nanoparticle and our initial studies on its utility as a defined semi-synthetic platform. Finally, we describe several proof-of-concept studies illustrating its potential in therapeutic settings and in future applications.

Bacteriophage Lambda

The lambda development pathway is generally conserved in the large double-stranded DNA viruses, from bacteriophages to the herpesviruses [8, 11, 33, 67, 68]. The virus attaches to an *Escherichia coli* bacterial cell wall using a "tail" structure through which DNA exits the capsid and is introduced into the cell (Fig. 13.1). The 48.5 kbp linear genome circularizes via 12 base complementary single strand ends (the **cohesive ends**) and is used as a DNA replication substrate that ultimately yields linear concatemers of multiple genomes covalently linked in a head-to-tail fashion (immature DNA). Late gene expression affords structural proteins, including capsid proteins and tail proteins, that self assemble into functional procapsid and tail structures, respectively.

The next step, **genome packaging**, is catalyzed by an enzyme known as **terminase** and represents the intersection of the procapsid assembly and DNA replication pathways (Fig. 13.2) [10, 65, 14, 15]. Terminase assembles at the cohesive end site (***cos***) in a concatemer and site-specifically nicks the duplex to generate the 12-base single stranded mature left end (D_L) of the first genome to be packaged (*cos*

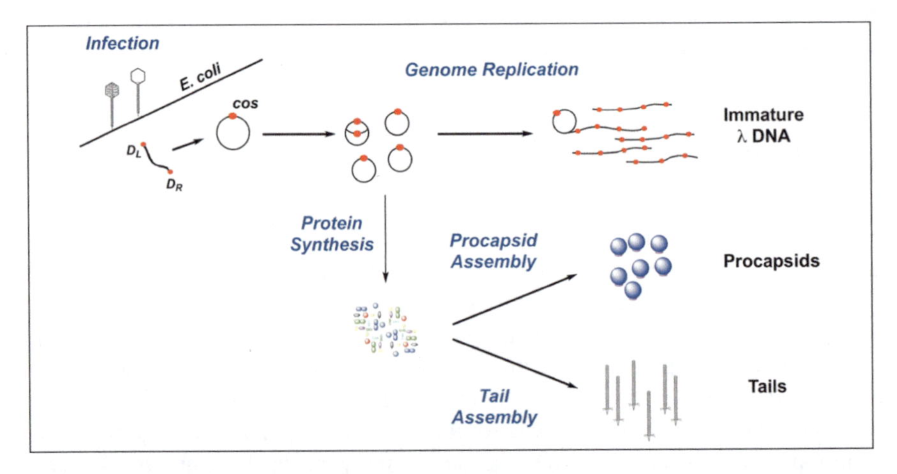

Fig. 13.1 Bacteriophage Lambda Development. The D_L and D_R 12-base cohesive ends of the mature λ genome (linear monomer) anneal upon injection into the *E. coli* cell to engender a circular genome containing the cohesive end site (***cos***, red dot). This sequence represents the junction between genomes in concatemeric (immature) viral DNA. Details provided in the text. This figure was taken from [13], with permission

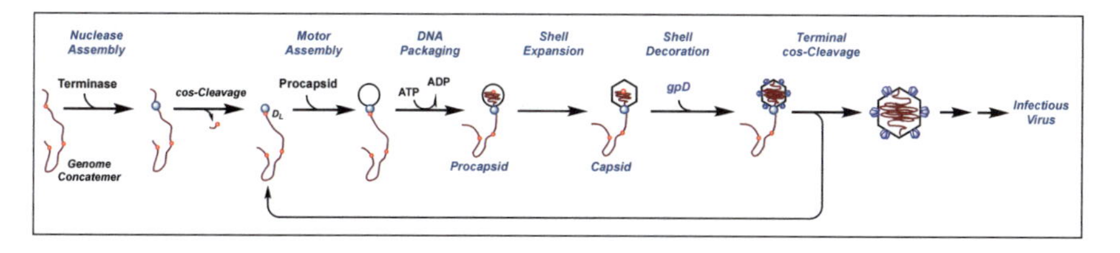

Fig. 13.2 Genome Packaging. The terminase enzyme cycles between a stable nuclease complex and a dynamic motor complex to processively excise and package multiple matured genomes from concatemeric λ DNA. These reactions are strongly conserved in all of the large dsDNA viruses. Details provided in the text. This figure was modified from [13], with permission

cleavage reaction). This stable nucleoprotein complex then binds to the "portal ring" situated at a unique vertex of the icosahedral procapsid; the portal provides a conduit for DNA packaging into the shell interior, and for genome exit during infection. The enzyme then switches from a stable nuclease complex to a dynamic DNA packaging motor that inserts DNA into the shell powered by ATP hydrolysis.

Upon packaging ~30% genome length (15 kbp), the *procapsid* shell undergoes a remarkable conformational change that affords a larger, angularized, and thinner *capsid* shell [23, 33, 81]. This exposes nucleation sites for the **gpD decoration protein** that adds as trimer spikes at the 140 three-fold axes on the shell exterior (Fig. 13.2) [46, 70]. Terminase continues to package DNA until the next downstream *cos* sequence is reached (the genome end) and the enzyme switches back to a nuclease complex that again introduces nicks into the duplex; this affords a decorated shell filled with a single *mature* lambda genome. Finishing proteins and a pre-assembled tail add to the nucleocapsid to yield an infectious phage particle while the ejected terminase•concatemer complex (minus one genome) binds to another procapsid to initiate a second round of processive genome packaging [10], [65, 14, 15].

Procapsid Assembly in Vivo

Lambda procapsid assembly is a stepwise process that initiates with self-association of twelve portal proteins into ring-like structure, chaperoned by host groELS and the lambda scaffolding protein (**SP**) (Fig. 13.3) [29, 34]. The portal nucleates co-polymerization of the major capsid protein (**CP**) and SP into a thick, spherical icosahedral shell; the scaffolding protein acts as a chaperone for high-fidelity shell assembly and in its absence CP aggregates into a variety aberrant and non-functional structures [56, 62, 91]. The lambda capsid protease (**gpC**), which is a fusion composed of a N-terminal protease domain and a C-terminal SP domain, is also incorporated into the shell to afford an "immature" procapsid composed of a portal ring situated at a unique vertex of an icosahedral shell that is made up of 415 copies of CP and containing 10–12 copies of gpC and 100–200 copies of SP at the interior [25, 29, 34, 55, 57]. The protease is autoproteolytic, it degrades the

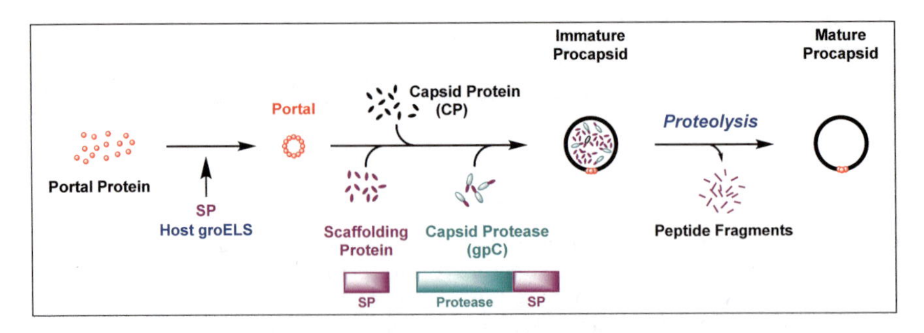

Fig. 13.3 Lambda Procapsid Assembly In Vivo. The assembly and maturation reactions are strongly conserved in all of the large dsDNA viruses. The gpC protease is incorporated into the shell by co-polymerization with SP, mediated by the SP domain. Details provided in the text. This figure was modified from [13], with permission

scaffolding protein and in addition removes 20 N-terminal residues from the portal proteins. The peptide fragments exit the shell to afford an empty and packaging competent mature procapsid shell (Fig. 13.3) [29, 33].

Characterization of Capsid Assembly in Vitro

Our lab has interrogated the mechanisms of procapsid assembly, genome packaging, shell expansion, and shell decoration reactions in vitro using defined biochemical assay systems [15, 14, 28, 45, 60, 85, 86]. Pertinent to the present discussion, we have examined maturation of the procapsid shell by the gpC protease (see Fig. 13.3) [55]. Mutation of the catalytic serine 166 to alanine abrogates the catalytic activity of the enzyme and as a result, the protease and scaffolding proteins remain intact. It was presumed that proteolysis was required for their exit from the shell prior to DNA packaging and while gpC-S166A (45.9 kDa) indeed remains trapped, the full-length scaffolding protein (13.4 kDa) is slowly released.

We have also characterized the shell expansion reaction in vitro (see Fig. 13.2). Procapsid shells can be artificially expanded with urea and we found that the transition is reversible, highly cooperative, strongly temperature dependent (exothermic) and inhibited by salt [54]. The free energy of expansion is 10 kcal/mol (in the presence of physiological Mg^{2+} concentrations) and the data indicate that significant hydrophobic surface area is exposed in the expanded shell. We next characterized the shell decoration reaction in vitro (see Fig. 13.2). The lambda decoration protein (**gpD**) is a monomer in solution and does not interact with the procapsid shell, but adds to the expanded capsid surface as trimeric spikes at each of the 140 three-fold axes (420 copies total) (Figs. 13.2, 13.4a) [46, 54]. The reaction is non-reversible, cooperative and strongly temperature dependent (endothermic) [45, 85]. In sum, the thermodynamic data suggest a model wherein expansion of the procapsid shell exposes hydrophobic surface area at the three-fold axes of the capsid which serves to nucleate cooperative gpD trimer spike assembly.

Finally, we examined lambda capsid shell assembly in vitro. While the lambda scaffolding protein is intrinsically disordered [22, 90], the purified protein is biologically active and is required for high-fidelity polymerization of CP into an icosahedral phage-like particle (**PLP**) [56]. These particles differ from procapsids in that they

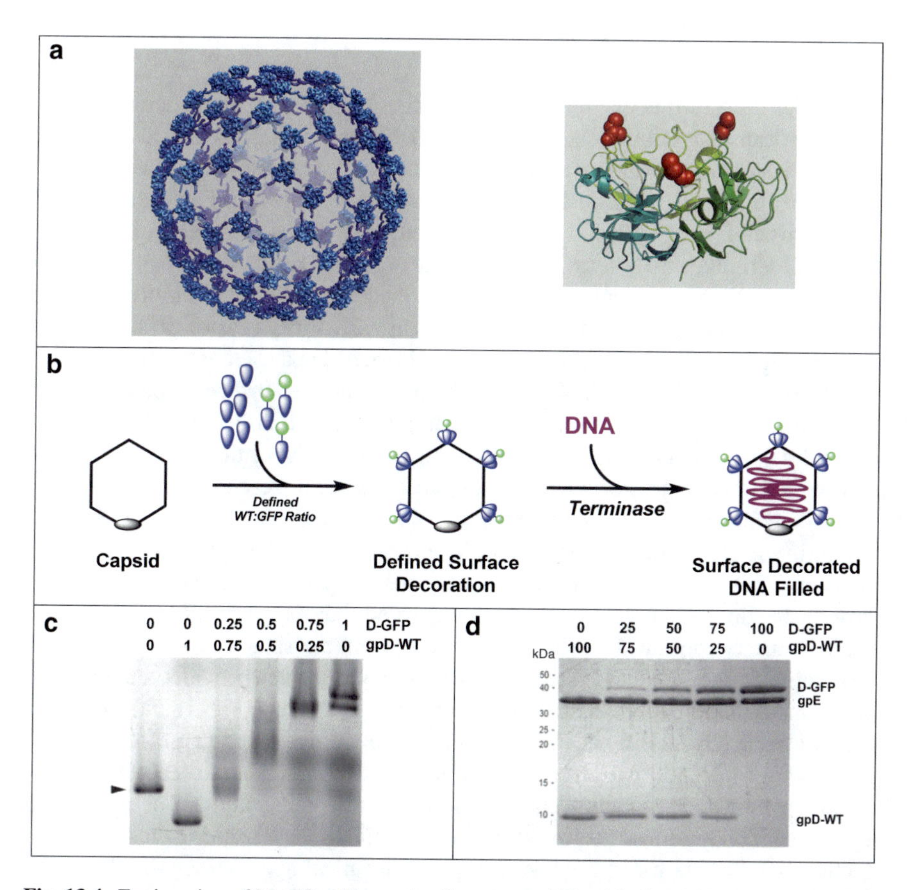

Fig. 13.4 Engineering of Modified Decoration Proteins and Tunable Capsid Decoration In Vitro. **a** Cryo-EM structure of a λ capsid decorated with gpD trimer spikes. Only the decoration protein is shown for clarity. The crystal structure of a gpD trimer (PDB ID# 1C5E) is shown at right with Ser42 shown as red spheres. The capsid surface is at the base of the spike and the residue projects away from the shell surface. **b** Strategy for tunable capsid decoration. The mole ratio of gpD-WT (blue) and D-GFP fusion protein (green) can be controlled in the decoration reaction, which allows precise capsid surface decoration. There are 420 gpD binding sites on the shell surface and we define ***percent surface density*** as (# Decoration Proteins/420)*100. **c** Agarose gel showing capsids decorated with D-GFP. The mole ratio of gpD-WT:D-GFP included in the decoration reaction is shown at top of the gel. **d** SDS-PAGE analysis of GFP-decorated particles purified by size exclusion chromatography. The percent surface density of ligands displayed on the shell surface are indicated at top. This figure was modified from [17], with permission

do not contain a portal. PLPs can also be isolated in high yield from *E. coli* cells that co-express capsid and scaffolding proteins [12]. As observed with capsids, the PLPs can be expanded and decorated under defined reaction conditions in vitro [54, 13].

Cooptation of Lambda as a Platform for Designer Nanoparticle Applications

The studies described above provided a fundamental, mechanistic understanding of lambda procapsid assembly, shell maturation, DNA packaging, procapsid expansion and capsid decoration reactions. We reasoned that these processes might be harnessed to engineer a tunable, rigorously defined nanoparticle platform. For instance, gpD has been used in phage display applications with peptides successfully displayed at both the C-terminal and N-terminal ends of the protein in the context of an infectious phage [24, 3, 27, 58, 92] and we recapitulated the phage display assembly reactions under defined biochemical reaction conditions in vitro. First, a fusion protein of gpD and green fluorescent protein (**D-GFP**) was constructed and used to decorate lambda capsids as outlined in Fig. 13.4b [17]. We found that the surface display density[1] can be tuned in a defined manner by adjusting the mole ratio of D-GFP: gpD-WT during the decoration reaction (Fig. 13.4c, d). The potential of these GFP-decorated particles for tracking and/or diagnostic applications did not escape our attention.

The demonstration that lambda capsids could be decorated with the D-GFP fusion protein in vitro provided support for its use as a tunable and defined nanoparticle platform; however, this approach limits the system to proteinaceous display ligands. Close inspection of the gpD trimer spike revealed that that serine 42 is positioned within a surface loop of the protein, projecting from the shell surface (Fig. 13.4a). We modified this residue to cysteine to afford **D(S42C)** which provides a unique site for chemical modification using simple maleimide chemistry (Fig. 13.5a). As a first test, we chemically modified D(S42C) with methoxy-polyethylene glycol maleimide (PEG average MW 5000) to afford **D-PEG** and used the construct to decorate lambda capsids in vitro [17]. As with D-GFP, the PEG surface density can be tuned in a defined manner by adjusting the mole ratio of D-PEG:gpD-WT during the decoration reaction (Fig. 13.5b, c). PEG is used in pharmaceutical applications as a "stealth" agent to improve immune and pharmacokinetic properties of therapeutics [61, 79]. Of note, particles decorated with 100% D-PEG migrate backwards in an agarose gel (Fig. 13.5b), indicating that the surface properties of the shell are significantly masked. This study sets the stage for tuning lambda PLP stealth properties as desired.

We next chemically modified D(S42C) with mannose-maleimide to afford **D-mannose** and used the synthetic glycoprotein to decorate capsids alone and in combination with GFP in defined surface densities (Fig. 13.5d) [17]. Importantly,

[1] There are 420 gpD binding sites on the shell surface and we define ***percent surface density*** as (# Decoration Proteins/420)*100.

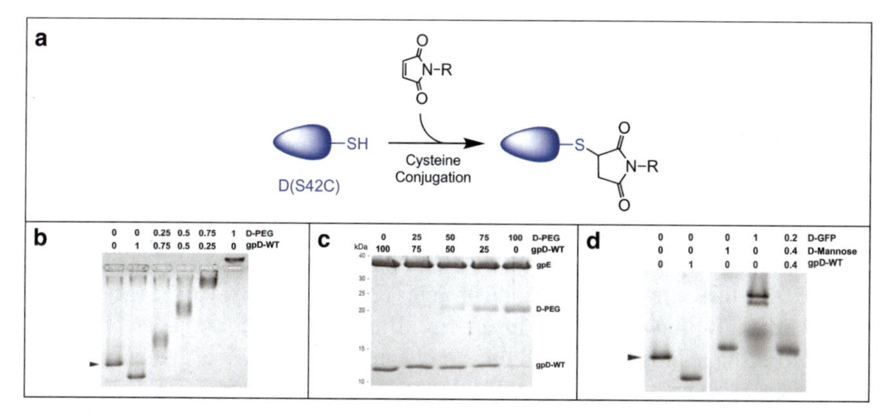

Fig. 13.5 Engineering of Semi-Synthetic Nanoparticles. **a** Strategy for site-specific modification of Cys42 with maleimide derivatives. **b** Agarose gel showing capsids decorated with D-PEG. The mole ratio of gpD-W:D-PEG included in the decoration reaction is shown at top of the gel. Note that PLPs decorated with 100% D-PEG migrate backwards in the gel, indicating significant shielding of the shell surface by the polymer. **c** SDS-PAGE analysis of PEG-decorated particles purified by size exclusion chromatography. The percent surface density of ligands displayed on the shell surface are indicated at top. **d** Agarose gel showing capsids decorated with D-GFP and D-mannose, alone and in combination. The mole ratio of each modified decoration protein included in the decoration reaction is shown at top. This figure was modified from [17], with permission

shells decorated with GFP, PEG, or mannose can be packaged with heterologous DNA in vitro to afford particles that display defined ligands on the shell surface, alone or in combination, and that simultaneously carry a protected DNA cargo, as depicted in Fig. 13.4b [17].

Application of the Lambda Platform

The studies described above demonstrated that the lambda shell can be decorated with multiple display ligands in a tunable manner to afford defined semi-synthetic nanoparticles. We next set out to engineer PLPs that displayed complex biological ligands with potential in therapeutic applications.

Targeted Intracellular Delivery of Therapeutic Antibodies

Trastuzumab (**Trz**) is a therapeutic antibody that specifically targets the extracellular domain of the Human Epidermal Growth Factor 2 (**HER2**). This receptor tyrosine kinase is overexpressed in a variety of cancers and is associated with a poor clinical prognosis [47, 4, 38]. While Trz has been used with success in HER2$^+$ breast cancers, systemic toxicity, intrinsic and acquired resistance, among other issues are of concern

[47, 2]. This necessitates fundamental investigations to develop additional technologies in this setting and we explored the potential of lambda PLPs as a platform to target HER2^{+} overexpressing breast cells in vitro.

We first engineered a construct consisting of D(S42C) crosslinked to trastuzumab (**D-Trz**) using the generalized approach outlined in Fig. 13.6a (Protein A = Trz) [12]. While chemical modification of D(S42C) is specific (only one cysteine), Trz contains 88 lysine residues, and we anticipated that a single antibody would likely contain multiple gpD adducts. Indeed, characterization of the cross-linked product indicates that there are 3–6 decoration proteins per antibody on average (Fig. 13.6b). Not surprisingly, this is similar to approved antibody–drug conjugate biologics generated using analogous cross-linking strategies [48, 77]. We also engineered **D-F5M** wherein D(S42C) is site specifically modified with fluorescein 5-maleimide (Fig. 13.5a; R = fluorescein). Purified PLPs were then decorated with 25% D-F5M for particle tracking, and 0.5% to 30% D-Trz for targeting of HER2^{+} overexpressing cells, as outlined in Fig. 13.6c.

We first characterized the physiochemical properties of the decorated particles using a battery of approaches, including agarose gel electrophoresis (**AGE**), SDS-PAGE, size exclusion chromatography (**SEC**), analytical ultracentrifugation (**AUC**), transmission electron microscopy (**TEM**), electrophoretic light scattering (**ELS**), and dynamic light scattering (**DLS**) [12]. In sum, the data indicate that the particles are decorated with ligands at the anticipated surface densities, that the shells retain

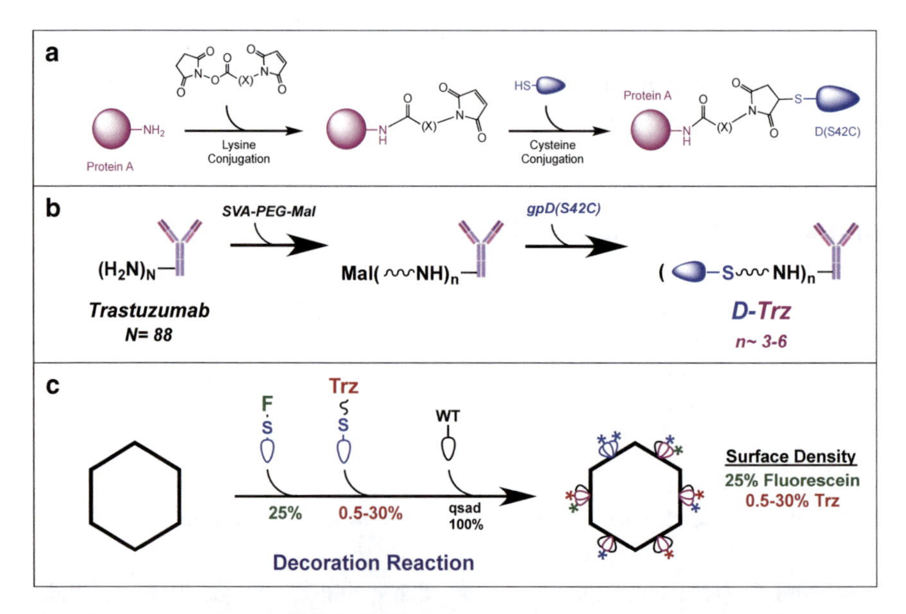

Fig. 13.6 Construction of Trastuzumab-Decorated Nanoparticles. **a** Strategy for engineering gpD-protein constructs using succinimidyl valerate maleimide (SVA-PEG-Mal) bi-functional chemical cross-linking. **b** Engineering of D-Trz. **c** Construction of Trz-PLP nanoparticles. This figure was modified from [12], with permission

structural and physical integrity and that they possess physiochemical properties compatible with pharmaceutical standards [12]. Of note, micrographs of particles decorated with Trz display projections from the shell surface in a surface-density dependent manner (Fig. 13.7a, b). This protein "corona" bears a morphology similar to that of free Trz and becomes denser as the surface density of D-Trz on the PLP is increased, indicating a tunable decoration reaction.

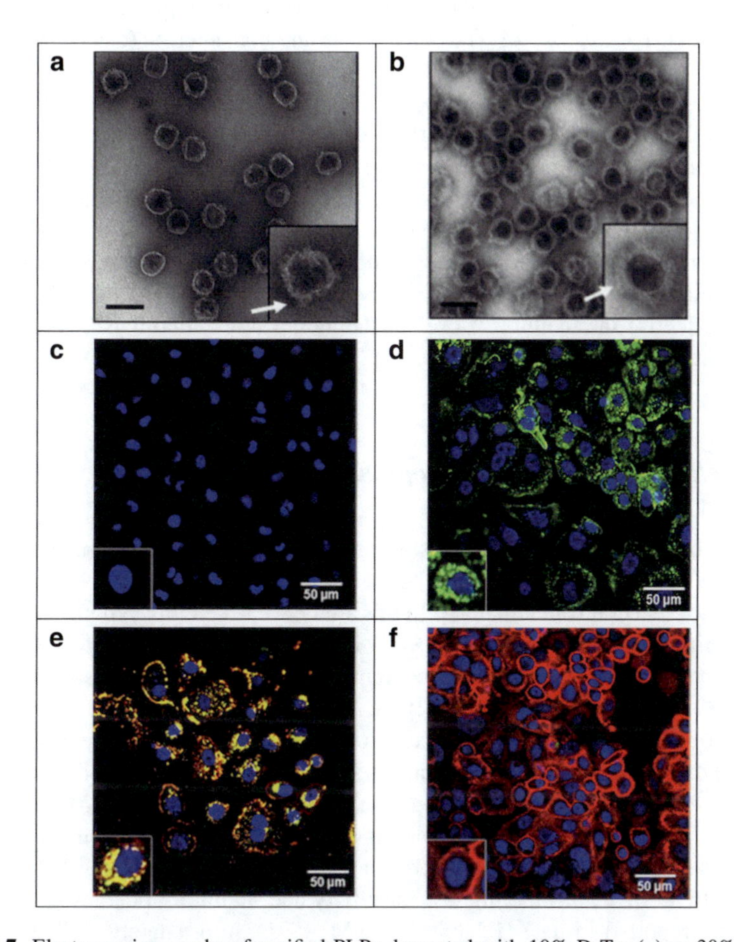

Fig. 13.7 Electron micrographs of purified PLPs decorated with 10% D-Trz (**a**) or 30% D-Trz (**b**) at 120,000X magnification; black scale bars represent 100 nm and inset shows an expanded view of a single particle; white arrows indicate the density attributed to D-Trz on the surface of thePLPs. Confocal fluorescence microscopy was performed on breast cancer cells after the following treatments. **c**. MDA-MB-231 cells ($HER2^-$) treated with 30% Trz-PLPs (2 nM). **d** SKBR3 cells ($HER2^+$) treated with 10% Trz-PLPs (2 nM). **e** SKBR3 cells ($HER2^+$) treated with 30% Trz-PLPs (2 nM). **f** SKBR3 cells ($HER2^+$) treated with 2 μM free (unmodified) Trz. All particles were simultaneously decorated with 25% D-F for particle tracking. Fluorescence signals: cell nuclei (blue); Trz and Trz-PLPs (red); Trz-PLPs (green). The internalized Trz and PLP fluorescent signals colocalize, and gradually turn yellow with increasing Trz surface decoration (compare **d** and **e**). Details are discussed in [12]. This figure was modified from [12], with permission

We next examined Trz-PLPs for their cell targeting behavior using human breast carcinoma cell lines. As anticipated, Trz-decorated PLPs bind to cells that overexpress the HER2 receptor, but not to $HER2^-$ cells (see Fig. 13.7c–e). Moreover, unlike the unmodified antibody that remains predominantly bound at the cell surface (Fig. 13.7f), Trz-PLPs are efficiently internalized by $HER2^+$ cells. Robust particle internalization is observed with Trz surface densities as low as 1% and the *number* of internalized particles does not significantly vary with higher surface densities. Nevertheless, the *amount* of Trz delivered to the cell interior (e.g., dose) can be controlled by manipulation of the PLP surface density (see Fig. 13.7d, e).

As a result of particle internalization, Trz-PLPs provide prolonged inhibition of cancer cell proliferation relative to the free antibody [12]. Moreover, multi-omics analysis (metabolomics and proteomics) demonstrate that the Trz-PLPs significantly modulate cellular programs associated with HER2 signaling, proliferation, metabolism and protein synthesis. In sum, this study demonstrates that (i) PLPs can be decorated with trastuzumab to specifically target HER2 overexpressing breast cancer cells, that (ii) the particles are efficiently internalized by the cells and that (iii) the effective intracellular "dose" of the therapeutic antibody can be controlled by varying the PLP surface density.

Intracellular Delivery of Biological Macromolecules

Based on our studies with Trz-PLPs, we reasoned that PLPs simultaneously decorated with Trz and a second protein could be used to specifically target a "biologic" to the cell interior in a controlled manner. In this study, we tested this hypothesis.

Ubiquitin-specific protease 7 (**USP7**) is a deubiquitinating enzyme that plays important roles in cell cycle control, tumor suppression, apoptosis, the immune response and in viral infection [63, 82]. Because of its association with viral infection and the observation that aberrant activation or overexpression of the enzyme may promote oncogenesis, it has become an attractive therapeutic target and several small molecule inhibitors of the enzyme have been developed [5, 52, 82]. More recently, Sidhu and co-workers employed a ubiquitin scaffold to develop protein based USP7 inhibitors using a phage display approach [26, 50, 87]. One "hit" (UbV.7.2, herein referred to as **UbV**) has the highest selectivity and inhibitory potency of all USP7 inhibitors reported to date ($IC_{50} = 16.2$ nM) [87]. Unfortunately, delivery of the protein biologic to the cell interior, wherein lies the enzyme target, remained a significant problem. We therefore sought to engineer PLPs that (i) specifically target $HER2^+$ cells and that (ii) deliver the UbV biologic for intracellular inhibition of USP7.

We first engineered a **D-UbV** construct using a "click" chemistry approach, as generally outlined in Fig. 13.8a [53]. UbV and D(S42C) each posses a sole cysteine residue and the proteins were site-specifically labeled with maleimide click reagents (Protein A = UbV). The modified proteins were then "clicked" together to afford the D-UbV cross-linked product. We next decorated PLPs with 25% D-F (particle tracking), 1% D-Trz (cell targeting) and varying surface densities of D-UbV (the "biologic") as outlined in Fig. 13.8b. Physiochemical characterization of the resulting particles as described for Trz-PLPs above confirmed that they are decorated with ligands at the anticipated surface densities (Fig. 13.8c), that the shells retain structural and physical integrity (Fig. 13.8d) and that they possess physiochemical properties compatible with pharmaceutical standards [53]. Co-immunoprecipitation studies confirmed that UbV bound to the PLP shell retained USP7 binding interactions and that the amount of USP7 pulled down by the particles increases in a UbV surface density (dose) dependent manner (Fig. 13.8e). As anticipated, Trz decoration is necessary and sufficient for specific targeting of $HER2^+$ cells and importantly, simultaneous decoration with UbV neither affects cell targeting nor particle internalization [53].

We next tested the hypothesis that the targeting PLPs can be used to deliver a second functional biologic to the cell interior. Mouse double minute 2 homolog (**MDM2**) ubiquitin ligase is an important negative regulator of the p53 tumor suppressor [87, 88]. Cellular levels of MDM2 are regulated by auto-ubiquitination, which marks the protein for proteasomal degradation, and by USP7-catalyzed de-ubiquitination, which in concert maintain steady state cellular concentrations, as outlined in Fig. 13.9a. Thus, we predicted that PLPs simultaneously decorated with Trz and UbV would be internalized by $HER2^+$ cells and deliver the biologic to the cytosol, resulting in inhibition of USP7 and thus, a decrease in MDM2 levels. Western blot analysis confirmed this prediction and further showed that MDM2 knockdown is increased with increased UbV display density (dose) (Fig. 13.9b, c). Further, the Trz:UbV particles have a significant impact on the proteomics and metabolomics landscape of $HER2^+$ cells that is distinct from the effects of Trz-PLPs alone and consistent with inhibition of USP7 [53]. Finally and as would be expected from MDM2 knockdown, the UbV decorated particles strongly inhibit cellular proliferation in a surface-density dependent manner [53].

Cellular Trafficking of the Designer Nanoparticles

We have demonstrated that the lambda designer nanoparticle can be decorated with multiple ligands for particle tracking, cell specific targeting and delivery of biologic agents to the cell interior in a defined dose; however, an important observation in the above studies is that internalization of Trz-decorated PLPs affords a punctate fluorescence patter throughout the cytoplasm, suggesting that the particles are taken into an intracellular compartment (see Fig. 13.7d and e) [53]. Indeed, endosomal escape is most often the rate-limiting step in the delivery of biologics to the cytoplasm [30, 43, 51]. Within this context, our initial MDM2 knockdown studies using the UbV-decorated particles failed to demonstrate an effect [53]. We reasoned that PLP release from the putative internal compartment was slow, which limited particle access to cytosolic USP7 during the short (3 h) study. We therefore employed chloroquine (**CLQ**), an endosomal disruption agent that has been shown to promote endosomal

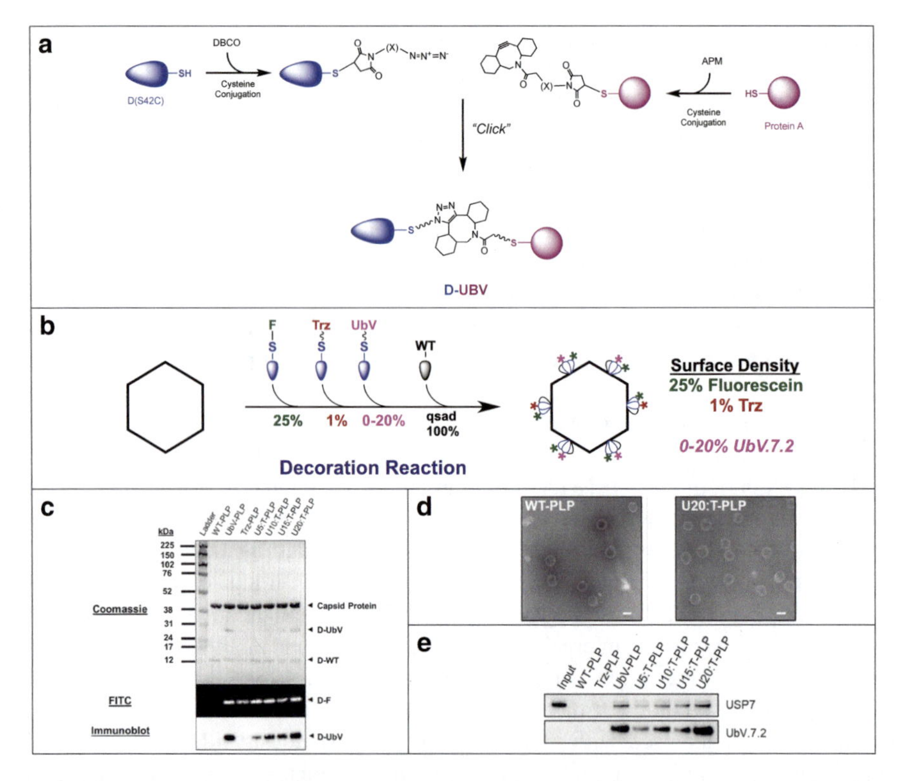

Fig. 13.8 Construction of Bivalent Trastuzumab-UbV Nanoparticles. **a** Strategy for engineering gpD-protein constructs using a "click" approach. *DBCO*, dibenzylcyclooctyne-PEG4-maleimide; *APM*, azido-PEG3-maleimide. In both cases, (X) denotes PEG linkers. Details are described in [53]. **b** Scheme for the construction of Trz-UbV nanoparticles. **c** SDS-PAGE Characterization of bivalent Trz:UbV-Decorated PLPs. *Top*, Coomassie-stained PVDF transfer membrane. Molecular weight standards are indicated at left and migration of the major capsid protein (38.2 kDa), D-WT (11.6 kDa) and D-UbV (22.7 kDa) are indicated at right. *Middle*, fluorescence imaging of the membrane; migration of D-F is indicated at right. *Bottom*, western immunoblot imaging of D-UbV. Migration of D-UbV is indicated at right. **d** Micrographs of WT-PLPs (*left*) and PLPs decorated with 20% UbV (*right*). White bar denotes 50 nm. **e** Co-immunoprecipitation of USP7 by decorated PLPs. The input control is a tenfold dilution of SKBR3 cell lysate. All particles were decorated with 25% D-F, 1% D-Trz (T) and the indicated surface density of D-UbV (U). This figure was modified from [53], with permission

escape of biological therapeutics [37, 78]. The data presented in Fig. 13.9b and c were obtained in the presence of CLQ and indicate that under these conditions a sufficient cytosolic concentration of UbV is achieved and strong inhibition of USP7 is observed. Importantly, Trz-PLPs and UbV:Trz-PLPs both inhibit cell growth, and both have significant impacts on the multi-omics profile of the cells in the *absence* of CLQ [12, 53], indicating that a significant fraction of the particles are naturally released. In sum, our data are consistent with a model wherein Trz-PLPs are taken into an endosomal compartment in $HER2^+$ cells from which they are slowly released.

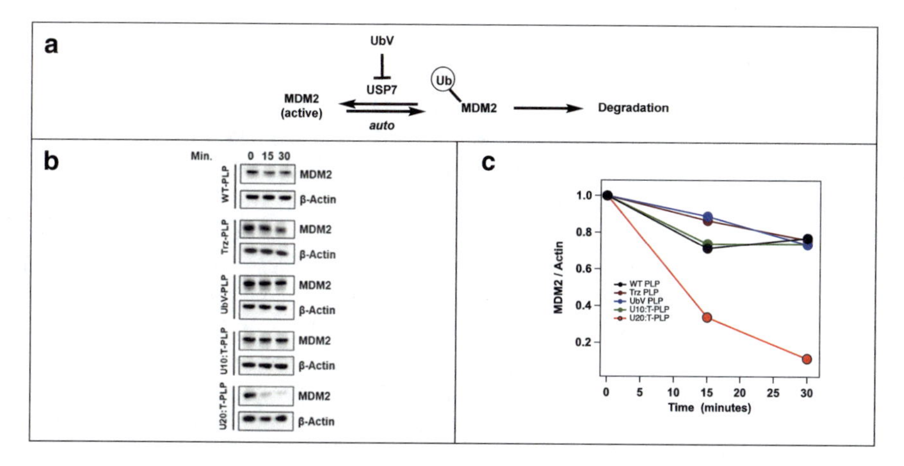

Fig. 13.9 Controlled Intracellular Delivery of Biologics. **a** Schematic diagram of USP7 regulation of MDM2. Inhibition of USP7 results in increased proteasomal degradation of MDM2. **b** UbV-Decorated PLPs Knockdown Cellular MDM2. Cells were treated with the indicated particles for three hours in the presence of chloroquine, at which point cycloheximide was added and MDM2 levels then quantified at the indicated times. **c** Quantitation of the data presented in **b** Details are described in [53]. This figure was modified from [53], with permission

Within this context, we recently demonstrated that PLPs decorated with cetuximab, a therapeutic antibody used for a number of cancers, specifically target bladder cancer cells that overexpress the epidermal growth factor receptor, and that the PLPs are efficiently internalized by these cells (Brown and Catalano, unpublished). In this case, however, the particles distribute throughout the cytosol to afford a diffuse fluorescent pattern (*not shown*), in contrast to the punctate pattern observed with Trz-PLPs. Thus, trafficking of internalized particles likely varies depending on the PLP surface decoration and/or the cell type.

Lambda PLPs as a Vaccine Platform

In addition to its potential as a targeted biologic delivery platform, we have investigated the use of lambda designer nanoparticles as a vaccine platform. A number of biological and synthetic nanoparticles and antigen coupling strategies have been evaluated for vaccine design [6, 19, 75]. Infectious phages and phage-derived nanoparticles have developed as vaccine platforms to display antigens from a number of pathogens including *Y. pestis* [76], *P. falciparum* [7], and SARS-CoV-2 [18, 72], among many others [35]. Each platform has its strengths and weaknesses and we considered that the semi-synthetic lambda designer nanoparticles possess some advantages and unique features that could be harnessed for vaccine development. Specifically, modification of the particle surface is fast, facile and can be tuned in a user-defined manner to allow rigorous and controlled antigen display in a timely manner. The icosahedral shell can be engineered to display multiple antigens (and adjuvants) in a defined geometry and in high density that may mimic a native antigen presentation to afford a robust immune response [40].

As a proof-of-concept study, we engineered lambda PLPs to simultaneously display the receptor binding domains from SARS-CoV-2 (RBD_{SARS}) and MERS-Cov (RBD_{MERS}) as a multi-valent vaccine platform [21]. We first constructed D-RBD_{SARS} and D-RBD_{MERS} decoration proteins as generally outlined in Fig. 13.6a, where Protein A represents the purified RBD proteins. These constructs were then used to decorate lambda PLPs in defined surface densities, alone (**RBD_{SARS}-PLP** and **RBD_{MERS}-PLP**) and in combination (**hCOV-RBD-PLP**) as outlined in Fig. 13.10a. Similar to the studies described above, the particles can be decorated with the multiple RBDs at the defined surface densities (Fig. 13.10b), the shells retain structural integrity and display surface projections associated with RBDs (Fig. 13.10c), and they possess physiochemical properties compatible with pharmaceutical standards [21]. Enzyme immunoassays confirmed that the antigenic properties of the unmodified RBDs are retained in the monovalent and bivalent RBD PLPs [21].

We next performed in vivo studies which demonstrated that mice immunized with RBD_{SARS}-PLPs or RBD_{MERS}-PLPs possess serum RBD-specific IgG endpoint titers and live virus neutralization titers that are maintained up to 6 months post-prime and that, in the case of SARS-CoV-2 are comparable to those detected in convalescent plasma from infected patients. Mice immunized with the bivalent hCoV-RBD PLPs afforded similar levels of RBD_{SARS}- and RBD_{MERS}-specific IgG, indicating a robust response to *both* antigens displayed on the particle. Finally, animals immunized with RBD_{SARS}-PLPs, RBD_{MERS}-PLPs, and hCoV-RBD PLPs were protected against SARS-CoV-2 and/or MERS-CoV lung infection and disease resulting in a durable immune response that provides effective protection against virulent virus challenge (Fig. 13.10d, e).

In sum, these data demonstrate that the designer PLP system provides a platform for facile and rapid generation of single and multi-target vaccines. The antigen presenting PLPs can be designed, engineered and constructed in roughly one month from the time the antigenic sequence has been identified. The particles afford a robust immune response to the presented antigens, while no adverse effects due to immunization are observed with undecorated PLPs, even after repeated doses or after challenge. Importantly, we have demonstrated that lambda PLPs decorated with model antigens can be formulated to afford temperature stable, timed release preparations that abrogate the onerous "cold-chain" issues associated with many vaccine preparations [83]. Overall, the lambda designer nanoparticle provides a platform with substantial flexibility and rigorous control in vaccine design, development and formulation, and is readily adaptable to any infectious agent. Continued efforts into the optimization of this platform could expand the diversity of protein-based, multi-valent vaccine technologies available and aid in the prevention of a variety of infectious diseases deleterious to global health.

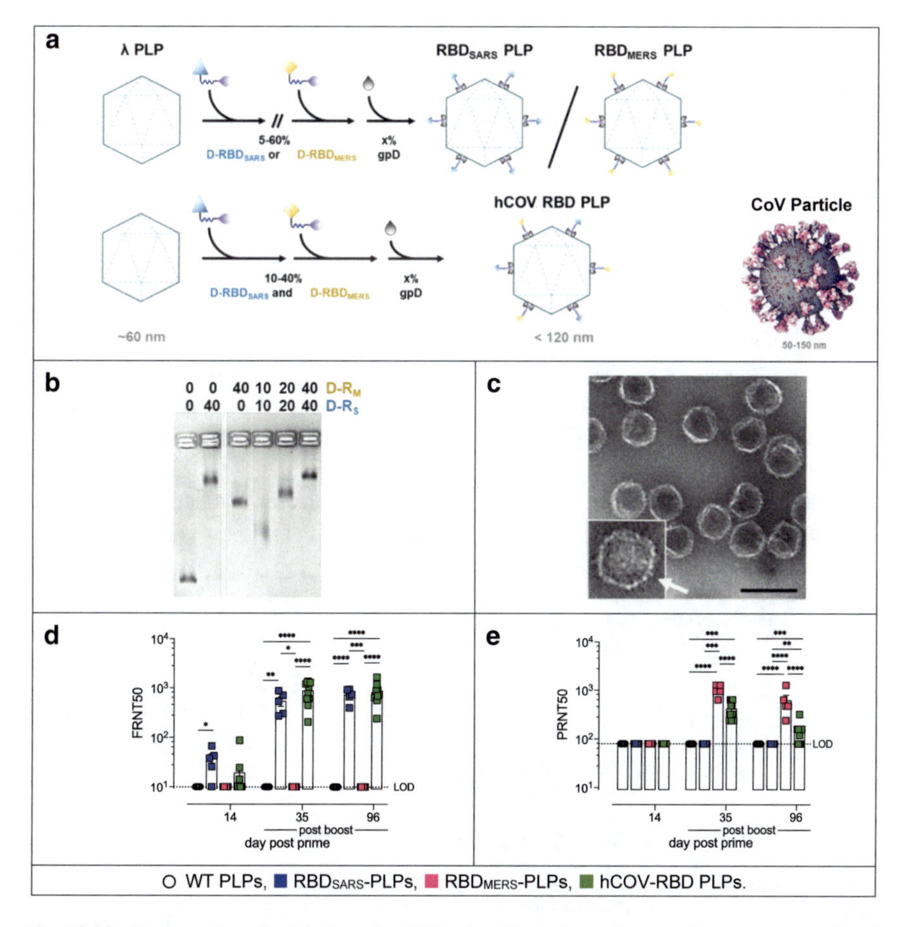

Fig. 13.10 Construction of a Bivalent Covid Vaccine Platform. **a** Strategy for the construction of the CoV vaccine platform. PLPs are decorated with SARS and MERS RBDs at the desired surface densities either independently (top) or in combination (bottom). A CoV virus particle is shown at right. **b** PLPs were decorated with RBD$_{SARS}$ (D-R$_S$, blue) and RBD$_{MERS}$ (D-R$_M$, yellow) at the indicated surface densities and the reaction mixtures were fractionated by AGE. **c** Electron micrograph of PLPs simultaneously decorated with 20% RBD$_{SARS}$ and 20% RBD$_{MERS}$ (magnification, 98,000x; scale bars, 100 nm). White arrow indicates the surface density attributed to the RBD proteins projecting from the PLP surface. Immunization of mice with mosaic hCoV-RBD-PLPs protects against both SARS-CoV-2 (**d**) and MERS-CoV challenge (**e**). Neutralizing activity was determined by a focus reduction neutralization test (FRNT). The dilution of serum that inhibited 50% of infectivity (FRNT50 titer) was calculated. Mann–Whitney test. $*P < 0.05$, $**P < 0.01$, $***P < 0.001$, $****P < 0.0001$. Details are described in [21]. This figure was modified from [21], with permission

Future Directions

Our studies to date have focused on decoration of the shell surface for particle tracking, optimization of Drug Metabolism/Pharmacokinetic (DMPK) characteristics, cell targeting, intracellular delivery of biologics and antigen presentation. An important aspect of the lambda designer nanoparticle platform is that heterologous DNA can be packaged into the decorated PLPs in vitro (see Fig. 13.4b) [17]. This provides an attractive option for targeted gene delivery to a cell and for DNA vaccine development, among other potential applications. Another option is genetic or chemical modification of the scaffolding protein, as outlined in Fig. 13.11a, to afford constructs analogous to the gpC protease (see Fig. 13.3). These fusion proteins could then be incorporated in a procapsid assembly reaction to package enzymes (bioreactor), tracking tags (GFP, quantum dots) and synthetic "drug binding" polymers into the shell interior as depicted in Fig. 13.11b. In aggregate, these particle assembly options provide a novel and customizable platform that can be adapted to a number and variety of applications in a user-defined manner.

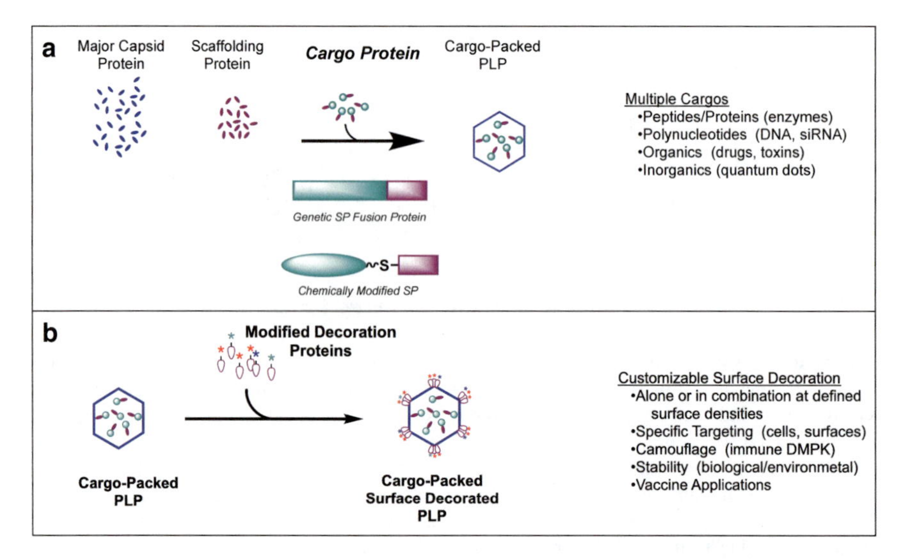

Fig. 13.11 Proposed Strategy for PLP Cargo Loading. **a** In vitro PLP Assembly. Purified capsid and scaffolding proteins assemble into PLPs under defined reaction conditions. Genetic fusion constructs containing a C-terminal SP-domain and proteins chemically cross-linked to SP can be engineered and included in the assembly reaction mixture. The constructs may then be incorporated into the PLP shell interior by co-polymerization with SP. This strategy allows proteinaceous, small molecule and synthetic cargo loading. **b** Cargo-laden PLPs assembled as in **a** can be decorated with multiple display ligands in defined surface densities. This figure was modified from [13], with permission

Conclusions

We have demonstrated that lambda PLPs can be decorated with a variety of chemical, biological and synthetic molecules in tunable surface densities, simultaneously integrating genetic and chemical modification strategies. This provides a robust and expansive set of integrated genetic and synthetic tools that can be harnessed in a user-defined manner. Rigorous characterization of a number and variety of decorated lambda nanoparticles has shown that they are monodisperse, physically and structurally stable, and that they possess physiochemical properties amenable to pharmaceutical development and formulation [12, 53, 83]. These features, which are not often considered in the early stages of nanoparticle development, are essential to address regulatory hurdles associated with therapeutic nanoparticle development [20], [59]. The ability to package DNA and the potential for selective incorporation of proteins and synthetic cargo into the shell interior, and to simultaneously decorate the surface with multiple ligands in a defined manner provide a nimble "designer" nanoparticle platform that can be rapidly adapted to a variety of theranostic applications.

References

1. Anselmo AC, Mitragotri S (2016) Nanoparticles in the clinic. Bioeng Transl Med 1:10–29
2. Ayoub NM, Al-Shami KM, Yaghan RJ (2019) Immunotherapy for HER2-positive breast cancer: recent advances and combination therapeutic approaches. Breast Cancer (Dove Med Press) 11:53–69
3. Beghetto E, Gargano N (2011) Lambda display: a powerful tool for antigen discovery. Molecules 16:3089–3105
4. Bertelsen V, Stang E (2014) The mysterious ways of ErbB2/HER2 trafficking. Membranes (Basel) 4:424–446
5. Bhattacharya S, Chakraborty D, Basu M, Ghosh MK (2018) Emerging insights into HAUSP (USP7) in physiology, cancer and other diseases. Signal Transduct Target Ther 3:17
6. Brune KD, Howarth M (2018) New routes and opportunities for modular construction of particulate vaccines: stick, click, and glue. Front Immunol 9:1432
7. Brune KD, Leneghan DB, Brian IJ, Ishizuka AS, Bachmann MF, Draper SJ, Biswas S, Howarth M (2016) Plug-and-display: decoration of Virus-Like Particles via isopeptide bonds for modular immunization. Sci Rep 6:19234
8. Calendar R, Abedon ST (2006) The Bacteriophages. Oxford University Press, New York
9. Carrico ZM, Farkas ME, Zhou Y, Hsiao SC, Marks JD, Chokhawala H, Clark DS, Francis MB (2012) N-Terminal labeling of filamentous phage to create cancer marker imaging agents. ACS Nano 6:6675–6680
10. Casjens SR (2011) The DNA-packaging nanomotor of tailed bacteriophages. Nat Rev Micro 9:647–657
11. Casjens SR, Hendrix RW (2015) Bacteriophage lambda: early pioneer and still relevant. Virology 0:310–330
12. Catala A, Dzieciatkowska M, Wang G, Gutierrez-Hartmann A, Simberg D, Hansen KC, D'Alessandro A, Catalano CE (2021) Targeted intracellular delivery of trastuzumab using designer phage lambda nanoparticles alters cellular programs in human breast cancer cells. ACS Nano 15:11789–11805

13. Catalano CE (2018) Bacteriophage lambda: the path from biology to theranostic agent. Wiley Interdisc Rev: Nanomed Nanobiotechnol 10:e1517
14. Catalano CE (2021) Enzymology of viral DNA packaging machines. In: Bamford DH, Zuckerman M (eds) Encyclopedia of Virology (4th edn). Academic Press, Oxford
15. Catalano CE, Morais MC (eds) (2021) Viral genome packaging machines: structure and enzymology. Academic Press, Cambridge, MA
16. Handrakala V, Aruna V, Angajala G (2022) Review on metal nanoparticles as nanocarriers: current challenges and perspectives in drug delivery systems. Emergent Mater 1–23
17. Chang JR, Song E-H, Nakatani-Webster E, Monkkonen L, Ratner DM, Catalano CE (2014) Phage lambda capsids as tunable display nanoparticles. Biomacromol 15:4410–4419
18. Chiba S, Frey SJ, Halfmann PJ, Kuroda M, Maemura T, Yang JE, Wright ER, Kawaoka Y, Kane RS (2021) Multivalent nanoparticle-based vaccines protect hamsters against SARS-CoV-2 after a single immunization. Commun Biol 4:597
19. Cohen AA, Gnanapragasam PNP, Lee YE, Hoffman PR, Ou S, Kakutani LM, Keeffe JR, Wu H-J, Howarth M, West AP, Barnes CO, Nussenzweig MC, Bjorkman PJ (2021) Mosaic nanoparticles elicit cross-reactive immune responses to zoonotic coronaviruses in mice. Science eabf6840
20. Cooper CJ, Khan Mirzaei M, Nilsson AS (2016) Adapting drug approval pathways for bacteriophage-based therapeutics. Front Microbiol 7:1–15
21. Davenport BJ, Catala A, Weston SM, Johnson RM, Ardanuy J, Hammond HL, Dillen C, Frieman MB, Catalano CE, Morrison TE (2022) Phage-like particle vaccines are highly immunogenic and protect against pathogenic coronavirus infection and disease. npj Vaccines, 7:57
22. Davis CR, Backos D, Morais MC, Churchill MEA, Catalano CE (2022) Characterization of a primordial major capsid-scaffolding protein complex in icosahedral virus shell assembly. J Mol Biol 434:167719
23. Dokland T, Murialdo H (1993) Structural transitions during maturation of bacteriophage lambda capsids. J Mol Biol 233:682–694
24. Dunn IS (1995) Assembly of functional bacteriophage lambda virions incorporating c-terminal peptide or protein fusions with the major tail protein. J Mol Biol 248:497–506
25. Earnshaw WC, Casjens SR (1980) DNA packaging by the double-stranded DNA bacteriophages. Cell 21:319–331
26. Ernst A, Avvakumov G, Tong J, Fan Y, Zhao Y, Alberts P, Persaud A, Walker JR, Neculai AM, Neculai D, Vorobyov A, Garg P, Beatty L, Chan PK, Juang YC, Landry MC, Yeh C, Zeqiraj E, Karamboulas K, Allali-Hassani A, Vedadi M, Tyers M, Moffat J, Sicheri F, Pelletier L, Durocher D, Raught B, Rotin D, Yang J, Moran MF, Dhe-Paganon S, Sidhu SS (2013) A strategy for modulation of enzymes in the ubiquitin system. Science 339:590–595
27. Garufi G, Minenkova O, Passo CL, Pernice I, Felici F (2005) Display libraries on bacteriophage lambda capsid. Biotechnol Annu Rev 11:153–190
28. Gaussier H, Yang Q, Catalano CE (2006) Building a virus from scratch: assembly of an infectious virus using purified components in a rigorously defined biochemical assay system. J Mol Biol 357:1154–1166
29. Georgopoulos C, Tilly K, Casjens S (1983) Lambdoid phage head assembly. In: Hendrix RW, Roberts JW, Stahl FW, Weisberg RA (eds) Lambda II. Cold Spring Harbor. Cold Spring Harbor Laboratory, NY
30. Gump JM, June RK, Dowdy SF (2010) Revised role of glycosaminoglycans in TAT protein transduction domain-mediated cellular transduction. J Biol Chem 285:1500–1507
31. Gupta AK, Gupta M (2005) Synthesis and surface engineering of iron oxide nanoparticles for biomedical applications. Biomaterials 26:3995–4021
32. Harish V, Tewari D, Gaur M, Yadav AB, Swaroop S, Bechelany M, Barhoum A (2022) Review on nanoparticles and nanostructured materials: bioimaging, biosensing, drug delivery, tissue engineering, antimicrobial, and agro-food applications. Nanomaterials (Basel), 12
33. Hendrix RW, Casjens S (2006) Bacteriophage lambda and its genetic neighborhood. In: Calendar R, Abedon ST (eds) The bacteriophages. 2nd ed. Oxford University Press, New York, N.Y.

34. Hendrix RW, Roberts, JW, Stahl, FW, Weisberg, RA (1983) Lamba II cold spring harbor. Cold Spring Harbor Laboratory, NY
35. Henry KA, Arbabi-Ghahroudi M, Scott JK (2015) Beyond phage display: non-traditional applications of the filamentous bacteriophage as a vaccine carrier, therapeutic biologic, and bioconjugation scaffold. Front Microbiol 6:755
36. Hoffman AS, Stayton PS (2020) 1.3.2G - Applications of "Smart Polymers" as Biomaterials. In Wagner WR, Sakiyama-Elbert SE, Zhang G, Yaszemski MJ (eds) Biomaterials science, 4th edn. Academic Press
37. Horn JM, Obermeyer AC (2021) Genetic and covalent protein modification strategies to facilitate intracellular delivery. Biomacromol 22:4883–4904
38. Iqbal N, Iqbal N (2014) Human epidermal growth factor receptor 2 (HER2) in cancers: overexpression and therapeutic implications. Mol Biol Int 2014:852748
39. Irvine DJ, Hanson MC, Rakhra K, Tokatlian T (2015) Synthetic nanoparticles for vaccines and immunotherapy. Chem Rev 115:11109–11146
40. Jegerlehner A, Tissot A, Lechner F, Sebbel P, Erdmann I, Kündig T, Bächi T, Storni T, Jennings G, Pumpens P, Renner WA, Bachmann MF (2002) A molecular assembly system that renders antigens of choice highly repetitive for induction of protective B cell responses. Vaccine 20:3104–3112
41. Joudeh N, Linke D (2022) Nanoparticle classification, physicochemical properties, characterization, and applications: a comprehensive review for biologists. J Nanobiotechnol 20:262
42. Kanamaru T, Sakurai K, Fujii S (2022) Impact of polyethylene glycol (PEG) conformations on the in vivo fate and drug release behavior of PEGylated core-cross-linked polymeric nanoparticles. Biomacromol 23:3909–3918
43. Kaplan IM, Wadia JS, Dowdy SF (2005) Cationic TAT peptide transduction domain enters cells by macropinocytosis. J Control Release 102:247–253
44. Karimi M, Mirshekari H, Moosavi Basri SM, Bahrami S, Moghoofei M, Hamblin MR (2016) Bacteriophages and phage-inspired nanocarriers for targeted delivery of therapeutic cargos. Adv Drug Del Rev 106:45–62
45. Lambert S, Yang Q, de Angeles R, Chang JR, Ortega M, Davis C, Catalano CE (2017) Molecular dissection of the forces responsible for viral capsid assembly and stabilization by decoration proteins. Biochemistry 56:767–778
46. Lander GC, Evilevitch A, Jeembaeva M, Potter CS, Carragher B, Johnson JE (2008) Bacteriophage lambda stabilization by auxiliary protein gpD: timing, location, and mechanism of attachment determined by cryo-EM. Structure 16:1399–1406
47. Larionov AA (2018) Current therapies for human epidermal growth factor receptor 2-positive metastatic breast cancer patients. Front Oncol 8:89
48. Lazar AC, Wang L, Blättler WA, Amphlett G, Lambert JM, Zhang W (2005) Analysis of the composition of immunoconjugates using size-exclusion chromatography coupled to mass spectrometry. Rapid Commun Mass Spectrom 19:1806–1814
49. Lee CS, Bishop ES, Zhang R, Yu X, Farina EM, Yan S, Zhao C, Zeng Z, Shu Y, Wu X, Lei J, Li Y, Zhang W, Yang C, Wu K, Wu Y, Ho S, Athiviraham A, Lee MJ, Wolf JM, Reid RR, He T-C (2017) Adenovirus-mediated gene delivery: Potential applications for gene and cell-based therapies in the new era of personalized medicine. Genes Diseases 4:43–63
50. Leung I, Jarvik N, Sidhu SS (2017) A highly diverse and functional naïve ubiquitin variant library for generation of intracellular affinity reagents. J Mol Biol 429:115–127
51. Lönn P, Kacsinta AD, Cui X-S, Hamil AS, Kaulich M, Gogoi K, Dowdy SF (2016) Enhancing endosomal escape for intracellular delivery of macromolecular biologic therapeutics. Sci Rep 6:32301
52. Lu J, Zhao H, Yu C, Kang Y, Yang X (2021) Targeting ubiquitin-specific protease 7 (USP7) in cancer: a new insight to overcome drug resistance. Front Pharmacol 12
53. McClary WD, Catala A, Zhang W, Gamboni F, Dzieciatkowska M, Sidhu SS, D'Alessandro A, Catalano CE (2022) A designer nanoparticle platform for controlled intracellular delivery of bioactive macromolecules: inhibition of ubiquitin-specific protease 7 in breast cancer cells. ACS Chem Biol 17:1853–1865

54. Medina E, Nakatani E, Kruse S, Catalano CE (2012) Thermodynamic characterization of viral procapsid expansion into a functional capsid shell. J Mol Biol 418:167–180
55. Medina E, Wieczorek DJ, Medina EM, Yang Q, Feiss M, Catalano CE (2010) Assembly and maturation of the bacteriophage lambda procapsid: gpC is the viral protease. J Mol Biol 401:813–830
56. Medina MM, Andrews BT, Nakatani E, Catalano CE (2011) The Bacteriophage lambda gpNu3 scaffolding protein is an intrinsically disordered and biologically functional procapsid assembly catalyst. J Mol Biol 412:723–736
57. Murialdo H, Becker A (1978) Head morphogenesis of complex double-stranded deoxyribonucleic acid bacteriophages. Microbiol Rev 42:529–576
58. Nicastro J, Sheldon K, El-Zarkout FA, Sokolenko S, Aucoin MG, Slavcev R (2013) Construction and analysis of a genetically tuneable lytic phage display system. App Microbiol Biotechnol 1–14
59. Nilsson AS (2019) Pharmacological limitations of phage therapy. Upsala J Med Sci 124:218–227
60. Nurmemmedov E, Castelnovo M, Catalano CE, Evilevitch A (2007) Biophysics of viral infectivity: matching genome length with capsid size. Q Rev Biophys 40:327–356
61. Pelegri-O'day EM, Lin E-W, Maynard HD (2014) Therapeutic protein-polymer conjugates: advancing beyond PEGylation. J Am Chem Soc 136:14323–14332
62. Prevelige PE, Fane BA (2012) Building the machines: scaffolding protein functions during bacteriophage morphogenesis. In: Rossmann MG, Rao VB (eds) Viral molecular machines. Springer US, Boston, MA
63. Qi S-M, Cheng G, Cheng X-D, Xu Z, Xu B, Zhang W-D, Qin J-J (2020) Targeting USP7-mediated deubiquitination of MDM2/MDMX-p53 pathway for cancer therapy: are we there yet? Front Cell Devel Biol 8:233–233
64. Raman V, van Dessel N, Hall CL, Wetherby VE, Whitney SA, Kolewe EL, Bloom SMK, Sharma A, Hardy JA, Bollen M, van Eynde A, Forbes NS (2021) Intracellular delivery of protein drugs with an autonomously lysing bacterial system reduces tumor growth and metastases. Nat Commun 12:6116
65. Rao VB, Feiss M (2015) Mechanisms of DNA packaging by large double-stranded DNA viruses. Ann Rev Virol 2:351–378
66. Rohevie MJ, Nagasawa M, Swartz JR (2016) Virus-like particles: next-generation nanoparticles for targeted therapeutic delivery. Bioeng Transl Med 2:43–57
67. Roizman B, Knipe DM, Whitley RJ (2007) Herpes simplex viruses. In: Knipe DM, Howley PM (eds) Fields Virology, 5 edn. Lippincott, Williams, and Wilkins, New York, NY
68. Roizman B, Palese P (1996) Multiplication of viruses: an overview. In: Fields BN, Knipe DM, Howley PM (eds) Fields virology, 3rd edn., Lippincott-Raven, New York, N.Y.
69. Schwarz B, Madden P, Avera J, Gordon B, Larson K, Miettinen HM, Uchida M, Lafrance B, Basu G, Rynda-Apple A, Douglas T (2015) Symmetry controlled, genetic presentation of bioactive proteins on the P22 virus-like particle using an external decoration protein. ACS Nano 9:9134–9147
70. Singh P, Nakatani E, Goodlett DR, Catalano CE (2013) A pseudo-atomic model for the capsid shell of bacteriophage lambda using chemical cross-linking/mass spectrometry and molecular modeling. J Mol Biol 425:3378–3388
71. Sponchioni M, Capasso Palmiero U, Moscatelli D (2019) Thermo-responsive polymers: applications of smart materials in drug delivery and tissue engineering. Mater Sci Eng C Mater Biol Appl 102:589–605
72. Staquicini DI, Tang FHF, Markosian C, Yao VJ, Staquicini FI, Dodero-Rojas E, Contessoto VG, Davis D, O'Brien P, Habib N, Smith TL, Bruiners N, Sidman RL, Gennaro ML, Lattime EC, Libutti SK, Whitford PC, Burley SK, Onuchic JN, Arap W, Pasqualini R (2021) Design and proof of concept for targeted phage-based COVID-19 vaccination strategies with a streamlined cold-free supply chain. Proc Natl Acad Sci 118:e2105739118
73. Steinmetz NF (2019) Biological and evolutionary concepts for nanoscale engineering. EMBO reports 20:e48806

74. Strable E, Finn MG (2009) Chemical modification of viruses and virus-like particles. In: Manchester M, Steinmetz NF (eds) Viruses and Nanotechnology, Springer, Berlin
75. Tan TK, Rijal P, Rahikainen R, Keeble AH, Schimanski L, Hussain S, Harvey R, Hayes JWP, Edwards JC, McLean RK, Martini V, Pedrera M, Thakur N, Conceicao C, Dietrich I, Shelton H, Ludi A, Wilsden G, Browning C, Zagrajek AK, Bialy D, Bhat S, Stevenson-Leggett P, Hollinghurst P, Tully M, Moffat K, Chiu C, Waters R, Gray A, Azhar M, Mioulet V, Newman J, Asfor AS, Burman A, Crossley S, Hammond JA, Tchilian E, Charleston B, Bailey D, Tuthill TJ, Graham SP, Duyvesteyn HME, Malinauskas T, Huo J, Tree JA, Buttigieg KR, Owens RJ, Carroll MW, Daniels RS, McCauley JW, Stuart DI, Huang K-YA, Howarth M, Townsend AR (2021) A COVID-19 vaccine candidate using SpyCatcher multimerization of the SARS-CoV-2 spike protein receptor-binding domain induces potent neutralising antibody responses. Nat Commun 12:542
76. Tao P, Mahalingam M, Kirtley ML, van Lier CJ, Sha J, Yeager LA, Chopra AK, Rao VB (2013) Mutated and bacteriophage T4 nanoparticle arrayed F1-V immunogens from yersinia pestis as next generation plague vaccines. PLoS Pathog 9:e1003495
77. Tsuchikama K, An Z (2018) Antibody-drug conjugates: recent advances in conjugation and linker chemistries. Protein Cell 9:33–46
78. Varkouhi AK, Scholte M, Storm G, Haisma HJ (2011) Endosomal escape pathways for delivery of biologicals. J Control Release 151:220–228
79. Veronese FM, Mero A (2008) The Impact of PEGylation on Biological Therapies. Biodrugs: Clinical Immunotherapeutics. Biopharmaceuticals Gene Therapy 22:315–329
80. Vieweger SE, Tsvetkova IB, Dragnea BG (2018) In vitro assembly of virus-derived designer shells around inorganic nanoparticles. Methods Mol Biol 1776:279–294
81. Wang C, Zeng J, Wang J (2022) Structural basis of bacteriophage lambda capsid maturation. Structure 30:637-645.e3
82. Wang Z, Kang W, You Y, Pang J, Ren H, Suo Z, Liu H, Zheng Y (2019) USP7: novel drug target in cancer therapy. Front Pharmacol 10
83. Witeof AE, McClary W, Rea LT, Yang Q, Davis MM, Funke H, Catalano CE, Randolph T (2022) Atomic-layer deposition processes applied to phage lambda and a phage-like particle platform yield thermostable. Single-Shot Vacc J Pharm Sci 111:1354–1362
84. Woźniak M, Płoska A, Siekierzycka A, Dobrucki LW, Kalinowski L, Dobrucki IT (2022) Molecular imaging and nanotechnology-emerging tools in diagnostics and therapy. Int J Mol Sci 23
85. Yang Q, Maluf NK, Catalano CE (2008) Packaging of a unit-length viral genome: the role of nucleotides and the gpD decoration protein in stable nucleocapsid assembly in bacteriophage lambda. J Mol Biol 383:1037–1048
86. Yang T-C, Ortiz D, Yang Q, Angelis RD, Sanyal SJ, Catalano CE (2017) Physical and functional characterization of nucleoprotein complexes along a viral assembly pathway. Biophys J 112:1551–1560
87. Zhang W, Sartori MA, Makhnevych T, Federowicz KE, Dong X, Liu L, Nim S, Dong A, Yang J, Li Y, Haddad D, Ernst A, Heerding D, Tong Y, Moffat J, Sidhu SS (2017) Generation and validation of intracellular ubiquitin variant inhibitors for USP7 and USP10. J Mol Biol 429:3546–3560
88. Zhao Y, Yu H, Hu W (2014) The regulation of MDM2 oncogene and its impact on human cancers. Acta Biochim Biophys Sin (Shanghai) 46:180–189
89. Zhu J, Ananthaswamy N, Jain S, Batra H, Tang W-C, Lewry DA, Richards ML, David SA, Kilgore PB, Sha J, Drelich A, Tseng C-TK, Chopra AK, Rao VB (2021) A universal bacteriophage T4 nanoparticle platform to design multiplex SARS-CoV-2 vaccine candidates by CRISPR engineering. bioRxiv, 2021.01.19.427310
90. Ziegelhoffer T, Yau P, Chandrasekhar GN, Kochan J, Georgopoulos C, Murialdo H (1992) The purification and properties of the scaffolding protein of bacteriophage lambda. J Biol Chem 267:455–461

91. Zlotnick A, Fane BA (2011) Mechanisms of icosahedral virus assembly. In Agbandje-Mckenna M, Mckenna R (eds) Structural Virology. The Royal Society of Chemistry, Cambridge, UK
92. Zucconi A, Dente L, Santonico E, Castagnoli L, Cesareni G (2001) Selection of ligands by panning of domain libraries displayed on phage lambda reveals new potential partners of synaptojanin 111Edited by. J Karn J Molecular Biol 307:1329–1339

Chapter 14
Therapeutic Interfering Particles (TIPs): Escape-Resistant Antiviral Against SARS-CoV-2

Sonali Chaturvedi

Abstract Despite identifying new antiviral targets and developing new drugs, resistance against drugs emerges in viruses regardless of their genome, making it imperative to identify therapies that will counter the problem of resistance. Recently, Therapeutic Interfering Particles (TIPs), a new class of antiviral therapy, was engineered against SARS-CoV-2, overcoming the problem of the emergence of resistance. TIPs conditionally replicate in the presence of SARS-CoV-2 to efficiently mobilize between cells with $R_0 > 1$ and interfere with virus replication. As TIPs steal replication and packaging machineries from the infectious virus, they keep up with virus evolution. In hamsters, a single administration of TIPs reduces virus titer by more than 100-fold, reduces proinflammatory genes associated with SARS-CoV-2, histopathology in lungs, and virus shedding from infected to uninfected hamsters. This proof-of-concept strategy can circumvent the general problem of the emergence of resistance and be instrumental in designing escape-resistant therapies against other viruses.

Keywords Defective interfering particles · Therapeutic interfering particles (TIPs) · SARS-CoV-2 · Lipid nanoparticles (LNPs) · Transmission

Unmet Need

The proliferation of new and resurgent viral diseases represents a formidable challenge to the global health system [2, 11]. While antivirals have been developed to combat these emerging viruses, their efficacy is severely compromised by the high genetic variability inherent in these pathogens. The likelihood of a virus evolving drug-resistant strains is dependent on a number of factors, including the existing

S. Chaturvedi (✉)
Gladstone|UCSF Center for Cell Circuitry, Gladstone Institute of Virology, Gladstone Institutes, San Francisco, CA 94158, USA
e-mail: sonali.chaturvedi@gladstone.ucsf.edu

M. Comas-Garcia and S. Rosales-Mendoza (eds.), *Physical Virology*, Springer Series in Biophysics 24, https://doi.org/10.1007/978-3-031-36815-8_14

genetic diversity of the pathogen, its mutation and recombination rates, and the size of the affected population [3–5, 14, 17, 23].

Consequently, there is an urgent need to develop innovative therapeutic approaches that are capable of overcoming resistance [3–5], offer the potential for single-dose administration to mitigate the issue of dose-limiting toxicity, and can be readily disseminated to high-risk populations. To this end, the recent development of Therapeutic Interfering Particles (TIPs), a therapy based on defective interfering particles (DIPs) for SARS-CoV-2, offers a promising solution [3, 5]. TIPs replicate only in the presence of an infectious virus and evolve alongside the virus, thereby countering drug-resistant strains. Furthermore, TIPs are capable of being administered in a single dose via lipid nanoparticles (LNPs) and could spread between cells at the same rate as the infectious virus, with a basic reproductive ratio (R_0) greater than 1 [18, 20, 21]. Moreover, TIPs is the only single-administration strategy reported to reduce virus shedding [3].

The TIPs Concept

History (DIP)

The first observations of the emergence of "incomplete virus particles" in the presence of a high multiplicity of infection (MOI) of the Influenza virus were made by Preben von Magnus in 1954 [26]. During the serial passaging of the infectious virus, fluctuations in the titer of the virus were noted and believed to result from the interaction between "incomplete virus" and wild-type virus, which established a predator–prey dynamic. These fluctuations became known as the "von Magnus effect," where the incomplete virus contested with the infectious virus for access to replication and packaging machinery. This phenomenon was later observed in other viral species, including Vesicular Stomatitis Virus (VSV), where the first direct evidence of the presence of incomplete visions was found [10]. A few years later, Alice Huang and David Baltimore gave these incomplete particles the name "Defective Interfering Particles" or DIPs [12].

TIP Definition and Mode of Action

Since the 1970s, there has been a growing interest in exploring the therapeutic capabilities of DIPs due to their capacity to impede the replication of viral pathogens. However, it wasn't until recently that their full therapeutic potential was understood. One issue faced by DIPs was their reliance on the presence of the native virus and their tendency to become diluted and eventually lost over time. In 2003, theoretical models demonstrated that an engineered DIP with a basic reproductive ratio (R_0)

greater than 1 could overcome these challenges, leading to the coinage of the term "Therapeutic Interfering Particle" or TIP to describe these engineered particles [18, 27]. Genetically, TIPs retain only cis-regulatory elements from the infectious virus and have trans elements deleted requiring them to intracellularly scavenge replication and packaging resources from infectious virus for their propagation [5]. As TIPs interfere with the replication of the infectious virus, they serve as obligate parasites with the ability to co-evolve with the pathogen and overcome the drug resistance that is a challenge for static therapies and vaccines.

TIP Versus DIP/DVG Mode of Action

Defective viral genomes (DVGs) have been shown to impede the spread of infectious viruses through a variety of mechanisms [5, 9, 15, 18, 22, 25]. At high multiplicities of infection, the formation of a substantial number of DVGs may occur through single-amino acid mutations or large deletion mutations resulting from recombination [8, 15]. Despite the fact that these mutations often negatively impact the viral fitness, a small proportion of DVGs persist in long-term cultures of infectious viruses. Some DVGs present in nature are capable of replication, but lack key structural elements, resulting in competition with the infectious virus for resources [13]. These DVGs give rise to defective interfering particles (DIPs), which are lacking significant portions of the viral genome and compete with the infectious virus for access to replication and packaging machinery. These DVGs and DIPs have been shown to elicit the expression of pro-inflammatory genes such as interferons, interleukins, and tumor necrosis factor, believed to be the primary antiviral mechanism [1, 16]. In contrast, therapeutic interfering particles (TIPs), which are engineered DIPs, interfere with the spread of infectious viruses by competing for replication and packaging resources while having the ability to transmit at $R_0 > 1$ [5, 18, 21, 27].

Theoretical Advantage of TIPs

The phenomenon of antivirals and vaccines losing their effectiveness due to the development of viral resistance is a widely observed occurrence across a diverse range of viral genome types [2, 11]. Traditional antivirals, designed to target rapidly evolving viral targets, inevitably become obsolete over time. The implementation of TIPs offers a solution to this issue, as they counteract the problem of resistance posed by static antivirals and vaccines. In theory, TIP is a one-time administration therapy that continuously replenishes itself in the presence of the infectious virus [18, 21, 27]. Additionally, as TIPs target both non-structural and structural viral proteins, they have the potential to evolve in parallel with the virus [5]. Furthermore, the theoretical transmission of TIPs with the virus presents a promising solution to the challenge of providing adequate protection to high-risk, hard-to-reach populations.

Identification of TIPs

Engineering of TIPs against any virus requires characterization of *cis*- and *trans*-regulatory elements of the infectious virus. TIP retains all the *cis*-regulatory elements and parasitize on infectious virus for *trans*-regulatory resources [5, 18, 27]. Along with the 5′ and 3′ untranslated region (UTRs) of infectious virus, packaging elements and replication initiation sites are few of the crucial *cis*-regulatory elements that are indispensable for the replication of the virus. Once characterized, *cis*-regulatory elements are engineered to the backbone of the virus and passaged in the presence of infectious virus to evaluate if the TIP candidate can: (i) conditionally replicate; (ii) interfere with infectious virus replication; and (iii) transmit with $R_0 > 1$ (Fig. 14.1) [27]. Two techniques can be exploited to characterize *cis*-regulatory elements in infectious virus: (a) a transposon based high-throughput method to characterize *cis*- and *trans*-regulatory elements, called RanDeL-Seq (random deletion library sequencing) [19]. This approach employs enzymatic various-size deletion by transposon integration followed by adding a barcode for tracking the libraries. RanDeL-Seq approach was employed to generate and screen deletion libraries for Human Immunodeficiency Virus-1 (HIV-1) and Zika virus (ZIKV) at a single-base resolution to identify *cis*- and *trans*-regulatory elements. Along with identifying previously known cis-regulatory elements for HIV-1, this approach identified central polypurine track (cPPT) as indispensable element. RanDeL-seq identified 300-bp region downstream of Rev response element critical for HIV-1 infection. Moreover, RanDeL-seq identified previously unidentified *cis*-regulatory elements in ZIKV. The second approach employs bioreactors and paired it with in-silico models to study HIV in vivo dynamics [24]. In vitro continuous culture bioreactor of cells infected with HIV-1 identified *cis*-regulatory elements for HIV-1 that guided the design of TIP for HIV-1. These TIPs interfered with the replication of infectious HIV-1, transmitted with $R_0 > 1$, and conditionally replicate in the presence of infectious virus.

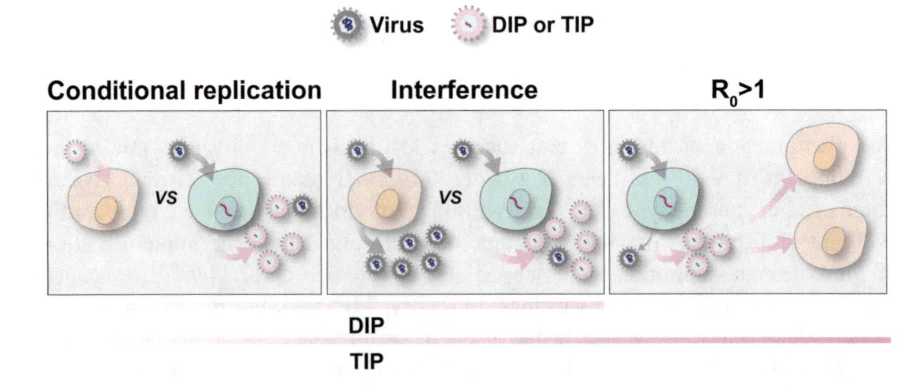

Fig. 14.1 Conditionally Replicating Therapeutic Interfering Particles (TIPs) demonstrate the capacity to replicate, impede the replication of the virus, and exhibit a reproductive number greater than one

Therapeutic Interfering Particles for SARS-CoV-2

The ongoing evolution of SARS-CoV-2 has led to a significant shift in the antiviral landscape for the virus [2, 11]. While small-molecule antivirals were designed to target SARS-CoV-2, their efficacy was quickly hampered by the emergence of drug-resistant strains. Vaccines, while effective in preventing severe disease and death, are less effective against the new variants [7]. Monoclonal antibodies have shown promise in prophylaxis but have not been successful in therapeutic administration [6]. Additionally, the above antiviral interventions have been largely ineffective in reducing SARS-CoV-2 transmission.

Given the theoretical advantage of TIPs, including the ability to be administered as a single dose and track variants, TIPs against SARS-CoV-2 were engineered to meet the requirements of conditional replication, transmission with $R_0 > 1$, and interference with infectious virus replication [3, 5]. Two TIP candidates, TIP-1 and TIP-2, were designed, encoding different portions of the SARS-CoV-2 genome, including 5′ and 3′ UTRs, part of ORF1a and the N gene, and an IRES-mCherry reporter. Control RNA (Ctrl RNA), of similar size, coding for firefly luciferase and IRES-mCherry was also included (Fig. 14.2). All RNAs were 5′ capped and poly-A tailed to reduce immune response. The antiviral activity of TIP candidates was evaluated in vitro by nucleofecting them into Vero-6 cells and infecting with SARS-CoV-2 at MOI = 0.05. Based on qRT-PCR and plaque assays, both TIP candidates reduced viral load by several Logs compared to the Ctrl RNA.

The feasibility of transmission of the TIP candidates with infectious SARS-CoV-2 virus was evaluated through a series of experiments. Cells were nucleofected with either TIP candidates or Ctrl RNA, followed by infection with the SARS-CoV-2 virus. The supernatant collected from infected wells one day post-infection was transferred to uninfected cells, and mCherry expression was quantified through flow cytometry at 24 h post-transfer to assess the amount of TIP or Ctrl RNA present in infectious particles. The results showed that both TIP candidates could transmit the virus, with greater than 16% of cells exhibiting mCherry expression compared to the Ctrl RNA.

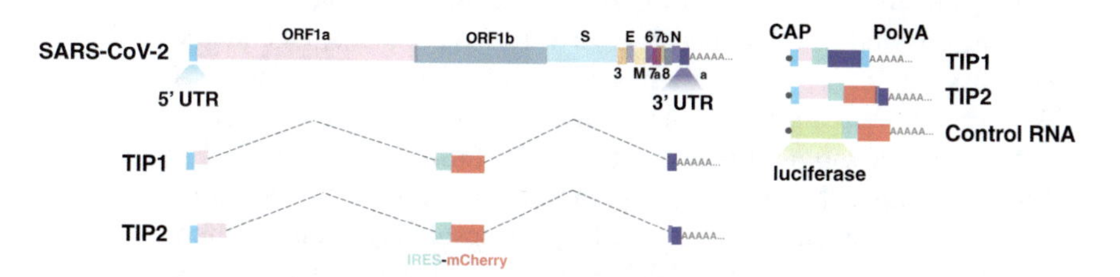

Fig. 14.2 A diagrammatic representation of the SARS-CoV-2 viral and TIP candidates genome. The TIP candidates are generated from the native genome of SARS-CoV-2 and comprise the 5′ untranslated region, a portion of the ORF1a gene, and the 3′ untranslated region. The TIP2 candidate additionally encompasses a segment of the N gene. A control is established using an RNA of comparable size (luciferase) and both the TIP and control RNA are capped at the 5′ end and tailed with a poly-A sequence to mitigate any immune response

The R_0 of the TIP candidates was also calculated. Briefly, GFP-positive cells were added at 20% confluency at 2 h post-infection to the cells nucleofected with TIP candidates and infected with SARS-CoV-2 (MOI = 0.05). The percentage of mCherry-positive cells in the GFP-positive population was quantified through flow cytometry at 12 h post-infection, revealing an R_0 of the TIP candidates to be greater than 1.

Further experiments demonstrated the efficacy of TIP candidates in overcoming the evolution of resistance in a long-term culture. A reconstitution assay was performed to gain insight into the antiviral activity of TIP candidates, where virus-like particles were synthesized using structural proteins (spike, matrix, nucleocapsid, and envelop) in the presence of TIP or Ctrl RNAs, and the relative mCherry level was quantified in the cells transfected with reconstituted virus-like particles. The results indicated that TIP candidates form functional virus-like particles.

An electromobility shift assay was performed to evaluate the direct interaction between TIP candidates with the nucleocapsid protein and RNA-dependent RNA polymerase (RdRp) of SARS-CoV-2. Both TIP candidates demonstrated direct interaction with the nucleocapsid protein and RdRp of the virus, suggesting a direct interaction between TIP candidates and both the structural and non-structural proteins of SARS-CoV-2. As TIP candidates compete with the infectious virus for access to structural and non-structural proteins, the therapy can keep pace with the evolution of the virus and effectively interfere with the replication of ancestral and variant strains, including WA-1, B.1.1.7, B.1.351, B.1.617.2, and others.

The immune response to TIPs, which are engineered DIPs, was meticulously quantified to verify that their mode of action is through competition for viral structural and non-structural proteins rather than eliciting an immune response. An evaluation of the relative gene expression of inflammatory genes in cells treated with TIP or Ctrl RNA revealed no significant expression of proinflammatory genes in the presence of TIP-treated samples. After thoroughly investigating the properties of TIP and establishing that its mode of action entails scavenging for *trans*-regulatory elements from the infectious virus, the efficacy of TIP candidates was tested using a donor-derived lung organoid model. The results showed that the TIP candidates significantly inhibited the replication of SARS-CoV-2 in the lung organoid model.

Having ensured that TIP met all its in vitro requirements, its efficacy was subsequently assessed in a live animal study utilizing Syrian golden hamsters. The delivery of TIP was optimized through encapsulation of RNA in lipid nanoparticles (LNPs) and the characterization of these LNPs was performed via dynamic light scattering (DLS) to determine their suitability for intranasal administration to animals. A proof-of-concept experiment was conducted by encapsulating firefly luciferase RNA in LNPs and testing their intranasal delivery in mice. The mice were treated with PBS, naked RNA, or LNP-encapsulated RNA, and the biodistribution of luciferase was quantified thereafter. Upon characterizing the LNPs, TIP1 and Ctrl RNA were encapsulated in LNPs and administered to hamsters without the presence of infection. Five days post-treatment, the lungs of the hamsters were harvested, RNA sequencing was performed on total RNA extracted from one lobe, and the other lobe was subjected to histopathological analysis. The histopathological examination and RNA sequencing results showed no significant increase in the pro-inflammatory immune response against TIPs in the hamster lungs in the absence of infection.

Furthermore, the potency of a single intranasal administration of TIP was evaluated both prophylactically and therapeutically in Syrian golden hamsters. To assess the prophylactic efficacy of TIP, a group of hamsters ($n = 5$) were treated with TIP or control LNPs, followed by an intranasal infection with SARS-CoV-2 (10^6 PFU). The lungs were harvested five days post-infection, the virus titer was quantified from the lungs, and the histopathology was scored. The results indicated a reduction in viral titer of more than 100-fold in the lungs of animals treated with TIPs compared to control LNPs. Additionally, the animals treated with TIP LNPs displayed fewer instances of alveolar edema and infiltrates compared to those treated with control LNPs. The expression of proinflammatory genes associated with SARS-CoV-2 infection, such as *ccl2*, *ccl6*, and *il6*, was also mitigated in animals treated with TIPs. To evaluate the therapeutic efficacy of TIP, animals were intranasally infected with SARS-CoV-2 (10^6 PFU) and treated with TIP or control LNPs at 12 h post-infection. The lungs were harvested five days post-infection, and the virus titer was quantified using a plaque assay and histopathology was performed to assess lung pathogenesis. The results showed a significant reduction in virus titer and alveolar edema in animals treated with TIP LNPs compared to those treated with control LNPs.

The examination of proinflammatory markers related to SARS-CoV-2 was conducted through the quantification of expression levels in the presence and absence of viral infection in the presence of TIPs. Even though no considerable variations in the expression of proinflammatory genes were detected in uninfected animals exposed to TIP LNPs, Ctrl LNPs, or PBS, in the presence of infection, animals treated with TIP LNPs demonstrated significant reduction in pro inflammatory response compared to Ctrl LNP treated animals. This evidence indicates that TIP serves as a safe antiviral strategy that does not provoke an immune response in the absence of infection and, in the presence of infection, reduces the expression of proinflammatory genes linked to SARS-CoV-2 as compared to Ctrl LNPs.

The potential of TIPs to curb the transmission of SARS-CoV-2, a driving force behind the COVID-19 pandemic, was evaluated in the lungs of hamsters [3]. To this end, a single intranasal administration of TIP or Ctrl LNPs was administered 12 h post-infection with the delta variant (B.1.617.2) of SARS-CoV-2 (10^6 PFU). Thirty-six hours post-infection, infected and treated animals were housed with uninfected and untreated animals for eight hours before being separated and housed individually. Nasal mucosa was harvested daily, and lungs were harvested at the endpoint. The virus titer and viral RNA levels were quantified from nasal washes of source and contact animals treated with TIP or Ctrl LNPs. Results showed a significant reduction in infectious virus and viral RNA levels in both TIP-treated source animals and their contacts, compared to Ctrl RNA-treated source and contact animals. Furthermore, there was a reduction in viral load by more than 3 Logs in the lungs of source animals treated with TIPs and their contacts, compared to those treated with Ctrl LNPs and their contacts. The TIP LNP-treated source animals and their contacts also exhibited a reduction in proinflammatory gene expression related to SARS-CoV-2, compared to Ctrl LNP-treated animals and their contacts.

Additionally, the lungs of hamsters treated with TIPs and their contacts exhibited reduced instances of alveolar edema, immune infiltrates, and an overall lower composite histopathology score compared to Ctrl LNP treated hamsters and their contacts. This data highlights that a single dose of TIP can effectively mitigate virus shedding and pathogenesis in hamsters.

Perspectives

The TIPs technology offers a unique and dynamic solution to the changing landscape of antivirals. As the prevalence of variants and breakthrough infections continue to pose a threat, a therapy that adapts to the evolving nature of the virus becomes increasingly necessary. The TIPs technology provides a formidable defense against viral challenges, as demonstrated by its efficacy in hamsters. Not only are TIPs effective

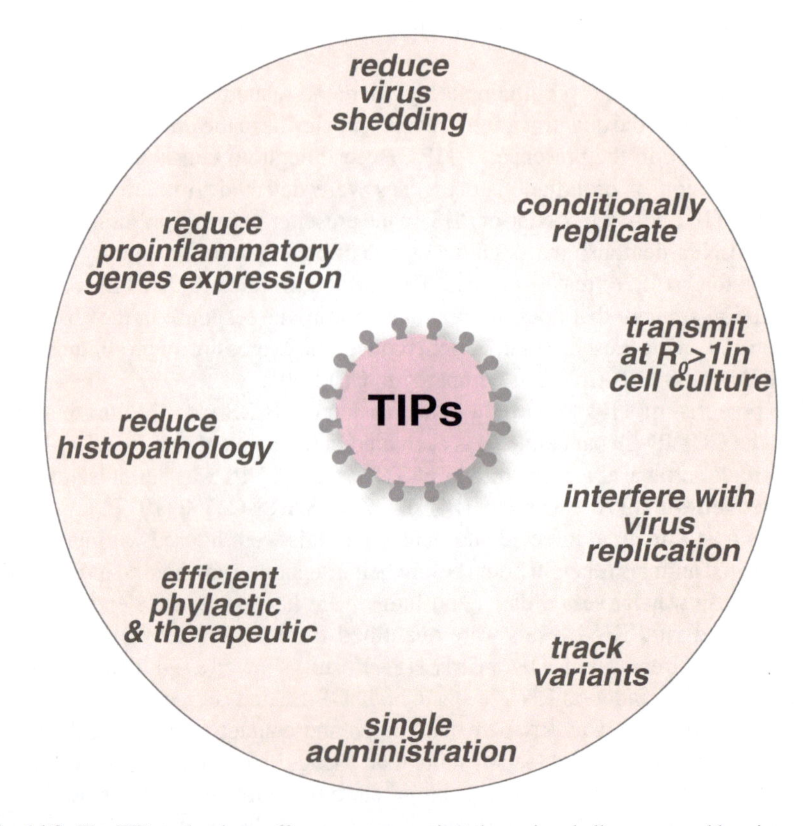

Fig. 14.3 The TIPs technology offers an armor against the major challenges posed by viruses

in downregulating virus replication even at high doses, they are also administered as a single dose, avoiding the need for a complex regimen. Additionally, TIPs have the ability to keep up with the evolving viruses, minimizing the risk of drug resistance. Furthermore, TIPs do not induce an immune response in the absence of infection and reduce the expression of proinflammatory genes associated with SARS-CoV-2 in the presence of infection (Fig. 14.3).

The TIPs technology can be expanded to target other viruses, including both RNA and DNA viruses, by removing the *trans*-regulatory elements that control viral replication. This flexibility and versatility make TIPs a valuable tool, particularly in cases where the cause of infection is unknown. With the capability to encapsulate TIPs for multiple viruses in a single administration, the TIPs technology presents a promising solution for addressing the ongoing challenges posed by viruses.

Acknowledgements We acknowledge Dr. Leor S. Weinberger, affiliated with the Gladstone Institutes and the University of California, San Francisco, and Robert Rodick from VxBiosciences Inc, for productive discussions. Additionally, we acknowledge the valuable editing assistance provided by Dr. Francoise Chanut of the Gladstone Institutes for the book chapter.

References

1. Barrett AD, Dimmock NJ (1984) Modulation of a systemic Semliki Forest virus infection in mice by defective interfering virus. J Gen Virol 65(Pt 10):1827–1831
2. Bekerman E, Einav S (2015) Infectious disease combating emerging viral threats. Science 348:282–283
3. Chaturvedi S, Beutler N, Vasen G, Pablo M, Chen X, Calia G, Buie L, Rodick R, Smith D, Rogers T et al (2022) A single-administration therapeutic interfering particle reduces SARS-CoV-2 viral shedding and pathogenesis in hamsters. Proc Natl Acad Sci U S A 119:e2204624119
4. Chaturvedi S, Pablo M, Wolf M, Rosas-Rivera D, Calia G, Kumar AJ, Vardi N, Du K, Glazier J, Ke R et al (2022) Disrupting autorepression circuitry generates "open-loop lethality" to yield escape-resistant antiviral agents. Cell 185(2086–2102):e2022
5. Chaturvedi S, Vasen G, Pablo M, Chen X, Beutler N, Kumar A, Tanner E, Illouz S, Rahgoshay D, Burnett J et al (2021) Identification of a therapeutic interfering particle-A single-dose SARS-CoV-2 antiviral intervention with a high barrier to resistance. Cell 184(6022–6036):e6018
6. Cohen MS, Nirula A, Mulligan MJ, Novak RM, Marovich M, Yen C, Stemer A, Mayer SM, Wohl D, Brengle B et al (2021) Effect of Bamlanivimab vs placebo on Incidence of COVID-19 among residents and staff of skilled nursing and assisted living facilities: a randomized clinical trial. JAMA 326:46–55
7. Cohn BA, Cirillo PM, Murphy CC, Krigbaum NY, Wallace AW (2022) SARS-CoV-2 vaccine protection and deaths among US veterans during 2021. Science 375:331–336
8. Fodor E, Mingay LJ, Crow M, Deng T, Brownlee GG (2003) A single amino acid mutation in the PA subunit of the influenza virus RNA polymerase promotes the generation of defective interfering RNAs. J Virol 77:5017–5020
9. Fuller FJ, Marcus PI (1980) Interferon induction by viruses. IV. Sindbis virus: early passage defective-interfering particles induce interferon. J Gen Virol 48:63–73
10. Hackett AJ (1964) A possible morphologic basis for the Autointerference phenomenon in vesicular stomatitis virus. Virology 24:51–59

11. Harrington WN, Kackos CM, Webby RJ (2021) The evolution and future of influenza pandemic preparedness. Exp Mol Med 53:737–749
12. Huang AS, Baltimore D (1970) Defective viral particles and viral disease processes. Nature 226:325–327
13. Kupke SY, Riedel D, Frensing T, Zmora P, Reichl U (2019) A novel type of influenza a virus-derived defective interfering particle with nucleotide substitutions in its genome. J Virol 93
14. Lorenzo-Redondo R, Fryer HR, Bedford T, Kim EY, Archer J, Pond SLK, Chung YS, Penugonda S, Chipman J, Fletcher CV et al (2016) Persistent HIV-1 replication maintains the tissue reservoir during therapy. Nature 530:51–56
15. Makino S, Yokomori K, Lai MM (1990) Analysis of efficiently packaged defective interfering RNAs of murine coronavirus: localization of a possible RNA-packaging signal. J Virol 64:6045–6053
16. Marcus PI, Sekellick MJ (1977) Defective interfering particles with covalently linked [+/-]RNA induce interferon. Nature 266:815–819
17. McLeod DV, Gandon S (2021) Understanding the evolution of multiple drug resistance in structured populations. Elife 10
18. Metzger VT, Lloyd-Smith JO, Weinberger LS (2011) Autonomous targeting of infectious superspreaders using engineered transmissible therapies. PLoS Comput Biol 7:e1002015
19. Notton T, Glazier JJ, Saykally VR, Thompson CE, Weinberger LS (2021) RanDeL-Seq: a high-throughput method to map viral cis- and trans-acting elements. mBio 12
20. Rast LI, Rouzine IM, Rozhnova G, Bishop L, Weinberger AD, Weinberger LS (2016) Conflicting selection pressures will constrain viral escape from interfering particles: principles for designing resistance-proof antivirals. PLoS Comput Biol 12:e1004799
21. Rouzine IM, Weinberger LS (2013) Design requirements for interfering particles to maintain coadaptive stability with HIV-1. J Virol 87:2081–2093
22. Snijder EJ, den Boon JA, Horzinek MC, Spaan WJ (1991) Characterization of defective interfering RNAs of Berne virus. J Gen Virol 72(Pt 7):1635–1643
23. Snoeck R, Andrei G, De Clercq E (1996) Patterns of resistance and sensitivity to antiviral compounds of drug-resistant strains of human cytomegalovirus selected in vitro. Eur J Clin Microbiol Infect Dis 15:574–579
24. Tanner EJJ, Jung S, Glazier J, Thompson C, Zhou Y, Martin B, Son H, Riley JL, Weinberger LS (2019) Discovery and Engineering of a Therapeutic Interfering Particle (TIP): a combination self-renewing antiviral. Biorxiv.
25. Vignuzzi M, Lopez CB (2019) Defective viral genomes are key drivers of the virus-host interaction. Nat Microbiol 4:1075–1087
26. Von Magnus P (1954) Incomplete forms of influenza virus. Adv Virus Res 2:59–79
27. Weinberger LS, Schaffer DV, Arkin AP (2003) Theoretical design of a gene therapy to prevent AIDS but not human immunodeficiency virus type 1 infection. J Virol 77:10028–10036

Index

M. Comas-Garcia and S. Rosales-Mendoza (eds.), *Physical Virology*, Springer Series in Biophysics 24, https://doi.org/10.1007/978-3-031-36815-8

Printed in the United States
by Baker & Taylor Publisher Services